Lecture Notes in Computer Science 10194

Commenced Publication in 1973
Founding and Former Series Editors:
Gerhard Goos, Juris Hartmanis, and Jan van Leeuwen

More information about this series at http://www.springer.com/series/7410

Said El Hajji · Abderrahmane Nitaj
El Mamoun Souidi (Eds.)

Codes, Cryptology and Information Security

Second International Conference, C2SI 2017
Rabat, Morocco, April 10–12, 2017, Proceedings
In Honor of Claude Carlet

 Springer

Editors
Said El Hajji
University Mohamed V in Rabat
Rabat
Morocco

El Mamoun Souidi
University Mohamed V in Rabat
Rabat
Morocco

Abderrahmane Nitaj
University of Caen Normandie
Caen
France

ISSN 0302-9743   ISSN 1611-3349 (electronic)
Lecture Notes in Computer Science
ISBN 978-3-319-55588-1   ISBN 978-3-319-55589-8 (eBook)
DOI 10.1007/978-3-319-55589-8

Library of Congress Control Number: 2017934218

LNCS Sublibrary: SL4 – Security and Cryptology

Printed on acid-free paper

This Springer imprint is published by Springer Nature
The registered company is Springer International Publishing AG
The registered company address is: Gewerbestrasse 11, 6330 Cham, Switzerland

Preface

This volume contains the papers accepted for presentation at C2SI-Carlet 2017, in honor of Professor Claude Carlet, from the University of Paris 8, France. C2SI-Carlet is an international conference on the theory and applications of cryptographic techniques, coding theory, and information security. One aim of this conference is to pay homage to Claude Carlet for his valuable contribution in teaching and disseminating knowledge in coding theory and cryptography worldwide, especially in Africa. The other aim of the conference is to provide an international forum for researchers from academia and practitioners from industry from all over the world for discussion of all forms of cryptology, coding theory, and information security.

The initiative of organizing C2SI-Carlet 2017 was initiated by the Moroccan Laboratory of Mathematics, Computing Sciences and Applications (LabMIA) at the Faculty of Sciences of the Mohammed V University in Rabat and performed by an active team of researchers from Morocco and France. The conference was organized in cooperation with the International Association for Cryptologic Research (IACR), and the proceedings are published in Springer's *Lecture Notes in Computer Science* series.

The first conference in this series was held at the same university during May 26–28, 2015, for which the proceedings were published in Springer's *Lecture Notes in Computer Sciences* as volume 9084.

The C2SI-Carlet 2017 Program Committee consisted of 49 members. There were 72 papers submitted to the conference. Each paper was assigned to two or three members of the Program Committee and was reviewed anonymously. The review process was challenging and the Program Committee, aided by reports from 26 external reviewers, produced a total of 164 reviews in all. After this period, 19 papers were accepted on January 28, 2017. Authors then had the opportunity to update their papers until February 6, 2017. The present proceedings include all the revised papers. We are indebted to the members of the Program Committee and the external reviewers for their diligent work.

The conference was honored by the presence of the invited speakers Mohammed Essaaidi, Caroline Fontaine, Maria Isabel Garcia Planas, Sylvain Guilley, and Tor Helleseth. They gave talks on various topics in cryptography, coding theory, and information security and contributed to the success of the conference.

We had the privilege to chair the Program Committee. We would like to thank all committee members for their work on the submissions, as well as all external reviewers for their support. We thank the authors of all submissions and all the speakers as well all the participants. They all contributed to the success of the conference.

We also would like to thank Professor Saaid Amzazi, Head of Mohammed V University in Rabat, for his unwavering support to research and teaching in the areas of cryptography, coding theory, and information security. We also want to thank Professor Mourad El Belkacemi, Dean of Faculty of Sciences in Rabat.

We are deeply grateful to Professor Claude Carlet for the great service in contributing to the establishment of a successful research group in coding theory, cryptography, and information security at the Faculty of Sciences of Mohammed V University in Rabat. We would like to take this opportunity to acknowledge his professional work.

Along with these individuals, we wish to thank our local colleagues and students who contributed greatly to the organization and success of the conference.

Finally, we heartily thank all the local Organizing Committee members, all sponsors, and everyone who contributed to the success of this conference. We are also thankful to the staff at Springer for their help with producing the proceedings and to the staff of EasyChair for the use of their conference management system.

April 2017   S. El Hajji
                                                                A. Nitaj
                                                                E.M. Souidi

Organization

C2SI-Carlet 2017 was organized by the Moroccan Laboratory of Mathematics, Computing Sciences and Applications (LabMIA) at the Faculty of Sciences of the Mohammed V University in Rabat.

Honorary Chairs

Saaid Amzazi	President of Mohammed V University in Rabat, Morocco
Claude Carlet	Paris 8 University, Paris, France

General Chair

Said El Hajji	Mohammed V University in Rabat, Morocco

Program Chairs

Said El Hajji	Mohammed V University in Rabat, Morocco
Abderrahmane Nitaj	University of Caen Normandie, France
El Mamoun Souidi	Mohammed V University in Rabat, Morocco

Organizing Committee

Said El Hajji (Chair)	LabMIA, Mohammed V University in Rabat, Morocco
El Mamoun Souidi (Co-chair)	LabMIA, Mohammed V University in Rabat, Morocco
Ghizlane Orhanou (Co-chair)	LabMIA, Mohammed V University in Rabat, Morocco
Abdelmalek Azizi	Mohammed I University, Morocco
Hicham Bensaid	INPT, Rabat, Morocco
Hafssa Benaboud	Mohammed V University in Rabat, Morocco
Redouane Benaini	Mohammed V University in Rabat, Morocco
Youssef Bentaleb	Ibn Tofail University, Kenitra, Morocco
Souad EL Bernoussi	Mohammed V University in Rabat, Morocco
Sidi Mohamed Douiri	Mohammed V University in Rabat, Morocco
Abelkrim Haqiq	Hassan I University, Settat, Morocco
Hicham Laanaya	Mohammed V University in Rabat, Morocco
Jalal Laassiri	Ibn Tofail University, Kenitra, Morocco
Mounia Mikram	Information Science School, Rabat, Morocco
Faissal Ouardi	Mohammed V University in Rabat, Morocco

Program Committee

Anas Aboulkalam	Cadi Ayyad University, Morocco
Amr Youssef	Concordia University, Canada
Muhammad Rezal Kamel Ariffin	University Putra Malaysia, Malaysia
François Arnault	University of Limoges, France
Hafssa Benaboud	Mohammed V University in Rabat, Morocco
Abdelmalek Azizi	Mohammed I University, Morocco
Youssef Bentaleb	Ibn Tofail University, Kenitra, Morocco
Thierry Berger	University of Limoges, France
Mohamed Bouhdadi	Mohammed V University in Rabat, Morocco
Mohammed Boulmalf	UIR, Rabat, Morocco
Lilya Budaghyan	University of Bergen, Norway
Anne Canteaut	Inria Rocquencourt, France
Claude Carlet	Paris 8 University, France
Pierre Louis Cayrel	University of Saint Etienne, France
Sherman S.M. Chow	The Chinese University of Hong Kong, SAR China
Pierre Dusart	University of Limoges, France
Nadia El Mrabet	SAS Ecole des Mines de Saint Etienne, Gardanne, France
Caroline Fontaine	Telecom Bretagne, Rennes, France
Philippe Gaborit	University of Limoges, France
Maria Isabel Garcia Planas	Catalonia University, Barcelona, Spain
Sanaa Ghouzali	King Saud University, Riyadh, Saudi Arabia
Guang Gong	University of Waterloo, Canada
Aline Gouget	Gemalto, France
Sylvain Guilley	TELECOM ParisTech and SecureIC S.A.S., France
Tor Helleseth	Bergen University, Norway
Mohammed Essaaidi	IEEE Section Morocco, Mohammed V University in Rabat, Morocco
Sidi Mohamed Douiri	Mohammed V University in Rabat, Morocco
Abelkrim Haqiq	Hassan I University, Settat, Morocco
Zoubida Jadda	Defense Department Vannes Coëtquidan, France
JonLark Kim	Sogang University, Seoul, South Korea
Salahddine Krit	IbnZohr University, Ouarzazate, Morocco
Jalal Laassiri	Ibn Tofail University, Kenitra, Morocco
Jean Louis Lanet	Inria Bretagne Atlantique, France
Sihem Mesnager	University of Paris 8, France
Mounia Mikram	Information Sciences School in Rabat, Morocco
Marine Minier	Laboratoire LORIA, University of Lorraine, Nancy, France
Ghizlane Orhanou	Mohammed V University in Rabat, Morocco
Faissal Ouardi	Mohammed V University in Rabat, Morocco
Ali Ouadfel	LabMIA, Mohammed V University in Rabat, Morocco
Francesco Sica	Nazarbayev University, Kazakhstan

Partrice Parraud   Defense Department Vannes Coëtquidan, France
Emmanuel Prouff   Safran Identity and Security and Université Pierre
                            et Marie Curie, Paris, France
Mohamed Rziza   Mohammed V University in Rabat, Morocco
Pantelimon (Pante) Stanica Naval Postgraduate School, USA
Joseph Tonien   University of Wollongong, Australia
Felix Ulmer   Université de Rennes 1, France
Damien Vergnaud   Ecole Normale Supérieure, Paris, France
Fouad Zinoun   Mohammed V University in Rabat, Morocco

Additional Reviewers

Amit Kumar Chauhan   Mohammed Benabdellah
Cedric Lauradoux   Nian Li
Chunlei Li   Nicolas Gama
David Pointcheval   Rafael Misoczki
Delphine Boucher   Raghvendra Rohit
Edoardo Persichetti   Riham Altawy
Essaid Chanigui   Said El Kafhali
Guillame Barbu   Siham Ezzouak
Guillaume Bouffard   Steve Szabo
Jean Belo Klamti   Thomas Debris-Alazard
Jiafan Wang   Wilfried Meidl
Kalikinkar Mandal   Xiuhua Wang
Matthew Parker   Yongjun Zhao

Invited Speakers

Mohammed Essaaidi   Mohammed V University in Rabat, Morocco
Caroline Fontaine   TELECOM Bretagne, France
Maria Isabel Garcia Planas UPC, Universitat Politècnica de Catalunya, Spain
Sylvain Guilley   TELECOM-Paris Tech, France
Tor Helleseth   University of Bergen, Norway

Sponsoring Institutions

Ministère de l'Enseignement Supérieur, de la Recherche Scientifique et de la Formation
des Cadres
Faculty of Sciences, Mohammed V University in Rabat, Morocco
Centre Marocain de Recherches Polytechniques et d'Innovation, Morocco
Laboratoire de Mathématiques, Informatique et Applications (LabMIA), Rabat,
Morocco
Ministère de l'Industrie, du Commerce, de l'Investissement et de l'Economie
Numérique, Morocco

Biography of Claude Carlet

Claude Carlet received in 1990 the Ph.D. degree from the University of Paris 6, France and in 1994 the Habilitation to Direct theses from the University of Amiens, France. He was associate professor in the Department of Computer Science at the University of Amiens from 1990 to 1994, and professor in the Department of Computer Science at the University of Caen, France, from 1994 to 2000 and in the department of Mathematics at the University of Paris 8, France, from 2000 to 2017. His research interests include Boolean functions (bent, correlation-immune, algebraic immune, SAC, etc.), vectorial functions (APN, etc.), cryptography (in particular, stream ciphers, block ciphers and side-channel attacks) finite fields and coding theory (in relationship with the domains above). He has participated as chapter author or editor to 11 books, (co-)written 100 journal papers, 60 papers in proceedings and 20 shorter international papers. He has been member of 70 program committees (7 as co-chair). He has been in charge of the French research group "codage-cryptographie C2" during ten years. He has been Associate Editor of IEEE Transactions on Information Theory and is currently editor in chief of the journal Cryptography and Communications (SPRINGER) and editor in the 4 journals DCC (SPRINGER), AMC (American Institute of Mathematical Sciences), IJCM-TCOM (Taylor & Francis) and IJOCT (Inderscience Publishers). He has supervised 13 students and is currently supervising 5. He has been plenary invited speaker in 20 international conferences and invited speaker in 25 other conferences and workshops.

Contents

Invited Papers

Some Results on the Known Classes of Quadratic APN Functions

Lilya Budaghyan, Tor Helleseth, Nian Li$^{(\boxtimes)}$, and Bo Sun

Department of Informatics, University of Bergen,
Postboks 7803, 5020 Bergen, Norway
{Lilya.Budaghyan,Tor.Helleseth,Nian.Li,Bo.Sun}@uib.no

Abstract. In this paper, we determine the Walsh spectra of three classes of quadratic APN functions and we prove that the class of quadratic trinomial APN functions constructed by Göloğlu is affine equivalent to Gold functions.

Keywords: APN function · Quadratic function · Walsh spectrum

1 Introduction

For given positive integers n and m, a function F from the finite field $\mathbb{F}_{2^n}$ to the finite field $\mathbb{F}_{2^m}$ is called a vectorial Boolean function or an (n, m)-function, and in the case when $m = 1$ it is simply called a Boolean function. When $m = n$ an (n, n)-function has a unique representation as a univariate polynomial over $\mathbb{F}_{2^n}$ of the form

$$F(x) = \sum_{i=0}^{2^n - 1} a_i x^i, \qquad a_i \in \mathbb{F}_{2^n}.$$

Boolean and vectorial Boolean functions have many applications in mathematics and information theory. In particular, they play an important role in cryptography.

In modern society, exchange and storage of information in an efficient, reliable and secure manner is of fundamental importance. Cryptographic primitives are used to protect information against eavesdropping, unauthorized changes and other misuse. In the case of symmetric cryptography ciphers are designed by appropriate composition of nonlinear Boolean functions. For example, the security of block ciphers depends on S-boxes which are (n, m)-functions. For most of cryptographic attacks on block ciphers there are certain properties of functions which measure the resistance of the S-box to these attacks. The differential attack introduced by Biham and Shamir is one of the most efficient cryptanalysis tools for block ciphers [2]. It is based on the study of how differences in an input can affect the resulting difference at the output. An (n, m)-function F is called differentially δ-uniform if the equation $F(x + a) - F(x) = b$ has at most δ

This work was supported by the Norwegian Research Council.

S. El Hajji et al. (Eds.): C2SI 2017, LNCS 10194, pp. 3–16, 2017.
DOI: 10.1007/978-3-319-55589-8_1

solutions for every nonzero element a of $\mathbb{F}_{2^n}$ and every b in $\mathbb{F}_{2^m}$. Functions with the smallest possible differential uniformity contribute an optimal resistance to the differential attack [34]. In this sense differentially 2^{n-m}-uniform functions, called perfect nonlinear (PN), are optimal. However, PN functions exist only for n even and $m \leq n/2$ [35]. An important case are differentially 2-uniform functions with $n = m$, called almost perfect nonlinear (APN), which have the smallest possible differential uniformity.

Another powerful attack on block ciphers is linear cryptanalysis by Matsui which is based on finding affine approximations to the action of a cipher [33]. The nonlinearity $NL(F)$ of an (n, m)-function F is the minimum Hamming distance between all the component functions of F (that is, the functions $\mathrm{tr}_1^m(vF(x))$ where

$$\mathrm{tr}_1^m(x) = x + x^2 + \cdots + x^{2^{m-1}}$$

denotes the absolute trace function of $\mathbb{F}_{2^m}$ and v is any nonzero element of $\mathbb{F}_{2^m}$) and all affine Boolean functions over $\mathbb{F}_{2^n}$. The nonlinearity quantifies the level of resistance of the function to the linear attack: the higher is the nonlinearity $NL(F)$ the better is the resistance of F [21]. The functions achieving the upper bound on nonlinearity are called bent functions. All bent functions are also PN and vice versa, that is, these functions have optimal resistance against both linear and differential attacks. As mentioned above, PN (or bent) functions do not exist when $m = n$. In this case, when also n is odd, functions with the best possible nonlinearity are called almost bent (AB). When n is even the upper bound on nonlinearity is still to be determined. All AB functions are APN, but the converse is not true in general. However, for n odd all quadratic APN functions are also AB.

The nonlinearity $NL(F)$ of an (n, m) function F can be expressed by means of the Walsh transform. The Walsh transform of F at $(\alpha, \beta) \in \mathbb{F}_{2^n} \times \mathbb{F}_{2^m}$ is defined by

$$W_F(\alpha, \beta) = \sum_{x \in \mathbb{F}_{2^n}} (-1)^{\mathrm{tr}_1^m(\beta F(x)) + \mathrm{tr}_1^n(\alpha x)},$$

and the Walsh spectrum of F is the set

$$\{W_F(\alpha, \beta) : \alpha \in \mathbb{F}_{2^n}, \beta \in \mathbb{F}_{2^m}^*\}.$$

Then

$$NL(F) = 2^{n-1} - \frac{1}{2} \max_{\alpha \in \mathbb{F}_{2^n}, \beta \in \mathbb{F}_{2^m}^*} |W_F(\alpha, \beta)|.$$

The Walsh spectrum of AB functions consists of three values $0, \pm 2^{\frac{n+1}{2}}$. The Walsh spectrum of a bent function is $\{\pm 2^{\frac{n}{2}}\}$.

There are several equivalence relations of functions for which differential uniformity and nonlinearity are invariant. Due to these equivalence relations, having only one APN (respectively, AB) function, one can generate a huge class of APN (respectively, AB) functions.

Two functions F and F' from $\mathbb{F}_{2^n}$ to $\mathbb{F}_{2^m}$ are called

- affine equivalent (or linear equivalent) if $F' = A_1 \circ F \circ A_2$, where the mappings A_1 and A_2 are affine (resp. linear) permutations of $\mathbb{F}_{2^m}$ and $\mathbb{F}_{2^n}$, respectively;
- extended affine equivalent (EA-equivalent) if $F' = A_1 \circ F \circ A_2 + A$, where the mappings $A : \mathbb{F}_{2^n} \to \mathbb{F}_{2^m}$, $A_1 : \mathbb{F}_{2^m} \to \mathbb{F}_{2^m}$, $A_2 : \mathbb{F}_{2^n} \to \mathbb{F}_{2^n}$ are affine, and where A_1, A_2 are permutations;
- Carlet-Charpin-Zinoviev equivalent (CCZ-equivalent) if for some affine permutation $\mathcal{L}$ of $\mathbb{F}_{2^n} \times \mathbb{F}_{2^m}$ the image of the graph of F is the graph of F', that is, $\mathcal{L}(G_F) = G_{F'}$ where $G_F = \{(x, F(x)) \mid x \in \mathbb{F}_{2^n}\}$ and $G_{F'} = \{(x, F'(x)) \mid x \in \mathbb{F}_{2^n}\}$.

Although different, these equivalence relations are connected to each other. It is obvious that linear equivalence is a particular case of affine equivalence, and that affine equivalence is a particular case of EA-equivalence. As shown in [20], EA-equivalence is a particular case of CCZ-equivalence and every permutation is CCZ-equivalent to its inverse. The algebraic degree of a function (if it is not affine) is invariant under EA-equivalence but, in general, it is not preserved by CCZ-equivalence.

There are six known infinite families of power APN functions. They are presented in Table 1. There are also eleven known infinite families of quadratic APN polynomilas CCZ-inequivalent to power functions. They are given in Table 2. Families 3, 4 and 11 in Table 2 are proven to be a part of a general binary construction of APN functions [18].

Classification of APN functions is complete for $n \leq 5$ [9]: for these values of n the only APN functions, up to CCZ-equivalence, are power APN functions, and up to EA-equivalence, are power APN functions and those APN functions constructed in [16]. For $n = 6$ classification is complete for quadratic APN functions: 13 quadratic APN functions are found in [10] and, as proven in [26], up to CCZ-equivalence, these are the only quadratic APN functions. The only known APN function CCZ-inequivalent to power functions and to quadratic functions was found in [9,27] for $n = 6$. For $n = 7$ and $n = 8$, as shown in a recent work [37], there are, respectively, more than 470 and more than a thousand

Table 1. Known APN power functions x^d on $\mathbb{F}_{2^n}$.

Functions	Exponents d	Conditions	$d^\circ(x^d)$	Proven
Gold	$2^i + 1$	$\gcd(i, n) = 1$	2	[28,34]
Kasami	$2^{2i} - 2^i + 1$	$\gcd(i, n) = 1$	$i + 1$	[31,32]
Welch	$2^t + 3$	$n = 2t + 1$	3	[23]
Niho	$2^t + 2^{\frac{t}{2}} - 1, \quad t$ even	$n = 2t + 1$	$(t+2)/2$	[22]
	$2^t + 2^{\frac{3t+1}{2}} - 1, t$ odd		$t + 1$	
Inverse	$2^{2t} - 1$	$n = 2t + 1$	$n - 1$	[1,34]
Dobbertin	$2^{4i} + 2^{3i} + 2^{2i} + 2^i - 1$	$n = 5i$	$i + 3$	[24]

Table 2. Known classes of quadratic APN polynomials CCZ-inequivalent to power functions on $\mathbb{F}_{2^n}$.

$N°$	Functions	Conditions	References
1-2	$x^{2^s+1} + \alpha^{2^k-1}x^{2^{ik}+2^{mk+s}}$	$n = pk$, $\gcd(k,p) = \gcd(s,pk) = 1$, $p \in \{3,4\}$, $i = sk \bmod p$, $m = p-i$, $n \geq 12$, α primitive in $\mathbb{F}_{2^n}^*$	[13]
3	$x^{2^{2i}+2^i} + bx^{q+1} + cx^{q(2^{2i}+2^i)}$	$q = 2^m$, $n = 2m$, $\gcd(i,m) = 1$, $\gcd(2^i+1, q+1) \neq 1$, $cb^q + b \neq 0$, $c \notin \{\lambda^{(2^i+1)(q-1)}, \lambda \in \mathbb{F}_{2^n}\}$, $c^{q+1} = 1$	[12]
4	$x(x^{2^i} + x^q + cx^{2^iq})$ $+ x^{2^i}(c^qx^q + sx^{2^iq}) + x^{(2^i+1)q}$	$q = 2^m$, $n = 2m$, $\gcd(i,m) = 1$, $c \in \mathbb{F}_{2^n}$, $s \in \mathbb{F}_{2^n} \setminus \mathbb{F}_q$, $X^{2^i+1} + cX^{2^i} + c^qX + 1$ is irreducible over $\mathbb{F}_{2^n}$	[12]
5	$x^3 + a^{-1}\mathrm{tr}_1^n(a^3x^9)$	$a \neq 0$	[14,15]
6	$x^3 + a^{-1}\mathrm{tr}_3^n(a^3x^9 + a^6x^{18})$	$3|n$, $a \neq 0$	[14]
7	$x^3 + a^{-1}\mathrm{tr}_3^n(a^6x^{18} + a^{12}x^{36})$	$3|n$, $a \neq 0$	[14]
8-10	$ux^{2^s+1} + u^{2^k}x^{2^{-k}+2^{k+s}}$ $+ vx^{2^{-k}+1} + wu^{2^k+1}x^{2^s+2^{k+s}}$	$n = 3k$, $\gcd(k,3) = \gcd(s,3k) = 1$, $v, w \in \mathbb{F}_{2^k}$, $vw \neq 1$, $3|(k+s)$, u primitive in $\mathbb{F}_{2^n}^*$	[3]
11	$\alpha x^{2^s+1} + \alpha^{2^k}x^{2^{k+s}+2^k}$ $+ \beta x^{2^k+1} + \sum_{i=1}^{k-1}\gamma_i x^{2^{k+i}+2^i}$	$n = 2k$, $\gcd(s,k) = 1$, s,k odd, $\beta \notin \mathbb{F}_{2^k}$, $\gamma_i \in \mathbb{F}_{2^k}$, α not a cube	[3,4]

CCZ-inequivalent quadratic APN functions. For n odd all power APN functions and the known APN binomials are permutations (see [13,19]). For n even the only known APN permutation is constructed in [11] for $n = 6$. Existence of APN permutations for even $n \geq 8$ is an open problem.

A class of APN functions over $\mathbb{F}_{2^n}$.

$$x^3 + \mathrm{tr}_1^n(x^9)$$

was constructed by Budaghyan, Carlet and Leander in [14]. Later, they generalized this result in [15] to the APN function $F_0(x)$ of the form

$$F_0(x) = x^3 + a^{-1}\mathrm{tr}_1^n(a^3x^9) \tag{1}$$

for any positive integer n and any nonzero element a in $\mathbb{F}_{2^n}$, and they also obtained two other classes of APN functions over $\mathbb{F}_{2^n}$

$$F_1(x) = x^3 + a^{-1}\mathrm{tr}_3^n(a^3x^9 + a^6x^{18}) \tag{2}$$
$$F_2(x) = x^3 + a^{-1}\mathrm{tr}_3^n(a^6x^{18} + a^{12}x^{36}) \tag{3}$$

for any positive integer n divisible by 3 and any nonzero element a in $\mathbb{F}_{2^n}$ and where $\mathrm{tr}_3^n(x) = \sum_{i=0}^{n/3-1} x^{2^{3i}}$ is the trace function from $\mathbb{F}_{2^n}$ to its subfield $\mathbb{F}_{2^3}$. When n is even each of the functions F_0, F_1 and F_2 defines two CCZ-inequivalent functions one for $a = 1$ and one for any $a \neq 1$, that is, all together they give six different functions. When n is odd each of the functions F_0, F_1 and F_2 defines only one function, up to CCZ-inequivalence, that is, all together they give three different functions [15]. In Table 2 the functions F_0, F_1 and F_2 correspond to families 5, 6 and 7, respectively. In the present paper we determine the Walsh spectra of the functions F_0, F_1 and F_2. The Walsh spectra of the remaining functions in Tables 1 and 2 have been determined in [6–8,17,28,30,36]. Note that the Walsh spectrum of the function F_0 with $a = 1$ was previously found in [5] and we generalize this result to any $a \neq 0$. The results on the Walsh spectra show that all the known families of quadratic APN functions have Gold like Walsh spectra. Note that there exists a quadratic APN function for $n = 6$ with Walsh spectrum different from Gold [10].

In 2015 a family of quadratic APN trinomials on $\mathbb{F}_{2^n}$

$$G(x) = x^{2^k+1} + \left(\mathrm{tr}_m^n(x)\right)^{2^k+1}, \tag{4}$$

with $\gcd(k, n) = 1$ and $n = 2m = 4t$, was constructed in [29]. It was claimed there to be CCZ-inequivalent to power functions. However, in the present paper we prove that this family is in fact affine equivalent to Gold power functions.

2 Walsh Spectra of F_1 and F_2

In this section, we determine the Walsh spectra of the APN functions F_1 and F_2. According to the definition, for any $b, c \in \mathbb{F}_{2^n}$, one gets

$$
\begin{aligned}
g_i(x) = \mathrm{tr}_1^n(bF_i(x) + cx) &= \mathrm{tr}_1^n(bx^3 + ba^{-1}\mathrm{tr}_3^n(a^3x^9 + a^6x^{18})^i + cx) \\
&= \mathrm{tr}_1^n(bx^3 + cx) + \mathrm{tr}_1^n(ba^{-1}\mathrm{tr}_3^n(a^3x^9 + a^6x^{18})^i) \\
&= \mathrm{tr}_1^n(bx^3 + cx) + \mathrm{tr}_1^3\mathrm{tr}_3^n(ba^{-1}\mathrm{tr}_3^n(a^3x^9 + a^6x^{18})^i) \\
&= \mathrm{tr}_1^n(bx^3 + cx) + \mathrm{tr}_1^3\mathrm{tr}_3^n(\mathrm{tr}_3^n(ba^{-1})(a^3x^9 + a^6x^{18})^i) \\
&= \mathrm{tr}_1^n(bx^3 + cx + \mathrm{tr}_3^n(ba^{-1})(a^3x^9 + a^6x^{18})^i)
\end{aligned}
$$

for $i \in \{1, 2\}$. For simplicity, denote $\mathrm{tr}_3^n(ba^{-1}) = \delta^2$. By a direct calculation, one obtains that

$$
\begin{aligned}
&g_i(x) + g_i(x + u) + g_i(u) \\
&= \mathrm{tr}_1^n(bx^2u + bxu^2 + \delta^2(a^3x^8u + a^3xu^8 + a^6x^2u^{16} + a^6x^{16}u^2)^i) \\
&= \mathrm{tr}_1^n\left(x((bu)^{2^{-1}} + bu^2 + (\delta^{2/i}a^3u)^{2^{-3}} + \delta^{2/i}a^3u^8 + \delta^{1/i}a^3u^8 + (\delta^{1/i}a^3u)^{2^{-3}})\right) \\
&= \mathrm{tr}_1^n\left(x((\delta^{2/i} + \delta^{1/i})a^3u^8 + bu^2 + (bu)^{2^{-1}} + ((\delta^{2/i} + \delta^{1/i})a^3u)^{2^{-3}})\right), \tag{5}
\end{aligned}
$$

which implies that

$$
\begin{aligned}
|W_{F_i}(b, c)|^2 &= \sum_{x \in \mathbb{F}_{2^n}} \sum_{u \in \mathbb{F}_{2^n}} (-1)^{g_i(x) + g_i(x+u)} \\
&= \sum_{u \in \mathbb{F}_{2^n}} (-1)^{g_i(u)} \sum_{x \in \mathbb{F}_{2^n}} (-1)^{\mathrm{tr}_1^n(xL_{a,b,\delta}^i(u))},
\end{aligned}
$$

where $L_{a,b,\delta}^i(u)$ is defined as

$$L_{a,b,\delta}^i(u) = (\delta^{2/i} + \delta^{1/i})a^3u^8 + bu^2 + (bu)^{2^{-1}} + ((\delta^{2/i} + \delta^{1/i})a^3u)^{2^{-3}}. \quad (6)$$

Note that $g_i(u) + g_i(u+v) + g_i(v) = \mathrm{tr}_1^n(vL_{a,b,\delta}^i(u))$ due to (5) and (6). This means that for any u satisfying $L_{a,b,\delta}^i(u) = 0$ and any $v \in \mathbb{F}_{2^n}$ we have

$$g_i(u+v) = g_i(u) + g_i(v)$$

which implies that

$$|W_{F_i}(b,c)|^2 = 0, \qquad \text{or} \qquad 2^n \cdot |\{x \in \mathbb{F}_{2^n} : L_{a,b,\delta}^i(u) = 0\}|. \quad (7)$$

In what follows, we discuss the number of solutions $u \in \mathbb{F}_{2^n}$ to $L_{a,b,\delta}^i(u) = 0$ by adopting Dobbertin's method [25], which also was used by Bracken et al. in [5] to determine the Walsh spectrum of $F_0(x)$ for the case of $a = 1$.

For simplicity, define $\theta_i = (\delta^{2/i} + \delta^{1/i})a^3$ for $i = 1, 2$. Then $L_{a,b,\delta}^i(u) = 0$ can be written as $\theta_i u^8 + bu^2 + (bu)^{2^{-1}} + (\theta_i u)^{2^{-3}} = 0$ and it can be readily verified that

$$uL_{a,b,\delta}^i(u) = \phi_i(u) + \phi_i(u)^{2^{-1}},$$

where $\phi_i(u)$ is given as

$$\phi_i(u) = bu^3 + \theta_i u^9 + \theta_i^{\frac{1}{2}}u^{\frac{9}{2}} + \theta_i^{\frac{1}{4}}u^{\frac{9}{4}}. \quad (8)$$

Then, if $L_{a,b,\delta}^i(u) = 0$, we must have $\phi_i(u) \in \mathbb{F}_2$.

Proposition 1. *Let $a, b \in \mathbb{F}_{2^n}$ with $ab \neq 0$ and $\delta^2 = \mathrm{tr}_3^n(ba^{-1})$. If $\delta^{2/i} + \delta^{1/i} \neq 0$, then $L_{a,b,\delta}^i(u) = 0$ if and only if $\phi_i(u) = 0$ for $i = 1, 2$.*

Proof. If $\phi_i(u) = 0$, we have $L_{a,b,\delta}^i(u) = 0$; and if $L_{a,b,\delta}^i(u) = 0$, we have $\phi_i(u) \in \mathbb{F}_2$. Thus, to complete the proof, we need to show that $L_{a,b,\delta}^i(u) = 0$ implies that $\phi_i(u) = 0$ for $i = 1, 2$. Suppose that $\phi_i(u) = 1$, one then gets $b = \theta_i u^6 + \theta_i^{1/2}u^{3/2} + \theta_i^{1/4}u^{-3/4} + u^{-3}$ which together with $\theta_i = (\delta^{2/i} + \delta^{1/i})a^3$ leads to

$$\frac{b}{a} = (\delta^{\frac{2}{i}} + \delta^{\frac{1}{i}})a^2u^6 + (\delta^{\frac{2}{i}} + \delta^{\frac{1}{i}})^{\frac{1}{2}}a^{\frac{1}{2}}u^{\frac{3}{2}} + (\delta^{\frac{2}{i}} + \delta^{\frac{1}{i}})^{\frac{1}{4}}a^{-\frac{1}{4}}u^{-\frac{3}{4}} + a^{-1}u^{-3}. \quad (9)$$

For convenience, define $\mathrm{tr}_3^n(a^2u^6) = t$ and $\mathrm{tr}_3^n(a^{-1}u^{-3}) = r$. Notice that $\delta^{\frac{1}{2}} = \delta^4$ and $\delta^{\frac{1}{4}} = \delta^2$ since $\delta \in \mathbb{F}_{2^3}$. Then by $\mathrm{tr}_3^n(ba^{-1}) = \delta^2$ and (9) one has that

$$\delta^2 = \mathrm{tr}_3^n(\frac{b}{a}) = (\delta^{\frac{2}{i}} + \delta^{\frac{1}{i}})t + (\delta^{\frac{1}{i}} + \delta^{\frac{1}{2i}})t^{\frac{1}{4}} + (\delta^{\frac{1}{2i}} + \delta^{\frac{1}{4i}})r^{\frac{1}{4}} + r. \quad (10)$$

Rewrite (10) with respect to the variable r we have

$$(\delta^{\frac{1}{2i}} + \delta^{\frac{1}{4i}})r^2 + r + (\delta^{\frac{2}{i}} + \delta^{\frac{1}{i}})t + (\delta^{\frac{1}{i}} + \delta^{\frac{1}{2i}})t^2 + \delta^2 = 0.$$

Note that $\delta^{\frac{1}{2i}} + \delta^{\frac{1}{4i}} \neq 0$ due to $\delta^{\frac{2}{i}} + \delta^{\frac{1}{i}} \neq 0$. Then the above equation has solution $r \in \mathbb{F}_{2^3}$ if and only if

$$\mathrm{tr}_1^3((\delta^{\frac{1}{2i}} + \delta^{\frac{1}{4i}})((\delta^{\frac{2}{i}} + \delta^{\frac{1}{i}})t + (\delta^{\frac{1}{i}} + \delta^{\frac{1}{2i}})t^2 + \delta^2)) = 0. \tag{11}$$

It can be readily verified that for $i = 1, 2$ we have

$$(\delta^{\frac{1}{2i}} + \delta^{\frac{1}{4i}})^2(\delta^{\frac{2}{i}} + \delta^{\frac{1}{i}})^2 = (\delta^{\frac{1}{2i}} + \delta^{\frac{1}{4i}})(\delta^{\frac{1}{i}} + \delta^{\frac{1}{2i}}),$$

which implies that (11) holds if and only if

$$\mathrm{tr}_1^3((\delta^{\frac{1}{2i}} + \delta^{\frac{1}{4i}})\delta^2) = 0.$$

Observe that $(\delta^{\frac{1}{2i}} + \delta^{\frac{1}{4i}})\delta^2 = (\delta^4 + \delta^2)\delta^2 = \delta^6 + \delta^4$ if $i = 1$, and it equals $(\delta^2 + \delta)\delta^2 = \delta^4 + \delta^3$ if $i = 2$. Thus, no matter which case we arrive at $\mathrm{tr}_1^3(\delta^3 + \delta) = 0$. By $\mathrm{tr}_1^3(\delta^3) = \mathrm{tr}_1^3(\delta)$ and $\delta^7 = 1$ we have $\delta^3 + \delta^6 + \delta^5 = \delta + \delta^2 + \delta^4$ which leads to $\delta = 0, 1$, a contradiction with $\delta^{2/i} + \delta^{1/i} \neq 0$. Therefore, if $\delta^{2/i} + \delta^{1/i} \neq 0$, then there is no solution $r \in \mathbb{F}_{2^3}$ to (10) and $L_{a,b,\delta}^i(u) = 0$ if and only if $\phi_i(u) = 0$. This completes the proof.

Proposition 2. *Let $a, b \in \mathbb{F}_{2^n}$ with $ab \neq 0$ and $\delta^2 = \mathrm{tr}_3^n(ba^{-1})$. Then $L_{a,b,\delta}^i(u) = 0$ defined by (6) has at most four roots in $\mathbb{F}_{2^n}$ for any $i \in \{1, 2\}$.*

Proof. If $\theta_i = 0$, i.e., $\delta^{2/i} + \delta^{1/i} = 0$, then (6) is reduced to $bu^2 + (bu)^{2^{-1}} = 0$ which has at most four roots in $\mathbb{F}_{2^n}$ for any nonzero b. Next we consider the case $\theta_i \neq 0$. By Proposition 1, for this case we have $L_{a,b,\delta}^i(u) = 0$ if and only if $\phi_i(u) = 0$. Thus, to complete the proof, it suffices to show that $\phi_i(u) = 0$ has at most four roots in $\mathbb{F}_{2^n}$ for any $i \in \{1, 2\}$. If $\phi_i(u) = 0$ has no nonzero solution for some θ_i and b, then the desired result follows. Now let v be any fixed nonzero solution of $\phi_i(u) = 0$, then for any u satisfying $\phi_i(u) = 0$ we have

$$u(u + v)\phi_i(v) + v(u + v)\phi_i(u) + uv\phi_i(u + v) = 0.$$

A direct calculation based on (8) gives

$$\theta_i^{\frac{1}{2}}(u^2v^{\frac{9}{2}} + v^2u^{\frac{9}{2}} + u^5v^{\frac{3}{2}} + v^5u^{\frac{3}{2}}) = \theta_i^{\frac{1}{4}}(u^2v^{\frac{9}{4}} + v^2u^{\frac{9}{4}} + u^3v^{\frac{5}{4}} + v^3u^{\frac{5}{4}}),$$

which can be written as

$$\theta_i^{\frac{1}{4}}(u^4v + uv^4)(u^{\frac{1}{2}}v + uv^{\frac{1}{2}}) = (u^2v + uv^2)(u^{\frac{1}{4}}v + uv^{\frac{1}{4}}) \tag{12}$$

since $\theta_i \neq 0$. Then, let $u = vz$, one obtains that

$$\theta_i^{\frac{1}{4}}v^{\frac{9}{4}}(z^4 + z)(z^{\frac{1}{2}} + z) = (z^2 + z)(z^{\frac{1}{4}} + z). \tag{13}$$

Note that v is a fixed nonzero element which means that z is uniquely determined by u. Thus, one can conclude that the number of solutions $z \in \mathbb{F}_{2^n}$ to (13) is no less than that of $u \in \mathbb{F}_{2^n}$ to $\phi_i(u) = 0$. Let $w = z^2 + z$ and rewrite (13) as

$$w\Omega_i(w) := \theta_i^{\frac{1}{4}}v^{\frac{9}{4}}(w^2 + w)w^{\frac{1}{2}} + w(w^{\frac{1}{2}} + w^{\frac{1}{4}}) = 0. \tag{14}$$

Observe that (12) holds for any u satisfying $\phi_i(u) = 0$ and the solution set of $\phi_i(u) = 0$ is an $\mathbb{F}_2$-linear space due to Proposition 1. Then, one can conclude that the solution sets of both (13) and (14) are $\mathbb{F}_2$-linear spaces. Assume that w_1, w_2 and $w_1 + w_2$ are solutions of (14), then we have

$$0 = \Omega_i(w_1) + \Omega_i(w_2) + \Omega_i(w_1 + w_2) = \theta_i^{\frac{1}{4}} v^{\frac{9}{4}} (w_1^{\frac{1}{2}} w_2 + w_2^{\frac{1}{2}} w_1)$$

since (14) holds if and only if $\Omega_i(w) = 0$, which leads to $w_1 w_2^2 + w_2^2 w_1 = w_1 w_2(w_1 + w_2) = 0$, i.e., $w_1 = 0$, $w_2 = 0$ or $w_1 = w_2$. This means that (14) has at most two distinct solutions in $\mathbb{F}_{2^n}$ and then (13) has at most four solutions in z since $w = z^2 + z$. This completes the proof.

Theorem 1. *The Walsh spectra of both functions F_1 and F_2 defined by (2) and (3) respectively are $\{0, \pm 2^{(n+1)/2}\}$ if n is odd and $\{0, \pm 2^{n/2}, \pm 2^{(n+2)/2}\}$ otherwise.*

Proof. The Walsh transform of F_i, $i = 1, 2$, takes values from $\{0, \pm 2^{(n+1)/2}\}$ if n is odd and takes values from $\{0, \pm 2^{n/2}, \pm 2^{(n+2)/2}\}$ if n is even. This follows from (7) and Proposition 2.

The Walsh transform takes all three values for n odd and all 5 values for n even since quadratic functions are plateaud and there exists no bent function from $\mathbb{F}_{2^n}$ to itself, while in case of n even quadratic APN functions have some bent components.

3   Walsh Spectrum of F_0

Bracken et al. in [5] had determined the Walsh spectrum of the APN function F_0 for the case of $a = 1$. In this section, we determine its Walsh spectrum for any nonzero element $a \in \mathbb{F}_{2^n}$ by using the same techniques. By the definition, for any $b, c \in \mathbb{F}_{2^n}$, one gets

$$\begin{aligned} \mathrm{tr}_1^n(bF_0(x) + cx) &= \mathrm{tr}_1^n(bx^3 + ba^{-1}\mathrm{tr}_1^n(a^3 x^9) + cx) \\ &= \mathrm{tr}_1^n(bx^3 + cx + \mathrm{tr}_1^n(ba^{-1})a^3 x^9). \end{aligned}$$

For simplicity, let $\mathrm{tr}_1^n(ba^{-1}) = \delta$ and $g_0(x) = \mathrm{tr}_1^n(bF_0(x) + cx)$. Then, by a direct calculation, one obtains that

$$\begin{aligned} & g_0(x) + g_0(x + u) + g_0(u) \\ &= \mathrm{tr}_1^n(bx^2 u + bxu^2 + \delta a^3 x^8 u + \delta a^3 x u^8) \\ &= \mathrm{tr}_1^n\left(x((bu)^{2^{-1}} + bu^2 + (\delta a^3 u)^{2^{-3}} + \delta a^3 u^8)\right), \end{aligned} \tag{15}$$

which implies that

$$\begin{aligned} |W_{F_0}(b, c)|^2 &= \sum_{x \in \mathbb{F}_{2^n}} \sum_{u \in \mathbb{F}_{2^n}} (-1)^{g_0(x) + g_0(x+u)} \\ &= \sum_{u \in \mathbb{F}_{2^n}} (-1)^{g_0(u)} \sum_{x \in \mathbb{F}_{2^n}} (-1)^{\mathrm{tr}_0^n(x L_{a,b,\delta}^0(u))}, \end{aligned}$$

where $L^0_{a,b,\delta}(u)$ is defined as

$$L^0_{a,b,\delta}(u) = (bu)^{2^{-1}} + bu^2 + (\delta a^3 u)^{2^{-3}} + \delta a^3 u^8. \tag{16}$$

Note that $g_0(u) + g_0(u+v) + g_0(v) = \mathrm{tr}^n_1(vL^0_{a,b,\delta}(u))$ due to (15) and (16). This means that for any u satisfying $L^0_{a,b,\delta}(u) = 0$ and any $v \in \mathbb{F}_{2^n}$ we have

$$g_0(u+v) = g_0(u) + g_0(v)$$

which implies that

$$|W_{F_0}(b,c)|^2 = 0, \text{ or } 2^n | |\{x \in \mathbb{F}_{2^n} : L^0_{a,b,\delta}(u) = 0\}|. \tag{17}$$

Next we aim to determine the number of solutions to $L^0_{a,b,\delta}(u) = 0$ in order to determine the possible values of the Walsh spectrum of $F_0(x)$. First, if $\delta = \mathrm{tr}^n_1(ba^{-1}) = 0$, then $L^0_{a,b,\delta}(u) = 0$ is reduced to $L^0_{a,b,0}(u) = (bu)^{2^{-1}} + bu^2 = 0$ which has at most 4 roots in $\mathbb{F}_{2^n}$. Now we assume that $\delta = \mathrm{tr}^n_1(ba^{-1}) = 1$, then $L^0_{a,b,\delta}(u) = 0$ is reduced to $L^0_{a,b,1}(u) = (bu)^{2^{-1}} + bu^2 + (a^3u)^{2^{-3}} + a^3u^8 = 0$, and it is straightforward to verify that

$$uL^0_{a,b,1}(u) = \phi_0(u) + \phi_0(u)^{2^{-1}}, \tag{18}$$

where $\phi_0(u)$ is defined by

$$\phi_0(u) = bu^3 + a^3u^9 + a^{\frac{3}{2}}u^{\frac{9}{2}} + a^{\frac{3}{4}}u^{\frac{9}{4}}.$$

Proposition 3. *Let $a,b \in \mathbb{F}_{2^n}$ with $\delta = \mathrm{tr}^n_1(ba^{-1}) = 1$. Then $L^0_{a,b,1}(u) = 0$ if and only if $\phi_0(u) = 0$.*

Proof. According to (18), we have $L^0_{a,b,1}(u) = 0$ if $\phi_0(u) = 0$; and $\phi_0(u) \in \mathbb{F}_2$ if $L^0_{a,b,1}(u) = 0$. Thus, to complete the proof, we need to show that $L^0_{a,b,1}(u) = 0$ implies that $\phi_0(u) = 0$. Suppose that $\phi_0(u) = 1$, one then gets $b = a^3u^6 + a^{3/2}u^{3/2} + a^{3/4}u^{-3/4} + u^{-3}$ which leads to

$$ba^{-1} = a^2u^6 + a^{\frac{1}{2}}u^{\frac{3}{2}} + a^{-\frac{1}{4}}u^{-\frac{3}{4}} + a^{-1}u^{-3}.$$

This contradicts with the condition that $\mathrm{tr}^n_1(ba^{-1}) = 1$. This completes the proof.

Proposition 4. *Let $a,b \in \mathbb{F}_{2^n}$ with $ab \neq 0$ and $\delta = \mathrm{tr}^n_1(ba^{-1})$. Then $L^0_{a,b,\delta}(u) = 0$ defined by (16) has at most four roots in $\mathbb{F}_{2^n}$.*

Proof. This can be proved completely the same as Proposition 2.

Theorem 2. *The Walsh spectrum of the function F_0 defined by (1) is $\{0, \pm 2^{(n+1)/2}\}$ if n is odd and $\{0, \pm 2^{n/2}, \pm 2^{(n+2)/2}\}$ otherwise.*

Proof. The Walsh transform of F_0 takes values from $\{0, \pm 2^{(n+1)/2}\}$ if n is odd and takes values from $\{0, \pm 2^{n/2}, \pm 2^{(n+2)/2}\}$ if n is even. This follows from (17) and Proposition 4. The Walsh transform takes all three values for n odd and all 5 values for n even by the same reasons as in Theorem 1.

4   Equivalence of Göloğlu's APN Trinomial to Gold Functions

In this section we prove that the function G defined by (4) is affine equivalent to the Gold function $x^{2^{m-k}+1}$. Note first that

$$G(x) = x^{2^m(2^k+1)} + x^{2^k+2^m} + x^{2^{m+k}+1},$$

and it is affine equivalent to the function

$$G'(x) = \left(G(x)\right)^{2^m} = x^{2^k+1} + x^{2^k+2^m} + x^{2^{m+k}+1}.$$

Linear functions

$$L_1(x) = \gamma^{2^k} x^{2^{m+k}} + \gamma x^{2^k},$$

$$L_2(x) = \gamma x^{2^m} + \gamma^{2^k} x,$$

where γ is a primitive element of $\mathbb{F}_{2^2}$, are permutations. Indeed, it is easy to see that the equations $L_1(x) = 0$ and $L_2(x) = 0$ have only 0 as a solution. Note that $L_1(x) = 0$ implies $L_1(x)^{2^m} = 0$ which give

$$\gamma^{2^k} x^{2^{m+k}} = \gamma x^{2^k},$$

$$\gamma^{2^{m+k}} x^{2^k} = \gamma^{2^m} x^{2^{m+k}}.$$

Hence, assuming $x \neq 0$ and multiplying both sides of the equalities above gives $\gamma^{2^k+2^{m+k}} = \gamma^{2^m+1}$ or $\gamma = \gamma^2$ (see explanation below) contradicting that γ is primitive in $\mathbb{F}_{2^2}$. The proof for L_2 being a permutation is similar.

Further we have

$$\left(L_1(x)\right)^{2^{m-k}+1} = \gamma^{2^m+2^k} x^{2^{m+k}+1} + \gamma^{2^{m-k}+1} x^{2^m+2^k} + \gamma^{2^{m-k}+2^k} x^{2^{m+k}+2^m} + \gamma^{2^m+1} x^{2^k+1}$$

$$= x^{2^{m+k}+1} + x^{2^m+2^k} + \gamma x^{2^{m+k}+2^m} + \gamma^2 x^{2^k+1}$$

and

$$L_2 \circ G'(x) = \gamma \left(x^{2^k+1} + x^{2^k+2^m} + x^{2^{m+k}+1}\right)^{2^m} + \gamma^{2^k}\left(x^{2^k+1} + x^{2^k+2^m} + x^{2^{m+k}+1}\right)$$

$$= \left(\gamma + \gamma^{2^k}\right)\left(x^{2^{m+k}+1} + x^{2^m+2^k}\right) + \gamma x^{2^{m+k}+2^m} + \gamma^{2^k} x^{2^k+1}$$

$$= x^{2^{m+k}+1} + x^{2^m+2^k} + \gamma x^{2^{m+k}+2^m} + \gamma^2 x^{2^k+1}$$

since $\gcd(k, n) = 1$, $n = 2m = 4t$, and then

$$\gamma + \gamma^{2^k} = \gamma + \gamma^2 = 1,$$
$$\gamma^{2^m+2^k} = \gamma^{2^{m-k}+1} = \gamma^3 = 1,$$
$$\gamma^{2^{m-k}+2^k} = \gamma^{2^k+2^{m+k}} = \gamma^4 = \gamma,$$
$$\gamma^{2^m+1} = \gamma^2.$$

Hence $\left(L_1(x)\right)^{2^{m-k}+1} = L_2 \circ G'(x) = L_2' \circ G(x)$ where $L_2'(x) = L_2(x^{2^m})$ is, obviously, a linear permutation. Therefore $x^{2^{m-k}+1}$ and G are affine equivalent.

Table 3. CCZ-inequivalent APN functions over $\mathbb{F}_{2^n}$ from the known APN classes ($6 \leq n \leq 11$ and a primitive in $\mathbb{F}_{2^n}$).

n	N°	Functions	Families from Tables 1 and 2	Relation to [27]
6	6.1	x^3	Gold	Table 5: $N^\circ 1.1$
	6.2	$x^6 + x^9 + a^7 x^{48}$	$N^\circ 3$	5: $N^\circ 1.2$
	6.3	$ax^3 + a^4 x^{24} + x^{17}$	$N^\circ 8$-10	5: $N^\circ 2.3$
7	7.1	x^3	Gold	Table 7 : $N^\circ 1.1$
	7.2	x^5	Gold	7 : $N^\circ 3.1$
	7.3	x^9	Gold	7 : $N^\circ 4.1$
	7.4	x^{13}	Kasami	7 : $N^\circ 5.1$
	7.5	x^{57}	Kasami	7 : $N^\circ 6.1$
	7.6	x^{63}	Inverse	7 : $N^\circ 7.1$
	7.7	$x^3 + \mathrm{tr}_1^7(x^9)$	$N^\circ 5$	7 : $N^\circ 1.2$
8	8.1	x^3	Gold	Table 9 : $N^\circ 1.1$
	8.2	x^9	Gold	9 : $N^\circ 1.2$
	8.3	x^{57}	Kasami	9 : $N^\circ 7.1$
	8.4	$x^3 + x^{17} + a^{48} x^{18} + a^3 x^{33} + ax^{34} + x^{48}$	$N^\circ 4$	9 : $N^\circ 2.1$
	8.5	$x^3 + \mathrm{tr}_1^8(x^9)$	$N^\circ 5$	9 : $N^\circ 1.3$
	8.6	$x^3 + a^{-1}\mathrm{tr}_1^8(a^3 x^9)$	$N^\circ 5$	9 : $N^\circ 1.5$
9	9.1	x^3	Gold	
	9.2	x^5	Gold	
	9.3	x^{17}	Gold	
	9.4	x^{13}	Kasami	
	9.5	x^{241}	Kasami	
	9.6	x^{19}	Welch	
	9.7	x^{255}	Inverse	
	9.8	$x^3 + \mathrm{tr}_1^9(x^9)$	$N^\circ 5$	
	9.9	$x^3 + \mathrm{tr}_3^9(x^9 + x^{18})$	$N^\circ 6$	
	9.10	$x^3 + \mathrm{tr}_3^9(x^{18} + x^{36})$	$N^\circ 7$	
10	10.1	x^3	Gold	
	10.2	x^9	Gold	
	10.3	x^{57}	Kasami	
	10.4	x^{339}	Dobbertin	
	10.5	$x^6 + x^{33} + a^{31} x^{192}$	$N^\circ 3$	
	10.6	$x^{72} + x^{33} + a^{31} x^{258}$	$N^\circ 3$	
	10.7	$x^3 + \mathrm{tr}_1^{10}(x^9)$	$N^\circ 5$	
	10.8	$x^3 + a^{-1}\mathrm{tr}_1^{10}(a^3 x^9)$	$N^\circ 5$	
11	11.1	x^3	Gold	
	11.2	x^5	Gold	
	11.3	x^9	Gold	
	11.4	x^{17}	Gold	
	11.5	x^{33}	Gold	
	11.6	x^{13}	Kasami	
	11.7	x^{57}	Kasami	
	11.8	x^{241}	Kasami	
	11.9	x^{993}	Kasami	
	11.10	x^{35}	Welch	
	11.11	x^{287}	Niho	
	11.12	x^{1023}	Inverse	
	11.13	$x^3 + \mathrm{tr}_1^{11}(x^9)$	$N^\circ 5$	

Proposition 5. *Let k, n, m, t be positive integers such that $\gcd(k, n) = 1$, $n = 2m = 4t$. Then the function G defined by (4) and the function $x^{2^{m-k}+1}$ over $\mathbb{F}_{2^n}$ are affine equivalent.*

Remark 1. *For $k = 1$ the APN function G and its equivalence to Gold functions were known from [14].*

We note that it is possible to check CCZ-equivalence of functions with a computer for n up to 12. However, since most of the known families of APN functions are defined for many different parameters and coefficients it becomes extremely difficult to check CCZ-equivalence of a given function to all of them. For this reason we tested all possible parameters and coefficients to produce Table 3 of all CCZ-inequivalent functions arising from the known infinite families of APN functions for $n \leq 11$.

References

1. Beth, T., Ding, C.: On almost perfect nonlinear permutations. In: Helleseth, T. (ed.) EUROCRYPT 1993. LNCS, vol. 765, pp. 65–76. Springer, Heidelberg (1994). doi:10.1007/3-540-48285-7_7
2. Biham, E., Shamir, A.: Differential cryptanalysis of DES-like cryptosystems. J. Cryptology **4**(1), 3–72 (1991)
3. Bracken, C., Byrne, E., Markin, N., McGuire, G.: A few more quadratic APN functions. Crypt. Commun. **3**(1), 43–53 (2011)
4. Bracken, C., Byrne, E., Markin, N., McGuire, G.: New families of quadratic almost perfect nonlinear trinomials and multinomials. Finite Fields Appl. **14**(3), 703–714 (2008)
5. Bracken, C., Byrne, E., Markin, N., McGuire, G.: On the Walsh spectrum of a new APN function. In: Galbraith, S.D. (ed.) Cryptography and Coding 2007. LNCS, vol. 4887, pp. 92–98. Springer, Heidelberg (2007). doi:10.1007/978-3-540-77272-9_6
6. Bracken, C., Byrne, E., Markin, N., McGuire, G.: On the Fourier spectrum of binomial APN functions. SIAM J. Discrete Math. **23**(2), 596–608 (2009)
7. Bracken, C., Byrne, E., Markin, N., McGuire, G.: Determining the nonlinearity of a new family of APN functions. In: Boztaş, S., Lu, H.-F.F. (eds.) AAECC 2007. LNCS, vol. 4851, pp. 72–79. Springer, Heidelberg (2007). doi:10.1007/978-3-540-77224-8_11
8. Bracken, C., Zha, Z.: On the Fourier spectra of the infinite families of quadratic APN functions. Finite Fields Appl. **18**(3), 537–546 (2012)
9. Brinkmann, M., Leander, G.: On the classification of APN functions up to dimension five. Des. Codes Crypt. **49**(1–3), 273–288 (2008)
10. Browning, A.K., Dillon, F.J., Kibler, E.R., McQuistan, T.M.: APN polynomials and related codes. J. Comb. Inf. Syst. Sci. **34**(1–4), 135–159 (2009). Special Issue in Honor of Prof. D.K Ray-Chaudhuri on the occasion of his 75th birthday
11. Browning, A.K., Dillon, F.J., McQuistan, T.M., Wolfe, J.A.: An APN permutation in dimension six. In: Post-proceedings of the 9-th International Conference on Finite Fields and Their Applications Fq 2009, Contemporary Math, AMS, vol. 518, pp. 33–42 (2010)
12. Budaghyan, L., Carlet, C.: Classes of quadratic APN trinomials and hexanomials and related structures. IEEE Trans. Inform. Theor. **54**(5), 2354–2357 (2008)

13. Budaghyan, L., Carlet, C., Leander, G.: Two classes of quadratic APN binomials inequivalent to power functions. IEEE Trans. Inform. Theor. **54**(9), 4218–4229 (2008)
14. Budaghyan, L., Carlet, C., Leander, G.: Constructing new APN functions from known ones. Finite Fields Appl. **15**(2), 150–159 (2009)
15. Budaghyan, L., Carlet, C., Leander, G.: On a construction of quadratic APN functions. In: IEEE Information Theory Workshop, pp. 374–378 (2009)
16. Budaghyan, L., Carlet, C., Pott, A.: New classes of almost bent and almost perfect nonlinear functions. IEEE Trans. Inform. Theor. **52**(3), 1141–1152 (2006)
17. Canteaut, A., Charpin, P., Dobbertin, H.: Binary m-sequences with three-valued crosscorrelation: a proof of Welch's conjecture. IEEE Trans. Inform. Theor. **46**(1), 4–8 (2000)
18. Carlet, C.: Relating three nonlinearity parameters of vectorial functions and building APN functions from bent functions. Des. Codes Crypt. **59**(1–3), 89–109 (2011)
19. Carlet, C.: Vectorial Boolean functions for cryptography. In: Crama, Y., Hammer, P. (eds.) Boolean Methods and Models. Cambridge University Press (2005–2006, to appear). Chapter of the Monography
20. Carlet, C., Charpin, P., Zinoviev, V.: Codes, bent functions and permutations suitable for DES-like cryptosystems. Des. Codes Crypt. **15**(2), 125–156 (1998)
21. Chabaud, F., Vaudenay, S.: Links between differential and linear cryptanalysis. In: Santis, A. (ed.) EUROCRYPT 1994. LNCS, vol. 950, pp. 356–365. Springer, Heidelberg (1995). doi:10.1007/BFb0053450
22. Dobbertin, H.: Almost perfect nonlinear power functions over $GF(2^n)$: the Niho case. Inform. Comput. **151**, 57–72 (1999)
23. Dobbertin, H.: Almost perfect nonlinear power functions over $GF(2^n)$: the Welch case. IEEE Trans. Inform. Theor. **45**, 1271–1275 (1999)
24. Dobbertin, H.: Almost perfect nonlinear power functions over $GF(2^n)$: a new case for n divisible by 5. In: Proceedings of Finite Fields and Applications Fq5, pp. 113–121 (2000)
25. Dobbertin, H.: Another proof of Kasami's theorem. Des. Codes Crypt. **17**, 177–180 (1999)
26. Edel, Y.: Quadratic APN functions as subspaces of alternating bilinear forms. In: Contact Forum Coding Theory and Cryptography III 2009, Belgium, pp. 11–24 (2011)
27. Edel, Y., Pott, A.: A new almost perfect nonlinear function which is not quadratic. Adv. Math. Commun. **3**(1), 59–81 (2009)
28. Gold, R.: Maximal recursive sequences with 3-valued recursive crosscorrelation functions. IEEE Trans. Inform. Theor. **14**, 154–156 (1968)
29. Göloğlu, F.: Almost perfect nonlinear trinomials and hexanomials. Finite Fields Appl. **33**, 258–282 (2015)
30. Hollmann, H., Xiang, Q.: A proof of the Welch and Niho conjectures on crosscorrelations of binary m-sequences. Finite Fields Appl. **7**, 253–286 (2001)
31. Janwa, H., Wilson, R.M.: Hyperplane sections of fermat varieties in P^3 in char. 2 and some applications to cyclic codes. In: Cohen, G., Mora, T., Moreno, O. (eds.) AAECC 1993. LNCS, vol. 673, pp. 180–194. Springer, Heidelberg (1993). doi:10.1007/3-540-56686-4_43
32. Kasami, T.: The weight enumerators for several classes of subcodes of the second order binary Reed-Muller codes. Inform. Control **18**, 369–394 (1971)
33. Matsui, M.: Linear cryptanalysis method for DES cipher. In: Helleseth, T. (ed.) EUROCRYPT 1993. LNCS, vol. 765, pp. 386–397. Springer, Heidelberg (1994). doi:10.1007/3-540-48285-7_33

34. Nyberg, K.: Differentially uniform mappings for cryptography. In: Helleseth, T. (ed.) EUROCRYPT 1993. LNCS, vol. 765, pp. 55–64. Springer, Heidelberg (1994). doi:10.1007/3-540-48285-7_6
35. Nyberg, K.: Perfect nonlinear S-boxes. In: Davies, D.W. (ed.) EUROCRYPT 1991. LNCS, vol. 547, pp. 378–386. Springer, Heidelberg (1991). doi:10.1007/3-540-46416-6_32
36. Tan, Y., Qu, L., Ling, S., Tan, C.H.: On the Fourier spectra of new APN functions. SIAM J. Discrete Math. **27**(2), 791–801 (2013)
37. Yu, Y., Wang, M., Li, Y.: A matrix approach for constructing quadratic APN functions. In: Pre-proceedings of the International Conference WCC 2013, Bergen, Norway (2013)

Families of Convolutional Codes over Finite Fields: A Survey

M. Isabel García-Planas[✉]

Dept. de Matemàtiques, Universitat Politècnica de Catalunya, Barcelona, Spain
maria.isabel.garcia@upc.edu

Abstract. The goal of this work is to give explicit interconnections between control theory and coding. It is well-known the existence of a closed relation between linear systems over finite fields and convolutional codes that allow to understand some properties of convolutional codes and to construct them. The connection between convolutional codes and linear systems permit to consider control as well as analyze observability of convolutional codes under linear systems point of view.

An accurate look at the algebraic structure of convolutional codes using techniques of linear systems theory as well a study of input-state-output representation control systems. A particular property considered in control systems theory called output-controllability property is analyzed and used for solve the decoding process of this kind of codes.

1 Introduction

At the origin, coding theory has been devoted mainly to information theory. In coding theory had, in fact, emerged from the need for better communication and better computer data storage. Concretely, convolutional codes are used on many occasions to transfer data with high demands on speed. To this end, we require potent codes of high rates. These codes are frequently implemented in composite with a hard-decision code, particularly Reed Solomon. Before turbo codes, such constructions were the most efficient, coming closest to the Shannon limit.

The convolutional codes are an alternative to the block codes because of their simplicity of generation with a little shift register. The main difference between them is the introduction of the concept of memory, that is, the coding at any given time will not depend only on the word to be coded, also on those previously coded. These codes have a great advantage over those of blocks in channels with high noise (high probability of error). Wireless communications or satellite communications stand out among their uses.

Convolutional codes were introduced by Elias [3] which suggests using a polynomial matrix $G(z)$ in the encoding process and allow the generation of the code line without using a previous buffer. G.D. Forney in [4] explained that the term "convolutional" is used because the output sequences can be regarded as the convolution of the input sequence with the sequences in the encoder.

There is a considerable amount of literature on the theory of convolutional codes over finite fields, see [1,3,5,9–11,13–17] or [21], for example. In particular,

© Springer International Publishing AG 2017
S. El Hajji et al. (Eds.): C2SI 2017, LNCS 10194, pp. 17–34, 2017.
DOI: 10.1007/978-3-319-55589-8_2

in [16] the author find an overview of the different approaches to the subject of convolutional code. In this work we use the definition of convolutional code as submodule of $\mathbb{F}[z]^n$ being interesting ir order to obtain a realization as linear system. First order and input-state-output representations can be found in [18, 19, 22, 23].

2   Convolutional Codes over Finite Fields

Let $\mathbb{F}_q$ be the finite field of $q = p^r$ elements, the set of the input alphabet channel. In the sequel, and if the confusion is not possible, we denote $\mathbb{F}_q$ simply as $\mathbb{F}$.

Definition 1. *A rate (n, k) convolutional code $\mathcal{C}$, over a finite field $\mathbb{F}$ is a finitely generated $\mathbb{F}[z]$-submodule of $\mathbb{F}^n[z]$ of rank k.*

A convolutional code $\mathcal{C}$ can be expressed in a matrix form (called generator matrix) as follows.

$$G(z) : \mathbb{F}[z]^\ell \longrightarrow \mathbb{F}[z]^n$$
$$u(z) \longrightarrow v(z) = G(z)u(z)$$

of order $n \times \ell$, $\ell \geq k$, whose columns collect a system of generators of the finitely generated submodule representing the code, that is to say $\mathcal{C} = \operatorname{Im} G(z)$.

Note that $\mathbb{F}[z]$ is a principal ideal domain and then a convolutional code $\mathcal{C}$ has a well-defined rank k and there exists a full-rank matrix $G(z)$ (of rank k) such that $\mathcal{C} = \operatorname{colsp}_{\mathbb{F}[z]} G(z)$.

So, it is possible to refine the definition of generator matrix considering the notion of *encoder*, (see [23], for more details).

Definition 2. *An encoder to $\mathcal{C}$ is a matrix*

$$G(z) : \mathbb{F}[z]^k \longrightarrow \mathbb{F}[z]^n$$
$$u(z) \longrightarrow v(z) = G(z)u(z)$$

such that $\operatorname{Im} G(z) = \mathcal{C}$ and $G(z)$ is injective.

If we assume that $G(z)$ is a $n \times k$ matrix with entries in $\mathbb{F}[z]$, the set

$$\mathcal{C} = \{v(z) \in \mathbb{F}^n[z] \mid \exists u(z) \in \mathbb{F}^k[z] \text{ such that } v(z) = G(z)u(z)\}$$

defines a submodule of $\mathbb{F}^n[z]$. Note that $\operatorname{Im}(G)$ is a finitely generated submodule.

The above definition implies that a $n \times k$ polynomial matrix is an encoder of $\mathcal{C}$ if its columns form a basis of the free module $\mathcal{C}$. In particular, an encoder is a generator matrix which $l = k$ and $G(z)$ is injective.

We denote by ν_i the maximum of all degrees of each of the polynomials of each column and we can assume that $\nu_1 \geq \nu_2 \geq \dots \geq \nu_k$ up to realignment. The number ν_1 is called the memory of the code and the collection of numbers ν_i are known as Forney's indices.

Remember that in convolutional codes, the coding of a word varies according to the words transmitted previously. And just the memory of the code ν_1 corresponds to the number of previous words on which the encoding depends. Notice that if $\nu_1 = 0$ the convolutional code is a block code.

Moreover, there exists another parameter related with convolutional codes and their encoders; that is, the complexity of both objects. The relation between these complexities is the key of the definition of a minimal encoder.

Definition 3. *(a) The complexity of the encoder (also called constraint length) is $c = \sum_{i=0}^{k} \nu_i$.*
(b) The degree or complexity of a convolutional code $\mathcal{C}$ is the highest degree of the full size minors of any encoder, and it is denoted by $\delta(\mathcal{C})$.

We ask if these two numbers ever coincide, the answer is "in general no", and for the case where they coincide we have the following definition.

Definition 4. *Let $\mathcal{C} \subset \mathbb{F}[z]^n$ be a (n, k)-convolutional code. An encoder matrix $G(z)$ of $\mathcal{C}$ is called minimal if and only if the complexity of the encoder coincides with the complexity of the code. That is to say $c = \delta(\mathcal{C})$*

It is well known that if we apply a basis change in $\mathbb{F}[z]^k$, it does not change the path of the map $G(z)$. Then, we have the following results relating minimal encoders:

Lemma 1. *Let $G(z)$ be an $n \times k$ polynomial matrix of rank k defining a convolutional code $\mathcal{C} = colsp_{\mathbb{F}[z]}G(z)$. Let $\widehat{G}(z)$ be an $n \times k$ polynomial matrix of rank k over $\mathbb{F}[z]$. The following statements are verified:*

1. *$G(z)$ and $\widehat{G}(z)$ define the same behaviour if and only if there exists a $k \times k$ unimodular matrix $U(z)$ such that $\widehat{G}(z) = G(z)U(z)$*
2. *There exists an unimodular matrix $U(z)$ such that $\widehat{G}(z) = G(z)U(z)$ is a minimal encoder.*
3. *If $G(z)$ and $\widehat{G}(z)$ are minimal encoders of $\mathcal{C}$ then they have the same column degrees.*

Definition 5. *The column degrees $(\kappa_1, ..., \kappa_k)$ of any minimal encoder $\widehat{G}(z)$ of $\mathcal{C}$ are known as the Kronecker or controllability indices of the code. We can reorder them if it is necessary such that $\kappa_1 \geq ... \geq \kappa_k$. The invariant $\delta = \sum_{i=1}^{k} \kappa_i$ is the degree of complexity of the code $\mathcal{C}$.*

(In some coding literature, δ is called the complexity of the code).

Note that the controllability indices of a convolutional code are unique and invariant of the code. If we consider a minimal encoder of a convolutional code then the controllability indices and Forney's indices are equal, and in this case, $\kappa_1 = \nu_1$ is the memory of the encoder.

We give some notions about observable convolutional codes that are useful in the following Chapter.

Definition 6. *Let $G(z)$ be an encoder of a (n, k) convolutional code C over $\mathbb{F}$. A syndrome former for the code C is a homomorphism of modules given by*

$$\psi : \mathbb{F}[z]^n \to \mathbb{F}[z]^{n-k}$$

with the property that $Im\ G(z) \subseteq Ker\ \psi$.

Definition 7. *Let $G(z)$ be an encoder of a (n, k) convolutional code C over $\mathbb{F}$. The convolutional code C is observable if and only if $G(z)$ is right-prime, i.e. all $k \times k$-minors are non-zero and they have non trivial common factors (z^ℓ, $\ell \in \mathbb{N}$ are trivial).*

Proposition 1. *Let $G(z)$ be an encoder of a (n, k) convolutional code C over $\mathbb{F}$. The convolutional code C is observable if and only if there exits an encoder $G(z)$ and a syndrome Former ψ such that the following sequence is exact*

$$0 \to \mathbb{F}[z]^k \xrightarrow{G(z)} \mathbb{F}[z]^n \xrightarrow{\psi} \mathbb{F}[z]^{n-k} \to 0$$

in other words, if a convolutional code C is observable there exists a polynomial matrix $H(z)$ (a syndrome former) with the property that $v \in C$ if and only if $H(z)v(z) = 0$.

The representation of a code among relatively different representations by means of a polynomial matrix is not unique, but we have the following proposition.

Proposition 2. *Two $n \times k$ rational encoders $G_1(z)$, and $G_2(z)$ define the same convolutional code, if and only if there exists a $k \times k$ unimodular matrix $U(z)$ such that $G_1(z)U(z) = G_2(z)$.*

Remember that a polynomial matrix $P(z) \in \mathbb{F}[z]$ is unimodular if there exists another matrix $Q(z)$ such that $P(z)Q(z) = I$.

After a suitable permutation of the rows, we can assume that the generator matrix $G(z)$ is in the form

$$G(z) = \begin{pmatrix} P(z) \\ Q(z) \end{pmatrix} \tag{1}$$

with right coprime polynomial factors (block of polynomials) $P(z) \in \mathbb{F}_{(n-k) \times k}$ and $Q(z) \in \mathbb{F}_{k \times k}$, respectively.

It is possible to consider the equivalent rational encoder where $Q(z) \neq 0$

$$\begin{pmatrix} P(z) \\ Q(z) \end{pmatrix} Q^{-1}(z) = \begin{pmatrix} P(z)Q^{-1}(z) \\ I \end{pmatrix}. \tag{2}$$

In the convolutional codes the Hamming distance can be defined as in block codes, the number of symbols in which two encoded bit sequences differ.

In convolutional codes the free distance $d_{free}(C)$ of a code C is defined as the minimum Hamming distance between two encoded bit sequences. This depends on the number of errors that the code is able to correct. As in block codes, the Hamming distance is calculated by comparing the outputs with the null input.

In a more formal form

Definition 8.

$$\mathrm{d}_{\text{free}}(\mathcal{C}) = \min\left\{wt(v(z)) \mid v(z) \in \mathcal{C} \text{ with } v(z) \neq 0\right\}.$$

where the weight $wt(v(z))$ of $v(z) = v_0 + v_1 z + \ldots + v_l z^l \in \mathbb{F}^n q[z]$ (with $l \geq 0$) is defined as the sum of the Hamming weights of all their coefficients, that is,

$$wt(v(z)) = \sum_{i=0}^{l} wt(v_i).$$

and Hamming weight $wt(v)$ of a vector $v_i \in \mathbb{F}^n$, is the number of its nonzero components.

The importance of free distance is because it determines the corrective capacity of the code.

3   Convolutional Codes and Linear Systems

In this section, we recall the systems theory tools by introducing the input-state-output representation; then, we will talk about convolutional codes using the linear systems theory; and also introduce the realization for the transition between codes and linear systems.

A discrete linear time-invariant system is described by the equations

$$\begin{cases} x(t+1) = Ax(t) + Bu(t) \\ \quad\; y(t) = Cx(t) + Du(t) \end{cases} \tag{3}$$

where $A \in M_\delta(\mathbb{F})$, $B \in M_{\delta \times k}(\mathbb{F})$, $C \in M_{p \times \delta}(\mathbb{F})$, $D \in M_{p \times k}(\mathbb{F})$ (in our particular setup $p = n - k$) are constant matrices over the field $\mathbb{F}$, and $u(t) \in \mathbb{F}^k$, $x(t) \in \mathbb{F}^\delta$, $y(t) \in \mathbb{F}^p$ are the input, state and output vectors, respectively.

We will denote a system simply as the quadruple of matrices (A, B, C, D).

With initial condition $x(0) = 0$, a solution of the Eq. (3) can be obtained by making use of the Z-transform. Let $u(z)$, $x(z)$, $y(z)$ be the Z-transforms of the variables u, x, y of a time-invariant linear system. Then by applying the Z-transform to the equations of the system we obtain

$$\begin{cases} zx(z) = Ax(z) + Bu(z) \\ \;\; y(z) = Cx(z) + Du(z) \end{cases} \tag{4}$$

and as a result we have

$$y(z) = (C(zI_\delta - A)^{-1}B + D)u(z), \tag{5}$$

called the transfer function of the system, and the rational matrix

$$C(zI_\delta - A)^{-1}B + D = \frac{1}{\det(zIA)}C\mathrm{adj}(zIA)B + D,$$

where adj(M) represents the adjoint matrix of M, is called the transfer matrix, (notice that the transfer matrix will always be a rational matrix).

The values $z_0 \in \overline{\mathbb{F}}$ (where $\overline{\mathbb{F}}$ denotes the algebraic closure of the field $\mathbb{F}$) such that $\det(z_0 I_\delta - A) = 0$ are called eigenvalues of A and the set of all eigenvalues is called spectrum of A and is denoted by $Spec(A)$.

The bridge between linear systems theory and convolutional codes is given by a duality between codes and sets input/state/output representations that are controllable state space systems.

Given a convolutional code, with a specific encoding matrix $G(z)$, we can find four matrices (A, B, C, D) of adequate sizes, corresponding to the size of the encoder, defining the system

$$\left. \begin{array}{l} x(t+1) = Ax(t) + Bu(t) \\ y(t) = Cx(t) + Du(t) \\ v(t) = \begin{pmatrix} y(t) \\ u(t) \end{pmatrix} \\ x(0) = 0 \end{array} \right\} . \tag{6}$$

where $x(t)$ is called state vector, $u(t)$ information vector, $y(t)$ parity vector and $v(t)$ the code vector or codeword. The linear system (A, B, C, D) associated to the encoder $G(z)$ is called a realization of $G(z)$. We are interested in minimal realizations.

In terms of the input-state-output representation of a convolutional code, we have the following characterization of the free distance.

Definition 9.

$$d_{\text{free}}(\mathcal{C}) = \min \left\{ \sum_{t=0}^{\infty} wt(u_t) + \sum_{t=0}^{\infty} wt(y_t) \right\}$$

Where the minimum is considered over all non-null code words.

Due to algebraic reasons, we assume throughout the paper that the code words are of finite weight.

Another well-studied concept in convolutional codes theory is that of column distances. The jth column distance of the code $\mathcal{C}$ is defined as the following manner

Definition 10.

$$d_j = \min \left\{ \sum_{t=0}^{j} wt(u_t) + \sum_{t=0}^{j} wt(y_t) \right\},$$

where the minimum is taken over all trajectories (u_t, y_t) of the system (6) with initial vector $u_0 \neq 0$.

It is clear that

$$d_0 \leq d_1 \leq d_2 \leq \dots$$

and hence there exists an integer r such that $d_r = d_{r+j}$ for all $j \leq 0$. This largest possible column distance is of central importance in coding theory.

Proposition 3.

$$d_{\text{free}} = \lim_{j \to \infty} d_j$$

Codes with a large free distance and the largest possible column distances are very desirable.

3.1   Realization

Linear systems for convolutional codes represent a mechanism to work on every little sub-piece of the encoding process. If we try to understand the physical control process, that goes along with the coding, the state of our encoding machine is modified by both the dynamics matrix and the input matrix.

Now, we present a method to obtain a realization.

Let $G(z)$ be a matrix generator of (n, k) convolutional code, in which we consider that is in the form $\left(\begin{smallmatrix} P(z) \\ Q(z) \end{smallmatrix} \right)$ with $Q(z)$ invertible and the degree δ of the polynomial $\det Q(z)$ being maximal among all minors of order k.

We decompose $P(z)Q(z)^{-1}$ into a polynomial matrix and a strictly proper matrix.

Let $p(z) = z^{\delta} + a_{\delta-1}z^{\delta-1} + \ldots + a_1 z + a_0$ the monic polynomial deduced from $\det Q(z)$. So, the matrix $P(z)Q(z)^{-1}$ is written in the form

$$\begin{pmatrix} d_{11} + \dfrac{q_{11}(z)}{p(z)} & \cdots & d_{1k} + \dfrac{q_{1k}(z)}{p(z)} \\ \vdots & & \vdots \\ d_{n-k1} + \dfrac{q_{n-k1}(z)}{p(z)} & \cdots & d_{n-kk} + \dfrac{q_{n-kk}(z)}{p(z)} \end{pmatrix}$$

$$q_{ij} = c_0^{ij} + c_1^{ij} z + \ldots + c_{\delta-1}^{ij} z^{\delta-1}$$

(by construction $d_{ij} \in \mathbb{F}$ and degree $q_{ij} < \delta$).

First of all and for simplicity, we analyze the case where $k = 1$.

We consider the following matrices

$$D = \begin{pmatrix} d_{11} \\ \vdots \\ d_{n-k1} \end{pmatrix} \in M_{(n-k)\times 1}(\mathbb{F}).$$

$$A = \begin{pmatrix} -a_{\delta-1} & -a_{\delta-2} & \cdots & -a_1 & -a_0 \\ 1 & 0 & \cdots & 0 & 0 \\ & & \ddots & & \\ 0 & 0 & \cdots & 1 & 0 \end{pmatrix} \in M_{\delta}(\mathbb{F})$$

$$B = \begin{pmatrix} 1 \\ 0 \\ \vdots \\ 0 \end{pmatrix} \in M_{\delta \times 1}(\mathbb{F})$$

$$C = \begin{pmatrix} c_{\delta-1}^{11} & \cdots & c_0^{11} \\ \vdots & & \vdots \\ c_{\delta-1}^{n-k1} & \cdots & c_0^{n-k1} \end{pmatrix} \in M_{(n-k)\times\delta}.$$

A simple calculation shows that $C(zI_\delta - A)^{-1}B + D = P(z)Q(z)^{-1}$.

Example 1. We consider the following code

$$G(z) = \begin{pmatrix} 1+z+z^2 \\ \alpha+z+\alpha^2 z^2 \\ \alpha^2+z+\alpha z^2 \end{pmatrix}$$

over the field $\mathbb{F}_4$,

$$G(z) = \begin{pmatrix} 1+z+z^2 \\ \alpha+z+\alpha^2 z^2 \\ \alpha^2+z+\alpha z^2 \end{pmatrix} = \begin{pmatrix} \dfrac{1+z+z^2}{\alpha^2+z+\alpha z^2} \\ \dfrac{\alpha+z+\alpha^2 z^2}{\alpha^2+z+\alpha z^2} \\ 1 \end{pmatrix} \alpha^2+z+\alpha z^2$$

$$\begin{pmatrix} \dfrac{1+z+z^2}{\alpha^2+z+\alpha z^2} \\ \dfrac{\alpha+z+\alpha^2 z^2}{\alpha^2+z+\alpha z^2} \\ 1 \end{pmatrix} = \begin{pmatrix} \alpha+1+\dfrac{1+\alpha+\alpha z}{\alpha^2+z+\alpha z^2} \\ \alpha+\dfrac{(1+\alpha)+(1+\alpha)z}{\alpha^2+z+\alpha z^2} \\ 1 \end{pmatrix} ;$$

$$P(z)Q(z)^{-1} = \begin{pmatrix} 1+\alpha+\dfrac{\alpha+z}{\alpha+(\alpha+1)z+z^2} \\ \alpha+\dfrac{\alpha+\alpha z}{\alpha+(\alpha+1)z+z^2} \end{pmatrix}.$$

Following as before we obtain the following realization (A, B, C, D) of the convolutional code where

$$D = \begin{pmatrix} \alpha+1 \\ \alpha \end{pmatrix}, B = \begin{pmatrix} 1 \\ 0 \end{pmatrix},$$
$$q_{11} = \alpha+z = c_0^{11}+c_1^{11}z$$
$$q_{21} = \alpha+\alpha z = c_0^{21}+c_1^{21}z$$
$$C = \begin{pmatrix} c_1^{11} & c_0^{11} \\ c_1^{21} & c_0^{21} \end{pmatrix} = \begin{pmatrix} 1 & \alpha \\ \alpha & \alpha \end{pmatrix}$$
$$p(z) = a_0+a_1 z+z^2 = \alpha+(1+\alpha)z+z^2$$
$$A = \begin{pmatrix} -a_1 & -a_0 \\ 1 & 0 \end{pmatrix} = \begin{pmatrix} 1+\alpha & \alpha \\ 1 & 0 \end{pmatrix}$$

A similar result holds for $k > 1$ case, the single input state-space models that correspond to the individual transfer functions from each input to each output, could be stacked into one large $k > 1$ state-space model.

Example 2. Let $G(z)$ be the following encoder matrix

$$G(z) = \begin{pmatrix} \dfrac{1+z}{z} & \dfrac{1}{1+z} \\ 1+z+z^2 & 0 \\ 0 & 1+z+z^2 \end{pmatrix} = \begin{pmatrix} P(z) \\ Q(z) \end{pmatrix}$$

So,

$$C(zI - A)^{-1}B + D = P(z)Q(z)^{-1} = \begin{pmatrix} \dfrac{1+z}{1+z+z^2} & \dfrac{1}{1+z+z^2} \\ \dfrac{z}{1+z+z^2} & \dfrac{1+z}{1+z+z^2} \end{pmatrix}$$

In this case $D = 0$ and $A = \begin{pmatrix} -1 & 0 & -1 & 0 \\ 0 & -1 & 0 & -1 \\ 1 & 0 & 0 & 0 \\ 0 & 1 & 0 & 0 \end{pmatrix}$, $B = \begin{pmatrix} 1 & 0 \\ 0 & 1 \\ 0 & 0 \\ 0 & 0 \end{pmatrix}$, $C = \begin{pmatrix} 1 & 1 & 0 & 1 \\ 1 & 0 & 1 & 1 \end{pmatrix}$.

An important concept in realization theory is the minimality.

Definition 11. *A realization (A, B, C, D) of a transfer matrix $G(z)$ is said to be minimal if no other realization of $G(z)$ has smaller dimension.*

In order to know the minimality of the realization we have the following result

Theorem 1. *Let (A, B, C, D) be a realization of $G(z)$. The following statements are equivalent:*

(1) (A, B, C, D) is minimal.
(2) The poles of $G(z)$ are the eigenvalues of A

Theorem 2. *Given a transfer matrix $G(z)$, all the minimal realizations of $G(z)$ are algebraically equivalent.*

The equivalence is in the following sense.

Definition 12. *Two systems (A, B, C, D) and (A', B', C', D') are equivalent if and only if there exist an invertible matrix P such that*

$$\begin{pmatrix} A' & B' \\ C' & D' \end{pmatrix} = \begin{pmatrix} P & \\ & I_p \end{pmatrix} \begin{pmatrix} A & B \\ C & D \end{pmatrix} \begin{pmatrix} P^{-1} & \\ & I_k \end{pmatrix}$$

Notice that this equivalence relation preserve the transfer matrix associate to the system:

$$\begin{aligned} C'(zI - A')^{-1}B' + D' &= CP^{-1}(zI - PAP^{-1})^{-1}PB + D \\ &= CP^{-1}(P(zI - A)P^{-1})^{-1}PB + D = CP^{-1}P(zI - A)^{-1}P^{-1}PB + D \\ &= C(zI - A)^{-1}B + D \end{aligned}$$

3.2   Control Properties of Convolutional Codes

We review the standard conditions about reachability (controllability from the origin) over the input-state-output representation of a convolutional code $\mathcal{C}$ over $\mathbb{F}$. First, we recall some results.

Definition 13. *Let (A, B, C, D) be matrices over $\mathbb{F}$ describing a linear system as in (3). The controllability (reachability) matrix was defined by*

$$\mathcal{C}(A, B) = \begin{pmatrix} B & AB & \dots & A^{\delta-2}B & A^{\delta-1}B \end{pmatrix} \tag{7}$$

It is well-Known that, a linear system (A, B, C, D) over a field $\mathbb{F}$ is reachable if its controllability matrix has full row rank; that is, $\operatorname{rank} \Phi_{\delta}(A, B) = \delta$. Or, equivalently, the Hautus test is verified.

$$\operatorname{rank} \mathcal{C}(A, B) = \delta \text{ if and only if } \operatorname{rank}(z_0 I + A \mid B) = \delta, \ \forall z_0 \in \overline{\mathbb{F}}$$

Remark 1. The controllability depends only on the state equation of the system.

Remark 2. Note that by construction, realization constructed is controllable.

Duality between convolutional codes and reachable state space realization is useful to construct observable convolutional codes: an input-state-output realization is always a reachable dynamical linear system. If it is also observable, then the following results allow us to get an associated observable convolutional code.

Rosenthal and York in [19] show that, starting from a minimal representation of a convolutional code, then this code is non-catastrophic if and only if the pair (A, C) is observable.

In terms of linear systems, let (A, B, C, D) be matrices over $\mathbb{F}$ describing a system. The observability matrix is defined by

$$\mathcal{O}(A, C) = \begin{pmatrix} C \\ CA \\ CA^2 \\ \vdots \\ CA^{\delta-1} \end{pmatrix} \tag{8}$$

Lemma 2. *The system (A, B, C, D) is observable if and only $\operatorname{rank} \mathcal{O}(A, C) = \delta$ or equivalently, by the Hautus Test, $\forall z_0 \in \overline{\mathbb{F}}$,*

$$\operatorname{rank} \begin{pmatrix} -z_0 I + A \\ C \end{pmatrix} = \delta$$

There are multiple realizations (A, B, C, D) for a given linear system. In particular, δ, the size of matrix A is not constant in the set of all realizations. Since δ is always a positive integer, it must reach a minimum value for certain realization. This minimum value of δ is called the McMillan degree of the system.

A realization (A, B, C, D) for which δ is equal to the degree of McMillan, we say that is a minimal realization. It is well known that the minimality property of a realization is related to the concepts of controllability and observability in the following sense.

Theorem 3 ([2]). *The realization (A, B, C, D) of a linear system is minimal if and only if (A, B) is a controllable pair and (A, C) is an observable pair.*

It is important to note that while in linear systems theory, a realization is minimal if and only if the pair (A, B) is controllable and the pair (A, C) is observable, for input-state-output representation of a convolutional code we do not have the same result. In fact, it is enough that the pair (A, B) be controllable so that the representation is minimal.

Related to the decodification of the encoders is the output-observability property.

Output-observability represents the possibility of an internal state, to be only defined by a finite set of outputs, for a finite number of steps. There are some literature about this topic, as for example [6–8].

Definition 14. *A system (A, B, C, D) is said to be output observable if the state sequence $x(0), \ldots, x(\ell)$ is uniquely determined by the knowledge of the output sequence $y(0), \ldots, y(\ell)$ for a finite number of steps $\ell \in \mathbb{N}$.*

Observe that $x(1), \ldots, x(\ell)$ are determined by the knowledge of $x(0)$ and $u(0), \ldots, u(\ell - 1)$ and the elements $x(0), u(0), \ldots,$ and $u(\ell)$ can be obtained solving the following system of matrix equations.

$$
\begin{cases}
y(0) = Cx(0) + Du(0) \\
y(1) = Cx(1) + Du(1) \\
\quad\;\; = CAx(0) + CBu(0) + Du(1) \\
\quad\;\; \vdots \\
y(\ell) = Cx(\ell) + Du(\ell) \\
\quad\;\; = CA^{\ell}x(0) + CA^{\ell-1}Bu(0) + \ldots + CBu(\ell - 1) + Du(\ell).
\end{cases}
\tag{9}
$$

Calling $T_{\ell}(A, B, C, D)$ (that we simply write T_{ℓ} if no confusion is possible) the matrix

$$
T_{\ell} = \begin{pmatrix}
C & D & & & \\
CA & CB & D & & \\
CA^2 & CAB & CB & D & \\
\vdots & & & \ddots & \ddots \\
CA^{\ell} & CA^{\ell-1}B & CA^{\ell-2}B & \ldots & CB \; D
\end{pmatrix}.
\tag{10}
$$

We have the following.

Proposition 4. *A system* (A, B, C, D) *is output observable if and only if the matrix* $\mathcal{T}_\ell$ *has full row rank for all* $\ell \in \mathbb{N}$.

Remark 3. If the number of rows is bigger than the number of columns, there are values of $y(0), \ldots, y(\ell)$, for which $(y(0), \ldots, y(\ell))$ is not a parity vector.

Corollary 1. *A necessary condition for output-observability of the system* (A, B, C, D) *is that the matrix* $(C \; D)$ *has full row rank.*

Therefore, we assume that the number of rows is less than or equal to the number of columns. It is well known that in this case and for each ℓ, the systems (9) have solution for all $y(0), \ldots, y(\ell)$, if and only if the systems have full rank.

Fixing the initial state $x(s) = 0$, the output-observability matrix allows us to describe a sequence of trajectories $\{v_s, \ldots, v_{s+\ell}\}$ in the following manner.

Theorem 4. *Let* (A, B, C, D) *be a representation of a convolutional code. Suppose that the initial state of the system is* $x(s) = 0$, *then*

$$\{v_s, \ldots, v_{s+\ell}\} = \mathrm{Ker}\, \mathcal{T}_\ell,$$

where

$$\mathcal{T}_\ell = \begin{pmatrix} D & -I & & & & & \\ CB & 0 & D & -I & & & \\ CAB & 0 & CB & 0 & D & -I & \\ \vdots & & & \ddots & & \ddots & \\ CA^{\ell-1}B & 0 & CA^{\ell-2}B & 0 & \ldots & CB & 0 & D & -I \end{pmatrix}$$

The output observability matrix is related with the syndrome former matrix used by Rosenthal and York [20], solving the decoding problem.

Let (A, B, C, D) be a realization of a convolutional code.

From the system

$$\begin{pmatrix} C & D & & & \\ CA & CB & D & & \\ CA^2 & CAB & CB & D & \\ \vdots & & & \ddots & \ddots \\ CA^\ell & CA^{\ell-1}B & CA^{\ell-2}B & \ldots & CB & D \end{pmatrix} \begin{pmatrix} x(s) \\ u(s) \\ \vdots \\ u(s+\ell) \end{pmatrix} = \begin{pmatrix} y(s) \\ y(s+1) \\ \vdots \\ y(s+\ell) \end{pmatrix} \quad (11)$$

we can deduce the syndrome former matrix for the given code.

Proposition 5. *Suppose that* $\ell \geq \delta$. *By making elementary transformations to matrix Eq. (11) we can deduce the syndrome former matrix for the convolutional code.*

Proof. The system (11) can be rewritten as

$$
\begin{pmatrix} C \\ CA \\ CA^2 \\ \vdots \\ CA^\ell \end{pmatrix} x(s) = \begin{pmatrix} D & & & & & I \\ CB & D & & & & & I \\ CAB & CB & D & & & & & I \\ \vdots & & & \ddots & \ddots & & & & \ddots \\ CA^{\ell-1}B & CA^{\ell-2}B & \dots & CB & D & & & & & I \end{pmatrix} \begin{pmatrix} -u(s) \\ -u(s+1) \\ \vdots \\ -u(s+\ell) \\ y(s) \\ y(s+1) \\ \vdots \\ y(s+\ell) \end{pmatrix} \tag{12}
$$

and making row elementary transformations, we obtain

$$
\left(\frac{\mathcal{O}(A,B)}{0} \right) (x(s)) = \left(\frac{M_1 \mid M_2}{M_3 \mid M_4} \right) \begin{pmatrix} -u(s) \\ \vdots \\ -u(s+\ell) \\ y(s) \\ \vdots \\ y(s+\ell) \end{pmatrix} \tag{13}
$$

Then, $\left(M_3 \mid M_4 \right)$ is the syndrome former matrix.

Example 3. In $\mathbb{F}_2$, we consider the system (A, b, C, D) with

$$
A = \begin{pmatrix} 0 & 1 \\ 1 & 0 \end{pmatrix}, \quad B = \begin{pmatrix} 1 \\ 0 \end{pmatrix}, \quad C = (1\ 0), \quad D = (1).
$$

Then, the system (12) for this particular case is

$$
\begin{pmatrix} 1 & 0 \\ 0 & 1 \\ 1 & 0 \\ 0 & 1 \\ 1 & 0 \\ 0 & 1 \\ 1 & 0 \end{pmatrix} (x(s)) = \begin{pmatrix} 1 & 0 & 0 & 0 & 0 & 0 & 1 & 0 & 0 & 0 & 0 & 0 & 0 \\ 1 & 1 & 0 & 0 & 0 & 0 & 0 & 1 & 0 & 0 & 0 & 0 & 0 \\ 0 & 1 & 1 & 0 & 0 & 0 & 0 & 0 & 1 & 0 & 0 & 0 & 0 \\ 1 & 0 & 1 & 1 & 0 & 0 & 0 & 0 & 0 & 1 & 0 & 0 & 0 \\ 0 & 1 & 0 & 1 & 1 & 0 & 0 & 0 & 0 & 0 & 1 & 0 & 0 \\ 1 & 0 & 1 & 0 & 1 & 1 & 0 & 0 & 0 & 0 & 0 & 1 & 0 \\ 0 & 1 & 0 & 1 & 0 & 1 & 1 & 0 & 0 & 0 & 0 & 0 & 1 \end{pmatrix} \begin{pmatrix} -u(0) \\ -u(1) \\ -u(2) \\ -u(3) \\ -u(4) \\ -u(5) \\ -u(6) \\ y(0) \\ y(1) \\ y(2) \\ y(3) \\ y(4) \\ y(5) \\ y(6) \end{pmatrix}
$$

Now, taking

$$P = \begin{pmatrix} 1 & 0 & 0 & 0 & 0 & 0 & 0 \\ 0 & 1 & 0 & 0 & 0 & 0 & 0 \\ -1 & 0 & 1 & 0 & 0 & 0 & 0 \\ 0 & -1 & 0 & 1 & 0 & 0 & 0 \\ -1 & 0 & 0 & 0 & 1 & 0 & 0 \\ 0 & -1 & 0 & 0 & 0 & 1 & 0 \\ -1 & 0 & 0 & 0 & 0 & 0 & 1 \end{pmatrix}$$

The system is reduced to

$$\begin{pmatrix} 1 & 0 \\ 0 & 1 \\ 0 & 0 \\ 0 & 0 \\ 0 & 0 \\ 0 & 0 \\ 0 & 0 \end{pmatrix} (x(s)) = \begin{pmatrix} 1 & 0 & 0 & 0 & 0 & 0 & 1 & 0 & 0 & 0 & 0 & 0 \\ 1 & 1 & 0 & 0 & 0 & 0 & 0 & 1 & 0 & 0 & 0 & 0 \\ -1 & 1 & 1 & 0 & 0 & 0 & -1 & 0 & 1 & 0 & 0 & 0 \\ 0 & -1 & 1 & 1 & 0 & 0 & 0 & -1 & 0 & 1 & 0 & 0 \\ -1 & 1 & 0 & 1 & 1 & 0 & 0 & -1 & 0 & 0 & 1 & 0 & 0 \\ 0 & -1 & 1 & 0 & 1 & 1 & 0 & 0 & -1 & 0 & 0 & 0 & 1 & 0 \\ -1 & 1 & 0 & 1 & 0 & 1 & 1 & -1 & 0 & 0 & 0 & 0 & 1 \end{pmatrix} \begin{pmatrix} -u(0) \\ -u(1) \\ -u(2) \\ -u(3) \\ -u(4) \\ -u(5) \\ -u(6) \\ y(0) \\ y(1) \\ y(2) \\ y(3) \\ y(4) \\ y(5) \\ y(6) \end{pmatrix}$$

So, the syndrome former matrix is

$$\begin{pmatrix} -1 & 1 & 1 & 0 & 0 & 0 & -1 & 0 & 1 & 0 & 0 & 0 \\ 0 & -1 & 1 & 1 & 0 & 0 & 0 & -1 & 0 & 1 & 0 & 0 \\ -1 & 1 & 0 & 1 & 1 & 0 & 0 & -1 & 0 & 0 & 1 & 0 & 0 \\ 0 & -1 & 1 & 0 & 1 & 1 & 0 & 0 & -1 & 0 & 0 & 0 & 1 & 0 \\ -1 & 1 & 0 & 1 & 0 & 1 & 1 & -1 & 0 & 0 & 0 & 0 & 1 \end{pmatrix}.$$

On the other hand, the following L-order block Toeplitz submatrix of the output-observability matrix

$$\mathfrak{T}_L = \begin{pmatrix} D & & & \\ CB & D & & \\ CAB & CB & D & \\ \vdots & & \ddots & \ddots \\ CA^{L-1}B & CA^{L-2}B & \dots & CB & D \end{pmatrix}. \tag{14}$$

allows us to obtain a characterization of the convolutional codes with maximum distance profile, in terms of its input-state-output representation.

Remember that (see [12]), an (n,k)-code], with column distances d_j and free distance d_{free}. has a maximum distance profile if

$$d_j = (n-k)(j+1) + 1 \text{ for } j = 0, \dots, L = \left\lfloor \frac{\delta}{k} \right\rfloor + \left\lfloor \frac{\delta}{n-k} \right\rfloor$$

Maximum distance profile convolutional codes are characterized by the property that two trajectories which start in the same state and proceed to a different state will have the maximum possible distance from each other relative to any other convolutional code of the same rate and degree.

Theorem 5 ([12]). *The matrices* (A, B, C, D) *generate a* (n, k)*-code with of maximum distance profile, (in terms of the input-state-output representation), if and only if the matrix* $\mathfrak{T}_L$*, verifies that any minor that is not trivially zero, is non-zero.*

Minor not trivially zero is understood in the following sense. We consider In this definition, we think of the nonzero entries of the block Toeplitz matrix $\mathfrak{T}_L$ as indeterminates of the polynomial ring $R = \mathbb{F}_q[x_{1,1}, \ldots, x_{1,(L+1)k}, \ldots, x_{(L+1)p,1}, \ldots, x_{(L+1)p,(L+1)k}]$. Specifically, if the entry (i, j) of the matrix is nonzero, we set it equal to $x_{i,j}$; otherwise, we leave it zero. So, a minor of $\mathfrak{T}_L$ is called trivially zero if it is zero viewed as an element of the ring R.

Example 4. In $\mathbb{F}_4$, the convolutional code (A, B, C, D) with

$$A = \begin{pmatrix} \alpha & \\ & 1 \end{pmatrix}, \ B = \begin{pmatrix} 1 \\ 1 \end{pmatrix}, \ C = (\alpha + 1 \ \alpha), \ D = (\alpha + 1)$$

where $\delta = 2$, $k = 1$, $p = 1$ then $L = 1$ and

$$\mathfrak{T}_L = \begin{pmatrix} \alpha + 1 & \\ 1 & \alpha + 1 \end{pmatrix}$$

So, the convolutional code has maximum distance profile.

4   Families of Convolutional Codes over Finite Fields

We are interested in convolutional codes where the matrices (A, B, C, D) or one of them, are not entirely defined having in certain positions parameters that can take any value from the field. So we can consider this parametric code as a family of convolutional codes.

These families of codes may be of interest when attempting to protect or hide certain information.

Anyway, we can not place parameters anywhere if we want to maintain certain properties of the code. In particular the structure of the matrix A, in this case and taking into account that the equivalence relation given in Definition 12 preserves this structure of matrices, we can consider classes of systems, and as representative of each class we find a system in which the matrix A is in some reduced form.

Example 5. In $\mathbb{F}_5$, we consider the following family of systems $(A(a), B(a), C(a), D(a))$ with

$$A(a) = \begin{pmatrix} 1 & a & 0 \\ 0 & 2 & 0 \\ 0 & 0 & 3 \end{pmatrix}, \ B(a) = \begin{pmatrix} 1 & 2 \\ 1 & 1 \\ 1 & 0 \end{pmatrix}, \ C(a) = \begin{pmatrix} 1 & 1 & 1 \\ 2 & 0 & 1 \end{pmatrix}, \ D(a) = \begin{pmatrix} 1 & 2 \\ 2 & 0 \end{pmatrix}$$

Taking the family of invertible matrices $P = \begin{pmatrix} 1 & a & 0 \\ 0 & 1 & 0 \\ 0 & 0 & 1 \end{pmatrix}$, this family is equivalent to $(A_1(a), B_1(a), C_1(a), D_1(a))$ with

$$A_1(a) = \begin{pmatrix} 1 & 0 & 0 \\ 0 & 2 & 0 \\ 0 & 0 & 3 \end{pmatrix}, \quad B_1(a) = \begin{pmatrix} 1-a & 2-a \\ 1 & 1 \\ 1 & 0 \end{pmatrix}, \quad C_1(a) = \begin{pmatrix} 1 & a+1 & 1 \\ 2 & 2a & 1 \end{pmatrix}, \quad D_1(a) = \begin{pmatrix} 1 & 2 \\ 2 & 0 \end{pmatrix}$$

So, for each $a \in \mathbb{F}_5$ we have a different system but all matrices $A(a)$ have the same structure. Obviously is not the same for the family $(\bar{A}(a), \bar{B}(a), \bar{C}(a), \bar{D}(a))$ with

$$\bar{A}(a) = \begin{pmatrix} 1+a & 0 & 0 \\ 0 & 2 & 0 \\ 0 & 0 & 3 \end{pmatrix}, \quad \bar{B}(a) = \begin{pmatrix} 1 & 2 \\ 1 & 1 \\ 1 & 0 \end{pmatrix}, \quad \bar{C}(a) = \begin{pmatrix} 1 & 1 & 1 \\ 2 & 0 & 1 \end{pmatrix}, \quad \bar{D}(a) = \begin{pmatrix} 1 & 2 \\ 2 & 0 \end{pmatrix}$$

where the matrix A in each member of the family has a different structure.

In $\mathbb{F}_q$ there are exactly $q^{\delta^2} \times q^{k\delta} \times q^{p\delta} \times q^{pk}$ different systems. In particular, if the matrix A is in such a way that in its reduced form is diagonal, we have

$$\frac{(\delta + q - 1)!}{\delta!(q-1)!} \times q^{k\delta} \times q^{p\delta} \times q^{pk}.$$

Taking into account that the cardinal of the set of invertible matrices $Gl(\delta, \mathbb{F}_q)$ is $\prod_{k=0}^{\delta-1}(q^\delta - q^k)$, it is possible to count the number of elements of each equivalent class and the number of classes.

For that, it suffices to define an action of the linear group over the set of systems $\mathcal{M} = \{(A, B, C, D)\}$:

$$\varphi : Gl(\delta, \mathbb{F}_q) \times \mathcal{M} \longrightarrow \mathcal{M}$$
$$(P, (A, B, C, D)) \longrightarrow (P^{-1}AP, P^{-1}B, CP, D)$$

Then, after to compute the stabilizer $\mathcal{S}_{(A,B,C,D)}$ of a fixed point $(A, B, C, D) \in \mathcal{M}$ defined as $\mathcal{S}_{(A,B,C,D)} = \{P \in Gl(\delta, \mathbb{F}_q) \mid \alpha(P, (A, B, C, D)) = (A, B, C, D) = \{P \in Gl(\delta, \mathbb{F}_q) \mid AP - PA = 0, PB - B = 0, CP - C = 0\}$
and now, it is easy to prove that there is a bijection between the set of equivalent systems to (A, B, C, D) and the quotient group $Gl(\delta, \mathbb{F}_q)/\mathcal{S}_{(A,B,C,D)}$.

Given a convolutional code (A, B, C, D), we are interested in to perturb it in order to improve their behaviour and control properties. That is, to find the values of the parameters for which our code has the appropriate or expected properties.

Example 6. In $\mathbb{F}_4$, let (A, B, C, D) be a convolutional code with

$$A = \begin{pmatrix} \alpha & \alpha+1 \\ \alpha+1 & 1 \end{pmatrix}, \quad B = \begin{pmatrix} \alpha+1 \\ \alpha \end{pmatrix}, \quad C = (\alpha+1 \; 1), \quad D = (1)$$

This code is no controllable because of:

$$\text{rank} \begin{pmatrix} z - \alpha & -(\alpha + 1) & \alpha + 1 \\ -(\alpha + 1) & z - 1 & \alpha \end{pmatrix} = 1 \text{ for } z = \alpha + 1$$

And not observable because of:

$$\text{rank} \begin{pmatrix} z - \alpha & -(\alpha + 1) \\ -(\alpha + 1) & z - 1 \\ \alpha + 1 & 1 \end{pmatrix} = 1 \text{ for } z = 0$$

Considering the family of convolutional codes $(A(a), B(a), C(a), D(a))$ be a convolutional code with

$$A = \begin{pmatrix} \alpha + a & \alpha + 1 \\ \alpha + 1 & 1 \end{pmatrix}, \ B = \begin{pmatrix} \alpha + 1 \\ \alpha \end{pmatrix}, C = (\alpha + 1 \ 1), \ D = (1)$$

The codes of the family are controllable and observable if and only if $a \neq 0$.

References

1. de Castro, N.: Feedback classification of linear systems and convolutional codes. applications in cybernetics, coding theory and cryptography. Ph.D. Universidad de León (2016)
2. Chen, C.T.: Introduction to Linear System Theory. Holt, Rinehart and Winston Inc., New York (1970)
3. Elias, P.: Coding for noisy channels. IRE Conv. Rec. **4**, 37–46 (1955)
4. Forney, G.D.: Convolutional codes: algebraic structure. IEEE Trans. Inf. Theor. **16**(6), 720–738 (1970)
5. Fragouli, C., Wesel, R.D.: Convolutional codes and matrix control theory. In: Proceedings of the 7th International Conference on Advances in Communications and Control, Athens, Greece (1999)
6. Garcia-Planas, M.I., Domínguez-Garcia, J.L.: A general approach for computing residues of partial-fraction expansion of transfer matrices. WSEAS Trans. Math. **12**(7), 647–756 (2013)
7. Garcia-Planas, M.I., Domínguez-Garcia, J.L.: Alternative tests for functional and pointwise output-controllability of linear time-invariant systems. Syst. Control Lett. **62**(5), 382–387 (2013)
8. García-Planas, M.I., Tarragona, S.: Output observability of time-invariant singular linear systems. In: PHYSCON 2011, Léon, Spain (2011)
9. García-Planas, M.I., Souidi, E.M., Um, L.E.: Convolutional codes under linear systems point of view. Analysis of output-controllability. Wseas Trans. Math. **11**(4), 324–333 (2010)
10. Gluesing-Luerssen, H.: On the weight distribution of convolutional codes. Linear Algebra Appl. **408**, 298–326 (2005)
11. Gluesing-Luerssen, H., Helmke, U., Iglesias Curto, J.I.: Algebraic decoding for doubly cyclic convolutional codes. Adv. Math. Commun. **4**, 83–99 (2010)
12. Hutchinsona, R., Rosenthal, J., Smarandacheb, R.: Convolutional codes with maximum distance prole. Syst. Control Lett. **54**(1), 53–63 (2005)

13. Kuijper, M., Pinto, R.: On minimality of convolutional ring encoders. IEEE Trans. Inf. Theor. **55**(11), 4890–4897 (2009)
14. Rosenthal, J.: Some interesting problems in systems theory which are of fundamental importance in coding theory. In: Proceedings of the 36th IEEE Conference on Decision and Control (1997)
15. Rosenthal, J.: An algebraic decoding algorithm for convolutional codes. In: Dynamical Systems, Control, Coding, Computer Vision; New Trends, Interfaces, and Interplay, Birkhäuser, Basel, pp. 343–360 (1999)
16. Rosenthal J.: Connections between linear systems and convolutional codes. In: Marcus B., Rosenthal J. (eds.) Codes, Systems, and Graphical Models. The IMA Volumes in Mathematics and its Applications, vol 123. Springer, New York (2001)
17. Rosenthal, J., Smarandache, R.: Maximum distance separable convolutional codes. Appl. Algebra Eng. Commun. Comput. **10**(1), 15–32 (1999)
18. Rosenthal, J., Schumacher, J.M., York, E.V.: On behaviors and convolutional codes. IEEE Trans. Inf. Theor. **42**(6), 1881–1891 (1996)
19. Rosenthal, J., York, E.V.: BCH convolutional codes. IEEE Trans. Inform. Theor. **45**(6), 1833–1844 (1999)
20. Rosenthal, J., York, E.V.: On behaviors and convolutional codes. IEEE Trans. Inform. Theor. **42**(6), 1881–1891 (1996)
21. Smarandache, R., Gluesing-Luerssen, H., Rosenthal, J.: Constructions of MDS-convolutional codes. IEEE Trans. Inf. Theor. **47**(5), 2045–2049 (2002)
22. Um, L.E.: A contribution to the theory of convolutional codes from systems theory point of view. Ph.D. dissertation. Université Mohammed V. Rabat (2015)
23. York, E.V.: Algebraic description and construction of error correcting codes, a systems theory point of view, Ph.D. dissertation, Univ. Notre Dame (1997)

Codes for Side-Channel Attacks and Protections

Sylvain Guilley[1,2(✉)], Annelie Heuser[3], and Olivier Rioul[2]

[1] Secure-IC S.A.S., 15 Rue Claude Chappe, Bât. B, 35 510 Cesson-Sévigné, France
sylvain.guilley@secure-ic.com
[2] LTCI, Télécom ParisTech, Université Paris-Saclay, 75 013 Paris, France
[3] IRISA, 263 Avenue Général Leclerc, 35 000 Rennes, France

Abstract. This article revisits side-channel analysis from the standpoint of coding theory. On the one hand, the attacker is shown to apply an optimal decoding algorithm in order to recover the secret key from the analysis of the side-channel. On the other hand, the side-channel protections are presented as a coding problem where the information is mixed with randomness to weaken as much as possible the sensitive information leaked into the side-channel. Therefore, the field of side-channel analysis is viewed as a struggle between a coder and a decoder. In this paper, we focus on the main results obtained through this analysis. In terms of attacks, we discuss optimal strategy in various practical contexts, such as type of noise, dimensionality of the leakage and of the model, etc. Regarding countermeasures, we give a formal analysis of some masking schemes, including enhancements based on codes contributed via fruitful collaborations with Claude Carlet.

1 Introduction

Digital information is handled by electronic devices, such as smartphones or servers. Some information, such as keys, is sensitive, in the sense that it shall remain confidential. In general, information is present in three states within devices: *at rest*, *in transit*, and *in computation*. The protection of information at rest can be ensured by on-chip encryption in the memories. The same technique applies to the data in transit: the buses can be encrypted (e.g., in a lightweight way, in which case one uses the term *scrambling*). Therefore, the protection of information during computation is the big issue to be dealt with. It is a real challenge, as a computing devices inadvertently leak some information about the data they manipulate. In this context, three questions are of interest:

1. How does an attacker best exploit the leaked information? The situation is similar to that of a decoding problem, and one aims at finding the optimal decoder.
2. Second, the designer (and the end user) aim at being protected against such attacks. Their goal is thus to try and weaken the side-channel. Randomization is one option, referred to as masking in the literature. We will illustrate that it can be seen as the use of code to optimally mix some random bits into the computations, with the possibility to eventually get rid off this entropy,

© Springer International Publishing AG 2017
S. El Hajji et al. (Eds.): C2SI 2017, LNCS 10194, pp. 35–55, 2017.
DOI: 10.1007/978-3-319-55589-8_3

e.g., at the end of the computation. Another interesting usage of codes is to detect faults in circuits. This dual use of codes is of interest in general security settings, where attacks can choose to be either passive or active. It is also very relevant in the case the circuit is trapped with a Hardware Trojan Horse.

3. Third, it is interesting to know in which respect the circuit leakage favors or not attacks. In particular, we will investigate the effect of glitches as a threat to masking schemes.

Outline. We start with the adversarial strategies in Sect. 2. Protection strategies, especially masking, are presented in Sect. 3. We will show how the circuit itself can contribute to the attack, through the analysis of glitches, in Sect. 4. Conclusions are in Sect. 5. Eventually, Appendix A gives some computation evidences why masking protection can be seen as reducing the signal-to-noise ratio, by increasing the noise.

2   Side-Channel Analysis as a Decoding Problem

In this section, we first describe the setup and the objective of the attacker. Second, we solve the objective of the attacker in various different setups.

2.1   Setup

We assume the device manipulates some data known by the attacker, such as a plaintext or a ciphertext, called T. This data is mixed with some secret, say a key k^*. The attacker manages to capture some noisy function of T and k^*, and attempts to extract k^*. For this purpose, he will enumerate (manageable) parts of the key (e.g., bytes), denoted k, and choose the key candidate $\hat{k}$ which is the most likely. Therefore, the attack resembles a communication channel, where the input is k^* and the output is $\hat{k}$. The attack is termed successful if $\hat{k} = k^*$.

Two kinds of leakage models are realistic in practice:

1. **direct probing** model, where the attacker uses some kind of probes, each being able to measure one bit,
2. **indirect measurement** of an aggregated function of the bits, using for instance an electromagnetic probe.

These two ways of capturing the signal are, by nature, very different. They are illustrated in Fig. 1.

The first one is noiseless. However, the bits in integrated circuits are nanometric, whereas probes are mesometric. Therefore, only few such probes can be used simultaneously. The security parameter is thus linked to the ability for the attacker to recover some useful information out of d probes (where d is typically 1, 2, 3 or 4). Besides, the probing requires a physical access to the wires, which is challenging, since it is possible that the contact breaks the bit to be probed.

(a) probing model   (b) bounded-moments model

Fig. 1. Settings for side-channel analysis. In the probing model (a), a few bits (here, $d = 3$) are measured with dedicated probes. In the bounded moments model (b), the attacker measures an integrated quantity of several bits.

Such attack is termed semi-invasive, since it leaves an evidence that the circuit has been tampered with (an opening is necessity to insert the probe).

The second one is noisy and also leaks some function of the bits. Therefore, the attacker needs to capture more than one trace to extract some information. This is why we model, in the sequel, traces by random variables. By convention, the variables are printed with capital letters, such as X, when designating a random variable, and with small letters, such as x, when designating the realization of random variables. We also denote by Q the number of queries ($=$ of measurements), and by $\mathbf{x} = (x_1, \ldots, x_Q)$ the vector of measurements. This attack will require a statistical analysis, which in general consists in the study of the leakage probability distribution. This starts in general by the analysis of the leakage moments.

We will link the two models in the case of RSM countermeasure (Sect. 3.5). The next Sect. 2.2 discusses the channel $k^\star \to \hat{k}$, for the second case.

2.2 Example of AWGN Channel

The key recovery setup is illustrated in Fig. 2 (see Fig. 1 in [24]). When the noise is Gaussian and independent from one measurement to others, it is referred to as AWGN (Additive white Gaussian noise). We write:

$$X = y(T, k^*) + N, \qquad \text{where } N \sim \mathcal{N}(0, \sigma^2). \tag{1}$$

The random variable $y(T, k^*)$ is the aggregated leakage model, and N is the noise (independent from Y). Let n the bitwidth of the key k and of the texts T. The function $y : \mathbb{F}_2^n \times \mathbb{F}_2^n \to \mathbb{R}$ is, in practice, the composition two functions $y = \varphi \circ f$, where:

- f is an algorithmic function called *sensitive variable*, such as $f(T, k^*) = S(T \oplus k^*)$, where S is a substitution box, and

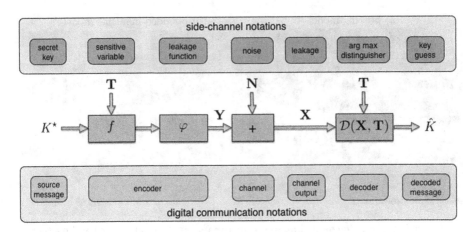

Fig. 2. Side-channel analysis as a communication channel

- $\varphi : \mathbb{F}_2^n \to \mathbb{R}$ accounts for the way the sensitive variable leaks, such as the Hamming weight $\varphi : z \mapsto w_H(z) = \sum_{i=1}^n z_i$.

2.3 Absence of Countermeasures

The optimal distinguisher is the key guess $\hat{k}$ which maximizes the success probability, that is the probability that $\hat{k}$ is actually k^*.

When there is no protection, all the uncertainty resides in the measurement noise. Thus, as the attacker knows T, he also knows $Y = Y(T, k)$ (for all key guess k).

Theorem 1 ([24, Theorem 4]). *In the AWGN setup, the optimal distinguisher is demonstrated to be equal to:*

$$\mathcal{D}_{opt}(\mathbf{x}, \mathbf{t}) = \mathrm{argmin}_k \|\mathbf{x} - \mathbf{y}(\mathbf{t}, k)\|_2^2 = \mathrm{argmax}_k \langle \mathbf{x} | \mathbf{y}(\mathbf{t}, k) \rangle - \frac{1}{2} \|\mathbf{y}(\mathbf{t}, k)\|_2^2, \quad (2)$$

where $\|\cdot\|_2$ is the Euclidean norm and $\langle \cdot | \cdot \rangle$ is the canonical scalar product.

2.4 Multivariate and Multimodel Setting

In the multivariate and multimodel case, the attacker is able to collect:

- not only one sample, but D (dimensionality) samples, and
- each function of the bits (e.g., $z \mapsto 1$, $z \mapsto z_i$ for $1 \le i \le n$, but also any selection of $z \mapsto \bigwedge_{i \in I} z_i$ where $I \subseteq \mathbb{F}_2^n$) has a different contribution.

We call S the number of models, and α the $D \times S$ matrix of the leakages, such that Eq. (1) is generalized as:

$$\mathbf{X} = \alpha \mathbf{y}(\mathbf{T}, k^*) + \mathbf{N}, \qquad \text{where } \mathbf{N} \sim \mathcal{N}(\mathbf{0}, \Sigma), \quad (3)$$

where $\mathbf{N}$ is multivariate normal of $D \times D$ covariance matrix Σ, and $\mathbf{Y} = \mathbf{y}(\mathbf{T}, k^*)$ is set of S models (e.g., $S = 1$ if the leakage model is the Hamming weight, or $S = n + 1$ if there is a non-zero offset (such offset is modeled by $z \mapsto 1$) and each bit $1 \leq i \leq n$ of the leakage model leaks differently). In this case also, **boldface** variables are vectorial (either *multivariate* or *multimodel*).

We have a generalization of Theorem 1:

Theorem 2 ([7, Theorem 1]). *Let us define* $\mathbf{x}' = \Sigma^{-1/2}\mathbf{x}$ *and* $\alpha' = \Sigma^{-1/2}\alpha$. *Then, in the multivariate and multimodel AWGN setup, the optimal distinguisher is demonstrated to be equal to:*

$$\mathcal{D}_{opt}^{D,S}(\mathbf{x}, \mathbf{t}) = \operatorname{argmin}_k \sum_{d=1}^{D} \|\mathbf{x}_d' - \alpha_d'\mathbf{y}((\mathbf{t}, k))\|_2^2$$

$$= \operatorname{argmax}_k \ \operatorname{tr}\left(\mathbf{x}'(\alpha'\mathbf{y}(\mathbf{t}, k))^{\mathsf{T}}\right) - \frac{1}{2}\|\alpha'\mathbf{y}(\mathbf{t}, k)\|_F^2,$$

where $\operatorname{tr}(\cdot)$ *is the trace operator of a square matrix and* $\|\cdot\|_F$ *is the Frobenius normal of a (rectangular) matrix.*

2.5   Collision

In some situations, the attacker does not know the leakage function $y = \varphi \circ f$, but knows that it is reused several times for different bytes, say $L > 1$. We denote by $x^{(\cdot)} = (x^{(1)}, \ldots, x^{(\ell)}, \ldots, x^{(L)})$ the L leakages. Therefore, the optimal attack consists in a collision attack where all the coefficients of the leakage function are regressed.

Theorem 3 ([5, Theorem 2.5]). *The optimal collision attack is:*

$$\mathcal{D}_{opt}^{L}(\mathbf{x}^{(\cdot)}, \mathbf{t}^{(\cdot)}) = \operatorname{argmax}_{k^{(\cdot)} \in (\mathbb{F}_2^n)^L} \sum_{u \in \mathbb{F}_2^n} \frac{\left(\sum_\ell \sum_{q/t_q^{(\ell)} \oplus k^{(\ell)} = u} x_q^{(\ell)}\right)^2}{\sum_\ell \sum_{q/t_q^{(\ell)} \oplus k^{(\ell)} = u} 1}.$$

Notice that in general, this attack allows to recover $(L - 1)$ n-bit keys when the collision is involving L samples with identical leakage model.

2.6   General Setting, with Countermeasures

In general, the device defends itself, by the implementation of protections. Masking is one of them. In the expression of y, in addition to T and k, another random variable M is introduced, called the mask, unknown to the attacker. It is usually assumed that it is uniformly distributed.

Theorem 4 ([8, Proposition 8]). *The optimal attack in case of masking countermeasure is:*

$$\mathcal{D}_{opt}^{M;L}(\mathbf{x}^{(\cdot)}, \mathbf{t}^{(\cdot)}) = \operatorname{argmax}_k \sum_{q=1}^{Q} \log\left\{\sum_m \exp\left\{\sum_{d=1}^{D} \frac{1}{\sigma^{(d)2}}\left(x_q^{(d)}y_q^{(d)} - \frac{1}{2}y_q^{(d)2}\right)\right\}\right\},$$

assuming that the noise at each sample d is normal of variance $\sigma^{(d)^2}$.

2.7 Link Between Success Probability, SNR and Leakage Function

The optimal distinguishers $\mathcal{D}_{opt}$ given in various scenarios ($\mathcal{D}_{opt}$ for nominal case in Sect. 2.3, $\mathcal{D}_{opt}^{D,S}$ for multivariate and multimodel case in Sect. 2.3, $\mathcal{D}_{opt}^{L}$ for the collision case in Sect. 2.5, and $\mathcal{D}_{opt}^{M;L}$ for the masked case in Sect. 2.6) allow to recover the secret key with the largest success rate (denoted as SR), but do not help in predicting the number of traces to reach a given success rate (or vice-versa).

Such relationship can be easily derived from the analysis of so-called *first-order exponents* [23]. Let us denote $\mathcal{A}_{opt}(\mathbf{x}, \mathbf{t}, k)$ the argument of maximization in either of $\mathcal{D}_{opt}$, $\mathcal{D}_{opt}^{D,S}$, $\mathcal{D}_{opt}^{L}$ or $\mathcal{D}_{opt}^{M;L}$. We have:

Theorem 5 ([23, Corollary 1]).

$$1 - \mathrm{SR}(\mathcal{D}) \approx e^{-Q \cdot \mathrm{SE}(\mathcal{D})} \tag{4}$$

where the first-order success exponent $\mathrm{SE}(\mathcal{D})$ *is equal to:*

$$\mathrm{SE}(\mathcal{D}) = \frac{1}{2} \min_{k \neq k^*} \frac{\left(\mathcal{A}_{opt}(\mathbf{x}, \mathbf{t}, k^*) - \mathcal{A}_{opt}(\mathbf{x}, \mathbf{t}, k) \right)^2}{\mathrm{Var}\left(\mathcal{A}_{opt}(\mathbf{x}, \mathbf{t}, k^*) - \mathcal{A}_{opt}(\mathbf{x}, \mathbf{t}, k) \right)}. \tag{5}$$

For the sake of the introduction of a signal-to-noise, we rewrite Eq. (1) as:

$$X = \alpha y(T, k^*) + N, \text{ where } \mathbb{E}(y(T, k^*)) = 0, \mathrm{Var}(y(T, k^*)) = 1 \text{ and } N \sim \mathcal{N}(0, \sigma^2).$$

Let us introduce generalized *confusion coefficients* [20]:

Definition 6 (General 2-way confusion coefficients [23, Definitions 8 and 10]). *For* $k \neq k^*$ *we define*

$$\kappa(k^*, k) = \mathbb{E}\left\{ \left(\frac{Y(k^*) - Y(k)}{2} \right)^2 \right\}, \tag{6}$$

$$\kappa'(k^*, k) = \mathbb{E}\left\{ \left(\frac{Y(k^*) - Y(k)}{2} \right)^4 \right\}. \tag{7}$$

For example, for the optimal distinguisher in the nominal case, the success exponent expression is:

Lemma 7 (SE for the optimal distinguisher, [23, Proposition 5]). *The success exponent for the optimal distinguisher takes the closed-form expression*

$$\mathrm{SE}(\mathcal{D}) = \frac{1}{2} \min_{k \neq k^*} \frac{\alpha^2 \kappa^2(k^*, k)}{\sigma^2 \kappa(k^*, k) + \alpha^2 (\kappa'(k^*, k) - \kappa(k^*, k)^2}. \tag{8}$$

This closed-form expression simplifies for high noise $\sigma \gg \alpha$ in a simple equation:

Corollary 8 ([23, Corollary 2]).

$$\mathrm{SE}(\mathcal{D}) \approx \frac{1}{2} \min_{k \neq k^*} \frac{\alpha^2 \kappa^2(k^*, k)}{\sigma^2 \kappa(k^*, k)} = \frac{1}{2} \cdot SNR \cdot \min_{k \neq k^*} \kappa(k^*, k), \tag{9}$$

where $\mathrm{SNR} = \alpha^2 / \sigma^2$ *is the signal-to-noise ratio (see [6] for the definition of SNR in the multivariate case).*

3   Side-Channel Protection

Side-channel attacks threaten the security of cryptographic implementations. Protections against such attacks can be devised using the coding theory. We illustrate in this section several techniques which randomize leakages in a view to decorrelate them from the internally manipulated data, and that (in some cases) also allow to detect malicious fault injections.

3.1   Strategies to Thwart Side-Channel Attacks

As discussed in Sect. 2.7 (especially in (9)), the success of an attack is all the larger as the leakage function has a higher confusion (6) and the SNR is high. However, the input of confusion is limited, since $0 \leq \min_{k \neq k^*} \kappa(k^*, k) \leq 1/2$ is bounded. Moreover, the defender cannot always change the algorithm nor its leakage model, that is $\min_{k \neq k^*} \kappa(k^*, k)$ is fixed. Thus, the defender is better off focusing on the reduction of the SNR.

This can be achieved in two flavors:

1. reduce the signal, as done in strategies aiming at flattening the leakage. This is easily achieved for some side-channels, such as timing: the execution time is made constant, e.g., by inserting dummy instructions or by balancing the code in each branch when the control flow forks. However, balancing an analogue quantity (such as power or electromagnetic field) is more challenging, let alone because of process variations, two identical gates or structures behave differently after fabrication. For instance, this is the working factor of physically unclonable functions (PUFs). Therefore, the quality of the protection depends on the ability of the fabrication plant to produce reproducible patterns. This fact naturally limits the quality of the designer's work, hence does not encourage to reach very high levels of security. In case this case, the second option is preferred;
2. increase the noise, by resorting to some extra random variables independent of that involved in the leakage function. Obviously, some artificial noise can be easily produced: one practical example consists in running an algorithm known to produce a lot of leakage (such as an asymmetrical engine, e.g., RSA) in parallel to the algorithm to protect. However, there remains the risk that the attacker manages, by a subtle placement of the probes, to limit or completely avoid the externally added noise; imagine an attacker with a very selective electromagnetic probe which would place its probe over the targetted algorithm, which is micrometers apart from the noise source (RSA). Therefore, it sounds wiser to entangle the computation and the random variables. This is what is achieved by so-called masking schemes. Appendix A explains why masking reduces the SNR.

Notice that the two strategies are orthogonal, that is, it is beneficial to employ them at the same time. Still, in the sequel, we will focus on masking, since it allows (at least in theory) to increase the noise at the maximal extent.

3.2   Masking Schemes

Masking schemes have been introduced to obfuscate the internals of a computation, in a view to make it more difficult to be attacked. The strategy in masking is based on randomization:

- for *data* (e.g., in algorithms with constant-execution flow, such as AES), and
- for *operations* (e.g., in algorithms where the sequence of operations leak some secrets, such as RSA).

In practice, a masking scheme consists in four algorithms, as depicted in Fig. 3.

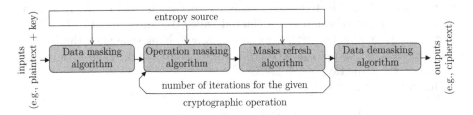

Fig. 3. Masking schemes

Initially, the input data must be masked, thanks to a first algorithm. Second, the masked data is manipulated, so as to implement the intended cryptographic operation. Many techniques exist. One way to envision masking is to see all the operations making up the cryptographic function as look-up tables. In this case, the masked look-up tables can be implemented as [37, Table 1]:

- new larger look-up tables, where the masking material is now part of the addressing strategy,
- table recomputation specifically for the current mask, or
- computation style which is able to operate on masked data.

After the operation has been computed, it can be necessary to refresh the masks. Indeed, if the value is intended to be used more than once, then some masks would be duplicated during the computation. It is thus wise to re-randomize the current masks. Eventually, at the end of the computation, the masked data shall be freed from its mask. Hence a demasking step. The first three algorithms require entropy, whereas the last one destroys entropy.

3.3   Security of Masking Schemes

It is easy to measure the amount of entropy consumed by a masking scheme (see top of Fig. 3). However, this does not obviously reflect its actual security level. Indeed, the entropy can be wasted, e.g., by being badly used: XORing together entropy reduces it, while bringing no additional difficulty for the attacker.

The first attempt to measure security arise from [1, Definition 1]. The order is defined as the minimum number of *intermediate values* an attacker must collect to recover part of the secret. In this framework, the overall security is that of the weakest link.

Still, the exact definition of an intermediate variable is unclear. The difficulty arises from the fact the designer would like to link the security to properties of its design. However, the intermediate variables encompass different notions depending on the refinement stage: after compilations, variables are mapped to internal resources. Thus, the granularity [1, Sect. 3] can change between the cryptographic algorithm specification, the source code, the machine code, and what is actually executed on the device.

Some early works considered intermediate values are bits, such as in *private circuits* [25, 26]. This makes sense for hardware circuits, for which (in general CMOS processes) an equipotential has only two licit values, that is carries one bit. However, private circuits have been extended to software implementations (see e.g. [40]), where intermediate variables become bitvectors of the machine word length. But after considering some new threats, such as glitches, a new trend has consisted in looking back to bit-oriented masking. This is typically the case of *threshold implementations* [35], where the granularity is again the bit.

In this article, we are interested with the lowest possible level of security analysis, hence we consider that intermediate variables are bits.

3.4   Orthogonal Direct Sum Masking (ODSM), a Masking Scheme Based on Codes

We illustrate in this section several masking schemes, and show in which respect they relate to coding theory.

We will show that the two security notions related to masking (*probing* and *bounded-moment* models) are equivalent when conducting analyses at bit-level. We model a circuit as a parallel composition of bits, seen as elements of $\mathbb{F}_2$. The exemple, when there are n wires in the circuit, we model the circuit state as an element of $\mathbb{F}_2^n$, that is the Cartesian product $\mathbb{F}_2 \times \ldots \times \mathbb{F}_2$.

At this stage, we use the following new notations. Let X a k-bit information word to be concealed. Let Y an $(n-k)$-bit mask used to protect X. The protected variable is $Z = XG + YH$, where:

- G is an $k \times n$ generating matrix of a code,
- H is an $(n-k) \times n$ generating matrix of a code of dual distance $d+1$,
- $+$ is the bitwise addition in $\mathbb{F}_2^n$, sometimes also denoted by $\oplus$.

The random variable YH is the mask. In practice, the bits making up Z can be manipulated in whatever order, i.e., they can even be scheduled to be manipulated one after the other, like in a bitslice implementation. We call Z an encoding with codes, or ODSM [3].

Then, we have the following twain theorems.

Theorem 9. *Encoding with codes is secure against probing of order d.*

Proof. By definition of a code of dual distance $d + 1$, any tuple of less than d coordinates is uniformly distributed [9]. Thus, if the attacker probes up to d (inclusive) wires, this word seen as an element of $\mathbb{F}_2^d$ is perfectly masked. Therefore, no information on X can be recovered.   □

Theorem 10. (Masking with codes is d-th order secure in the bounded-moments model). *For all pseudo-Boolean function $\psi : \mathbb{F}_2^n \to \mathbb{R}$ (leakage function, denoted $y = \varphi \circ f$ in Sect. 2.2) of degree $d^\circ(\psi) \leq d$, we have*

$$\mathrm{Var}(\mathbb{E}(\psi(XG + YH|X))) = 0. \tag{10}$$

Proof. Let ψ' the indicator of the code generated by H. Since H has dual-distance $d + 1$, we have that for all $z \in \mathbb{F}_2^n$, $0 < w_H(z) \leq d$, $\hat{\psi}'(z) = 0$, where $\hat{\psi}'(z) = \sum_{z' \in \mathbb{F}_2^n} \psi'(z)(-1)^{z' \cdot z}$. Now, owing to Lemma 1 in [4], we also know that for all $z \in \mathbb{F}_2^n$, $w_H(z) > d^\circ(\psi)$, $\hat{\psi}(z) = 0$.

Now, we must prove that $\mathrm{Var}(\mathbb{E}(\psi(XG + YH|X))) = 0$, that for all $x \in \mathbb{F}_2^k$, $\sum_{y \in \mathbb{F}_2^{n-k}} \psi(xG + yH) = \sum_{z \in \mathbb{F}_2^n} \psi(xG + z)\psi'(z) = (\psi \otimes \psi')(xG)$ is the same, where $\otimes$ is the convolution product.

Actually, we can prove more than that, namely that $\psi \otimes \psi'$ is constant on the full $\mathbb{F}_2^n$. This is equivalent to proving that $\widehat{\psi \otimes \psi'} = \hat{\psi}\hat{\psi}'$ is equal to zero on $\mathbb{F}_2^n \backslash \{0\}$. Indeed, let $z \in \mathbb{F}_2^n$, $z \neq 0$. If $w_H(z) > d^\circ(\psi)$, then $\hat{\psi}(z) = 0$. And if $w_H(z) \leq d^\circ(\psi) \leq d$, then $\hat{\psi}'(z) = 0$. So, in both cases, we have $\hat{\psi}(z)\hat{\psi}'(z) = 0$. □

Notice that the function $\psi : \mathbb{F}_2^n \to \mathbb{R}$ such that $\psi(x) = \sum_{i=0}^{n-1} x_i 2^i$, has degree one. It is sometimes (abusively) referred to as the identity function. Obviously, if the attacker gets to know $\psi(Z)$, then he can recover Z, hence deduce X by projection on subspace vector C. But this is not our security hypothesis. Our result from Theorem 10 (and in particular its Eq. (10)) is that the inter-class variance of $\psi(Z)$ knowing X is equal to zero, for all $d^\circ(\psi) \leq d$.

In Eq. (10), the degree of ψ can be accounted by two reasons:

1. High-order leakage in $y = \varphi \circ f$, owing to *glitches* (see Sect. 4), *capacitive coupling, IR drop*, etc. (refer to [18, Sect. 4.2]);
2. Combination function from the attacker, which can be: multivariate (which involved a product of shares), monovariate (hence necessarily high-order zero-offset).

As another remark, we notice that, although it is not strictly mandatory, the randomized variable Z can be manipulated by subwords, a bit like for classical masking, where the subwords coincide with shares.

Let us give the example of the look-up table, in the case $k = 8$ and $n = 16$. We know that we can reach 4-th order security [4]. But we can decide not to manipulate only Z as such, but to cut it into two parts, $Z = (Z_H, Z_L)$, where $Z_H, Z_L \in \mathbb{F}_2^8$. This cut is motivated by the adequation between the masking scheme and the machine architecture, where maybe the basic register size is 8 bits. Then, we also cut the T-table(s) into two tables, namely T_H and T_L, both of 256 bytes. The Algorithm 1 allows to evaluate the T-table using bytes only, i.e., without placing Z_H and Z_L side-by-side for all data Z.

Input :

- $(z_H, z_L) \in \mathbb{F}_2^8 \times \mathbb{F}_2^8$
- T_H, T_L, two tables of size 2^{16} bytes

Output : The result of the lookup $(T_H[z_H \times 2^8 + z_L], T_L[z_H \times 2^8 + z_L])$

1 Initialize $z'_H \in \mathbb{F}_2^8$ and $z'_L \in \mathbb{F}_2^8$ to zero
2 **for** $h = 0$ to $2^8 - 1$ **do**
3   **for** $l = 0$ to $2^8 - 1$ **do**
4          $z'_H \leftarrow z'_H \oplus T_H[h \times 2^8 + l] \wedge (h = z_H) \wedge (l = z_L)$
5          $z'_L \leftarrow z'_L \oplus T_L[h \times 2^8 + l] \wedge (h = z_H) \wedge (l = z_L)$
6   **end**
7 **end**
8 **return** (z'_H, z'_L)

Algorithm 1. S-box evaluation by block, without ever using a 16-bit word

3.5 Illustration for Some Coding-Based Masking Schemes

In the previous section, we have shown with Theorems 9 and 10 that the two models (bit-level probing and bounded moments) are equivalent, which motivates to consider the probing model at bit level (as opposed to at word level, as done in many papers (to cite a few: [16,19]). We give hereafter some examples of masking with codes at bit-level.

Perfect Masking. The masks M_1, M_2, etc. are chosen uniformly in $\mathbb{F}_2^k$. We assume here that $k|n$. It is possible to see perfect masking as a special case of ODSM [3], where:

$$G = \begin{pmatrix} I_k \ 0 \ 0 \dots 0 \end{pmatrix} \qquad \text{and} \qquad H = \begin{pmatrix} I_k \ I_k \ 0 \ \dots \ 0 \\ I_k \ 0 \ I_k \ \dots \ 0 \\ \vdots \ 0 \ 0 \ \ddots \ 0 \\ I_k \ 0 \ 0 \ \dots \ I_k \end{pmatrix}. \tag{11}$$

Rotating Substitution-Box Masking (RSM [32]). Let us illustrate RSM on $n = 8$ bits. The mask M is chosen uniformly in:

- the set $\mathcal{C}_0 = \{\texttt{0x00}\}$ for no resistance,
- the set $\mathcal{C}_1 = \{\texttt{0x00}, \texttt{0xff}\}$ for resistance to first-order attacks,
- the set $\mathcal{C}_2$, a non-linear code of length 8, size 12 and dual distance $d_{\mathcal{C}_2}^\perp = 3$,
- the set $\mathcal{C}_3$, a linear code of length 8, dimension 4 and dual distance $d_{\mathcal{C}_3}^\perp = 4$. This code is fully described in [15]. It is a self-dual code of parameters $[8, 4, 4]$.

The case $\mathcal{C}_3$ is interesting since there are sixteen masks, hence (in hardware), the sixteen Substitution-boxes (S) of an algorithm such as AES can be implemented masked. When $\varphi = w_H$ and $Z = f(T, k^*) = S(T \oplus k^*)$, then the leakage distributions $X = \varphi(Z \oplus M)$ are represented in Fig. 4.

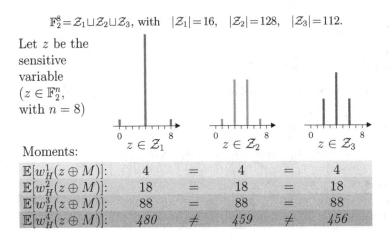

$$\mathbb{F}_2^8 = \mathcal{Z}_1 \sqcup \mathcal{Z}_2 \sqcup \mathcal{Z}_3, \text{ with } |\mathcal{Z}_1| = 16, \quad |\mathcal{Z}_2| = 128, \quad |\mathcal{Z}_3| = 112.$$

Let z be the sensitive variable ($z \in \mathbb{F}_2^n$, with $n = 8$)

Moments:

	$z \in \mathcal{Z}_1$		$z \in \mathcal{Z}_2$		$z \in \mathcal{Z}_3$
$\mathbb{E}[w_H^1(z \oplus M)]$:	4	=	4	=	4
$\mathbb{E}[w_H^2(z \oplus M)]$:	18	=	18	=	18
$\mathbb{E}[w_H^3(z \oplus M)]$:	88	=	88	=	88
$\mathbb{E}[w_H^4(z \oplus M)]$:	*480*	$\neq$	*459*	$\neq$	*456*

Fig. 4. Leakage distribution of RSM using $M \sim \mathcal{U}(\mathcal{C}_3)$ on $n = 8$ bits

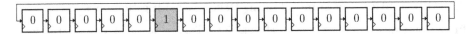

Fig. 5. Example of *one-hot* counter (out of 16), used to designate the round index position

RSM involves a random index, that is the choice of the initial codeword in $\mathcal{C}_d$, for a protection order of d. This choice can be done in a leak-free manner by using a *one-hot* representation. In the case of $\mathcal{C}_3$, sixteen such indices can be selected. The *one-hot* representation is given in Fig. 5. The random index is selected at random initially; then, from round to round, it is simply shifted.

Leakage Squeezing (LS). In leakage squeezing, the shares are like for perfect masking, except that some bijective functions are applied to the them, thereby mixing bits better [10,12,13,17].

Results. For the illustration of the *bounded moment* model, we use for our illustrations the *Hamming weight* leakage model. Notice that any other first-order leakage model would yield comparable results.

Also, we illustrate the leakage based on two extreme plaintexts, that is 0x00 and 0xff. However, in some situations, these two plaintexts lead to the same leakage (e.g., for symmetry reasons).

In all the presented schemes, security holds only provided there is no *high-order leakage*. Said differently, it is possible to consider that there is a high-order leakage. For instance, in recap Fig. 6, the indicated security order is the attack total order. The total attack order is the sum of multiplicative contribution

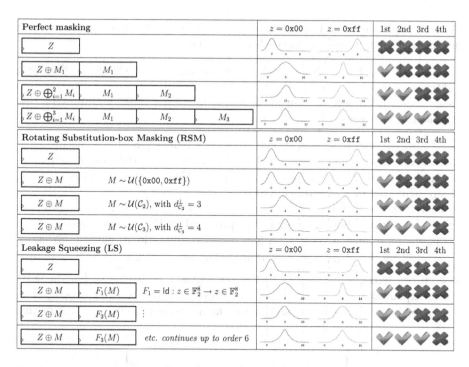

Fig. 6. Security level of several masking schemes. The order $d = 1, 2, 3, 4$ corresponds both to the number of probes (see Fig. 1(a)) used by the attacker and to the moment of leakage when the attacker uses an integrating probe (see Fig. 1(b))

from the hardware and the operations carried out by the attacker. That is, poor hardware which couples bits contributes to facilitates attacks by combining bits.

3.6   Masking and Faults Detection

Codes are also suitable tools when both side-channel leakage must be masked and faults must be detected. This need is general in cryptography, and has specific applications when thwarting Hardware Trojan Horses (HTH) [11,33,34]. Indeed, the *activation part* of a HTH is impeded by masking, whereas the *payload part* is caught red-handed by a detection code.

4   Leakage Model, and Glitches

The term *glitch* refers to a non-functional transition(s) occurring in combinational logic. They exist because combinational gates are non-synchronizing, i.e., they evaluate as soon as one input arrive. In terms of *hardware description languages* (VHDL, Verilog, etc.), they are modelled as processes where all inputs belong to the sensitivity list. Thus, for the vast majority of gates with many inputs, there is the possibility of a race between the inputs. Therefore, some gates can evaluate several times within one clock period. Actually, the deeper the combinational gates, the more likely it is that:

- there is a large timing difference between the inputs, thereby generating new glitches, and
- some input is already the output of a glitching gate, thereby amplifying the number of glitches.

It is known that glitches can defeat masking schemes [28–30]. Some masking schemes which somehow *tolerate* [21,22,35,39] or *avoid* glitches [27,31] have been put forward. However, the real negative effect of glitches on security is usually perceived in a *qualitative* manner.

Therefore, we would like to account *quantitatively* for the effect of glitches. Let us start by an illustrative example, provided in Fig. 7. The upper part of this figure represents a pipeline, where some combinational gates (AND gates represented by $\neg\!\!\!D\!\!-$ and XOR gate represented by $\neg\!\!\!\!D\!\!-$) form a partial netlist between two barriers of flip-flops (DFF gates represented by $\neg\!\!\Box$)). For the sake of this explanation, all the gates are assumed to have the same propagation time, namely 1 ns. The lower part of this figure gives the chronograms of the

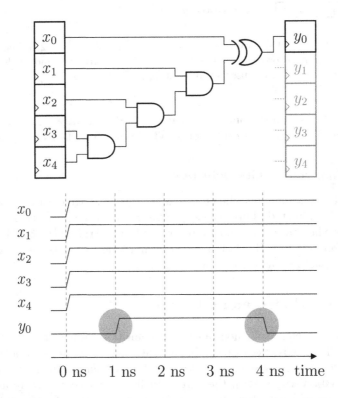

Fig. 7. Example of 4th-order glitch occurring upon 4th-order conjunction $\bigwedge_{i=1}^{i=4} x_i$

execution of this netlist, when initially all signals are set to zero. It appears that, owing to the difference of paths between the two inputs of the final XOR gate, this gate generates a glitch, highlighted with symbol ●, which lasts 3 ns, between time 1 and 4 ns within the depicted clock period. The condition for this glitch to appear is the following: $x_1 \wedge x_2 \wedge x_3 \wedge x_4$. This means that this glitch is a 4th-order leakage. So, if the masking scheme is only 3rd-order resistant, the setup of Fig. 7 would generate a glitch which compromises the security in a 1st-order side-channel attack. That is, the circuit itself contributes to the attack, in combining the bits on behalf of the attacker.

Assume now a setup slightly more simple than that of Fig. 7, where there is only one AND gate behind the second input of the XOR gate. However, we assume such pattern is present twice, once computing $y_0 = x_0 \oplus (x_1 \wedge x_2)$, and another time computing $y_5 = x_5 \oplus (x_4 \wedge x_3)$. Then, in this case depicted in Fig. 8,

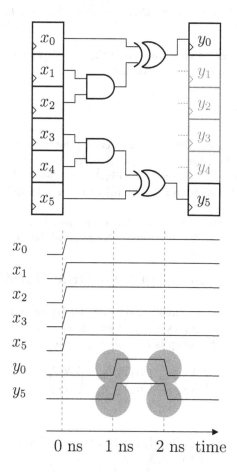

Fig. 8. Example of two 2nd-order glitches occurring upon 2th-order conjunctions $\bigwedge_{i=1}^{i=2} x_i$ and $\bigwedge_{i=3}^{i=4} x_i$

the leakage incurred by the glitches at the output of the XOR gates would only combine two bits amongst the x_i (namely x_1 & x_2, and x_3 & x_4). Therefore, it suffices for the attacker to conduct a 2nd-order attack on the glitchy traces to succeed a $2 \times 2 = 4$th order attack on the masking scheme. The circuit and the attacker collaborate in the objective of realizing a 4th-order attack: half of the combination is carried out by the circuit $((x_1 \wedge x_2)$ and $(x_3 \wedge x_4))$, while the other half is left remaining to the attacker. Indeed, by raising the traces to the second power, the attacker obtains a term $(x_1 \wedge x_2) \times (x_3 \wedge x_4)$, which coincides with the leakage condition of Fig. 7, that is $\bigwedge_{i=1}^{i=4} x_i$.

To conclude on the leakage model complexification, we underline that it has a negative impact on two situations:

– on low-entropy masking schemes, where the *individual shares* are not protected at the maximum order (see for instance RSM in Sect. 3.5), and
– on any masking schemes, where *shares interact between themselves* by some combinational logic.

In those two cases, a great care must be taken; tools as that described in [18] can help check the design is secure (or not).

5   Conclusion

Throughout this paper, we have seen how coding and side-channel analysis can benefit one from another, for attack as well as for protection.

This is a nice example of cross fertilization between disciplines, in which Claude Carlet played a decisive role. Thanks to you, Claude!

Acknowledgements. Part of this work has been funded by the ANR CHIST-ERA project SECODE (*Secure Codes to thwart Cyber-physical Attacks*).

A   SNR in the Presence of First Order Masking

Let us consider a first-order masking scheme [1]. By design, a first-order side-channel attack fails. However, a second-order side-channel attack, combining two samples, can succeed. The setup is the following: the leakage is:

$$\begin{pmatrix} X_1 \\ X_2 \end{pmatrix} = \begin{pmatrix} \alpha_1 Y_1^\star \\ \alpha_2 Y_2^\star \end{pmatrix} + \begin{pmatrix} N_1 \\ N_2 \end{pmatrix},$$

where:

– $N_1 \sim \mathcal{N}(0, \sigma_1^2)$ and $N_2 \sim \mathcal{N}(0, \sigma_2^2)$ are two independent noise sources,
– α_1 and α_2 are the amount of leakage,
– $Y_1^\star$ and $Y_2^\star$ are leakage functions (assumed normalized, that is $\mathbb{E}(Y_i^\star) = 0$ and $\mathsf{Var}(Y_i^\star) = 1$, for $i \in \{1, 2\}$).

In the Boolean masking where the attacker target the pair (mask, masked substitution box S), the leakage model is:

- $Y_1 = \frac{2}{\sqrt{n}} \left(w_H(S(T \oplus k) \oplus M) - \frac{n}{2} \right) = -\frac{1}{\sqrt{n}} \sum_{b=1}^{n} (-1)^{S_b(T \oplus k) \oplus M_b}$ and
- $Y_2 = \frac{2}{\sqrt{n}} \left(w_H(M) - \frac{n}{2} \right) = -\frac{1}{\sqrt{n}} \sum_{b=1}^{n} (-1)^{M_b}$.

The notation M_b means bit $b \in \{1, \dots, n\}$ in bitvector $M \in \mathbb{F}_2^n$.

As the masking is first-order perfect, we indeed have that $\mathbb{E}(Y_i | T = t)$ does not depend on the key, for each share $i \in \{1, 2\}$. However, the attacker is inclined to *combine* the two leakages by a centered product, since the expectation of this combination $Y_c = Y_1 Y_2$ depends on the key, despite the masking with the uniform $M \sim \mathcal{U}(\mathbb{F}_2^n)$. Precisely, let $t \in \mathbb{F}_2^n$ one realization of T. We have that:

$$\mathbb{E}(Y_c | T = t) = \frac{1}{2^n} \sum_{m \in \mathbb{F}_2^n} \frac{1}{n} \sum_{b,b'} (-1)^{S_b(T \oplus k) \oplus m_b \oplus m_{b'}}$$

$$= \frac{1}{n2^n} \sum_{m \in \mathbb{F}_2^n} \sum_{b} (-1)^{S_b(T \oplus k)} \qquad \text{(because m is uniform on $\mathbb{F}_2^n$)}$$

$$= -\frac{1}{2\sqrt{n}} \left(w_H(S(T \oplus k)) - \frac{n}{2} \right), \qquad (12)$$

which happens to be proportional to the leakage model of the substitution box when the masking is disabled ($M = 0$). Indeed, one can derive from Eq. (12) that:

$$\mathbb{E}(Y_c | T = t) = -\frac{1}{2\sqrt{n}} \mathbb{E}(Y_1 | T = t, M = 0).$$

The second-order attack thus consists in applying the regular correlation power analysis (CPA [2]):

- targeting $X_c = X_1 X_2$ instead of X_1 or X_2,
- using as leakage model $\mathbb{E}(Y_c | T)$, where we recall that $Y_c = Y_1 Y_2$ [38].

Thus, the new leakage to analyse is:

$$X_c = X_1 X_2 = (\alpha_1 Y_1^\star + N_1)(\alpha_2 Y_2^\star + N_2)$$
$$= \underbrace{\alpha_1 \alpha_2 Y_1^\star Y_2^\star}_{\text{signal}} + \underbrace{\alpha_1 Y_1^\star N_2 + \alpha_2 Y_2^\star N_1 + N_1 N_2}_{\text{noise}}.$$

Indeed, the term $Y_1^\star Y_2^\star$ conditionally to the known plaintext T depends on the key (recall Eq. (12)), whereas the other terms $\alpha_1 Y_1^\star N_2 + \alpha_2 Y_2^\star N_1 + N_1 N_2$ do not.

Therefore, the SNR in the case of the second-order attack is:

$$\text{SNR(2o)} = \frac{\text{Var}(\alpha_1 \alpha_2 Y_1^\star Y_2^\star)}{\text{Var}(\alpha_1 Y_1^\star N_2 + \alpha_2 Y_2^\star N_1 + N_1 N_2)}. \qquad (13)$$

Proposition 11. *The SNR in the case of the second-order attack is:*

$$SNR(2o) = \frac{SNR_1 \cdot SNR_2}{1 + SNR_1 + SNR_2},$$

where $SNR_i = \alpha_i^2 / \sigma_i^2$ for $i \in \{1, 2\}$.

Proof. We have:

$$\mathbb{E}_{T,M}(Y_1^\star Y_2^\star) = \frac{1}{2^{2n}} \sum_{t \in \mathbb{F}_2^n, m \in \mathbb{F}_2^n} Y_1^\star Y_2^\star$$

$$= \frac{1}{2^{2n}} \left(\frac{2}{\sqrt{n}}\right)^2 \sum_m \left(w_H(m) - \frac{n}{2}\right) \sum_t \left(w_H(S(t \oplus k^\star) \oplus m) - \frac{n}{2}\right)$$

$$= \frac{1}{2^{2n}} \left(\frac{2}{\sqrt{n}}\right)^2 \sum_m \left(w_H(m) - \frac{n}{2}\right) \sum_z \left(w_H(z) - \frac{n}{2}\right) \tag{14}$$

$$= 0 \times 0 = 0.$$

At line (14), we used the fact that S is a bijection of $\mathbb{F}_2^n$ (as is SubBytes in AES [36]).

Besides, we also have:

$$\mathbb{E}_{T,M}\left((Y_1^\star Y_2^\star)^2\right) = \frac{1}{2^{2n}} \sum_{t \in \mathbb{F}_2^n, m \in \mathbb{F}_2^n} (Y_1^\star)^2 (Y_2^\star)^2$$

$$= \frac{1}{2^{2n}} \left(\frac{2}{\sqrt{n}}\right)^4 \sum_m \left(w_H(m) - \frac{n}{2}\right)^2 \sum_t \left(w_H(S(t \oplus k^\star) \oplus m) - \frac{n}{2}\right)^2$$

$$= \frac{1}{2^{2n}} \left(\frac{2}{\sqrt{n}}\right)^4 \sum_m \left(w_H(m) - \frac{n}{2}\right)^2 \sum_z \left(w_H(z) - \frac{n}{2}\right)^2 \tag{15}$$

$$= 1 \times 1 = 1 \quad \text{(as per the normalization of } Y_1^\star \text{ and } Y_2^\star).$$

Therefore, the variance of the signal is equal to $\alpha_1^2 \alpha_2^2$.

Regarding the noise part, we have:

$$\mathbb{E}(\alpha_1 Y_1^\star N_2 + \alpha_2 Y_2^\star N_1 + N_1 N_2) = 0,$$

by independence between N_1, N_2 and $Y_i^\star$ for $i \in \{1, 2\}$. We also have:

$$\mathsf{Var}(\alpha_1 Y_1^\star N_2 + \alpha_2 Y_2^\star N_1 + N_1 N_2) = \mathbb{E}\left((\alpha_1 Y_1^\star N_2 + \alpha_2 Y_2^\star N_1 + N_1 N_2)^2\right) - 0$$
$$= \alpha_1^2 \sigma_2^2 + \alpha_2^2 \sigma_1^2 + \sigma_1^2 \sigma_2^2.$$

As a result, we have:

$$SNR(2o) = \frac{\alpha_1^2 \alpha_2^2}{\alpha_1^2 \sigma_2^2 + \alpha_2^2 \sigma_1^2 + \sigma_1^2 \sigma_2^2} = \frac{SNR_1 \cdot SNR_2}{1 + SNR_1 + SNR_2}. \qquad \square$$

Corollary 12 (Limit of SNR(2o) in the presence of large noise). *When the noise is large, that is $SNR_i \ll 1$ for $i \in \{1, 2\}$, then*

$$SNR(2o) \approx SNR_1 \cdot SNR_2 \approx SNR^2 \qquad (if \ SNR_1 \approx SNR_2 = SNR). \quad (16)$$

Proof. Immediate first-order simplification of SNR(2o) as given in Proposition 11. □

References

1. Blömer, J., Guajardo, J., Krummel, V.: Provably secure masking of AES. In: Handschuh, H., Hasan, M.A. (eds.) SAC 2004. LNCS, vol. 3357, pp. 69–83. Springer, Heidelberg (2004). doi:10.1007/978-3-540-30564-4_5
2. Brier, E., Clavier, C., Olivier, F.: Correlation power analysis with a leakage model. In: Joye, M., Quisquater, J.-J. (eds.) CHES 2004. LNCS, vol. 3156, pp. 16–29. Springer, Heidelberg (2004). doi:10.1007/978-3-540-28632-5_2
3. Bringer, J., Carlet, C., Chabanne, H., Guilley, S., Maghrebi, H.: Orthogonal direct sum masking: a smartcard friendly computation paradigm in a code, with Builtin protection against side-channel and fault attacks. In: Naccache, D., Sauveron, D. (eds.) WISTP 2014. LNCS, vol. 8501, pp. 40–56. Springer, Heidelberg (2014). doi:10.1007/978-3-662-43826-8_4
4. Bringer, J., Carlet, C., Chabanne, H., Guilley, S., Maghrebi, H.: Orthogonal direct sum masking: a smartcard friendly computation paradigm in a code, with Builtin protection against side-channel and fault attacks. Cryptology ePrint Archive, Report 2014/665 (2014). http://eprint.iacr.org/2014/665/
5. Bruneau, N., Carlet, C., Guilley, S., Heuser, A., Prouff, E., Rioul, O.: Stochastic Collision Attack. In: IEEE Transactions on Information Forensics and Security (2016)
6. Bruneau, N., Guilley, S., Heuser, A., Marion, D., Rioul, O.: Less is more: dimensionality reduction from a theoretical perspective. In: Güneysu, T., Handschuh, H. (eds.) CHES 2015. LNCS, vol. 9293, pp. 22–41. Springer, Heidelberg (2015). doi:10.1007/978-3-662-48324-4_2
7. Bruneau, N., Guilley, S., Heuser, A., Marion, D., Rioul, O.: Optimal side-channel attacks for multivariate leakages and multiple models. J. Crypt. Eng. (2016, to appear). http://www.proofs-workshop.org/program.html
8. Bruneau, N., Guilley, S., Heuser, A., Rioul, O.: *Masks will fall off*: higher-order optimal distinguishers. In: Sarkar, P., Iwata, T. (eds.) ASIACRYPT 2014. LNCS, vol. 8874, pp. 344–365. Springer, Heidelberg (2014). doi:10.1007/978-3-662-45608-8_19
9. Carlet, C.: Boolean functions for cryptography and error correcting codes, chapter of the monography. In: Crama, Y., Hammer, P. (eds.) Boolean Models and Methods in Mathematics, Computer Science, and Engineering, pp. 257–397. Cambridge University Press, Cambridge (2010). Preliminary version, http://www.math.univ-paris13.fr/~carlet/chap-fcts-Bool-corr.pdf
10. Carlet, C.: Correlation-immune boolean functions for leakage squeezing and rotating S-Box masking against side channel attacks. In: Gierlichs, B., Guilley, S., Mukhopadhyay, D. (eds.) SPACE 2013. LNCS, vol. 8204, pp. 70–74. Springer, Heidelberg (2013). doi:10.1007/978-3-642-41224-0_6

11. Carlet, C., Daif, A., Danger, J.-L., Guilley, S., Najm, Z., Ngo, X.T., Porteboeuf, T., Tavernier, C.: Optimized linear complementary codes implementation for hardware Trojan prevention. In: European Conference on Circuit Theory and Design, ECCTD, Trondheim, Norway, pp. 1–4. IEEE, 24–26 August 2015

12. Carlet, C., Danger, J.-L., Guilley, S., Maghrebi, H.: Leakage squeezing of order two. In: Galbraith, S., Nandi, M. (eds.) INDOCRYPT 2012. LNCS, vol. 7668, pp. 120–139. Springer, Heidelberg (2012). doi:10.1007/978-3-642-34931-7_8

13. Carlet, C., Danger, J.-L., Guilley, S., Maghrebi, H.: Leakage squeezing: optimal implementation and security evaluation. J. Math. Crypt. **8**(3), 249–295 (2014)

14. Carlet, C., Guilley, S.: Side-channel indistinguishability. In: HASP, pp. 9:1–9:8. ACM, New York, 13–14 June 2013

15. Carlet, C., Guilley, S.: Side-channel indistinguishability. On HAL, 19 July 2014. Extended version of [14] with more results in appendix, http://hal. archives-ouvertes.fr/hal-00826618

16. Coron, J.-S.: Higher order masking of look-up tables. In: Nguyen, P.Q., Oswald, E. (eds.) EUROCRYPT 2014. LNCS, vol. 8441, pp. 441–458. Springer, Heidelberg (2014). doi:10.1007/978-3-642-55220-5_25

17. Danger, J.-L., Guilley, S.: Protection des modules de cryptographie contre les attaques en observation d'ordre élevé sur les implémentations à base de masquage. Brevet Français FR09/50341, assigné à l'Institut TELECOM, 20 January 2009

18. Danger, J.-L., Guilley, S., Nguyen, P., Nguyen, R., Souissi, Y.: Analyzing security breaches of countermeasures throughout the refinement process in hardware design flow. In: DATE, Lausanne, Switzerland, 27–31 March 2017

19. Duc, A., Faust, S., Standaert, F.-X.: Making masking security proofs concrete: or how to evaluate the security of any leaking device. In: Oswald, E., Fischlin, M. (eds.) EUROCRYPT 2015. LNCS, vol. 9056, pp. 401–429. Springer, Heidelberg (2015). doi:10.1007/978-3-662-46800-5_16

20. Fei, Y., Luo, Q., Ding, A.A.: A statistical model for DPA with novel algorithmic confusion analysis. In: Prouff, E., Schaumont, P. (eds.) CHES 2012. LNCS, vol. 7428, pp. 233–250. Springer, Heidelberg (2012). doi:10.1007/978-3-642-33027-8_14

21. Fischer, W., Gammel, B.M.: Masking at gate level in the presence of glitches. In: Rao, J.R., Sunar, B. (eds.) CHES 2005. LNCS, vol. 3659, pp. 187–200. Springer, Heidelberg (2005). doi:10.1007/11545262_14

22. Gomathisankaran, M., Tyagi, A.: Glitch resistant private circuits design using HORNS. In: IEEE Computer Society Annual Symposium on VLSI, ISVLSI, Tampa, FL, USA, pp. 522–527, 9–11 July 2014

23. Guilley, S., Heuser, A., Rioul, O.: A key to success: success exponents for side-channel distinguishers. In: Biryukov, A., Goyal, V. (eds.) INDOCRYPT 2015. LNCS, vol. 9462, pp. 270–290. Springer, Cham (2015). doi:10.1007/ 978-3-319-26617-6_15

24. Heuser, A., Rioul, O., Guilley, S.: Good is not good enough: deriving optimal distinguishers from communication theory. In: Batina, L., Robshaw, M. (eds.) CHES 2014. LNCS, vol. 8731, pp. 55–74. Springer, Heidelberg (2014). doi:10.1007/ 978-3-662-44709-3_4

25. Ishai, Y., Prabhakaran, M., Sahai, A., Wagner, D.: Private circuits II: keeping secrets in tamperable circuits. In: Vaudenay, S. (ed.) EUROCRYPT 2006. LNCS, vol. 4004, pp. 308–327. Springer, Heidelberg (2006). doi:10.1007/11761679_19

26. Ishai, Y., Sahai, A., Wagner, D.: Private circuits: securing hardware against probing attacks. In: Boneh, D. (ed.) CRYPTO 2003. LNCS, vol. 2729, pp. 463–481. Springer, Heidelberg (2003). doi:10.1007/978-3-540-45146-4_27

27. Lin, K.J., Fan, S.C., Yang, S.H., Lo, C.C.: Overcoming glitches, dissipation timing skews in design of DPA-resistant cryptographic hardware. In: IEEE Computer Society Proceedings of the Conference on Design, Automation and Test in Europe, DATE 2007, Nice, France, pp. 1265–1270. EDA Consortium, San Jose, 16–20 April 2007. doi:10.1109/DATE.2007.364471

28. Mangard, S., Popp, T., Gammel, B.M.: Side-channel leakage of masked CMOS gates. In: Menezes, A. (ed.) CT-RSA 2005. LNCS, vol. 3376, pp. 351–365. Springer, Heidelberg (2005). doi:10.1007/978-3-540-30574-3_24

29. Mangard, S., Pramstaller, N., Oswald, E.: Successfully attacking masked AES hardware implementations. In: Rao, J.R., Sunar, B. (eds.) CHES 2005. LNCS, vol. 3659, pp. 157–171. Springer, Heidelberg (2005). doi:10.1007/11545262_12

30. Mangard, S., Schramm, K.: Pinpointing the side-channel leakage of masked AES hardware implementations. In: Goubin, L., Matsui, M. (eds.) CHES 2006. LNCS, vol. 4249, pp. 76–90. Springer, Heidelberg (2006). doi:10.1007/11894063_7

31. Moradi, A., Mischke, O.: Glitch-free implementation of masking in modern FPGAs. In: HOST, pp. 89–95. IEEE Computer Society, Moscone Center, San Francisco, 2–3 June 2012. doi:10.1109/HST.2012.6224326

32. Nassar, M., Souissi, Y., Guilley, S., Danger, J.-L.: RSM: a small and fast countermeasure for aes, secure against first- and second-order zero-offset SCAs. In: DATE (TRACK A: "Application Design", TOPIC A5: "Secure Systems"), Dresden, Germany, pp. 1173–1178. IEEE Computer Society, 12–16 March 2012

33. Ngo, X.T., Bhasin, S., Danger, J.-L., Guilley, S., Najm, Z.: Linear complementary dual code improvement to strengthen encoded circuit against hardware Trojan horses. In: IEEE International Symposium on Hardware Oriented Security and Trust, HOST 2015, Washington, DC, USA, pp. 82–87. IEEE, 5–7 May 2015

34. Ngo, X.T., Guilley, S., Bhasin, S., Danger, J.-L., Najm, Z.: Encoding the state of integrated circuits: a proactive and reactive protection against hardware trojans horses. In: Proceedings of the 9th Workshop on Embedded Systems Security, WESS 2014, pp. 7:1–7:10. ACM, New York (2014)

35. Nikova, S., Rijmen, V., Schläffer, M.: Secure hardware implementation of nonlinear functions in the presence of glitches. J. Crypt. **24**(2), 292–321 (2011)

36. NIST/ITL/CSD: Advanced Encryption Standard (AES). FIPS PUB 197, November 2001. http://csrc.nist.gov/publications/fips/fips197/fips-197.pdf

37. Prouff, E., Rivain, M.: A generic method for secure SBox implementation. In: Kim, S., Yung, M., Lee, H.-W. (eds.) WISA 2007. LNCS, vol. 4867, pp. 227–244. Springer, Heidelberg (2007). doi:10.1007/978-3-540-77535-5_17

38. Prouff, E., Rivain, M., Bevan, R.: Statistical analysis of second order differential power analysis. IEEE Trans. Comput. **58**(6), 799–811 (2009)

39. Prouff, E., Roche, T.: Higher-order glitches free implementation of the AES using secure multi-party computation protocols. In: Preneel, B., Takagi, T. (eds.) CHES 2011. LNCS, vol. 6917, pp. 63–78. Springer, Heidelberg (2011). doi:10.1007/978-3-642-23951-9_5

40. Rivain, M., Prouff, E.: Provably secure higher-order masking of AES. In: Mangard, S., Standaert, F.-X. (eds.) CHES 2010. LNCS, vol. 6225, pp. 413–427. Springer, Heidelberg (2010). doi:10.1007/978-3-642-15031-9_28

An Overview of the State-of-the-Art of Cloud Computing Cyber-Security

H. Bennasar[1], A. Bendahmane[2], and M. Essaaidi[1(✉)]

[1] ENSIAS, Mohammed V University in Rabat, Rabat, Morocco
essaaidi@ieee.org
[2] ENS, Abdelmalek Essaadi University, Tétouan, Morocco

Abstract. We presented an overview of the state-of-the-art of cloud computing security which covers its essential challenges through the main different cyber-security threats, the main different approaches, algorithms and techniques developed to address them, as well as the open problems which define the research directions in this area. The bottom line is that the state of maturity of cloud computing security is very promising and there are many research directions still open and which promise continued improvements of cloud security and privacy.

1 Introduction

Cloud computing is the use of computing resources that are delivered as a service via Internet [1] to provide a secure, and on demand network access to shared pool of configurable resources and different kind of services, such as, Software as a Service (SaaS), Platform as a Service (PaaS), and Infrastructure as a service (IaaS). During the last decade, there has been an increasing demand and adoption of cloud computing systems, technologies, applications and services. This is owing mainly to the many advantages this technology offers for businesses and organizations such as its high resources elasticity and scalability which provide important savings in terms of investment and manpower. However, Cyber-security is still considered among the most important issues and concerns limiting the widespread adoption of cloud computing. Among the major issues related with Cloud Computing security we can mention data security, intrusions attacks, confidentiality and data integrity Cloud computing provides several advantages allowing to have new business opportunities. However, it also involves potential cyber-security risks and vulnerabilities. For instance, storing data in the cloud may expose them to serious cyber-security attacks. The main objective of this paper is to present an up-to-date overview of cloud computing cyber-security issues. This will allow to identify the major research challenges in this increasingly important area. The remainder of this paper is organized as follows. In Sect. 2 we provide an overview of cloud computing, Sect. 3 is dedicated to the state of the art of cloud computing challenges, the current approaches used to circumvent them and a comparative study of related approaches.

© Springer International Publishing AG 2017
S. El Hajji et al. (Eds.): C2SI 2017, LNCS 10194, pp. 56–67, 2017.
DOI: 10.1007/978-3-319-55589-8_4

2   Cloud Computing

A. Definition

According to the National Institute of Standards and Technology (NIST) [2]: "Cloud Computing is a model for enabling ubiquitous, convenient, on demand network access to a shared pool of configurable computing resources (e.g., networks, servers, storage, applications and services) that can be rapidly provisioned and released with minimal management effort or service provider interaction.

B. Cloud Computing Characteristics

The main characteristics and features of Cloud Computing can be summarized in the following:

(1) Multi-Tenancy [3] which refers to having more than one occupants of the cloud, living and sharing other occupants of the provider's infrastructures, including computational resources, storage, services, and applications. By multi-tenancy, clouds provide simultaneous, secure hosting of services for various clients or customers using the same cloud infrastructure resources. It is an exclusive characteristic to resource sharing in clouds.

(2) Elasticity [4] is another important feature of cloud computing and it implies that the user is able to scale up or down resources assigned to services or resources based on the current demand. For providers, scaling up and down of a tenant's resources give a prospect to other tenants to use the tenant previously assigned resources.

(3) Availability of Information based on the Service level Agreement (SLA) [6] which is a trust bond between the cloud provider and the customer. This defines the maximum time for which the network resources or applications may not be available for the customer. Due to the complex nature of the customer demands, a simple measure and trigger process may not work for SLA enforcement.

(4) Multiple Stakeholder in the cloud Computing model means that there are different Stakeholders involved [5], such as the cloud provider (an entity that delivers infrastructures to the cloud's customers), the service provider (an entity that uses the cloud infrastructure to deliver applications/services to end users), and the customer (an entity that uses services hosted in the cloud infrastructure). Each stakeholder has its own security management systems/processes and its own requirements and capabilities distributed from/to other stakeholders.

(5) Third-Party Control [7] which is considered to be the major security challenge, that is, the owner of the data has no control on their processing. The biggest change for Information Technology (IT) department of an organization using cloud computing will be reduced control even as it is being tasked to tolerate increased responsibility for the confidentiality and compliance of computing practices in the organization.

C. Service Models

Cloud Computing offers services that can be grouped into three categories, as shown in Fig. 1

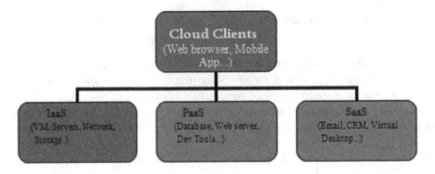

Fig. 1. Cloud computing service model

(1) Infrastructure-as-a-Service (IaaS) [1] through which the cloud providers deliver computation, storage and network resources. In this model, customers do not need to maintain huge servers; they just need to choose their required infrastructure using a web browser and they will be provided with all sorts of hardware infrastructure by the cloud service provider (CSP). As an examples of IaaS vendors, we can cite Citrix, 3tera, VMware, HP, and Dell.

(2) Platform-as-a-service (PaaS) [1] for which Cloud providers deliver platform, tools and business services to develop, deploy and manage their applications. PaaS facilitates the customer organization in developing software applications, without investing huge amounts of money on infrastructure, which will be delivered to the users over Internet on-demand and rent basis (i.e. pay-as-you-use). Web servers, application servers, development environment, and runtime environment are some example components with respect to PaaS. In this model, customers need not maintain underlying infrastructure including servers, cooling, operating systems, storage, etc. As examples of PaaS vendors, we can mention Google AppEngine, force.com, Microsoft Windows.

(3) Software-as-a-Service (SaaS) [1] for which Cloud computing providers offer applications hosted in the cloud infrastructures for application implementation. Example components for SaaS are office suites (docs), online games, email applications, online readers, online movie players, etc. In this model, customers need not maintain heavy investment on system configuration to run all these applications; they just require Internet access and a web browser. Salesforce.com, Amazon, Zoho, Microsoft Dynamics CRM, and Google are some examples of SaaS vendors.

D. Service Deployement

A cloud deployment model means a specific type of cloud computing environment, characterized by several features such as ownership, size, and access mode. As shown in Fig. 2, there are three common cloud deployment models, namely, private cloud, public cloud, and hybrid cloud.

(1) *Private cloud* [1] is for the only use of a single company/organization and its customers. This setup may reside inside or outside the customer's premises.

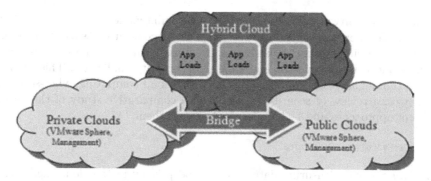

Fig. 2. Cloud computing deployment models

This cloud setup could be controlled, maintained or maneuvered by a third party or the organization itself or a combination of them.

(2) *Public cloud* [1] is for open use by the general public i.e. individuals or organizations. It resides on the premises of the CSP. This cloud setup could be controlled, maintained or maneuvered by different government organizations, corporate organizations, academic institutions or a combination of them to the extent permitted by the CSP.

(3) *Hybrid cloud* [1] is a combination of two or more distinct and unique cloud setups (private, community, or public) which are tied together by standardized or registered technology that ensures and allows data and application portability.

E. Cloud Actors

There are five cloud actors which are concisely explained below.

(1) *Cloud consumer* [8] is a person or an organization that maintains a business relationship with the cloud providers and uses their services.

(2) *Cloud provider* [8] is a person or an organization that is in charge for making a service available to other parties.

(3) *Cloud auditor* [8] is a party that performs independent evaluation of cloud services, information system operations, performance, and security.

(4) *Cloud broker* [8] is an entity that supervises the use, performance, and delivery of cloud services and which negotiates the relationships between cloud providers and cloud consumers.

(5) *Cloud carrier* [8] is an intermediary that provides connectivity and transport of cloud services from cloud providers to cloud consumers.

3   Cloud Computing Cyber-Security

Cloud computing attracts different users owing to its high resources elasticity and scalability which provide important savings in terms of investment and manpower. Cloud minimizes the need for user involvement by masking technical details such as software upgrades, licenses, and maintenance from its customers. However, the new concepts introduced by cloud computing, such as computation outsourcing, resource sharing, and external data warehousing, increase security and privacy concerns and create new security challenges. This section gives a thorough presentation and discussion of cloud computing cyber-security challenges, a review of security threats, and a comparative study of the latest different approaches and techniques used against them.

A. Cyber-security challenges

(1) *Data Security:* Ensuring data security and privacy in the cloud means the ability to ensure the principle key features of security, namely, confidentiality, integrity and availability. The main requirements for information security is data integrity that refers to the guarantee that users' data are not modified without authorization [10,13], in other words, data can be modified only by authorized users. In order to provide data integrity from both the provider and subscriber perspectives, secure encryption algorithms are generally used. However, encryption alone does not guarantee that data are not maliciously modified [12]. Due to the dynamic, shared and distributed nature of the cloud there is another important challenge for cloud users, namely, confidentiality. This refers to data privacy and accuracy which allows protecting private and sensitive data. To provide data confidentiality, one simple approach consists to save encrypted data in the cloud servers. As regards data availability, it refers to the ability of cloud users to access and use data any time and from anywhere. This means that the cloud system should be accessible and useful to authorized users anytime and anywhere [12,13]. There are several cyber-security threats that may face the cloud services availability. These are network based attacks such as Distributed Denial of Service (DDoS) attacks [10]. To ensure the safety and the availability of data, cloud providers should maintain an appropriate action plan for risk management to deal efficiently with these threats and to guarantee the cloud based services continuity [9].

(2) *Cloud Network Infrastructure Security:* A cloud service provider should be able to accept trustful network traffic, and to block malicious network traffic [9]. The cloud network infrastructure security should be able to block and protect against Denial of Service (DoS) attacks, to detect and prevent intrusions and to allow logging and notification. DoS defenses are based on network security, which should effectively filter queries and identify invaders to prevent malicious attacks [14]. The IDS/IPS systems detect or block known malware attacks, virus signatures and spam signatures but are also subject to false positives. Logging and notification allows cloud users to have some insight into the network's cyber-security health [9].

(3) *Cloud Applications Security:* Businesses and organizations should protect their cloud based applications from all sorts of cyber-security threats. Moreover, cloud applications security is similar to web applications security when hosted in data centers. Many organizations propose single sign on (SSO) as a solution to allow users to access multiple individual cloud services [14].

However, it is hard to implement SSO solutions correctly. In addition, many authentication methods require a secure software layer. To ensure cloud applications (APIs) security, there are different action items proposed in [9], namely:

- A design phase is used to carefully plan how the components of the cloud service will interact. Determine if the APIs can be restricted so that only trusted hosts can call them. Ensure that inter-service communication is securely authenticated.
- Ensure that the tools used are appropriate for APIs and can target the deployed technologies.
- Use testing to validate security monitoring and alerting capabilities. Ensure that any successfully exploited vulnerability was logged and appropriate alerting occurred.

B. Cyber-security threats

In this section, a classification of cloud computing cyber-security threats are detailed. In 2013, cloud security Alliance organized a panel of industry experts in order to present the Nine Cloud Computing Top Threats. Table presents a summary of Cloud security threats, proposed approaches to circumvent them, and a comparison with other approaches.

C. Cloud Cyber-security Techniques

In this section, we are going to make a zoom on some of the cloud cyber-security techniques presented in Table 1 above. These techniques may be classified into three categories:

- Data Integrity
- Authentication & Authorization
- Denial Of Service

Data Integrity (Data Loss & Data Breaches): In Table 1 above, several techniques have been proposed in order to deal with data integrity threats. These techniques use data encryption algorithms to give the data owner verifiable guarantees that their data remain trustful.

(1) Encryption algorithms: Plain RSA, AES, FDE, and Fully Homomorphic encryption (FHE).

(a) *Fully Homomorphic Encryption is* the most widely used encryption technique [22] in the literature. It means that the cloud provider can run the corresponding code a client requests, while not obtaining access neither to the argument data nor to the result data. Homomorphic encryption is an encryption algorithm proposing cloud computing data security scheme based on cloud data security problem. This encryption scheme includes four algorithms, namely, key generation algorithm, encryption algorithm, decryption algorithm and Additional Evaluation algorithm. The main idea behind this encryption scheme is the conversion of data into cipher-text that can be analyzed and worked with as if it were still in its original form [23]. FHE ensures the transmission of data between the cloud and the user safely, while the data stored in the cloud is still protected. However, FHE suffers so far from a problem related with huge computation requirements. This is still an open problem.

(b) *Field Programmable Gate Array (FPGA) [24]:* This is another approach which can also be used to ensure data integrity. FPGA is an integrated circuit designed to be configured by a customer or a designer after manufacturing - hence "field-programmable". J. M. Mondol proved that with the use of FPGA, four different types of solutions are given to ensure user authentication and user data security [25], namely:

- Trusted cloud computing platform ensuring computational trust.
- User enabled security groups for data collaboration.
- Data Security.
- Verifiable Attestation.

All these solutions guarantee that cyber-security is enabled by the Client who is the owner of the data. However, FPGA suffers from huge implementation complexity.

(2) Authentication and Authorization: Account hijacking, malicious intruders, and insecure Applications, are all threats resulting from authentication and authorization problems in cloud computing. *As solutions to these cyber-security threats, we present in* Table 1 *above the approaches and algorithms, such as Message Authentication code (MAC), key-hashed Message Authentication code (HMAC), Federated identity management (FIDM), Kerberos, Transport Layer Security (TLS), Trusted Third Party, Service Level Agreement (SLA) and cloud Security Management Framework. All these solutions could mitigate cloud computing security threats.*

The most widely used approaches, namely, Kerberos and the Service Level Agreement are explained below

(a) *Service Level Agreement (SLA)* [21] is a document that identifies the terms, conditions and it is able to create negotiations between the user and the provider. SLA is characterized by the following features.

- Minimum of performance level that the provider should provide
- Counteractive actions
- Consequences in case of breach of the agreement between user and provider.

The users have to be very obvious about security requirements for their property and all the requirement should be methodically agreed upon in the SLA. In case of doubts, it is harder to declare the defeat at a provider. In order to manage SLAs in a cloud computing environment, references [30, 32] suggest

Table 1. Summary of threats to cloud and solution directions

Threats	Affected Cloud services	Description	Proposed Solutions	Advantages/Drawbacks
Data Breaches	SaaS, PaaS, Iaas	A Security incident in which sensitive, protected or confidential data is copied, transmitted, viewed, stolen or used by an individual unauthorized to do so [16,27]	Plain RSA Advanced Encryption Standard (AES) Field Programmable Gate Array (FPGA)	(+) Strong Algorithms ensuring data confidentiality (-) Possibility to lose Encryption key(Losing data) (+)Ensures user authentication and user data security
Data Loss	SaaS, PaaS, IaaS	An error condition in information systems in which information is destroyed by failures or neglect in storage, transmission, or processing.[17,27]	Disaster recovery Full Disk Encryption (FDE) Homomorphic Encryption	(+) Backup (-) Possibility to lose Encryption key (Losing data) (+) Encrypting data before sending to providers (+) Strong security for data in process (-) High computation problem, needs further research
Account Hijacking	SaaS, PaaS, IaaS	A type of identity theft in which the hacker uses the stolen account information to carry out malicious or unauthorized activity.[18]	Message Authentication code (MAC) Federated identity management (FIDM) Kerberos	(+) Ensures data integrity and authenticity (+) Ensure strong security authentication and data access authorization. (+) Ensures the correctness of users data in cloud data storage and correctness of users (+) provides a centralized authentication server (-) Authentication mechanisms weakness
Insecure APIs	SaaS, PaaS, IaaS	Web and cloud services allow third-party access by exposing application programming interfaces, and many developers and customers do not adequately secure the keys to the cloud and their data[19]	Keyed-Hash Message Authentication code (HMAC) Transport Layer Security (TLS) Cloud Security Management Framework	(+) Ensures data integrity and authenticity (+) Strong traffic encryption to secure data in transit (+) Improves collaboration between cloud providers, service providers and service (-) Needs further extension of the automation of the security controls implementation phase customers
Denial of Service	IaaS	An attempt to make a machine or network ressource unavailable to its intended users.[20]	Firewall Intrusion Detection and Prevention systems	(+) Filter authorized traffic defined by security policies (-) traditional and not efficient (+) Monitor the usage of systems and detect insecure data (-) traditional and not efficient
Malicious Intruders	SaaS, PaaS, IaaS	Insider malicious activity bypassing firewall and other security model.	Trusted third Party Service level agreement (SLA)	(+) A strong mechanism to provide authorization, authentication, data confidentiality, data integrity (-) Huge complexity in implementation (+) Easy to implement (-) More sustained efforts towards standardization are required (-) Requires Advanced details and informations
Abuse of cloud Services	PaaS, Iaas	Allows interloper to start stronger attacks due to unidentified signup, lack of justification, and service fraud [12]	Fully Homomorphic Encryption	(+) Strong security for data in process (-) High computation problem, needs further research
Abuse of cloud Services	PaaS, Iaas	Allows interloper to start stronger attacks due to unidentified signup, lack of justification, and service fraud [12]	Fully Homomorphic Encryption	(+) Strong security for data in process (-) High computation problem, needs further research
Insufficient Due Diligence	SaaS, PaaS, Iaas	Organizations adopting the cloud without fully understanding the associated risks, they increase many operational, architectural and contractual issues over responsibility and transparency.[12]	Public key Infrastructure (PKI)	(+) Achieves strong authentication data confidentiality, data integrity and non-repudiation [6] (-) Huge implementation complexity.
Shared Technology Issues	Iaas	Allows one user to hinder other users' services by compromising hypervisor.	Isolation Advanced Cloud Protection System (ACPS)	(+) Strong Security (-) Requires advanced information, still a research problem (+) Ensures security of distributed computing middleware

Web Service Level Agreement (WSLA), a very flexible architecture for managing SLAs between providers and users. WSLA is designed to capture service level agreements in a formal way, but it suffers from some computation problems which need more investigations. The other approach used to provide Authentication and Authorization security is Kerberos [28].

(b) *Kerberos* is a computer network protocol which works on the basis of 'tickets' to allow nodes communicating over a non-secure network to prove their identity to one another in a secure manner. Cloud data storage security and user's data management in the cloud based upon Kerberos authentication service is proposed in [29]. In order to ensure the correctness of users' data in cloud, data storage and the users who can access the Cloud server, an effective and flexible distributed scheme with explicit dynamic data support, including Kerberos authentication service and third party, was proposed. Kerberos provides a centralized authentication server whose function is to authenticate the user to the cloud server and the cloud server to the user. To access the cloud server, all users should make the profile and set a password, then they can use the cloud server with some restrictions imposed by kerberos.

(3) Denial of Service: There are two main approaches proposed to deal with this cyber-security threat, namely, Firewalls and Intrusion detection Systems (IDPS).

(a) Firewalls [31] are utilized to reject or permit protocols, ports or IP addresses. Since firewalls detect the network packets at the limit of a network, intruders' attacks cannot be detected by traditional firewalls. Only some DoS or Distributed DOS (DDoS) attacks are also too complex to detect using traditional firewalls. For instance, if there is an attack on port 80 (web service), firewalls cannot distinguish good traffic from DoS attack traffic. Another solution is to add in IDS or IPS to the Cloud.

(b) *IDPS* [26,31] provides a real-time intrusion detection method and system. The IDS automatically and dynamically builds user profile data (known as a signature) for each user (or alternatively, a class of users) that can be used to determine normal actions for each user to reduce the occurrence of false alarms and to improve detection. The user profile data (signature) is saved and updated every time the user logs on and off Intrusion Prevention system [31] with the help of IDS, monitors network traffic and system activities to detect possible intrusions and dynamically responds to intrusions by blocking the traffic or quarantine it.

(c) *Intrusion Detection and Prevention Systems [31]:* Having their own strengths and weaknesses, individual IDS and IPS are not able to provide efficient security. It is very effective to use a combination of IDS and IPS, which is called IDPS. Apart from identifying possible intrusions, IDPS stops and reports them to security administrators. Proper configuration and management of IDS and IPS combination can improve Cloud security. NIST explained how intrusion detection and prevention can be used together to strengthen security, and it also proposed different ways to design, configure, and manage IDPS. However, IDPS drawbacks are still an open problem.

(d) Reputation-Based Voting (RBV) approach [33] for tolerating collusive computing resources in large-scale grid computing systems, which can be seen as a cloud computing service, which is used for business development has a strong potential for other Cloud computing services. To overcome the high overhead and the performance degradation by replication with voting mechanisms in the presence of collusion attacks, the voting method has been improved by combining it with reputation system. The voting decision of the task is generated based on the reputation value of the computing resources that participate in the computation of such task. This approach can provide a lower error rate with better performance in terms of overhead compared to the m-first voting and credibility-based voting techniques. This tolerance scheme for detecting collusive computing resources is more accurate and more reliable. Therefore, this approach can help improve the efficiency of voting-based techniques, to tolerate collusive computing resources, and to increase the security level of cloud computing. However, this approach considers only the case of an attacks model which represents a single group of collusive computing resources distributed in several Virtual Organizations (VOs). Such resources always return the same wrong results with a fixed collusion probability.

4   Conclusions

In this chapter, we presented an overview of the state-of-the-art of cloud computing security which covers its essential challenges through the main different cyber-security threats, the main different approaches, algorithms and techniques developed to address them, as well as the open problems which define the research directions in this area. The bottom line is that the state of maturity of cloud computing security is very promising and there are many research directions still open and which promise continued improvements of cloud security and privacy.

References

1. Armbrust, M., et al.: Above the clouds: a Berkeley view of cloud computing. UC Berkeley Technical Report (2009)
2. Mell, P.: The NIST Definition of Cloud, Reports on Computer Systems Technology, p. 7, September 2011
3. Microsoft multi tenant data architecture. http://msdn.microsoft.com/en-us/library/aa479016.aspx
4. Sosinsky, B.: Cloud Computing Bible (2015)
5. Tianfield, H.: Security issues in cloud computing school of engineering and built environment. Glasgow Caledonian University, United Kingdom (2012)
6. Service Level Agreement and Master Service Agreement. http://www.softlayer.com/sla.html. Accessed 05 Apr 2009

7. Kandukuri, B.R., Ramakrishna Paturi, V., Rakshit, A.: Cloud security issues. In: 2009 IEEE International Conference on Services Computing. Advanced Software Technologies International Institute of Information Technology (2009)

8. Karajeh, H., Maqableh, M., Masa'deh, R.: Security of Cloud Computing Environment (2014)

9. Security for cloud computation: Ten steps to ensure success, version 2, March 2015

10. Rabai, L.B.A., Jouini, M., Aissa, A.B., Mili, A.: A cyber security model in cloud computing environments. J. King Saud Univ. Comput. Inf. Sci. **25**, 63–75 (2013)

11. Zissis, D., Lekkas, D.: Addressing cloud computing security issues. Future Generation Comput. Syst. **21**, 513–592 (2012)

12. Wei, L., Zhu, H., Cao, Z., Dong, X., Jia, W., Chen, Y., Vasilakos, A.V.: Security and privacy for storage and computation in cloud computing. Inf. Sci. **258**(10), 371–386 (2014)

13. Turban, E., King, D.: Electronic Commerce, Global edn. Person, Upper Saddle River (2012)

14. Sullivan, B., Tabet, S.: Practices for Secure Development of Cloud Applications (2013)

15. Scarfone, K., Mell, P.: Guide to Intrusion Detection and Prevention Systems (IDPS). Recommendations of the National Institute of Standards and Technology (2007)

16. Security Breach - Explore the Internet - urlfo.com, dreached while scanning argument. www.urlfo.com/phrase/security-breach

17. Data loss - Wikipedia, the free encyclopedia. en.wikipedia.org/wiki/Data_loss

18. Claburn, T.: Google Study Finds Widespread Account Hijacking, February 2014

19. InsecureAPImplementationT. http://www.darkreading.com/cloud/insecure-api-implementations-threaten-cloud/d/d-id/1137550?

20. Ali, M., Khan, S.U.: Security in cloud computing approaches and solutions (2015)

21. Balachandra, K.R., Ramakrishna, V.P., Rakshit, A.: Cloud security issues. In: Proceedings of the IEEE International Conference on Services Computing, SCC 2009, pp. 517–520 (2009)

22. Gentry, C.: Fully homomorphic encryption using ideal lattices. In: STOC 2009, pp. 169–178. Association for Computing Machinery, New York (2009)

23. Ryan, M.D.: Cloud computing security: the scientific challenge, and a survey of solutions. J. Syst. Softw. **86**, 2263–2268 (2013)

24. http://en.wikipedia.org/wiki/Field-programmable_gate_array

25. Mondol, J.-A.M., IEEE Member: Cloud security using Solutions using FPGA (2012)

26. Leu, F.Y., Lin, J.C., Li, M.C., Yang, C.T., Shih, P.C.: Integrating grid with intrusion detection. In: Proceedings of the 19th International Conference on Advanced Information Networking and Applications, AINA 2005, vol. 1, pp. 304–309 (2005)

27. Lin, D., Squicciarini, A.: Data protection models for service provisioning in the cloud. In: Proceeding of the ACM Symposium on Access Control Models and Technologies, SACMAT 2010 (2010)

28. http://en.wikipedia.org/wiki/Kerberos_(protocol)

29. Hojabri, M., Venkat Rao, K.: Innovation in cloud computing: implementation of Kerberos version5 in cloud computing in order to enhance the security issues

30. Ludwig, H., Keller, A., Dan, A., King, R., Franck, R.: Web Service Level Agreement (WSLA) language specification. IBM Corporation (2003)

31. Modi, C., Patel, D., Borisaniya, B., Patel, H., Patel, A., Rajarajan, M.: A survey of intrusion detection techniques in cloud. J. Netw. Comput. Appl. **36**, 42–57 (2013)

32. Patel, P., Ranabahu, A.H., Sheth, A.P.: Service Level Agreement in Cloud Computing (2008)
33. Bendahman, A., Essaaidi, M., El Moussoaui, A.: The effectiveness of reputation-based voting for collusion tolerance in large-scale grids. IEEE Trans. Dependable Secure Comput. **12**(6), 665–674 (2015)

Somewhat/Fully Homomorphic Encryption: Implementation Progresses and Challenges

Guillaume Bonnoron[1,2], Caroline Fontaine[2(✉)], Guy Gogniat[3],
Vincent Herbert[2,4], Vianney Lapôtre[3], Vincent Migliore[3],
and Adeline Roux-Langlois[5]

[1] Chair of Naval Cyber Defense, Ecole Navale - CC600, 29240 Brest Cedex 9, France
`guillaume.bonnoron@ecole-navale.fr`
[2] CNRS and IMT Atlantique, UMR 6285, Lab-STICC,
CS 83818, 29238 Brest cedex 3, France
`caroline.fontaine@imt-atlantique.fr`
[3] Univ. Bretagne-Sud, UMR 6285, Lab-STICC, 56100 Lorient, France
`{guy.gogniat,vianney.lapotre,vincent.migliore}@univ-ubs.fr`
[4] CEA LIST, Point Courrier 172, 91191 Gif-sur-Yvette Cedex, France
`vincent.herbert@cea.fr`
[5] CNRS - IRISA, Campus universitaire de Beaulieu, 35042 Rennes, France
`adeline.roux-langlois@irisa.fr`

Abstract. The proposed article aims, for readers, to learn about the existing efforts to secure and implement Somewhat/Fully Homomorphic Encryption ((S/F)HE) schemes and the problems to be tackled in order to progress toward their adoption. For that purpose, the article provides, at first, a brief introduction regarding (S/F)HE. Then, it focuses on some practical issues related to the adoption of (S/F)HE schemes, i.e. the security parameters, the existing implementations and their limitations, and the management of the huge complexity caused by homomorphic calculation. These issues are analyzed with the help of recent related work published in the literature, and with the experience gained by the authors through their experiments.

Keywords: Homomorphic Encryption · Data privacy · Confidentiality · Security · Real world

1 Introduction

Homomorphic Encryption (HE) is a recent promising tool in modern cryptography, that allows to carry out operations on encryptYAed data. The key idea is that performing some operations on encrypted data will provide the same result after decryption as if the computation would have been performed on the original plain data. Then, with such a tool one could outsource storage and/or computation without endangering data's privacy. Figure 1 illustrates different client/server scenario benefiting from homomorphic encryption. Some examples of applications can be found in the literature, *e.g.* [NLV11, GLN12,

© Springer International Publishing AG 2017
S. El Hajji et al. (Eds.): C2SI 2017, LNCS 10194, pp. 68–82, 2017.
DOI: 10.1007/978-3-319-55589-8_5

LLN14, BPB09, CGGI16b]. While usual cryptographic schemes sometimes have homomorphic properties, for addition [Pai99] or multiplication [ElG85] operations (see also [FG07] for a survey on such partially homomorphic schemes), an important breakthrough has been made in 2009 according to the work of Gentry [Gen09b, Gen09a]. Based on hard lattice problems, he proposed the first *Fully Homomorphic Encryption* (FHE) scheme, enabling to perform an unlimited number of additions and multiplications secretly. Unfortunately, this scheme was too complex to be used in practice. Nevertheless, it introduced an interesting structure as well as a nice trick called *bootstrapping* to reduce the inherent noise that accompany the running of additions and multiplications. Following this seminal work, several FHE have been proposed, but none of them were usable in practice. It is interesting to notice that some FHE are related to more practical schemes called *Somewhat Homomorphic Encryption (SHE) schemes*, that enable an arbitrary number of additions but a bounded number of multiplications. In fact, with the help of bootstrapping, one can design FHE schemes from SHE schemes. Nevertheless, this bootstrapping step adds an extra cost to an already quite heavy process.

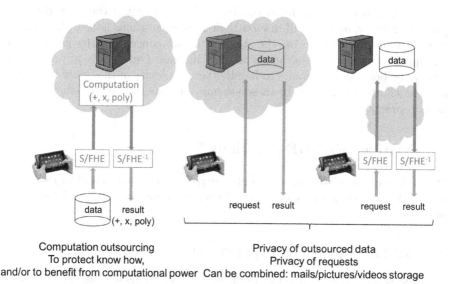

Fig. 1. A need for processing encrypted data, to ensure privacy in outsourced computation, outsourced storage and databases requests. Green areas show what is encrypted. (Color figure online)

To address a particular use case, one must have in mind several constraints that will be crucial for choosing the right homomorphic encryption solution. First, using (S/F)HE schemes will lead to a huge ciphertext expansion (say from 2.000 to 500.000 or even 1.000.000 according to the scheme and the targeted security level). This is due to the fact that homomorphic schemes must be probabilistic to ensure semantic security, and to the particular underlying mathematical structures. Second, as we will see later in this paper, we will need to consider only worse case complexity for the algorithms that will be run on the encrypted data; also considering that the underlying operations are intrinsically expansive, this will drastically penalize the global running time. These are the most important constraints we have to address. The underlying common point behind these remarks concerns the targeted security level, as it determines the parameters used to use the mathematical appropriate structure that will enable a coherent computation, and then the ciphertext expansion and the running time. Another important parameter concerns the strategy of the developer in terms of flexibility.

Hence, two strategies can be followed to drive the choice. On one hand one can fix a maximum multiplicative depth for the Boolean circuit to be evaluated on the encrypted data. This may take into account a small or at least bounded flexibility for future modifications of the circuit to be evaluated. In this case, SHE is the best choice. On the other hand one may want to be able to use any Boolean circuit and then to handle an unbounded multiplicative depth. In this case, FHE is the only choice.

In this paper, we will focus on SHE schemes because they are the most promising today, and we will discuss their implementations issues. We will also provide some information on the state-of-the-art concerning the use of bootstrapping to modify these schemes intro FHE ones. Our goal is to provide the reader the best starting points to handle the complexity of the issues to address and the efforts made to make these schemes become sufficiently efficient to be used in practice.

2   Existing (F/S)HE Solutions

Due to lack of space, we do not provide any precise description and let the reader refer to the mentioned papers to get all the mathematical details related to the schemes.

2.1   SHE from Classical Crypto World

The first scheme that enabled to perform both additions and multiplications is due to Boneh *et al.* [BGN05]. This pairing-based SHE enables to perform as many additions as wanted, but only one multiplication. Hence, it is not really flexible.

We have worked recently on the design of a variant of this scheme that allows to handle multiplicative depth 2 (instead of 1). Our solution employs together two improvements of the original scheme, based on [Fre10, CF15]. We will refer to it as BGN2 in the rest of the document [HF17].

2.2 Lattice-Based (S/F)HE

Whereas the first FHE scheme has been proposed by Gentry in [Gen09a], the first SHE based on lattices has been propose by Aguilar *et al.* in [AMGH10]. These schemes have been followed by many others. First generation encompass the older ones, like [vDGHV10,SV10,GHS12a,GH11]. Second generation [BGV12,CNT12,BGV12,Bra12,FV12] started with leveled SHE schemes based on modulus switching, which have then been improved to remain scale invariant, etc. They were followed by third generation schemes like [GSW13,BLLN13, BV14,KGV15,DS16]. In fact several of these schemes consist of improvement of previous ones, and the genealogy in rich of cross-fertilization. For example, SHIELD [GSW13] and F-NTRU [DS16] are the third generation schemes equivalent to, respectively, FV [FV12] and YASHE' [BLLN13]. With larger costs for the first homomorphic multiplications, these schemes have a much better asymptotic behavior.

3 Customization and Optimization of Both Program and Data

The issues discussed in this section have been addressed in few papers only [AMFF+13,FSF+13,CDS15] for the moment, but they are really important to handle real deployment of this technology.

3.1 Program Management

To establish a proper link between the program we want to run and (S/F)HE schemes, one must rely on the fact that any program can be expressed as a Boolean circuit involving XOR and AND gates, and that XOR and AND are precisely the addition and multiplication operators over bits. This being said, the game is to express the program through such a Boolean circuit, and to optimize it. The last step will then be to evaluate this optimized Boolean circuit over the encrypted data while replacing XOR and AND gates by the corresponding encrypted operators provided by a (S/F)HE library.

The optimization step of the Boolean circuit is crucial, as it will give us the multiplicative depth we will have to handle with the (S/F)HE scheme. This will have an impact on the parameters of this scheme (*e.g.* size of the lattice, modulus, etc.), and then on the ciphertexts size and on the resulting computation time. Usually, we focus on this multiplicative depth, but we also have to be careful that a very large amount of additions may also lead to heavy computations. Also, sometimes additions following multiplications may have a different impact than additions occurring at lower multiplicative depth. Also, when using leveled (S/F)HE using modulus switching techniques like [BGV12], one has to optimize at which level each ciphertext stands, to avoid any extra and costly modulus switching. At last, the noise growth is usually symmetric between the left and right operands, but for some particular recent schemes like [GSW13] the noise growth is asymmetric over the multiplications. In this case, it is important to optimize which operand will be left or right. All these aspects should be taken into account. This optimization issue has not been sufficiently addressed in the literature, and a lot of work remains to be done to set theoretical bounds and practical strategies to handle such an optimality.

Another issue related with the customization of the Boolean circuit is that it may contain `if-then-else` or `repeat...until` expressions leading to branches with dynamic size depending on the processed data. To handle such statements properly, on must have in mind that the processed data they are depending on are encrypted. Moreover, it is crucial that during the running process no information about the real value of the underlying data may leak. To handle this is easy but costly. In fact, the best approach is to rewrite such parts of the program with the help of Boolean expressions. If we look at the `if-then-else` example, one can rewrite `if c then x=a else x=b` as `x = (c AND a) XOR ((NOT c) AND b)`. The whole expression will then be evaluated over the encrypted data, and the final encrypted result will be the good one without revealing anything on the plain value of `c`. Unfortunately the price to be paid is high, as we always have to evaluated the whole expression, meaning that we always need to evaluate the deepest branch of the tree. Hence, computing over encrypted data always requires the worse-case complexity. This means that when choosing an algorithm to perform a particular computation over encrypted data, one must choose the algorithm that provides the best worse-case complexity (whereas usually we look for the best average-case complexity).

Finally, as (S/F)HE remain costly today, on must be really careful not to perform non necessary heavy computations. According to a particular applicative scenario, one must try to perform as much classical encryption as possible, and to think of some pre-processing over the plaintexts that may help to enlight the computation over the ciphertexts.

3.2 Data Management

Some schemes work only over integers, and some over bits. And some can manage both. If the application scenario does not require data depend behavior, then working over integers is usually the best choice. But if the application scenario requires some `if-then-else` or `repeat...until` like statements, then we need to work at the bit level to be able to perform <,> and = operators. In this case, we have to provide home maid basic operators for integers additions and multiplications, floats management,... but this is the price to pay to handle such dynamic behaviors properly.

Some schemes may also be compliant with batching ability, then enabling to group several pieces of data on the same plaintext, typically a polynomial, and to process all these pieces of data at the same time. This drastically reduces the costs in terms of space (memory) and running time, as several plaintexts (resp. ciphertexts) are processed in one shot. In the literature, batching is usually addressed through SIMD and RNS, see for example [SV14,BEHZ16].

4 Flexibility of (S/F)HE Schemes in Terms of Multiplicative Depth Management

4.1 SHE from Classical Crypto World

SHE schemes coming from classical crypto world are not flexible. Boneh-Goh-Nissim [BGN05] scheme only enables the evaluation only of degree 2 polynomials, whereas our extension BGN2 enables the evaluation of degree 4 polynomials.

4.2 Lattice-Based (S/F)HE Schemes

SHE schemes based on these lattice problems are generally much more flexible and can be turned into fully homomorphic schemes allowing computation with arbitrary depth. The downside for this flexibility is the increased size of ciphertexts and keys, leading to heavier computations. For these schemes, one have to choose the maximum multiplicative depth we want to be able to handle before deployment, as this multiplicative depth will drive the parameters determining the underlying lattice. Once the scheme is deployed, it cannot be modified. Moreover, as we have to handle "the higher the multiplicative depth, the higher the memory and time costs" the parameters much be chosen very carefully to maintain an acceptable cost while ensuring sufficient flexibility for future process.

FHE schemes are generally heavier to deploy, as they need the bootstrapping step to be sure to enhance the noise growth properly whatever the multiplicative depth. Hence, they should be used only if one does not know in advance the multiplicative depth we would like to handle in the future in the targeted application scenario. Good recent works to better understand the limits of this solution are [PV15], which provides a way to optimize the bootstraps management (for any FHE), and [CGGI16a] that proposes the more efficient current way to execute bootstrapping (especially for FHE based on [GSW13]).

5 Security of (S/F)HE Schemes

By nature, these schemes have been designed to enhance privacy and security. Hence, the analysis of their security is crucial. The best security level that has been achieved for such schemes is IND-CPA. We know that IND-CCA2 in not achievable, and the design of efficient schemes achieving IND-CCA1 is still open. Now, we will split the security study according to the underlying mathematical rationales.

5.1 SHE from Classical Crypto World

Schemes which are based on mathematical objects which have already been deeply studied by the cryptographic community are better understood, and their security is easier to manage. Among such schemes, one can mention partially homomorphic schemes as ElGamal, Paillier, and their variants, which are out of the scope of this article as they cannot handle at the same time additions and multiplications. As mentioned in Sect. 2.2, one can also mention Boneh-Goh-Nissim scheme [BGN05], a pairing-based SHE scheme which enables to perform as many additions as wanted, but only one multiplication, and our scheme BGN2 [HF17] which can handle one more multiplication depth.

The security of BGN2 is based on the generalized subgroup decision assumption[1]. This problem is derived from the decision Diffie-Hellman assumption [Bon98]. Two possible choices to instantiate groups are to select either an elliptic curve or an hyperelliptic curve. We place ourselves in the first case, where the security assumption reduces to the elliptic curve discrete logarithm problem. The recommended group size is given by different academic and private organizations at www.keylength.com according standard security levels. The order of ciphertext expansion in BGN2 is thousands.

5.2 Lattice-Based (S/F)HE

The *Learning With Errors* problem is about solving a system of several *noisy* linear equations. Its ring variant, *Ring-LWE*, allows to design more efficient schemes with faster computations and smaller keys and ciphertexts. These problems attract large attention, beyond HE, because they are believed to be quantum resistant: no known quantum algorithms perform better than the classical ones against them.

These schemes were introduced in the previous decade, i.e. quite recently in the time scale of cryptography, and one must tell that there is still a gap between the theoretical hardness proofs and the practical behavior of the known attacks.

Formal Proofs. The first FHE scheme of Gentry [Gen09b] was based on two assumptions, the *Bounded Distance Decoding problem* and the *Sparse Subset Sum problem* which are not standard in lattice-based cryptography. The first scheme proven secure under the LWE assumption is proposed by Brakerski and Vaikuntanathan in 2011 [BV11]. This result was followed by numerous constructions based on LWE and Ring-LWE (as [Bra12,BGV12]...).

The LWE and Ring-LWE problem are proven to be at least as hard as well-known hard problems (in their worst-case) on lattices (respectively on ideal-lattices). Another advantage of recent lattice-based schemes is that their security is proven under those assumptions.

Experimental Security Evaluation. Even at this point, concrete security behind homomorphic encryption is still moving. Thus, extracting realistic parameters for a given security level is a main challenge of (S/F)HE scenario.

---

[1] We choose to employ asymetric pairings to compute homomomorphic product of fresh ciphertexts. The use of symmetric pairings would change the computational hardness assumption [Fre10, p. 46].

The standard approach so far is to focus on concrete attack means. Building on surveys about existing attacks against LWE [APS15] and Ring-LWE [Pei16], experiments must be pushed further to provide experimental results of secured parameters for most promising homomorphic schemes.

Among these most promising schemes one can mention YASHE' [BLLN13], FV [FV12] and SHIELD [GSW13]. YASHE' and FV have been published almost at the same time, but YASHE' took benefit from a strong lobby and became more popular in the proposed implementations. However, confidence in this scheme has been recently damaged by the subfield/sublattice attack [ABD16, KF16]. Making YASHE' immune to these attacks would lead to oversize its parameters, far too much for practical use [DGBL+15]. This is why among second generation schemes FV is now the real challenger, and has received a lot of attention during the past months [LCP16, Cry16, BEHZ16].

6   Existing Implementations of (S/F)HE Schemes

6.1   Software Implementations

We implemented BGN2, which have been presented and briefly discussed above. In this cryptosystem, the homomorphic multiplication asks to compute pairings. We chose to compute an optimal Ate pairing over an elliptic curve in the Barreto-Naehrig curve family [BN05] using a program called DCLXVI [NNS10]. This work is not yet published, but will be available soon both as an article and as a software.

Concerning lattice-based (S/F)HE schemes, several implementations have emerged since their introduction in 2009. Due to the pace of evolution in the theoretical field, some are now outdated but served as good proof of concept in the early days of (S/F)HE [Bre, Cor, Lep]. Other libraries [DGBL+15, Hal, Cry16, LCP16] implement the latest techniques described in [SV10, FV12, BGV12, BLLN13]. There are also private implementations like those used in [AMFF+13, FSF+13, CDS15, BEHZ16]. Most of the libraries aim at providing tools for experiments to the academic community, except [DGBL+15, LCP16, CDS15] which can be used as building blocks for more industrialized developments.

Today, no complete benchmark is available to provide a fair and complete overviews of the efficiency of all these schemes. The reader can refer to [LN14] for a comparison of FV and YASHE', and to [MBF16] for a first discussion and comparison of FV, SHIELD and F-NTRU in terms of pros and cons, and parameters setting.

6.2   Hardware Implementations

Hardware implementation is one of the two principal ways to accelerate FHE/SHE schemes with dedicated components. The second one is the GPU acceleration. Even if performances using GPU are very scheme-dependent, it can be a good alternative to set up an homomorphic server quickly with acceptable performances [KGV15].

Hardware accelerators focus on accelerating the most complex operation of Homomorphic Encryption Schemes, the multiplication of homomorphic operands. Depending on the scheme, a million-bit integer multiplier or a polynomial of degree $n \in [4096, 32768]$ with coefficients of size $\log_2 q \in [125, 1228]$ is required. In [DOS13], an ASIC implementation of million-bits multiplier performs the multiplication operation in 7.74 ms using NTT algorithm. This computation time corresponds to the computation time on a Xeon, but can be implemented as a co-processor and thus requiring a much smaller area. For polynomial based homomorphic encryption, due to the fact that one must address various size of polynomials, different architectures have been investigated. To our knowledge, all implementations are based on NTT algorithm too, except in [MMRL+17] for small size polynomials which implements Karatsuba algorithm instead. In [PNPM], a usual but optimized NTT implementation is presented for two parameter sets. The proposed accelerator performs an homomorphic multiplication in 6.5 ms for $n = 4096$ and $\log_2 q = 125$ bits, and 48 ms for parameters $n = 16384$ and $\log_2 q = 512$ bits. Authors of [PNPM] implemented 512×512 bits multipliers with a small modular reduction by selecting a Solinas prime modulus [Sol11]. Due to the size of polynomials and coefficients, a cache is implemented to connect the external memory used to store intermediate coefficients. They also reported a bottleneck due to the divide and rounding required by YASHE' scheme, especially for large integers. That is why in [SRJV+] a precomputation is performed on polynomials to reduce the size of coefficients. They split a ciphertext into a few polynomials by using the *Chinese Reminder Theorem* (CRT) on each coefficient. The overall architecture is based on an array of crypto-units, which gives some flexibility to process several residue polynomials in parallel. For parameters $n = 32768$ and $\log_2 q = 1228$ bits, their accelerator performs an homomorphic multiplication in 121 ms including 25 ms spent for CRT.

Table 1. Timing results for the hardware implementation of Homomorphic Encryption.

Integer based Homomorphic encryption					
Scheme	Algorithm	Size		Homomorphic Encryption	Homomorphic Multiplication	Work
Gentry-Halevy	NTT	1 M bits		2.09 s	7.74 ms	[DOS13]
DHGV		19 M bits		3.9 s	*no results*	[CMO+]
Polynomial based Homomorphic encryption						
Scheme	Algorithm	n	$\log_2 q$	Homomorphic Encryption	Homomorphic Multiplication	Work
YASHE'	Karatsuba	2560	124 bits	*not implemented*	4.73 ms	[MMRL+17]
	NTT	4096	125 bits		6.5 ms	[PNPM]
		16384	512 bits		48 ms	
		32768	1228 bits		121 ms	[SRJV+]

Table 1 summarize the different hardware implementations available for both integer based and polynomial based Homomorphic Encryption Schemes.

7   How to Handle Such a Huge Complexity and Expansion?

(S/F)HE schemes defined on Elliptic Curves and Pairing present a smaller complexity and expansion, and their security level is quite clear, but are very limited in terms of multiplicative depth. For some applications this may be sufficient, and particularly interesting, this is why we discussed them here.

Nevertheless, the biggest hope for the future comes from lattice-based schemes, which promise to handle larger multiplicative depths processing. Once their security will be better understood, their main drawbacks are their algorithmic complexity and the related ciphertext expansion. This is why it is important to pursue designing new schemes and to look for lighter solutions. Current ciphertext expansions can go from $10,000$ up to $1,000,000$, depending on the schemes and on the parameters that have been chosen (and which are directly related with the targeted security level). The encrypted data must be uploaded from the client device to the server, then processed on the server, and finally the encrypted result must be downloaded from the server to the client device. Hence, its size is critical.

To reduce the size of the uploaded data on the first step, [NLV11] proposed to combine the (S/F)HE scheme with symmetric encryption schemes. The main idea is that the data to be uploaded will be encrypted by the symmetric encryption scheme, and then sent to the server without any size expansion. Hence, the server will trans-crypt this encrypted data to produce a new ciphertext which corresponds to the encryption of the same data with the (S/F)HE that will be used on the server to perform the desired computation. This trans-cryption step will require the decryption circuit of the underlying symmetric encryption scheme to be evaluated by the (S/F)HE scheme. This step's complexity is critical, and will lead the choice of the symmetric cipher to use. Following this idea, several symmetric encryption schemes have been investigated. We will first mention on-the-shelf block ciphers like AES [GHS12b, CCK+13, DHS14], and the lightweight block ciphers Simon [LN14] and Prince [DSES14]. But the evaluation of these ciphers by the (S/F)HE remains too complex, and recent papers proposed new ciphers designed specifically for this purpose (*i.e.* with a low multiplicative depth): the block cipher Low-MC [ARS+15], which has been broken [DLMW15] and patched [Rec16]; the stream cipher Kreyvium [CCF+16], whose security is the same as the well-studied stream cipher Trivium; and a more recent stream cipher proposal called FLIP [MJSC16], which should be used carefully [DLR16]. These papers include experimental material and results. More exotic solutions are discussed in [FHK16], but without experimental data in the paper.

The second way to reduce ciphertexts weight is to pack several inputs on the same plaintext structure through batching. This has been briefly discussed in Sect. 3.2

8 Conclusion

There is still a lot of work to be done, but everything is moving fast and recent progress are quite impressive. Hence, for some applications which are not too critical in terms of memory and time costs it is time to adopt a practical approach to make dream become reality. How to choose the good (S/F)HE scheme for a given application scenario, and how to set it up in the best way? The answer is not trivial at all, and this article provides some hints concerning the implementation issues of these promising but still tricky and heavy schemes. Our goal was to share our discussions and reflections about all the identified issues, and to provide ad-hoc references to help the reader in his exploration of a very prolific and dense literature. It is clear that more comparisons should be performed between the most promising schemes. A few papers compared two schemes at a time, like [LN14], and a first attempt to provide a wider analysis can be found in [MBF16]. But it is clear that it should be pushed further, and that fair benchmarks based on available implementations have to be driven. Moreover, a fair and precise comparison should also be driven to properly compare SHE schemes coming from classical crypto world with lattice-based ones when targeting a small multiplicative depth, in terms of time and space complexity. At the same time, open issues remain concerning a precise evaluation of the practical security of lattice-based schemes, as well as on the optimization of the Boolean circuits we want to evaluate over the encrypted data.

Acknowledgement. This project has received funding from the European Union's Horizon 2020 research and innovation programme under grant agreement No 643964.

References

[ABD16] Albrecht, M., Bai, S., Ducas, L.: A subfield lattice attack on overstretched ntru assumptions: cryptanalysis of some fhe and graded encoding schemes. Cryptology ePrint Archive, Report 2016/127 (2016)

[AMFF+13] Aguilar-Melchor, C., Fau, S., Fontaine, C., Gogniat, G., Sirdey, R.: Recent advances in homomorphic encryption: a possible future for signal processing in the encrypted domain. IEEE Sig. Process. Mag. **30**(2), 108–117 (2013)

[AMGH10] Melchor, C.A., Gaborit, P., Herranz, J.: Additively homomorphic encryption with d-operand multiplications. In: Rabin, T. (ed.) CRYPTO 2010. LNCS, vol. 6223, pp. 138–154. Springer, Heidelberg (2010). doi:10.1007/978-3-642-14623-7_8

[APS15] Albrecht, M.R., Player, R., Scott, S.: On the concrete hardness of learning with errors. J. Math. Cryptology **9**(3), 169–203 (2015)

[ARS+15] Albrecht, M.R., Rechberger, C., Schneider, T., Tiessen, T., Zohner, M.: Ciphers for MPC and FHE. In: Oswald, E., Fischlin, M. (eds.) EUROCRYPT 2015. LNCS, vol. 9056, pp. 430–454. Springer, Heidelberg (2015). doi:10.1007/978-3-662-46800-5_17

[BEHZ16] Bajard, J.-C., Eynard, J., Hasan, A., Zucca, V.: A full RNS variant of FV like somewhat homomorphic encryption schemes. Cryptology ePrint Archive, Report 2016/510 (2016). http://eprint.iacr.org/2016/510

[BGN05] Boneh, D., Goh, E.-J., Nissim, K.: Evaluating 2-DNF formulas on ciphertexts. In: Kilian, J. (ed.) TCC 2005. LNCS, vol. 3378, pp. 325–341. Springer, Heidelberg (2005). doi:10.1007/978-3-540-30576-7_18

[BGV12] Brakerski, Z., Gentry, C., Vaikuntanathan, V.: (leveled) fully homomorphic encryption without bootstrapping. In: Proceedings of the 3rd Innovations in Theoretical Computer Science Conference - ITCS 2012, pp. 309–325. ACM (2012)

[BLLN13] Bos, J.W., Lauter, K., Loftus, J., Naehrig, M.: Improved security for a ring-based fully homomorphic encryption scheme. In: Stam, M. (ed.) IMACC 2013. LNCS, vol. 8308, pp. 45–64. Springer, Heidelberg (2013). doi:10.1007/978-3-642-45239-0_4

[BN05] Barreto, P.S.L.M., Naehrig, M.: Pairing-friendly elliptic curves of prime order. In: Preneel, B., Tavares, S. (eds.) SAC 2005. LNCS, vol. 3897, pp. 319–331. Springer, Heidelberg (2006). doi:10.1007/11693383_22

[Bon98] Boneh, D.: The decision Diffie-Hellman problem. In: Buhler, J.P. (ed.) ANTS 1998. LNCS, vol. 1423, pp. 48–63. Springer, Heidelberg (1998). doi:10.1007/BFb0054851

[BPB09] Bianchi, T., Piva, A., Barni, M.: On the implementation of the discrete Fourier transform in the encrypted domain. IEEE Trans. Inf. Forensics Secur. 4(1), 86–97 (2009)

[Bra12] Brakerski, Z.: Fully homomorphic encryption without modulus switching from classical GapSVP. In: Safavi-Naini, R., Canetti, R. (eds.) CRYPTO 2012. LNCS, vol. 7417, pp. 868–886. Springer, Heidelberg (2012). doi:10.1007/978-3-642-32009-5_50

[Bre] Brenner, M.: Hcrypt project. http://www.hcrypt.com

[BV11] Brakerski, Z., Vaikuntanathan, V.: Efficient fully homomorphic encryption from (Standard) LWE. In: Proceedings of FOCS, pp. 97–106 (2011)

[BV14] Brakerski, Z., Vaikuntanathan, V.: Lattice-based fhe as secure as pke. In: Proceedings of the 5th Conference on Innovations in Theoretical Computer Science - ITCS 2014, pp. 1–12. ACM (2014)

[CCF+16] Canteaut, A., Carpov, S., Fontaine, C., Lepoint, T., Naya-Plasencia, M., Paillier, P., Sirdey, R.: Stream ciphers: a practical solution for efficient homomorphic-ciphertext compression. In: Peyrin, T. (ed.) FSE 2016. LNCS, vol. 9783, pp. 313–333. Springer, Heidelberg (2016). doi:10.1007/978-3-662-52993-5_16

[CCK+13] Cheon, J.H., Coron, J.-S., Kim, J., Lee, M.S., Lepoint, T., Tibouchi, M., Yun, A.: Batch fully homomorphic encryption over the integers. In: Johansson, T., Nguyen, P.Q. (eds.) EUROCRYPT 2013. LNCS, vol. 7881, pp. 315–335. Springer, Heidelberg (2013). doi:10.1007/978-3-642-38348-9_20

[CDS15] Carpov, S., Dubrulle, P., Sirdey, R.: Armadillo: a compilation chain for privacy preserving applications. In: Proceedings of the 3rd International Workshop on Security in Cloud Computing, pp. 13–19. ACM (2015)

[CF15] Catalano, D., Fiore, D.: Using linearly-homomorphic encryption to evaluate degree-2 functions on encrypted data. In: Ray, I., Li, N., Kruegel, C. (eds.) Proceedings of the 22nd ACM SIGSAC Conference on Computer and Communications Security, Denver, CO, USA, 12–16 October, pp. 1518–1529. ACM (2015)

[CGGI16a] Chillotti, I., Gama, N., Georgieva, M., Izabachène, M.: Faster fully homomorphic encryption: bootstrapping in less than 0.1 seconds. In: Cheon, J.H., Takagi, T. (eds.) ASIACRYPT 2016. LNCS, vol. 10031, pp. 3–33. Springer, Heidelberg (2016). doi:10.1007/978-3-662-53887-6_1

[CGGI16b] Chillotti, I., Gama, N., Georgieva, M., Izabachène, M.: A homomorphic LWE based e-voting scheme. In: Takagi, T. (ed.) PQCrypto 2016. LNCS, vol. 9606, pp. 245–265. Springer, Cham (2016). doi:10.1007/978-3-319-29360-8_16

[CMO+] Cao, X., Moore, C., O'Neill, M., Hanley, N., O'Sullivan, E.: High-speed fully homomorphic encryption over the integers. In: Böhme, R., Brenner, M., Moore, T., Smith, M. (eds.) FC 2014. LNCS, vol. 8438. Springer, Heidelberg (2014). doi:10.1007/978-3-662-44774-1_14

[CNT12] Coron, J.-S., Naccache, D., Tibouchi, M.: Public key compression and modulus switching for fully homomorphic encryption over the integers. In: Pointcheval, D., Johansson, T. (eds.) EUROCRYPT 2012. LNCS, vol. 7237, pp. 446–464. Springer, Heidelberg (2012). doi:10.1007/978-3-642-29011-4_27

[Cor] Coron, J.-S.: An implementation of the DGHV fully homomorphic scheme. https://github.com/coron/fhe

[Cry16] CryptoExperts: FV-NFLlib (2016). https://github.com/CryptoExperts/FV-NFLlib

[DGBL+15] Dowlin, N., Gilad-Bachrach, R., Laine, K., Lauter, K., Naehrig, M., Wernsing, J.: Manual for using homomorphic encryption for bioinformatics. Technical report MSR-TR-2015-87, November 2015

[DHS14] Doröz, Y., Yin, H., Sunar, B.: Homomorphic AES evaluation using NTRU. IACR Cryptology ePrint Archive 2014:39 (2014)

[DLMW15] Dinur, I., Liu, Y., Meier, W., Wang, Q.: Optimized interpolation attacks on LowMC. IACR Cryptology ePrint Archive 2015:418 (2015)

[DLR16] Duval, S., Lallemand, V., Rotella, Y.: Cryptanalysis of the FLIP family of stream ciphers. IACR Cryptology ePrint Archive (271) (2016)

[DOS13] Doroz, Y., Ozturk, E., Sunar, B.: Evaluating the hardware performance of a million-bit multiplier. In: Proceedings of Euromicro Conference on Digital System Design – DSD 2013 (2013)

[DS16] Doröz, Y., Sunar, B.: Flattening ntru for evaluation key free homomorphic encryption. Cryptology ePrint Archive, Report 2016/315 (2016)

[DSES14] Doröz, Y., Shahverdi, A., Eisenbarth, T., Sunar, B.: Toward practical homomorphic evaluation of block ciphers using prince. In: Böhme, R., Brenner, M., Moore, T., Smith, M. (eds.) FC 2014. LNCS, vol. 8438, pp. 208–220. Springer, Heidelberg (2014). doi:10.1007/978-3-662-44774-1_17

[ElG85] ElGamal, T.: A public key cryptosystem and a signature scheme based on discrete logarithms. IEEE Trans. Inf. Theor. $31(4)$, 469–472 (1985)

[FG07] Fontaine, C., Galand, F.: A survey of homomorphic encryption for nonspecialists. EURASIP J. Inf. Secur. $2007(1)$, 1–15 (2007)

[FHK16] Fouque, P.-A., Hadjibeyli, B., Kirchner, P.: Homomorphic evaluation of lattice-based symmetric encryption schemes. In: Dinh, T.N., Thai, M.T. (eds.) COCOON 2016. LNCS, vol. 9797, pp. 269–280. Springer, Cham (2016). doi:10.1007/978-3-319-42634-1_22

[Fre10] Freeman, D.M.: Converting pairing-based cryptosystems from composite-order groups to prime-order groups. In: Gilbert, H. (ed.) EUROCRYPT 2010. LNCS, vol. 6110, pp. 44–61. Springer, Heidelberg (2010). doi:10.1007/978-3-642-13190-5_3

[FSF+13] Fau, S., Sirdey, R., Fontaine, C., Aguilar-Melchor, C., Gogniat, G.: Towards practical program execution over fully homomorphic encryption schemes. In: Eighth International Conference on P2P, Parallel, Grid, Cloud and Internet Computing (3PGCIC), pp. 284–290. IEEE (2013)

[FV12]   Fan, J., Vercauteren, F.: Somewhat practical fully homomorphic encryption. IACR Cryptology ePrint Archive 2012:144 (2012)

[Gen09a]   Gentry, C.: A fully homomorphic encryption scheme. Ph.D. thesis, Stanford University (2009)

[Gen09b]   Gentry, C.: Fully homomorphic encryption using ideal lattices. In: STOC, vol. 9, pp. 169–178 (2009)

[GH11]   Gentry, C., Halevi, S.: Fully homomorphic encryption without squashing using depth-3 arithmetic circuits. In: IEEE 52nd Annual Symposium on Foundations of Computer Science (FOCS 2011), pp. 107–109. IEEE (2011)

[GHS12a]   Gentry, C., Halevi, S., Smart, N.P.: Fully homomorphic encryption with polylog overhead. In: Pointcheval, D., Johansson, T. (eds.) EUROCRYPT 2012. LNCS, vol. 7237, pp. 465–482. Springer, Heidelberg (2012). doi:10. 1007/978-3-642-29011-4_28

[GHS12b]   Gentry, C., Halevi, S., Smart, N.P.: Homomorphic evaluation of the AES circuit. In: Safavi-Naini, R., Canetti, R. (eds.) CRYPTO 2012. LNCS, vol. 7417, pp. 850–867. Springer, Heidelberg (2012). doi:10.1007/978-3-642-32009-5_49

[GLN12]   Graepel, T., Lauter, K., Naehrig, M.: ML confidential: machine learning on encrypted data. In: Kwon, T., Lee, M.-K., Kwon, D. (eds.) ICISC 2012. LNCS, vol. 7839, pp. 1–21. Springer, Heidelberg (2013). doi:10.1007/978-3-642-37682-5_1

[GSW13]   Gentry, C., Sahai, A., Waters, B.: Homomorphic encryption from learning with errors: conceptually-simpler, asymptotically-faster, attribute-based. In: Canetti, R., Garay, J.A. (eds.) CRYPTO 2013. LNCS, vol. 8042, pp. 75–92. Springer, Heidelberg (2013). doi:10.1007/978-3-642-40041-4_5

[Hal]   Halevi, S.: HElib. https://github.com/shaih/HElib

[HF17]   Herbert, V., Fontaine, C.: Software Implementation of 2-Depth Pairing-based Homomorphic Encryption Scheme, Cryptology ePrint Archive, Report 2017/091 (2017). http://eprint.iacr.org/2017/091

[KF16]   Kirchner, P., Fouque, P.-A.: Comparison between subfield and straightforward attacks on NTRU. Cryptology ePrint Archive, 2016/717 (2016)

[KGV15]   Khedr, A., Gulak, G., Vaikuntanathan, V.: SHIELD: scalable homomorphic implementation of encrypted data-classifiers. IEEE Trans. Comput. **PP**(99), 1 (2015)

[LCP16]   Laine, K., Chen, H., Player, R.: Simple encrypted arithmetic library - seal (v2.1). Technical report, September 2016

[Lep]   Lepoint, T.: A proof-of-concept implementation of the homomorphic evaluation of SIMON using FV and YASHE. https://github.com/tlepoint/homomorphic-simon

[LLN14]   Lauter, K., López-Alt, A., Naehrig, M.: Private computation on encrypted genomic data. In: Aranha, D.F., Menezes, A. (eds.) LATINCRYPT 2014. LNCS, vol. 8895, pp. 3–27. Springer, Cham (2015). doi:10.1007/978-3-319-16295-9_1

[LN14]   Lepoint, T., Naehrig, M.: A comparison of the homomorphic encryption schemes FV and YASHE. In: Pointcheval, D., Vergnaud, D. (eds.) AFRICACRYPT 2014. LNCS, vol. 8469, pp. 318–335. Springer, Cham (2014). doi:10.1007/978-3-319-06734-6_20

[MBF16]   Migliore, V., Bonnoron, G., Fontaine, C.: Determination and exploration of practical parameters for the latest somewhat homomorphic encryption (SHE) schemes. Working paper or preprint, October 2016

[MJSC16] Méaux, P., Journault, A., Standaert, F.-X., Carlet, C.: Towards stream
          ciphers for efficient FHE with low-noise ciphertexts. In: Fischlin, M.,
          Coron, J.-S. (eds.) EUROCRYPT 2016. LNCS, vol. 9665, pp. 311–343.
          Springer, Heidelberg (2016). doi:10.1007/978-3-662-49890-3_13

[MMRL+17] Migliore, V., Real, M.M., Lapotre, V., Tisserand, A., Fontaine, C., Gog-
          niat, G.: Hardware/software co-design of an accelerator for FV homomor-
          phic encryption scheme using Karatsuba algorithm. IEEE Trans. Comput.
          (2017, accepted)

[NLV11]   Naehrig, M., Lauter, K.E., Vaikuntanathan, V.: Can homomorphic encryp-
          tion be practical? In: ACM CCSW, pp. 113–124. ACM (2011)

[NNS10]   Naehrig, M., Niederhagen, R., Schwabe, P.: New software speed records for
          cryptographic pairings. In: Abdalla, M., Barreto, P.S.L.M. (eds.) LATIN-
          CRYPT 2010. LNCS, vol. 6212, pp. 109–123. Springer, Heidelberg (2010).
          doi:10.1007/978-3-642-14712-8_7

[Pai99]   Paillier, P.: Public-key cryptosystems based on composite degree residu-
          osity classes. In: Stern, J. (ed.) EUROCRYPT 1999. LNCS, vol. 1592, pp.
          223–238. Springer, Heidelberg (1999). doi:10.1007/3-540-48910-X_16

[Pei16]   Peikert, C.: How (not) to instantiate ring-LWE. In: Zikas, V., Prisco, R.
          (eds.) SCN 2016. LNCS, vol. 9841, pp. 411–430. Springer, Cham (2016).
          doi:10.1007/978-3-319-44618-9_22

[PNPM]   Pöppelmann, T., Naehrig, M., Putnam, A., Macias, A.: Accelerating homo-
          morphic evaluation on reconfigurable hardware. In: Güneysu, T., Hand-
          schuh, H. (eds.) CHES 2015. LNCS, vol. 9293, pp. 143–163. Springer, Hei-
          delberg (2015). doi:10.1007/978-3-662-48324-4_8

[PV15]   Paindavoine, M., Vialla, B.: Minimizing the number of bootstrappings in
          fully homomorphic encryption. In: Dunkelman, O., Keliher, L. (eds.) SAC
          2015. LNCS, vol. 9566, pp. 25–43. Springer, Cham (2016). doi:10.1007/
          978-3-319-31301-6_2

[Rec16]   Rechberger, C.: The FHEMPCZK-Cipher Zoo. Presented at the FSE
          Rump Session (2016)

[Sol11]   Solinas, J.A.: Generalized mersenne prime. In: van Tilborg, H.C.A., Jajo-
          dia, S. (eds.) Encyclopedia of Cryptography and Security, pp. 509–510.
          Springer, New York (2011)

[SRJV+]   Sinha Roy, S., Järvinen, K., Vercauteren, F., Dimitrov, V., Verbauwhede,
          I.: Modular hardware architecture for somewhat homomorphic func-
          tion evaluation. In: Güneysu, T., Handschuh, H. (eds.) CHES 2015.
          LNCS, vol. 9293, pp. 164–184. Springer, Heidelberg (2015). doi:10.1007/
          978-3-662-48324-4_9

[SV10]   Smart, N.P., Vercauteren, F.: Fully homomorphic encryption with rela-
          tively small key and ciphertext sizes. In: Nguyen, P.Q., Pointcheval, D.
          (eds.) PKC 2010. LNCS, vol. 6056, pp. 420–443. Springer, Heidelberg
          (2010). doi:10.1007/978-3-642-13013-7_25

[SV14]   Smart, N.P., Vercauteren, F.: Fully homomorphic simd operations. Des.
          Codes Crypt. 71(1), 57–81 (2014)

[vDGHV10] van Dijk, M., Gentry, C., Halevi, S., Vaikuntanathan, V.: Fully homo-
          morphic encryption over the integers. In: Gilbert, H. (ed.) EUROCRYPT
          2010. LNCS, vol. 6110, pp. 24–43. Springer, Heidelberg (2010). doi:10.
          1007/978-3-642-13190-5_2

Regular Papers

Two-Source Randomness Extractors for Elliptic Curves for Authenticated Key Exchange

Abdoul Aziz Ciss[1](✉) and Djiby Sow[2]

[1] Laboratoire de Traitement de l'Information et Systèmes Intelligents,
École Polytechnique de Thiès, Thiès, Senegal
aaciss@ept.sn
[2] Département de Mathématiques et Informatique,
Université Cheikh Anta Diop de Dakar, Dakar, Senegal
sowdjibab@ucad.edu.sn

Abstract. This paper studies the task of two-sources randomness extractors for elliptic curves defined over finite fields K, where K can be a prime or a binary field. In fact, we introduce new constructions of functions over elliptic curves which take in input two random points from two different subgroups. In other words, for a given elliptic curve E defined over a finite field $\mathbb{F}_q$ and two random points $P \in \mathcal{P}$ and $Q \in \mathcal{Q}$, where $\mathcal{P}$ and $\mathcal{Q}$ are two subgroups of $E(\mathbb{F}_q)$, our function extracts the least significant bits of the abscissa of the point $P \oplus Q$ when q is a large prime, and the k-first $\mathbb{F}_p$ coefficients of the abscissa of the point $P \oplus Q$ when $q = p^n$, where p is a prime greater than 5. We show that the extracted bits are close to uniform.

Our construction extends some interesting randomness extractors for elliptic curves, namely those defined in [7,9,10], when $\mathcal{P} = \mathcal{Q}$. The proposed constructions can be used in any cryptographic schemes which require extraction of random bits from two sources over elliptic curves, namely in key exchange protocol, design of strong pseudo-random number generators, etc.

Keywords: Elliptic curves · Randomness extractor · Key derivation · Bilinear sums

1 Introduction

A deterministic randomness extractor for an elliptic curve is a function which allows to produce close to uniform random bit-string from a random point of the elliptic curve. The main difficulty of extracting randomness in elliptic curve points is to find suitable and explicit constructions for such function, i.e. computable in polynomial time by a Turing Machine.

The task of randomness extraction from a point of an elliptic curve has several cryptographic applications. For example, it can be used in key derivation functions, in key exchange protocols like Diffie-Hellman [11] and to design cryptographically secure pseudorandom number generators [21].

© Springer International Publishing AG 2017
S. El Hajji et al. (Eds.): C2SI 2017, LNCS 10194, pp. 85–95, 2017.
DOI: 10.1007/978-3-319-55589-8_6

For instance, by the end of Diffie-Hellman key exchange protocol [11], Alice and Bob agree on a common secret $K_{AB} \in G$, where G is a cryptographic cyclic group, which is indistinguishable from another element of G under the decisional Diffie-Hellman assumption [6]. The secret key used for encryption or authentication of data has to be indistinguishable from a uniformly random bit-string. Hence, the common secret K_{AB} cannot be directly used as a session key.

A classical solution is the use of a hash function to map an element of the group G onto a uniformly random bit-string of fixed length. However, the indistinguishability cannot be proved under the decisional Diffie-Hellman assumption. In this case, it is necessary to appeal to the Random Oracle or to other technics. Many results in this direction can be found in [12,16]. An alternative to hash function is to use a deterministic extractor when G is the group of points of an elliptic curve [7–10,13–15]. These constructions use exponential sums to bound the statistical distance.

In this paper, we introduce two new constructions of two-sources randomness extractors for elliptic curves defined over finite field. More precisely, we deal with finite fields $\mathbb{F}_p$ for large prime p and finite fields $\mathbb{F}_q$ where $q = p^n$. Consider an elliptic curve E defined over a finite field $\mathbb{F}_p$, with $p > 5$, and $\mathcal{P}$ and $\mathcal{Q}$ be two distinct subgroups of $E(\mathbb{F}_q)$. For given two points $P \in \mathcal{P}$ and $Q \in \mathcal{Q}$, the first extractor outputs the k-least significant bits of the abscissa of the point $P \oplus Q$. We show that the extracted bits are indistinguishable from a random bit-string of length k. In fact, we use bilinear exponential sums, recently proposed by Ahmadi and Shparlinski [1] to bound the statistical distance.

We use the same technique to define a two-source randomness extractor for elliptic curves defined over finite fields $\mathbb{F}_q$, where $q = p^n$. The proposed function extracts the k-first $\mathbb{F}_p$ coefficients of the abscissa of the point $P \oplus Q$.

We organize the paper as follows: the next section recalls some basic notion on theory of randomness extraction, namely tools for measuring randomness: collision probability, statistical distance, min-entropy, exponential character sums over finite fields and elliptic curves, in particular we recall fundamental results on bilinear exponential sums over elliptic curves we use in this paper. We also give some previous results related to the randomness extraction in elliptic curves when working only one subgroup. Section 3 introduces our first contribution, i.e. a new construction of a two-source deterministic randomness extractor for elliptic curves defined over prime fields. An analogue of this extractor for elliptic curves defined over $\mathbb{F}_{p^n}$ is given in Sect. 4.

2 Preliminaries

2.1 Deterministic Extractor

Definition 1 (Collision probability). *Let S be a finite set and X be an S-valued random variable. The collision probability of X, denoted by $Col(X)$, is the probability*

$$Col(X) = \sum_{s \in S} \Pr[X = s]^2$$

If X and X' are identically distributed random variables on S, the collision probability of X is interpreted as $Col(X) = \Pr[X = X']$

Definition 2 (Statistical distance). *Let X and Y be S-valued random variables, where S is a finite set. The statistical distance $\Delta(X, Y)$ between X and Y is*

$$\Delta(X, Y) = \frac{1}{2} \sum_{s \in S} |\Pr[X = s] - \Pr[Y = s]|$$

Let U_S be a random variable uniformly distributed on S. Then a random variable X on S is said to be δ-uniform if

$$\Delta(X, U_S) \leq \delta$$

An equivalent definition is that $|X(A) - Y(A)| \leq \delta$ for every event $A \subseteq S$, which means that the two distributions are almost indistinguishable.

Lemma 1. *Let S be a finite set and let $(\alpha_x)_{x \in S}$ be a sequence of real numbers. Then,*

$$\frac{(\sum_{x \in S} |\alpha_x|)^2}{|S|} \leq \sum_{x \in S} \alpha_x^2. \tag{1}$$

Proof. This inequality is a direct consequence of Cauchy-Schwarz inequality:

$$\sum_{x \in S} |\alpha_x| = \sum_{x \in S} |\alpha_x| . 1 \leq \sqrt{\sum_{x \in S} \alpha_x^2} \sqrt{\sum_{x \in S} 1^2} \leq \sqrt{|S|} \sqrt{\sum_{x \in S} \alpha_x^2}.$$

The result can be deduced easily.

If X is an S-valued random variable and if we consider that $\alpha_x = \Pr[X = x]$, then

$$\frac{1}{|S|} \leq Col(X), \tag{2}$$

since the sum of probabilities is 1 and since $Col(X) = \sum_{x \in S} \Pr[X = x]^2$.

The following lemma gives an explicit relation between the statistical distance and collision probability.

Lemma 2. *Let X be a random variable over a finite S of size $|S|$ and $\delta = \Delta(X, U_S)$ be the statistical distance between X and U_S, the uniformly distributed random variable over S. Then,*

$$Col(X) \geq \frac{1 + 4\delta^2}{|S|}$$

Proof. If $\delta = 0$, then the result is an easy consequence of Eq. 2. Let suppose that $\delta \neq 0$ and define

$$q_x = |\Pr[X = x] - 1/|S||/2\delta.$$

Then $\sum_x q_x = 1$ and by Eq. 1, we have

$$\frac{1}{|S|} \leq \sum_{x \in S} q_x^2 = \sum_{x \in S} \frac{(\Pr[X = x] - 1/|S|)^2}{4\delta^2} = \frac{1}{4\delta^2} \left(\sum_{x \in S} \Pr[X = x]^2 - 1/|S| \right)$$

$$\leq \frac{1}{4\delta^2} (Col(X) - 1/|S|).$$

The lemma can be deduced easily.

Definition 3 (Min-entropy). *The min-entropy of a distribution X on a set S denoted by $H_\infty(x)$ is defined by :*

$$H_\infty(x) = \min_{x \in S} \log_2 \frac{1}{\Pr[X = x]}$$

In other words, a distribution has a min-entropy at least k if the probability of each element is bounded by 2^{-k}. Intuitively, such a distribution contains k random bits.

Definition 4 (Extractor). *Let S and T be two finite sets. A (k, ϵ)-extractor is a function*

$$Ext : S \longrightarrow T$$

such that for every distribution X on S with $H_\infty(x) \geq k$ the distribution $Ext(X)$ is ϵ-close to the uniform distribution on $\{0,1\}^m$

Definition 5 (Two-sources-extractor). *Let R, S and T be finite sets. The function $Ext : R \times S \longrightarrow T$ is a two-sources-extractor if the distribution $Ext(X_1, X_2)$ is δ-close to the uniform distribution U_T for every uniformly distributed random variables X_1 in R and X_2 in S.*

For more information on extractors, see [20].

2.2   Character Sums in Finite Fields

In the following, we denote by e_p the character on $\mathbb{F}_p$ such that, for all $x \in \mathbb{F}_p$

$$e_p(x) = e^{\frac{2i\pi x}{p}} \in \mathbb{C}^*.$$

If I is an interval of integers, it's well known [7] that

$$\sum_{x \in \mathbb{F}_p} \left| \sum_{\theta \in I} e_p(\theta x) \right| \leq p \log_2(p).$$

Denote by $\Psi = \mathrm{Hom}(\mathbb{F}_{p^n}, \mathbb{C}^*)$, the group of additive characters on $\mathbb{F}_{p^n}$ that can be described by the set

$$\Psi = \{\psi, \psi(z) = e_p(\mathrm{Tr}(\alpha z)), \text{ for } \alpha \in \mathbb{F}_{p^n}\}$$

where $\mathrm{Tr}(x)$ is the trace of $x \in \mathbb{F}_{p^n}$ to $\mathbb{F}_p$ (see [19]).

Lemma 3. *Let V be an additive subgroup of $\mathbb{F}_{p^n}$. Then,*

$$\sum_{\psi \in \Psi} \left| \sum_{z \in V} \psi(z) \right| \leq p^n.$$

Proof. See [22] for the proof.

2.3 Character Sums with Elliptic Curves

Let q be a prime power and let E be an elliptic curve defined over a finite field $\mathbb{F}_q$ of q elements of characteristic $p \geq 5$ given by an affine Weierstrass equation

$$E : y^2 = x^3 + ax + b$$

with $a, b \in \mathbb{F}_q$, see [18]. The set of all points on E forms an abelian group with neutral element $\mathcal{O}$. Let $\oplus$ denote the group law operation. For a point $P \neq \mathcal{O}$ on E we write $P = (\mathrm{x}(P), \mathrm{y}(P))$. Let ψ be a nonprincipal additive character of $\mathbb{F}_q$ and let $\mathcal{P}$ and $\mathcal{Q}$ be two subsets of $E(\mathbb{F}_q)$. For arbitrary complex functions $\rho(P)$ and $\vartheta(Q)$ supported on $\mathcal{P}$ and $\mathcal{Q}$ we consider the bilinear sums of additive type:

$$V_{\rho,\vartheta}(\psi, \mathcal{P}, \mathcal{Q}) = \sum_{P \in \mathcal{P}} \sum_{Q \in \mathcal{Q}} \rho(P)\vartheta(Q)\psi(\mathrm{x}(P \oplus Q)).$$

In the following, we write $f \ll g$ if $f = o(g)$ for given functions f and g and we recall the following interesting result of [1].

Lemma 4. *Let E be an elliptic curve defined over $\mathbb{F}_q$ and let*

$$\sum_{P \in \mathcal{P}} |\rho(P)|^2 \leq R \quad and \quad \sum_{Q \in \mathcal{Q}} |\vartheta(Q)|^2 \leq T.$$

Then, uniformly over all nontrivial additive character ψ of $\mathbb{F}_q$,

$$|V_{\rho,\vartheta}(\psi, \mathcal{P}, \mathcal{Q})| \ll \sqrt{qRT}.$$

Proof. See [1].

Previous Works. For $q = p$ a prime number > 5 let's recall the extractor of Chevalier *et al.* in [7].

Definition 6. *Let E be an elliptic curve defined over a finite field $\mathbb{F}_p$, for a prime $p > 2$. Let G be a subgroup of $E(\mathbb{F}_p)$ and let k be a positive integer. Define the function*

$$\mathcal{L}_k : G \longrightarrow \{0,1\}^k$$
$$P \longmapsto \mathrm{lsb}_k(\mathrm{x}(P)),$$

where $\mathrm{lsb}_k(n)$ is the function which outputs the k-least significant bits of the integer n.

The following lemmas state that $\mathcal{L}_k$ is a deterministic randomness extractor for the elliptic curve E

Lemma 5. *Let p be a n-bit prime, G a subgroup of $E(\mathbb{F}_p)$ of cardinality q generated by a point P_0, q being an l-bit prime, U_G a random variable uniformly distributed in G and k a positive integer. Then*

$$\Delta(\mathcal{L}_k(U_G), U_k) \leq 2^{(k+n+\log_2(n))/2+3-l},$$

where U_k is the uniform distribution in $\{0,1\}^k$.

Proof. See [7].

Corollary 1. *Let e be a positive integer and suppose that*

$$k \leq 2l - (n + 2e + \log_2(n) + 6).$$

Then $\mathcal{L}_k$ is a $(U_G, 2^{-e})$-deterministic extractor.

Consider now the finite field $\mathbb{F}_{p^n}$, where $p > 5$ is prime and n is a positive integer. Then $\mathbb{F}_{p^n}$ is a n-dimensional vector space over $\mathbb{F}_p$. Let $\{\alpha_1, \alpha_2, \ldots, \alpha_n\}$ be a basis of $\mathbb{F}_{p^n}$ over $\mathbb{F}_p$. That means, every element x of $\mathbb{F}_{p^n}$ can be represented in the form $x = x_1\alpha_1 + x_2\alpha_2 + \ldots + x_n\alpha_n$, where $x_i \in \mathbb{F}_{p^n}$. Let E be the elliptic curve over $\mathbb{F}_{p^n}$ defined by the Weierstrass equation

$$y^2 + (a_1 x + a_3)y = x^3 + a_2 x^2 + a_4 x + a_6.$$

The extractor $\mathcal{D}_k$, where k is a positive integer less than n, for a given point P on $E(\mathbb{F}_{p^n})$, outputs the k first $\mathbb{F}_p$-coordinates of the abscissa of the point P.

Definition 7. *Let G be a subgroup of $E(\mathbb{F}_{p^n})$ and k a positive integer less than n. Define the function $\mathcal{D}_k$*

$$\mathcal{D}_k : G \longrightarrow \mathbb{F}_{p^k}$$
$$P = (x, y) \longmapsto (x_1, x_2, \ldots, x_k)$$

where $x \in \mathbb{F}_{p^n}$ is represented as $x = x_1\alpha_1 + x_2\alpha_2 + \ldots + x_n\alpha_n$, and $x_i \in \mathbb{F}_{p^n}$.

Lemma 6. *Let E be an elliptic curve defined over $\mathbb{F}_q$, whit $q = p^n$ and let G be a subgroup of $E(\mathbb{F}_{p^n})$. Let $\mathcal{D}_k$ be the function defined above. Then,*

$$\text{Col}(\mathcal{D}_k(U_G)) \leq \frac{1}{p^k} + \frac{4\sqrt{q}}{|G|^2}$$

and

$$\Delta(\mathcal{D}_k(U_G), U_{\mathbb{F}_{p^k}}) \leq \frac{2\sqrt{p^{n+k}}}{|G|}$$

where U_G is uniformly distributed in G and $U_{\mathbb{F}_{p^k}}$ is the uniform distribution in $\mathbb{F}_{p^k}$.

Proof. See [10].

In the following, $|i|$ represents the bit size of the integer i.

Lemma 7. *Let $p > 2$ be a prime and $E(\mathbb{F}_{p^n})$ be an elliptic curve over $\mathbb{F}_{p^n}$ and $G \subset E(\mathbb{F}_{p^n})$ be a multiplicative subgroup of order r with $|r| = t$ bits and $|p| = m$ bits and let U_G be the uniform distribution in G. If $e > 1$ is an integer and $k > 1$ is an integer such that*

$$k \leq \frac{2t - 2e - nm - 4}{m},$$

then $\mathcal{D}_k$ is a $(\mathbb{F}_p^k, 2^{-e})$-deterministic randomness extractor over the elliptic curve $E(\mathbb{F}_{p^n})$.

Proof. See [10].

3   Randomness Extractors for $E(\mathbb{F}_p)$

Definition 8. *Let E be an elliptic curve defined a finite field $\mathbb{F}_q$, with $q = p$ a prime greater than 5, and let $\mathcal{P}$ and $\mathcal{Q}$ be two subgroups of $E(\mathbb{F}_q)$ with $\#\mathcal{P} = r$ and $\#\mathcal{Q} = t$. Define the function*

$$Ext_1 : \mathcal{P} \times \mathcal{Q} \longrightarrow \{0,1\}^k$$
$$(P, Q) \longmapsto \mathrm{lsb}_k(\mathrm{x}(P \oplus Q))$$

Theorem 1. *Let E be an elliptic curve defined over $\mathbb{F}_p$ and let $\mathcal{P}$ and $\mathcal{Q}$ be two subgroups of $E(\mathbb{F}_p)$, with $\#\mathcal{P} = r$ and $\#\mathcal{Q} = t$. Let $U_\mathcal{P}$ and $U_\mathcal{Q}$ be two random variables uniformly distributed in $\mathcal{P}$ and $\mathcal{Q}$ respectively and let U_k be the uniform distribution in $\{0,1\}^k$. Then,*

$$\Delta(Ext_1(U_\mathcal{P}, U_\mathcal{Q}), U_k) \ll \sqrt{\frac{2^{k-1} p \log(p)}{rt}}$$

Proof. Let $\alpha = 2^k$ and let $\theta_0 = \mathrm{msb}_{n-k}(p - 1)$. Define the set

$$\mathcal{A} = \{(P, Q), (R, S) \in \mathcal{P} \times \mathcal{Q} \mid \exists\, \theta \leq \theta_0, \mathrm{x}(P \oplus Q) - \mathrm{x}(R \oplus S) - \alpha\theta = 0 \bmod p\}.$$

Consider the double character sum $V_{\rho,\vartheta}(\psi, \mathcal{P}, \mathcal{Q})$, with $\rho(P) = 1 \;\; \forall\, P$ and $\vartheta(Q) = 1 \;\; \forall\, Q$. Then,

$$\mathrm{Col}(Ext_1(U_\mathcal{P}, U_\mathcal{Q})) = \frac{\#\mathcal{A}}{(rt)^2}$$

$$= \frac{1}{r^2 t^2 p} \sum_{P \in \mathcal{P}} \sum_{Q \in \mathcal{Q}} \sum_{R \in \mathcal{P}} \sum_{S \in \mathcal{Q}} \sum_{\theta \leq \theta_0} \sum_{\psi \in \Psi} \psi(\mathrm{x}(P \oplus Q) - \mathrm{x}(R \oplus S) - \alpha\theta)$$

$$= \frac{1}{2^k} + \frac{1}{r^2 t^2 p} \sum_{P \in \mathcal{P}} \sum_{Q \in \mathcal{Q}} \sum_{R \in \mathcal{P}} \sum_{S \in \mathcal{Q}} \sum_{\theta \leq \theta_0} \sum_{\psi \neq \psi_0} \psi(\mathrm{x}(P \oplus Q) - \mathrm{x}(R \oplus S) - \alpha\theta)$$

$$\leq \frac{1}{2^k} + \frac{1}{r^2 t^2 p} \left| \sum_{P \in \mathcal{P}} \sum_{Q \in \mathcal{Q}} \psi(\mathrm{x}(P \oplus Q)) \right| \left| \sum_{R \in \mathcal{P}} \sum_{S \in \mathcal{Q}} \psi(-\mathrm{x}(R \oplus S)) \right| \left| \sum_{\theta \leq \theta_0} \sum_{\psi \neq \psi_0} \psi(-\alpha\theta) \right|$$

$$\ll \frac{1}{2^k} + \frac{V^2}{r^2 t^2 p} \sum_{\theta \leq \theta_0} \left| \sum_{\psi \neq \psi_0} \psi(-\alpha\theta) \right|$$

$$\ll \frac{1}{2^k} + \frac{p \log(p)}{rt}$$

Therefore,

$$\Delta(Ext_1(U_{\mathcal{P}}, U_{\mathcal{Q}}), U_k) \ll \sqrt{\frac{2^{k-1} p \log(p)}{rt}}$$

Corollary 2. *Let m and l be the bit size of r and t respectively and let e be a positive integer. If k is a positive integer such that*

$$k \leq m + l - (n + 2e + \log_2(n) + 1),$$

then Ext_1 is a $(k, O(2^{-e}))$-deterministic extractor for $\mathcal{P} \times \mathcal{Q}$.

Corollary 3. *Let $\mathcal{P} = \mathcal{Q}$, m be the bit size of r and let e be a positive integer. If k is a positive integer such that*

$$k \leq 2m - (n + 2e + \log_2(n) + 1),$$

then Ext_1 is a $(k, O(2^{-e}))$-deterministic extractor for $\mathcal{P} \times \mathcal{P}$.

4   Randomness Extractor for $E(\mathbb{F}_{p^n})$, with $p > 5$

Definition 9. *Let E be an elliptic curve defined over the finite field $\mathbb{F}_{p^n}$, where p is a prime greater than 5 and $n > 1$. Consider two subgroups $\mathcal{P}$ and $\mathcal{Q}$ of $E(\mathbb{F}_q)$. Define the function*

$$Ext_2 : \mathcal{P} \times \mathcal{Q} \longrightarrow \mathbb{F}_p^k$$

$$(P, Q) \longmapsto (x_1, x_2, \dots, x_k)$$

where $\mathrm{x}(P \oplus Q) = (x_1, x_2, \dots, x_k, x_{k+1}, \dots, x_n)$. In other words, the function Ext_2 output the k first $\mathbb{F}_p$-coefficients of the point $P \oplus Q$.

Theorem 2. *Let E be an elliptic curve defined over $\mathbb{F}_{p^n}$ and let $\mathcal{P}$ and $\mathcal{Q}$ be two subgroup of $E(\mathbb{F}_{p^n})$ with $\#\mathcal{P} = r$ and $\#\mathcal{Q} = t$. Denote by $U_{\mathcal{P}}$ and $U_{\mathcal{Q}}$ two random variables uniformly distributed on $\mathcal{P}$ and $\mathcal{Q}$ respectively. Then,*

$$\Delta(Ext_2(U_{\mathcal{P}}, U_{\mathcal{Q}}), U_{\mathbb{F}_p^k}) \ll \sqrt{\frac{p^{n+k}}{4rt}}$$

Sketch of Proof. Consider the sets

$$\mathcal{M} = \{(x_{k+1}\alpha_{k+1} + x_{k+2}\alpha_{k+2} + \dots + x_n\alpha_n), x_i \in \mathbb{F}_p\} \subset \mathbb{F}_{p^n}$$

and

$$\mathcal{A} = \{(P, Q), (R, S) \in \mathcal{P} \times \mathcal{Q} \mid \exists \lambda \in \mathcal{M}, \mathrm{x}(P \oplus Q) - \mathrm{x}(R \oplus S) = \lambda\}.$$

Then,

$$\mathrm{Col}(Ext_2(U_{\mathcal{P}}, U_{\mathcal{Q}})) = \frac{\#\mathcal{A}}{(rt)^2}.$$

Use the technique of the proof of Theorem 1 and Lemmas 3 and 4 to complete the proof.

5   Application to Key Agreement

In this section, we give an application of our new extractor to key agreement schemes.

The unified model (UM) proposed by Ankney *et al.* in [2] is an authenticated key agreement in the standards ANSI X9.63 [3], ANSI X9.42 [4] and the IEEE P1363-2000 [17]. The UM has a very simple design and interesting security properties such as forward secrecy and key confirmation. It relies on the Diffie-Hellman key exchange. At the end of the protocol, the two parties involved agree on a shared session key which is the concatenation of two Diffie-Hellman keys.

More precisely, Alice and Bob share two points Z_s and Z_e on the elliptic curve from a two-pass Diffie-Hellman key exchange. The session key K is then obtained by applying $K = \mathrm{kdf}(Z_s || Z_e)$, where kdf is a key derivation function. In the following, we show how to use Ext_1 as the kdf; i.e. $K = Ext_1(Z_e, Z_s)$.

In practice, the following parameters the base can be used at the 80 bit-security level. Note that the prime sizes indicated in Table 1 are those recommended by the NIST [5] for Elliptic Curves Cryptography.

Table 1. Parameters for $Ext_1(Z_e, Z_s)$

Symmetric key size	Bit size of the finite field : $|p|_2$	Bit size of $\mathcal{P}$: $|m|_2$
$|k|_2 = 64$: DES-64	521	378
	384	309
	256	245
$|k|_2 = 128$: AES-128	521	410
	384	340
$|k|_2 = 256$: AES-256	521	474

6   Conclusion

We have successfully introduced new randomness extractors for elliptic curves, namely two-source extractors. The proposed functions take in input two random points from two different subgroups and output the k-least significant bits of the abscissa of the sum of these two points. We have shown that the bit-string extracted is close to uniform. These results extend also some interesting randomness extractors for elliptic curves in the literature, namely those defined in [7,9,10]. Future works includes extension of the proposed extractors to Jacobian of hyperelliptic curves.

Acknowledgments. The authors acknowledge support from the Simons Foundation through the Pole of Research in Mathematics and their Applications to Information Security in Subsaharan Africa (PRMAIS) and the LIRIMA-MACISA project.

References

1. Ahmadi, O., Shparlinski, I.E.: Exponential Sums over Points of Elliptic Curves. arXiv preprint arXiv:1302.4210 (2013)
2. Ankney, R., Honson, D., Matyas, M.: The Unified Model. Contribution to X9F1, October 1995
3. ANSI X9.42, Agreement of Symmetric Algorithm Keys using Diffie-Hellman, Working draft, July 1998
4. ANSI X9.63, Elliptic Curve Key Agreement and Key Transport Protocols, Working draft, July 1998
5. Barker, E.B., Chen, L., Roginsky, A., Smid, M.E.: Recommendation for Pair-Wise Key Establishment Schemes Using Discrete Logarithm Cryptography, NIST Special Publication 800–56A Revision 2, May 2013
6. Boneh, D.: The decision Diffie-Hellman problem. In: Buhler, J.P. (ed.) ANTS 1998. LNCS, vol. 1423, pp. 48–63. Springer, Heidelberg (1998). doi:10.1007/BFb0054851
7. Chevalier, C., Fouque, P.-A., Pointcheval, D., Zimmer, S.: Optimal randomness extraction from a Diffie-Hellman element. In: Joux, A. (ed.) EUROCRYPT 2009. LNCS, vol. 5479, pp. 572–589. Springer, Heidelberg (2009)
8. Ciss, A.A.: Arithmétique et Extracteurs déterministes sur les courbes elliptiques. Thèse de doctorat unique (2012)
9. Ciss, A.A., Sow, D.: Randomness extraction in elliptic curves and secret key derivation at the end of Diffie-Hellman protocol. Int. J. Appl. Cryptol. $2(4)$, 360–365 (2012)
10. Ciss, A.A., Sow, D.: On randomness extraction in elliptic curves. In: Nitaj, A., Pointcheval, D. (eds.) AFRICACRYPT 2011. LNCS, vol. 6737, pp. 290–297. Springer, Heidelberg (2011)
11. Diffie, W., Hellman, M.: New directions in cryptography. IEEE Trans. Inf. Theor. $22(6)$, 644–654 (1976)
12. Dodis, Y., Gennaro, R., Håstad, J., Krawczyk, H., Rabin, T.: Randomness extraction and key derivation using the CBC, cascade and HMAC modes. In: Franklin, M. (ed.) CRYPTO 2004. LNCS, vol. 3152, pp. 494–510. Springer, Heidelberg (2004)
13. Farashahi, R.R., Pellikaan, R.: The quadratic extension extractor for (hyper)elliptic curves in odd characteristic. In: Carlet, C., Sunar, B. (eds.) WAIFI 2007. LNCS, vol. 4547, pp. 219–236. Springer, Heidelberg (2007)
14. Farashahi, R.R., Sidorenko, A., Pellikaan, R.: Extractors for binary elliptic curves. Des. Codes Crypt. 94, 171–186 (2008)
15. Gürel, N.: Extracting bits from coordinates of a point of an elliptic curve, Cryptology ePrint Archive, Report 2005/324 (2005). http://eprint.iacr.org/
16. Håstad, J., Impagliazzo, R., Levin, L., Luby, M.: A pseudorandom generator from any one-way function. SIAM J. Comput. $28(4)$, 1364–1396 (1999)
17. IEEE P1363, Standard specification for public key cryptography, Working draft, July 1998
18. Koblitz, N.: Guide to Elliptic Curve Cryptography. Springer, Heidelberg (2004)
19. Kohel, D.R., Shparlinski, I.E.: On exponential sums and group generators for elliptic curves over finite fields. In: Bosma, W. (ed.) ANTS 2000. LNCS, vol. 1838, pp. 395–404. Springer, Heidelberg (2000)
20. Shaltiel, R.: An introduction to randomness extractors. In: Aceto, L., Henzinger, M., Sgall, J. (eds.) ICALP 2011. LNCS, vol. 6756, pp. 21–41. Springer, Heidelberg (2011)

21. Trevisan, L.: Extractors and pseudorandom generators. J. ACM **48**(4), 860–879 (2001)
22. Winterhof, A.: Incomplete additive character sums and applications. In: Jungnickel, D., Niederreiter, H. (eds.) Finite Fields and Applications, pp. 462–474. Springer, Heidelberg (2001)

Generalization of BJMM-ISD Using May-Ozerov Nearest Neighbor Algorithm over an Arbitrary Finite Field $\mathbb{F}_q$

Cheikh Thiécoumba Gueye[1], Jean Belo Klamti[1(✉)], and Shoichi Hirose[2]

[1] Faculté des Sciences et Techniques, DMI, LACGAA,
Université Cheikh Anta Diop, Dakar, Senegal
{cheikht.gueye,jeanbelo.klamti}@ucad.edu.sn
[2] Graduate School of Engineering, University of Fukui, Fukui, Japan
hrs_shch@u-fukui.ac.jp

Abstract. The security of McEliece cryptosystem heavily relies on the hardness of decoding a random linear code. The best known generic decoding algorithms are derived from the Information-Set Decoding (ISD) algorithm. The ISD algorithm was proposed in 1962 by *Prange* and improved in 1989 by *Stern* and later in 1991 by *Dumer*. Since then, there have been numerous works improving and generalizing the ISD algorithm: *Peters* in 2009, *May, Meurer* and *Thomae* in 2011, *Becker, Joux, May* and *Meurer* in 2012, *May* and *Ozerov* in 2015, and *Hirose* in 2016. Among all these improvement and generalization only those of *Peters* and *Hirose* are over $\mathbb{F}_q$ with q an arbitrary prime power. In *Hirose*'s paper, he describes the *May-Ozerov* nearest-neighbor algorithm generalized to work for vectors over the finite field $\mathbb{F}_q$ with arbitrary prime power q. He also applies the generalized algorithm to the decoding problem of random linear codes over $\mathbb{F}_q$. And he observed by a numerical analysis of asymptotic time complexity that the *May-Ozerov* nearest-neighbor algorithm may not contribute to the performance improvement of Stern's ISD algorithm over $\mathbb{F}_q$ with $q \geq 3$. In this paper, we will extend the *Becker, Joux, May*, and *Meurer*'s ISD using the *May-Ozerov* algorithm for Nearest-Neighbor problem over $\mathbb{F}_q$ with q an arbitrary prime power. We analyze the impact of May-Ozerov algorithm for Nearest-Neighbor Problem over $\mathbb{F}_q$ on *the Becker, Joux, May* and *Meurer*'s ISD.

Keywords: Code-based cryptography · Information-Set Decoding (ISD) algorithm · Linear code · Nearest neighbor

1 Introduction

Code-based cryptography introduced by McEliece [29] is one of the most promising solution for designing secure cryptosystems against quantum attacks. The McEliece public-key encryption scheme, based on binary Goppa codes, has so far successfully resisted all cryptanalysis efforts. But it is not used in real life because

© Springer International Publishing AG 2017
S. El Hajji et al. (Eds.): C2SI 2017, LNCS 10194, pp. 96–109, 2017.
DOI: 10.1007/978-3-319-55589-8_7

of the key length problem. In order to decrease the public-key size, some variants were proposed by concentrating on subclasses of alternant/Goppa codes which admit very compact public matrices, typically quasi-cyclic (QC), quasi-dyadic (QD), or quasi-monoidic (QM) matrices [2,14,18,27,28,30,36]. The security of the McEliece cryptosystem relies on the fact that the public key does not have any known structure. The attacker is faced with the problem of decoding a random code. A way to do this decoding is to use the Information-Set Decoding (ISD) algorithm. The ISD algorithm was introduced by Prange in 1962 [38]. Its principle is to find an information set where there are no errors positions. Its target is to answer to the Computational Syndrome Decoding (CSD) Problem.

In this paper, we will extend the best version of the ISD attack algorithm to arbitrary code over $\mathbb{F}_q$ and analyze the security of such codes to this new improved version. It is important to note that Peters used the ISD attack to prove the security of arbitrary codes over $\mathbb{F}_q$ [37], later *Ayoub et al.* introduced a polynomial attack against Wild McEliece over quadratic extensions and their attack is a structural attack [9]. Recently, *Hirose* applied the *May-Ozerov* algorithm for Nearest-Neighbor problem over $\mathbb{F}_q$ to generalize Stern's ISD version and he observed that the *May-Ozerov* algorithm for Nearest-Neighbor problem may not contribute to improve Stern's ISD [19]. The contribution of our paper is the generalization of Becker, Joux, May, and Meurer's ISD using the May-Ozerov algorithm for Nearest-Neighbor problem [32] over $\mathbb{F}_q$ with q an arbitrary prime power. We analyze the contribution of the *May-Ozerov* algorithm for Nearest-Neighbor problem over an arbitrary finite field $\mathbb{F}_q$ to the performance of *Becker, Joux, May, and Meurer*'s ISD. And we analyze the security over an arbitrary finite field $\mathbb{F}_q$.

q-ary Computational Syndrome Decoding (CSD) Problem

Input: $\mathbf{H} \in \mathbb{F}_q^{(n-k)\times n}$, $s \in \mathbb{F}_q^{n-k}$ and an integer $\omega > 0$.
Output: Find $e \in \mathbb{F}_q^n$ of weight $\leq \omega$ such that $\mathbf{H}e^T = s$

Information-Set Decoding (ISD) Algorithm. The best known attacks against the classical McEliece code-based cryptosystem are generic decoding attacks that treat McEliece's hidden binary Goppa codes as random linear codes. Introduced by Prange in 1962 (see [38]), the ISD algorithm is a generic decoding attack algorithm. Its target is to solve the CSD problem taking only as inputs a basis of the code and a noisy codeword. Improvements of this form of ISD were developed by Lee and Brickell [25], Stern [40], May, Meurer and Thomae [31], Becker, Joux, May and Meurer (BJMM-ISD) [4], later by May and Ozerov [32] used the nearest neighbor algorithm to improve the BJMM-ISD.

Organisation of Paper. The paper is organized as follows: in Sect. 2, we give some definitions and notations on coding theory, in Sect. 3 we give a summary of previous and recent results on ISD algorithm over an arbitrary finite fields $\mathbb{F}_q$. In Sect. 4, we give the version of BJMM-ISD using the May-Ozerov Nearest Neighbor algorithm. And in Sect. 5, we give the asymptotic complexity of our algorithm.

2 Coding Theory Background

2.1 Definitions and Notations

Let $\mathbb{F}_q$ be a finite field ($q = p^m$, p is prime). A q-ary linear code $\mathcal{C}$ of length n and dimension k over $\mathbb{F}_q$ is a vectorial subspace of dimension k of the full vectorial space $\mathbb{F}_q^n$. It can be specified by a full rank matrix $\mathbf{G} \in \mathbb{F}_q^{k \times n}$ called generator matrix of $\mathcal{C}$ whose rows span the code. Namely, $C = \{ \boldsymbol{x}\mathbf{G} \; such \; that \; \boldsymbol{x} \in \mathbb{F}_q^k \}$.

A linear code can be also defined by the right kernel of matrix $\mathbf{H}$ called parity-check matrix of $\mathcal{C}$ as follows:

$$\mathcal{C} = \{ \boldsymbol{x} \in \mathbb{F}_q^n \; such \; that \; \mathbf{H}\boldsymbol{x}^T = \boldsymbol{0} \}.$$

The *Hamming distance* between two codewords is the number of positions (coordinates) where they differ. The minimal distance of a code is the minimal distance of all codewords. The *weight* of a word $\boldsymbol{x} \in \mathbb{F}_q^n$ denote by $wt(\boldsymbol{x})$ is the number of its nonzero positions. Then the minimal *weight* of a code $\mathcal{C}$ is the minimal *weight* of all codewords. If a code $\mathcal{C}$ is linear, the minimal distance is equal to the minimal *weight* of the code.

Let $\mathcal{C}$ be a q-ary linear code of length n, dimension k and generator matrix $\mathbf{G} = (\boldsymbol{g}_0, \boldsymbol{g}_1, ..., \boldsymbol{g}_{n-1})$ with $\boldsymbol{g}_i \in \mathbb{F}_q^n$ for all $i \in \{0, 1, ..., n-1\}$. Let $I \subset \{0, 1, ..., n-1\}$ with $|I| = k$. We call I an *information set* if and only if the matrix $\mathbf{G}_I = (\boldsymbol{g})_{i \in I}$ is invertible.

A vector $\boldsymbol{u} \in \mathbb{F}_q^\ell$ is called a balanced vector if the number of its coordinates equal to x is ℓ/q for all $x \in \mathbb{F}_q$.

For $\boldsymbol{x} = (x_1, ..., x_n) \in \mathbb{F}_q^n$ and a non zero integer $j < n$, let $x_{[j]} = (x_1, ..., x_j)$ and $\boldsymbol{x}^{[j]} = (x_{n-j+1}, ..., x_n)$.

We denote the q-ary entropy function by:

$$H_q(x) = x \log(q - 1) - x \log(x) - (1 - x) \log(1 - x)$$

For all integer n, let $[n] = \{1, ..., n\}$. If I is a subset of $[n]$, for all vector $\boldsymbol{x} = (x_1, ..., x_n)$, let $\boldsymbol{x}_I = (x_i)_{i \in I}$.

2.2 McEliece's Cryptosystem

McEliece's cryptosystem is a public-key encryption scheme introduced in 1978 by *McEliece*. The original version used the Goppa binary code remained unbroken. It can also be used with any class of codes which has an efficient decoding algorithm.

Secret keys: A matrix $\mathbf{G} \in \mathbb{F}_2^{k \times n}$, $\mathbf{S} \in \mathbb{F}_2^{k \times k}$ (an invertible matrix), $\mathbf{P} \in \mathbb{F}_2^{n \times n}$ (a random permutation matrix).
Public keys: The matrix $\tilde{\mathbf{G}} = \mathbf{SGP}$ and the corrector capacity t.
Encryption: Let $\boldsymbol{m}$ be a plaintext then the ciphertext $\boldsymbol{c}$ is given by:

$$\boldsymbol{c} = \boldsymbol{m}\tilde{\mathbf{G}} + \boldsymbol{e}$$

with $\boldsymbol{e}$ a q-ary vector of length n and weight t.

Decryption: Compute

$$\tilde{c} = m\tilde{\mathbf{G}}\mathbf{P}^{-1} + e\mathbf{P}^{-1}$$

and use the decoding algorithm to find $\tilde{m} = m\mathbf{S}$ and finally find m by computing $m = \tilde{m}\mathbf{S}^{-1}$.

2.3   Nearest-Neighbor Problem

The nearest-neighbor (NN) problem over the binary field defined in [32] is generalized over other finite fields in [19].

Neartest Neighbor Problem over $\mathbb{F}_q$: Let q be a prime power. Let m be a positive integer. Let $0 < \gamma < 1/2$ and $0 < \lambda < 1$. Then (m, γ, λ)-NN problem is defined by:

Input: The constant γ and two lists $U \subset \mathbb{F}_q^m$, $V \subset \mathbb{F}_q^m$ of size $|U| = |U| = q^{\lambda n}$ with uniform and pairwise independent vectors.
Output: $\mathcal{C} \subset U \times V$ which has (u, v) such that $wt(u - v) = \gamma m$ with $wt(u - v)$ is the weight of $u - v$.

3   Preview Work on Information-Set Decoding over $\mathbb{F}_q$

We denote in the rest of the paper the concatenation of two vectors x and y (respectively of two matrices $\mathbf{A}$ and $\mathbf{B}$) by $(x|y)$ (respectively $(\mathbf{A}|\mathbf{B})$).

In this section we give a survey of the generalization of ISD algorithm over an arbitrary finite field.

Peters: In 2009, *Peters* was the first to propose a generalization of the ISD algorithm over an arbitrary finite field $\mathbb{F}_q$. In her paper [37], she proposed the generalization of Stern-ISD which all of the ISD improvements are based on.

Cayrel et al.: In 2010 just few months after *Peters*'s paper, *Cayrel et al.* [34] improved the performance of the ISD over an arbitrary finite field by giving a lower bound of ISD algorithm and they generalized the formula of the lower bound introduced by *Finiasz et al.* in [15].

Meurer: In 2012 just after their ISD algorithm in the binary case in [4,31], *Meurer* proposed a new generalization of the ISD algorithm over an arbitrary finite field in his dissertation thesis [33] based on these two papers.

Hirose: In 2016 *Hirose* gave a generalization of the *nearest-neighbor* algorithm introduced by *May-Ozerov* [32] to generalize the Stern-ISD algorithm. And he analyzed the contribution of the *May-Ozerov*'s nearest-neighbor algorithm over an arbitrary finite field to the performance of Stern-ISD algorithm over an arbitrary finite field.

The following tables give us a summary complexity results on the ISD algorithm generalization previous work. We denote the ISD algorithm generalization given by*Peters* by q-Stern-ISD, *Hirose*'s generalization by q-Hirose-ISD.

Table 1. Complexity of ISD algorithm over an arbitrary finite field given in [19].

q	q-Stern-ISD	q-Hirose-ISD
	Half distance	Half distance
2	0.05563	0.05498
3	0.05217	0.05242
4	0.04987	0.05032
5	0.04815	0.04864
7	0.04571	0.04614
8	0.04478	0.04519
8	0.04266	0.04299

Table 2. Complexity of ISD algorithm over an arbitrary finite field given by Meurer in [33].

q	q-Meurer-ISD	
	BReps	XBReps
2	0.1053	-
4	0.1033	0.1014
8	0.0989	0.0969
16	0.0929	0.0918
32	0.0867	0.0863
64	0.0808	0.0806

In *Meurer*'s dissertation thesis, he gave two variants of ISD algorithm general-ization then we denote the basic variant by BReps and the extended variant by XBReps.

4 Becker, Joux, May and Meurer ISD Using May-Ozerov Nearest-Neighbor Algorithm over $\mathbb{F}_q$

The Becker, Joux, May and Meurer ISD using May-Ozerov Nearest-Neighbor algorithm over an arbitrary finite field $\mathbb{F}_q$ is presented in *Algorithm* 1.

In this algorithm we construct Base Lists over $\mathbb{F}_q$ like in [4]. For all $j = 0, 1$ we denote the Base Lists by $\mathcal{B}_{j,1}^{\mathcal{L}_j}$, $\mathcal{B}_{j,2}^{\mathcal{L}_j}$, $\mathcal{B}_{j,1}^{\mathcal{R}_j}$ and $\mathcal{B}_{j,2}^{\mathcal{R}_j}$. We define $\mathcal{B}_{j,1}^{\mathcal{L}_j}$ as follows:

Let $\mathcal{P}_{j,1}^{\mathcal{L}_j}$ and $\mathcal{P}_{j,2}^{\mathcal{L}_j}$ be be a partition of $[k + \ell] = \{1, ..., k + \ell\}$ such that $\left|\mathcal{P}_{j,1}^{\mathcal{L}_j}\right| = \left|\mathcal{P}_{j,2}^{\mathcal{L}_j}\right| = \dfrac{k+\ell}{2}$ then

$$\mathcal{B}_{j,1}^{\mathcal{L}_j} = \left\{ x \in \mathbb{F}_q^{k+\ell} \times \left\{ 0^{n-k-\ell} \right\} \ s.t \ wt(x) = \frac{p}{8} + \frac{\epsilon_1}{4} + \frac{\epsilon_2}{2} \ with \ x_{P_{j,2}^{\mathcal{L}_j}} = (0, 0, ..., 0) \right\}$$

Where p, ϵ_1 and ϵ_2 are the parameters of the algorithm such that $0 \leq p < k + \ell$, $0 < \epsilon_1 < k + \ell - p$, $0 < \epsilon_2 < k + \ell - \dfrac{p}{2} - \epsilon_1$. The construction of $\mathcal{B}_{j,2}^{\mathcal{L}_j}$, $\mathcal{B}_{j,1}^{\mathcal{R}_j}$ and $\mathcal{B}_{j,2}^{\mathcal{R}_j}$ is similar.

We use these Base Lists to compute a vector $e \in \mathbb{F}_q^{k+\ell} \times \{0^{n-k-\ell}\}$ such that $wt\left(e_{[k+\ell]}\right) = p$ and $e = e_1 - e_2$ with e_1, $e_2 \in \mathbb{F}_q^{k+\ell} \times \{0^{n-k-\ell}\}$ and $wt(e_1) = wt(e_2) = \dfrac{p}{2} + \epsilon_1$.

Proposition 1 *[33]. Let $0 \leq p \leq k + \ell$ be an integer and $e \in \mathbb{F}_q^{k+\ell} \times \{0^{n-k-\ell}\}$ be a vector such that $wt(e) = p$. For all integer ϵ such that $0 \leq \epsilon < k + \ell - p$, denote $\vartheta\left(k, \ell, \epsilon, p, q\right)$ the number of pairs (e_1, e_2) such that $e = e_1 - e_2$ with e_1, $e_2 \in \mathbb{F}_q^{k+\ell} \times \{0^{n-k-\ell}\}$ and $wt(e_1) = wt(e_2) = \dfrac{p}{2} + \epsilon$. It holds*

$$\vartheta\left(k, \ell, \epsilon, p, q\right) = \sum_{i=0}^{\min\left(\frac{p}{2}, \epsilon\right)} \binom{p - 2i}{\frac{p}{2} - i} (q - 2)^{2i} \binom{k + \ell - p}{\epsilon - i} (q - 1)^{\epsilon - i}$$

Then $\vartheta\left(k, \ell, \epsilon, p, q\right) \geq \binom{p}{\frac{p}{2}}\binom{k+\ell-p}{\epsilon}(q - 1)^{\epsilon}$.

And asymptoticalLy by using the inequality $\log_q 2 < H_q\left(\frac{1}{2}\right)$, we implicity lower bound $\log_q \vartheta\left(k, \ell, \epsilon, p, q\right) \geq p\log_q 2 + (k + \ell - p) H_q\left(\frac{\epsilon}{k+\ell-p}\right)$ [33]. This brief analysis will allow us to give a constraint on some parameters of our algorithm.

Algorithm 1. q-BJMM-MO algorithm over $\mathbb{F}_q$

Constants: Let n, k, d and ω be nonzero integers such that $k \leq n$ and $\omega = \lfloor \frac{d-1}{2} \rfloor$ with $d = H_q^{-1}\left(1 - \dfrac{k}{n}\right)$

Parameters: Integers p, ℓ, r_1, ϵ_1 and ϵ_2 such that $0 \leq p \leq \min\{k + \ell, \omega\}$, $0 < r_1 < \ell \leq \min\{n - k - \omega + p, n - k\}$, $0 < \epsilon_1 < k + \ell - p$ and $0 < \epsilon_2 < k + \ell - \frac{p}{2} - \epsilon_1$.

Input: two nonzero integers n and k, a matrix $\mathbf{H} \in \mathbb{F}_q^{(n-k)\times n}$, and a nonzero vector $x \in \mathbb{F}_q^n$.

Output: A vector $e \in \mathbb{F}_q^n$ of weight $wt(e) = \omega$ such that $\mathbf{H}e^T = \mathbf{H}x^T$.

```
1 : Procedure: BJMM-MO(n, k, H, x)
2 :     s ⟵ Hx^T
3 :     d ⟵ nH^{-1}(1 - k/n)
4 :     ω ⟵ ⌊(d-1)/2⌋
5 :     Choose parameters p, ε_1, ε_2, 0 < r_1 < ℓ < n - k.
6 :     Repeat:
7 :         π ⟵ a random permutation on {1, 2, ..., n}.
8 :         (Q_1|Q_2) ⟵ π(H) with Q_2 ∈ F_q^{(n-k)×(n-k)} and Q_1 ∈ F_q^{(n-k)×k}
9 :         While Q_2 is not invertible:
```

10 : $\pi \longleftarrow$ a random permutation on $\{1, 2, ..., n\}$.

11 : $(\mathbf{Q_1}|\mathbf{Q_2}) \longleftarrow \pi(\mathbf{H})$

12 : $\tilde{\mathbf{H}} \longleftarrow \mathbf{Q_2}^{-1}\pi(\mathbf{H})$ and $\tilde{s} \longleftarrow \mathbf{Q_2}^{-1}s$

13 :   Choose randomly $t_{\mathcal{L}} \in \mathbb{F}_2^{\ell}$ and $t_{\mathcal{L}_0}, t_{\mathcal{R}_0} \in \mathbb{F}_q^{r_1}$

14 :   Compute $t_{\mathcal{R}} = t_{\mathcal{L}} - \tilde{s}_{[\ell]}$, $t_{\mathcal{L}_1} = t_{\mathcal{L}_0} - (t_{\mathcal{L}})_{[r_1]}$ and

                     $t_{\mathcal{R}_1} = t_{\mathcal{R}_0} - (t_{\mathcal{R}})_{[r_1]}$.

15 :   Compute Base Lists $\mathcal{B}_{i,1}^{\mathcal{L}_i}$, $\mathcal{B}_{i,2}^{\mathcal{L}_i}$, $\mathcal{B}_{i,1}^{\mathcal{R}_i}$ and $\mathcal{B}_{i,1}^{\mathcal{R}_i}$, with $i = 0, 1$ and:

16 : $\mathcal{L}_i \longleftarrow \left\{ u = a - b \ s.t \ a \in \mathcal{B}_{i,1}^{\mathcal{L}_i}, \ b \in \mathcal{B}_{i,2}^{\mathcal{L}_i} \ with \ wt(u) = \frac{p}{4} + \frac{\epsilon_1}{2} + \epsilon_2, \right.$

          $\left. and \ \left(\tilde{\mathbf{H}}u^T\right)_{[r_1]} = t_{\mathcal{L}_i} \right\}$

17 : $\mathcal{R}_i \longleftarrow \left\{ u = a - b \ s.t \ a \in \mathcal{B}_{i,1}^{\mathcal{R}_i}, \ b \in \mathcal{B}_{i,2}^{\mathcal{R}_i} \ with \ wt(u) = \frac{p}{4} + \frac{\epsilon_1}{2} + \epsilon_2, \right.$

          $\left. and \ \left(\tilde{\mathbf{H}}u^T\right)_{[r_1]} = t_{\mathcal{R}_i} \right\}$

18 : $\mathcal{L} \longleftarrow \left\{ \left(\tilde{\mathbf{H}}z^T\right)^{[n-k-\ell]} \quad s.t \ z = u - v \ and \ (u, v) \in \mathcal{L}_0 \times \mathcal{L}_1 \right.$

          $\left. with \ \left(\tilde{\mathbf{H}}z^T\right)_{[\ell]} = t_{\mathcal{L}} \right\}$

19 : $\mathcal{R} \longleftarrow \left\{ \left(\tilde{\mathbf{H}}z^T + \tilde{s}\right)^{[n-k-\ell]} \quad s.t \ z = u - v \ and \ (u, v) \in \mathcal{R}_0 \times \mathcal{R}_1 \right.$

          $\left. with \ \left(\tilde{\mathbf{H}}z^T\right)_{[\ell]} = t_{\mathcal{R}} \right\}$

20 :   In 18 and 19 we keep only elements with $wt(z) = \frac{p}{2} + \epsilon_1$

21 : $\mathcal{C} \longleftarrow$ MO-NN $\left(\mathcal{L}, \mathcal{R}, \frac{\omega - p}{n - k - \ell}\right)$

22 :   **For all** $(u, v) \in \mathcal{C} \cap \mathcal{L} \times \mathcal{R}$:

23 :   Find (e_1, e_2) s.t $u = \left(\tilde{\mathbf{H}}e_1^T\right)^{[n-k-\ell]}$ and

                        $v = \left(\tilde{\mathbf{H}}e_2^T + \tilde{s}\right)^{[n-k-\ell]}$

24 :   **If** $wt(e_1 - e_2) = p$:

25 :   **Return** $\pi^{-1}\left(e_1 - e_2 - \left(0^{k+\ell}|u - u\right)\right)$

26 : **End Procedure**

The complexity the q-BJMM-MO is given by:

Theorem 1. *Let $\varepsilon > 0$ be a real. The q-BJMM-MO algorithm solves the Syndrome Decoding problem of random $[n, k]$-linear code over $\mathbb{F}_q$ with overwhelming probability in time*

$$\tau(q, n, k, p, \omega, h_x, \varepsilon) = \tilde{\mathcal{O}}\left(q^{n\tau_1}\left(q^{n\tau_2} + q^{2n\tau_2 - r_1} + q^{4n\tau_2 - r_1 - \ell} + q^{n\mu} + q^{(y+\varepsilon)(n-k-\ell)}\right)\right)$$

where

$$\tau_1 = \left(H\left(\frac{\omega}{n}\right) - \left(\frac{k+\ell}{n}\right)H\left(\frac{p}{k+\ell}\right) - \left(1 - \frac{k+\ell}{n}\right)H\left(\frac{\omega - p}{n - k - \ell}\right)\right)\log_q 2,$$

$$\tau_2 = \frac{k+\ell}{2n}H_q\left(\frac{\frac{p}{4} + \frac{\epsilon_1}{2} + \epsilon_2}{k+\ell}\right) \quad and \quad \mu = \frac{k+\ell}{n}H_q\left(\frac{\frac{p}{2} + \epsilon_1}{k+\ell}\right) - \frac{\ell}{n}$$

with

$$y = (1 - \gamma) \left(H_q(\beta) - \frac{1}{q} \sum_{x \in \mathbb{F}_q} H_q \left(\frac{qh_x - \gamma}{1 - \gamma} \beta \right) \right), \quad \gamma = \frac{\omega - p}{n - k - p}, \quad 0 < \beta < 1,$$

$$\max \{0, \omega + k + \ell - n\} \le p \le \min \{k + \ell, \omega\}, \quad \sum_{x \in \mathbb{F}_q} h_x = 1,$$

$$\frac{\gamma}{q} < h_x < \frac{\gamma}{q} + \frac{1 - \gamma}{q\beta} \quad for \quad each \quad x \in \mathbb{F}_q,$$

$$\ell = p \log_q 2 + (k + \ell - p) H_q \left(\frac{\epsilon_1}{k + \ell - p} \right) \quad and \quad \ell \le \min \{n - k - \omega + p, n - k\}$$

$$r_1 = \left(\frac{p}{2} + \epsilon_1 \right) \log_q 2 + \left(k + \ell - \frac{p}{2} - \epsilon_1 \right) H_q \left(\frac{\epsilon_2}{k + \ell - \frac{p}{2} - \epsilon_1} \right)$$

$$\lambda = \frac{n\mu}{n - k - \ell} \le H_q(\beta) - \frac{1}{q} \sum_{x \in \mathbb{F}_q} H_q(qh_x\beta).$$

Proof. Recall that

$$\mathcal{T}(q, n, k, p, \omega, h_x, \varepsilon) = \frac{1}{\mathbb{P}(\pi_{succ})} \mathcal{C}_{in}$$

where $\mathbb{P}(\pi_{succ})$ is a the probability to have the good permutation (permutation allowing to have a success decoding) and $\mathcal{C}_{in}$ is the cost of each iteration with:

$$\mathbb{P}(\pi_{succ}) = \tilde{\mathcal{O}} \left(\frac{\binom{k+\ell}{p}\binom{n-k-\ell}{\omega-p}}{\binom{n}{\omega}} \right) \implies \frac{1}{\mathbb{P}(\pi_{succ})} = \tilde{\mathcal{O}} \left(\frac{\binom{n}{\omega}}{\binom{k+\ell}{p}\binom{n-k-\ell}{\omega-p}} \right).$$

Using the equality

$$\binom{n}{k} = 2^{nH\left(\frac{k}{n}\right)}$$

with H the binary entropie function.

$$\mathbb{P}(\pi_{succ}) = \tilde{\mathcal{O}} \left(2^{n\left(H(\frac{\omega}{n}) - \frac{k+\ell}{n}H(\frac{p}{k+\ell}) - (1 - \frac{k+\ell}{n})H(\frac{\omega-p}{n-k-\ell})\right)} \right)$$

$$= \tilde{\mathcal{O}} \left(q^{n\left(H(\frac{\omega}{n}) - \frac{k+\ell}{n}H(\frac{p}{k+\ell}) - (1 - \frac{k+\ell}{n})H(\frac{\omega-p}{n-k-\ell})\right)\log_q 2} \right)$$

$$= \tilde{\mathcal{O}}\left(q^{n\tau_1} \right).$$

Let us examine the complexity of each iteration. First we construct Base Lists and the cardinality of each Base List is given by, for each $i = 1, 2$ and $j = 1, 2$

$$|\mathcal{B}_{j,i}^{\mathcal{L}_j}| = \binom{\frac{k+\ell}{2}}{\frac{p}{8} + \frac{\epsilon_1}{4} + \frac{\epsilon_2}{2}} (q - 1)^{\frac{p}{8} + \frac{\epsilon_1}{4} + \frac{\epsilon_2}{2}}.$$

Then by using the equality

$$\binom{n}{k}(q-1)^k = \tilde{\mathcal{O}}\left(q^{nH_q\left(\frac{k}{n}\right)}\right),$$

the complexity to compute Base Lists is given by

$$\tilde{\mathcal{O}}\left(q^{n\left(\frac{k+\ell}{2n}H_q\left(\frac{\frac{p}{4}+\frac{\epsilon_1}{2}+\epsilon_2}{k+\ell}\right)\right)}\right) = \tilde{\mathcal{O}}\left(q^{n\tau_2}\right).$$

Second we use Base Lists to make a filtering to compute $\mathcal{L}_i$ and $\mathcal{R}_i$ for each $i = 1, 2$ and the cost of this filtering is given by:

$$\tilde{\mathcal{O}}\left(\frac{|\mathcal{B}_{i,1}^{\mathcal{L}_i}||\mathcal{B}_{i,2}^{\mathcal{L}_j}|}{q^{r_1}}\right) = \tilde{\mathcal{O}}\left(q^{2n\tau_2 - r_1}\right).$$

Third we compute the lists $\mathcal{L}$ and $\mathcal{R}$ with a filtering and the cost of this filtering is given by

$$\tilde{\mathcal{O}}\left(\frac{|\mathcal{L}_i||\mathcal{L}_j|}{q^{\ell-r_1}}\right) = \tilde{\mathcal{O}}\left(q^{4n\tau_2 - r_1 - \ell}\right).$$

Line 20 only gives the upper bound on $|\mathcal{L}| = |\mathcal{R}|$.

$$\tilde{\mathcal{O}}\left(\frac{\binom{k+\ell}{\frac{p}{2}+\epsilon_1}(q-1)^{\frac{p}{2}+\epsilon_1}}{q^\ell}\right) = \tilde{\mathcal{O}}\left(q^{n\left(\frac{k+\ell}{n}H_q\left(\frac{\frac{p}{2}+\epsilon_1}{k+\ell}\right)-\frac{\ell}{n}\right)}\right) = \tilde{\mathcal{O}}\left(q^{\mu n}\right).$$

And finally we make a last filtering using the May-Ozerov Nearest Neighbor algorithm and the cost of this filtering is given by:

$$\tilde{\mathcal{O}}\left(q^{(y+\varepsilon)(n-k-\ell)}\right).$$

We have $|\mathcal{L}| = |\mathcal{R}| = q^{\mu n}$. Thus MO-NN is given an instance of (m, γ, λ)-NN problem with:

$$m = n - k - \ell, \quad \gamma = \frac{\omega - p}{n - k - \ell}$$

and

$$\lambda = \frac{\mu n}{n - k - \ell}.$$

According to *Lemma* 3 in [19] we must have

$$\lambda \leq H_q(\beta) - \frac{1}{q}\sum_{x\in\mathbb{F}_q} H_q\left(qh_x\beta\right).$$

5 Numerical Analysis of Time Complexity

We give in this section a optimization numerical time complexity of our algorithm in the half distance decoding using the code's parameters given in [19] and in the full distance decoding using the code's parameters given in [33]. We give these complexities for $q \geq 3$ because the case $q = 2$ is already done in [4, 31–33]

Table 3. Complexity of the q-BJMM-MO algorithm in the half distance decoding for parameters in [19].

q	q-BJMM-MO					
	c_k	c_ℓ	c_p	h	β	Half dist.
3	0.4545	0.06273	0.015678	0.104457	0.081899	0.04427
4	0.4625	0.05936	0.012787	0.109280	0.065891	0.04194
5	0.4727	0.05664	0.010710	0.119404	0.059101	0.03955
7	0.4812	0.05383	0.009768	0.103261	0.042989	0.03706
8	0.4891	0.05232	0.008728	0.116760	0.039019	0.03593
11	0.4959	0.05045	0.009829	0.093971	0.029929	0.03335

Table 4. Complexity of the q-BJMM-MO algorithm in the full distance decoding for parameters in [33].

q	q-BJMM-MO					
	c_k	c_ℓ	c_p	h	β	Full dist.
4	0.4259	0.047749	0.015721	0.113254	0.058929	0.09951
8	0.4529	0.036823	0.009021	0.123717	0.019890	0.09388
16	0.4729	0.029908	0.008021	0.049354	0.021199	0.09012
32	0.4829	0.025151	0.007521	0.031235	0.014109	0.08264
64	0.4929	0.021496	0.006521	0.012637	0.013109	0.07861

6 Conclusion

The *May-Ozerov*'s Nearest Neighbor algoritm allows us to improve the generalization of BJMM-ISD. We show in the Tables 1 and 3 that our generalization is faster than *Hirose*'s generalization in the half distance decoding and in addition by comparing the Tables 2 and 4 we show that is faster than *Meurer*'s generalization.

Acknowlegment. This work was carried out with financial support of CEA-MITIC for CBC project and financial support of the government of Senegal's Ministry of Hight Education and Research for ISPQ project. The third author was supported in part by JSPS KAKENHI Grant Number JP16H02828.

Appendix

Nearest-Neighbor Algorithm over an Arbitrary Finite Field $\mathbb{F}_q$

We give in this section the May-Ozerov Nearest-Neighbor algorithm over $\mathbb{F}_q$ proposed by *Hirose* in [19]

Algorithm 2. May-Ozerov Nearest-Neighbor algorithm over $\mathbb{F}_q$

1: **Procedure:**MO-NN($\mathcal{L}$, $\mathcal{R}$, γ)

2: $\quad\quad y \longleftarrow (1-\gamma)\left(H_q(\beta) - \frac{1}{q}\sum_{x \in \mathbb{F}_q} H_q\left(\frac{qh_x - \gamma}{1-\gamma}\beta\right)\right)$

3: $\quad\quad$ Choose $\epsilon > 0$

4: $\quad\quad t \longleftarrow \left\lceil \frac{\log_2(y-\lambda+\frac{\epsilon}{2}) - \log_2(\frac{\epsilon}{2})}{\log_2(y) - \log_2(\lambda)}\right\rceil$

5: $\quad\quad \alpha_1 \longleftarrow \frac{y-\lambda+\frac{\epsilon}{2}}{y}$

6: $\quad\quad \alpha_j \longleftarrow \frac{\lambda}{y}\alpha_{j-1}$ **for** $2 \le j \le t$

7: $\quad\quad$ **For** $m^{\mathcal{O}(1)}$ **times:**

8: $\quad\quad\quad\quad$ Choose a permutation π on $\mathbb{F}_q^m$ uniformly at random

10: $\quad\quad\quad\quad$ Choose a vector $\mathbf{r} = (\mathbf{r}_1, \ldots, \mathbf{r}_t) \in \mathbb{F}_q^{\alpha_1 m} \times \ldots \times \mathbb{F}_q^{\alpha_t m} = \mathbb{F}_q^m$ uniformly at random s.t $\mathbf{r}_i$ is balanced for all $1 \le i \le t$

11: $\quad\quad\quad\quad \tilde{\mathcal{L}} \longleftarrow \{\tilde{\mathbf{u}} = (\tilde{\mathbf{u}}_1, \ldots, \tilde{\mathbf{u}}_t) \text{ s.t } \tilde{\mathbf{u}} = \pi(\mathbf{u}) + \mathbf{r} \text{ with } \mathbf{u} \in \mathcal{L}$ and $\tilde{\mathbf{u}}_j$ is balanced for every $1 \le j \le t\}$

12: $\quad\quad\quad\quad \tilde{\mathcal{R}} \longleftarrow \{\tilde{\mathbf{v}} = (\tilde{\mathbf{v}}_1, \ldots, \tilde{\mathbf{v}}_t) \text{ s.t } \tilde{\mathbf{v}} = \pi(\mathbf{v}) + \mathbf{r} \text{ with } \mathbf{u} \in \mathcal{R}$ and $\tilde{\mathbf{v}}_j$ is balanced for every $1 \le j \le t\}$

13: $\quad\quad\quad\quad$ **Return** MO-NNR($\tilde{\mathcal{L}}$, $\tilde{\mathcal{R}}$, m, t, γ, λ, $\alpha_1, \ldots, \alpha_t, y, \epsilon, 1$)

14: **End Procedure**

The complexity of May-Ozerov Nearest Neighbor algorithm is given by:

Theorem 2 [19]. *Let q be a prime power. Let γ, β, $\epsilon > 0$ and λ be reals such that $0 < \gamma < \frac{1}{2}$, $0 < \beta < 1$, $\varepsilon > 0$ and $\lambda \le H_q(\beta) - \frac{1}{q}\sum_{x \in \mathbb{F}_q} H_q(q\beta h_x)$ with*

$$\sum_{x \in \mathbb{F}_q} h_x = 1 \text{ and for each } x \in \mathbb{F}_q, \frac{\gamma}{q} < h_x < \frac{\gamma}{q} + \frac{1-\gamma}{q\beta}.$$

Let $y = (1-\gamma)\left(H_q(\beta) - \frac{1}{q}\sum_{x \in \mathbb{F}_q} H_q\left(\frac{qh_x-\gamma}{1-\gamma}\beta\right)\right)$. Then the MO-NN algorithm solves the $(m, \gamma, \lambda)NN$ problem over $\mathbb{F}_q$ with overwhelming probability in time

$$\tilde{\mathcal{O}}\left(q^{(y+\varepsilon)m}\right).$$

Algorithm 3. May-Ozerov NearestNeighborRec algorithm over $\mathbb{F}_q$

1: **Procedure:**MO-NNR($\mathcal{L}$, $\mathcal{R}$, m, t, γ, λ, $\alpha_1, \ldots, \alpha_t$, y, ϵ, i)

2: $\quad\quad$ **If** $i = t+1$:

3: $\quad\quad\quad\quad \mathcal{C} \longleftarrow \{(\mathbf{u}, \mathbf{v}) \in \mathcal{L} \times \mathcal{R} \text{ s.t } wt(\mathbf{u} - \mathbf{v}) = \gamma m\}$

4: $\quad\quad$ **For** $\mathcal{O}(q^{y\alpha_i m})$ **times:**

5: $\quad\quad\quad\quad$ Choose $A_i \subset \{(\alpha_1 + \cdots + \alpha_{i-1})m + 1, \ldots, (\alpha_1 + \cdots + \alpha_i)m\}$ uniformly at random s.t $|A_i| = \beta\alpha_i m$ with

$(\alpha_1 + \cdots + \alpha_{i-1})m = 0$ if $i = 1$

6:      $\mathcal{L}' \longleftarrow \{\mathbf{u} \in \mathcal{L}$ s.t the number of each $x \in \mathbb{F}_q$ on A_i is $h_x \beta \alpha_i m\}$

7:      $\mathcal{R}' \longleftarrow \{\mathbf{v} \in \mathcal{L}$ s.t the number of every $x \in \mathbb{F}_q$ on A_i is $h_x \beta \alpha_i m\}$

8:      **If** $|\mathcal{L}'| = |\mathcal{R}'| = \tilde{\mathcal{O}}\left(q^{\left(\lambda\left(1 - \sum\limits_{j=1}^{i} \alpha_j\right) + \frac{\epsilon}{2}\right)m}\right)$:

9:      $\mathcal{C} \longleftarrow \mathcal{C} \cup$ MO-NNR$(\mathcal{L}',\mathcal{R}',m,t,\gamma,\lambda, \alpha_1,\ldots,\alpha_t,y,\epsilon, i+1)$

10   **Return** $\mathcal{C}$

11: **End Procedure**

References

1. Andoni, A., Indyk, P., Nguyen, H.L., Razenshteyn, I.: Beyond locality-sensitive hashing. In: SODA, pp. 1018–1028 (2014)
2. Berger, T.P., Cayrel, P.-L., Gaborit, P., Otmani, A.: Reducing key length of the McEliece cryptosystem. In: Preneel, B. (ed.) AFRICACRYPT 2009. LNCS, vol. 5580, pp. 77–97. Springer, Heidelberg (2009). doi:10.1007/978-3-642-02384-2_6
3. Berlekamp, E., McEliece, R., van Tilborg, H.: On the inherent intractability of certain coding problems. IEEE Trans. Inf. Theor. **24**(3), 384–386 (1978)
4. Becker, A., Joux, A., May, A., Meurer A.: Decoding random binary linear codes in $2n, 20$: how $1 + 1 = 0$ improves information set decoding. In: Eurocrypt 2012 (2012)
5. Bernstein, D.J., Lange, T., Peters, C.: Smaller decoding exponents: ball-collision decoding. In: Rogaway, P. (ed.) CRYPTO 2011. LNCS, vol. 6841, pp. 743–760. Springer, Heidelberg (2011). doi:10.1007/978-3-642-22792-9_42
6. Chabot, C., Legeay, M.: Using permutation group for decoding. In: Proceedings of Algebraic and Combinatorial Coding Theory 2010, pp. 86–92 (2010)
7. Coffey, J.T., Goodman, R.M.: The complexity of Information-Set Decoding (ISD). IEEE Trans. Inf. Theor. **36**(5), 1031–1037 (1990)
8. Cohen, G., Wolfmann, J. (eds.): Coding Theory and Applications. LNCS, vol. 388. Springer, Heidelberg (1989)
9. Couvreur, A., Otmani, A., Tillich, J.-P.: Polynomial time attack on wild McEliece over quadratic extensions. Cryptology ePrint Archive 2014/112 (2014)
10. Dubiner, M.: Bucketing coding and information theory for the statistical high-dimensional nearest-neighbor problem. IEEE Trans. Inf. Theor. **56**(8), 4166–4179 (2010)
11. Dumer, I.: On minimum distance decoding of linear codes. In: Proceedings 5th Joint Soviet-Swedish International Workshop Information Theory, Moscow, pp. 50–52 (1991)
12. Faugère, J.-C., Otmani, A., Perret, L., Tillich, J.-P.: Algebraic cryptanalysis of McEliece variants with compact keys. In: Gilbert, H. (ed.) EUROCRYPT 2010. LNCS, vol. 6110, pp. 279–298. Springer, Heidelberg (2010)
13. Faugére, J.-C., Otmani, A., Perret, L., de Portzamparc, F., Tillich, J.-P.: Structural cryptanalysis of McEliece schemes with compact keys. Cryptology ePrint Archive: Report 2014/210 (2014)

14. Faugére, J.C., Otmani, A., Perret, L., de Portzamparc, F., Tillich, J.P.: Folding alternant and Goppa codes with non-nrivial automorphism groups. arXiv:1405.5101v1 [cs.IT], 20 May 2014
15. Finiasz, M., Sendrier, N.: Security bounds for the design of code-based cryptosystems. In: Matsui, M. (ed.) ASIACRYPT 2009. LNCS, vol. 5912, pp. 88–105. Springer, Heidelberg (2009)
16. Johansson, T., Löndahl, C.: An Improvement to Stern's Algorithm
17. Heyse, S.: Implementation of McEliece based on quasi-dyadic goppa codes for embedded devices. In: Yang, B.-Y. (ed.) PQCrypto 2011. LNCS, vol. 7071, pp. 143–162. Springer, Heidelberg (2011). doi:10.1007/978-3-642-25405-5_10
18. Gaborit, P.: Shorter keys for code based cryptography. In: Proceedings of the 2005 International Workshop on Coding and Cryptography (WCC 2005), Bergen, Norway, pp. 81–91, March 2005
19. Hirose, S.: May-Ozerov algorithm for nearest-neighbor problem over $\mathbb{F}_q$ and its application to information set decoding. Cryptology ePrint Archive: Report 2016/237 (2016)
20. Har-Peled, S., Indyk, P., Motwani, R.: Approximate nearest neighbor: towards removing the curse of dimensionality. Theor. Comput. 8(1), 321–350 (2012)
21. Howgrave-Graham, N., Joux, A.: New generic algorithms for hard knapsacks. In: Gilbert, H. (ed.) EUROCRYPT 2010. LNCS, vol. 6110, pp. 235–256. Springer, Heidelberg (2010)
22. Kobara, K.: Flexible quasi-dyadic code-based public-key encryption and signature. Cryptology ePrint Archive, Report 2009/635 (2009)
23. Legeay, M.: Permutation decoding: towards an approach using algebraic properties of the σ-subcode. In: Augot, D., Canteaut, A. (eds.) WCC 2011, pp. 193–202 (2011)
24. Legeay, M.: Utilisation du groupe de permutations d'un code correcteur pour améliorer l'éfficacité du décodage. Université de Rennes 1, Année (2012)
25. Lee, P.J., Brickell, E.F.: An observation on the security of McEliece's public-key cryptosystem. In: Barstow, D., et al. (eds.) EUROCRYPT 1988. LNCS, vol. 330, pp. 275–280. Springer, Heidelberg (1988). doi:10.1007/3-540-45961-8_25
26. Leon, J.S.: A probabilistic algorithm for computing minimum weights of large error-correcting codes. IEEE Trans. Inf. Theor. 34, 1354–1359 (1988)
27. Misoczki, R., Barreto, P.S.L.M.: Compact McEliece keys from Goppa codes. In: Jacobson, M.J., Rijmen, V., Safavi-Naini, R. (eds.) SAC 2009. LNCS, vol. 5867, pp. 376–392. Springer, Heidelberg (2009)
28. Misoczki, R., Tillich, J.P, Sendrier, N., Barreto, P.S.L.M.: MDPC-McEliece: new McEliece variants from moderate density parity-check codes. In: ISIT 2013, pp. 2069–2073 (2013)
29. McEliece, R.: A public-key cryptosystem based on algebraic coding theory. DSN Prog. Rep., Jet Propulsion Laboratory, California Institute of Technology, Pasadena, CA, pp. 114–116, January 1978
30. Barreto, P.S.L.M., Lindner, R., Misoczki, R.: Monoidic codes in cryptography. In: Yang, B.-Y. (ed.) PQCrypto 2011. LNCS, vol. 7071, pp. 179–199. Springer, Heidelberg (2011)
31. May, A., Meurer, A., Thomae, E.: Decoding random linear codes in $\tilde{\mathcal{O}}(2^{0.054n})$. In: Lee, D.H., Wang, X. (eds.) ASIACRYPT 2011. LNCS, vol. 7073, pp. 107–124. Springer, Heidelberg (2011). doi:10.1007/978-3-642-25385-0_6
32. May, A., Ozerov, I.: On computing nearest neighbors with applications to decoding of binary linear codes. In: Oswald, E., Fischlin, M. (eds.) EUROCRYPT 2015. LNCS, vol. 9056, pp. 203–228. Springer, Heidelberg (2015). doi:10.1007/978-3-662-46800-5_9

33. Meurer, A.: A coding-theoretic approach to cryptanalysis. Dissertation thesis, Universität Bochum Ruhr, Novenber 2012
34. Niebuhr, R., Persichetti, E., Cayrel, P.-L., Bulygin, S., Buchmann, J.: On lower bounds for information set decoding over $\mathbb{F}_q$ and on the effect of partial knowledge
35. Niederreiter, H.: Knapsack-type cryptosystems and algebraic coding theory. Probl. Control Inf. Theor. **15**, 159–166 (1986)
36. Persichetti, E.: Compact McEliece keys based on quasi-dyadic Srivastava codes. J. Math. Cryptology **6**(2), 149–169 (2012)
37. Peters, C.: Information-set decoding for linear codes over $\mathbb{F}_q$. Cryptology ePrint Archive 2009/589 (2009)
38. Prange, E.: The use of Information-Sets in decoding cyclic codes. IEEE Trans. **IT–8**, S5–S9 (1962)
39. Repka, M., Zajac, P.: Overview of the McEliece cryptosystem and its security. Tatra Mountains Math. Publ. **60**, 57–83 (2014). doi:10.2478/tmmp-2014-0025
40. Stern, J.: A method for finding codewords of small weight. In: Cohen, G., Wolfmann, J. (eds.) Coding Theory 1988. LNCS, vol. 388, pp. 106–113. Springer, Heidelberg (1989). doi:10.1007/BFb0019850
41. Umana, V.G., Leander, G.: Practical key recovery attacks on two McEliece variants. In: International Conference on Symbolic Computation and Cryptography SCC 2010, vol. 2010, p. 62 (2010)

Parameters of 2-Designs from Some BCH Codes

Cunsheng Ding[1]([✉]) and Zhengchun Zhou[2]

[1] Department of Computer Science and Engineering,
The Hong Kong University of Science and Technology,
Clear Water Bay, Kowloon, Hong Kong, China
cding@ust.hk
[2] School of Mathematics, Southwest Jiaotong University,
Chengdu 610031, China
zzc@home.swjtu.edu.cn
http://www.cse.ust.hk/faculty/cding/
http://userweb.swjtu.edu.cn/Userweb/zczhou/index.htm

Abstract. It has been known for decades that the codewords of a fixed weight in a code may hold a t-design. However, only a small amount of progress on the construction of t-designs from codes has been made so far. It was also proven that the automorphism groups of the extended codes of the narrow-sense primitive BCH codes over finite fields are doubly transitive and these extended codes hold 2-designs. But little is known about the parameters of these 2-designs. The objective of this extended abstract is to present the parameters of some 2-designs held in these extended codes of some classes of narrow-sense primitive BCH codes.

Keywords: BCH codes · Cyclic codes · t-designs · Weight distribution

1 Introduction

Let $\mathcal{P}$ be a set of $v \geq 1$ elements, and let $\mathcal{B}$ be a set of k-subsets of $\mathcal{P}$, where k is a positive integer with $1 \leq k \leq v$. Let t be a positive integer with $t \leq k$. The pair $\mathbb{D} = (\mathcal{P}, \mathcal{B})$ is called a t-(v, k, λ) *design*, or simply t-*design*, if every t-subset of $\mathcal{P}$ is contained in exactly λ elements of $\mathcal{B}$. The elements of $\mathcal{P}$ are called points, and those of $\mathcal{B}$ are referred to as blocks. We usually use b to denote the number of blocks in $\mathcal{B}$. A t-design is called *simple* if $\mathcal{B}$ does not contain repeated blocks. In this extended abstract, we consider only simple t-designs. A t-design is called *symmetric* if $v = b$. It is clear that t-designs with $k = t$ or $k = v$ always exist. Such t-designs are *trivial*. In this extended abstract, we consider only t-designs with $v > k > t$. A t-(v, k, λ) design is referred to as a *Steiner system* if $t \geq 2$ and $\lambda = 1$, and is denoted by $S(t, k, v)$.

The interplay between codes and t-designs has gone in two directions. In one direction, the incidence matrix of any t-design generates a linear code over any

---

C. Ding's research was supported by the Hong Kong Research Grants Council, Project No. 16300415.

Z. Zhou's research was supported by the NSF of China under grant 61672028.

S. El Hajji et al. (Eds.): C2SI 2017, LNCS 10194, pp. 110–127, 2017.
DOI: 10.1007/978-3-319-55589-8_8

finite field GF(q). A lot of progress in this direction has been made and documented in the literature (see, for examples, [1,6,20,21]). In the other direction, the codewords of a fixed Hamming weight in a linear or nonlinear code may hold a t-design. Some linear and nonlinear codes were employed to construct t-designs [1,11,13,15,17,19–21]. Binary and ternary Golay codes of certain parameters give 4-designs and 5-designs with fixed parameters. However, the largest t for which an infinite family of t-designs is derived directly from codes is $t = 3$ [20,21], to the best of our knowledge. According to [1,14,20,21], not much progress on the construction of t-designs from codes has been made so far, while many other constructions of t-designs are documented in the literature ([4,5,14,18]).

It was known for a long time that the extended codes of narrow-sense primitive BCH codes hold 2-designs. But little is known about the parameters of these 2-designs. The objective of this extended abstract is to document the parameters of some of the 2-designs held in the extended codes of several families of narrow-sense primitive BCH codes. The total number of 2-designs presented in this extended abstract are exponential. In addition, the block size of the designs can vary in a large range.

2 The Classical Construction of t-Designs from Codes

We assume that the reader is familiar with the basics of linear codes and cyclic codes, and proceed to introduce the classical construction of t-designs from codes directly. Let C be a $[v, \kappa, d]$ linear code over GF(q). Let $A_i := A_i(C)$, which denotes the number of codewords with Hamming weight i in C, where $0 \le i \le v$. The sequence $(A_0, A_1, \cdots, A_v)$ is called the *weight distribution* of C, and $\sum_{i=0}^{v} A_i z^i$ is referred to as the *weight enumerator* of C. For each k with $A_k \ne 0$, let $\mathcal{B}_k$ denote the set of the supports of all codewords with Hamming weight k in C, where the coordinates of a codeword are indexed by $(0, 1, 2, \cdots, v-1)$. Let $\mathcal{P} = \{0, 1, 2, \cdots, v-1\}$. The pair $(\mathcal{P}, \mathcal{B}_k)$ may be a t-(v, k, λ) design for some positive integer λ, which is called a *support design* of the code. In such a case, we say that the code C holds a t-(v, k, λ) design. Throughout this paper, we denote the dual code of C by $C^{\perp}$, and the extended code of C by $\overline{C}$.

2.1 Designs from Linear Codes via the Assmus-Mattson Theorem

The following theorem, developed by Assumus and Mattson, shows that the pair $(\mathcal{P}, \mathcal{B}_k)$ defined by a linear code is a t-design under certain conditions, [2,10, p. 303].

Theorem 1 (Assmus-Mattson Theorem). *Let C be a $[v, k, d]$ code over* GF(q). *Let $d^{\perp}$ denote the minimum distance of $C^{\perp}$. Let w be the largest integer satisfying $w \le v$ and*

$$w - \left\lfloor \frac{w + q - 2}{q - 1} \right\rfloor < d.$$

Define $w^\perp$ analogously using $d^\perp$. Let $(A_i)_{i=0}^v$ and $(A_i^\perp)_{i=0}^v$ denote the weight distribution of C and $C^\perp$, respectively. Fix a positive integer t with $t < d$, and let s be the number of i with $A_i^\perp \neq 0$ for $0 \leq i \leq v - t$. Suppose $s \leq d - t$. Then

- *the codewords of weight i in C hold a t-design provided $A_i \neq 0$ and $d \leq i \leq w$, and*
- *the codewords of weight i in $C^\perp$ hold a t-design provided $A_i^\perp \neq 0$ and $d^\perp \leq i \leq \min\{v - t, w^\perp\}$.*

The Assmus-Mattson Theorem is a very useful tool in constructing t-designs from linear codes, and has been recently employed to construct infinitely many 2-designs and 3-designs in [7,9].

2.2 Designs from Linear Codes via the Automorphism Group

In this section, we introduce the automorphism approach to obtaining t-designs from linear codes. To this end, we have to define the automorphism group of linear codes. We will also present some basic results about this approach.

The set of coordinate permutations that map a code C to itself forms a group, which is referred to as the *permutation automorphism group* of C and denoted by $\text{PAut}(C)$. If C is a code of length n, then $\text{PAut}(C)$ is a subgroup of the *symmetric group* Sym_n.

A *monomial matrix* over $\text{GF}(q)$ is a square matrix having exactly one nonzero element of $\text{GF}(q)$ in each row and column. A monomial matrix M can be written either in the form DP or the form PD_1, where D and D_1 are diagonal matrices and P is a permutation matrix.

The set of monomial matrices that map C to itself forms the group $\text{MAut}(C)$, which is called the *monomial automorphism group* of C. Clearly, we have

$$\text{PAut}(C) \subseteq \text{MAut}(C).$$

The *automorphism group* of C, denoted by $\text{Aut}(C)$, is the set of maps of the form $M\gamma$, where M is a monomial matrix and γ is a field automorphism, that map C to itself. In the binary case, $\text{PAut}(C)$, $\text{MAut}(C)$ and $\text{Aut}(C)$ are the same. If q is a prime, $\text{MAut}(C)$ and $\text{Aut}(C)$ are identical. In general, we have

$$\text{PAut}(C) \subseteq \text{MAut}(C) \subseteq \text{Aut}(C).$$

By definition, every element in $\text{Aut}(C)$ is of the form $DP\gamma$, where D is a diagonal matrix, P is a permutation matrix, and γ is an automorphism of $\text{GF}(q)$. The automorphism group $\text{Aut}(C)$ is said to be t-transitive if for every pair of t-element ordered sets of coordinates, there is an element $DP\gamma$ of the automorphism group $\text{Aut}(C)$ such that its permutation part P sends the first set to the second set.

A proof of the following theorem can be found in [10, p. 308].

Theorem 2. *Let C be a linear code of length n over $\text{GF}(q)$ where $\text{Aut}(C)$ is t-transitive. Then the codewords of any weight $i \geq t$ of C hold a t-design.*

This theorem gives another sufficient condition for a linear code to hold t-designs. To apply Theorem 2, we have to determine the automorphism group of C and show that it is t-transitive. It is in general very hard to find out the automorphism group of a linear code. Even if we known that a linear code holds t-(v, k, λ) designs, determining the parameters k and λ could be extremely difficult. This difficulty will be seen later. All the 2-designs presented in this extended abstract are obtained from this automorphism group approach.

3   The Parameters of Some 2-Designs from the Extended Codes of Some Narrow-Sense Primitive BCH Codes

Let b denote the number of blocks in a t-(v, k, λ) design. It is easily seen that

$$b = \lambda \frac{\binom{v}{t}}{\binom{k}{t}}. \tag{1}$$

In this section, we determine the parameters of some 2-designs from the extended codes of several families of narrow-sense primitive BCH codes.

We will need the following lemma in subsequent sections, which is a variant of the MacWilliam Identity [22, p. 41].

Theorem 3. *Let* C *be a* $[v, \kappa, d]$ *code over* $\mathrm{GF}(q)$ *with weight enumerator* $A(z) = \sum_{i=0}^{v} A_i z^i$ *and let* $A^{\perp}(z)$ *be the weight enumerator of* $C^{\perp}$. *Then*

$$A^{\perp}(z) = q^{-\kappa} \Big(1 + (q-1)z\Big)^v A\Big(\frac{1-z}{1+(q-1)z}\Big).$$

3.1   General Results About 2-Designs from the Extended Codes of Some Narrow-Sense Primitive BCH Codes

Let $\mathbb{Z}_n$ denote the set $\{0, 1, 2, \cdots, n-1\}$. Let s be an integer with $0 \le s < n$. The q-*cyclotomic coset of* s *modulo* n is defined by

$$C_s = \{s, sq, sq^2, \cdots, sq^{\ell_s - 1}\} \bmod n \subseteq \mathbb{Z}_n,$$

where ℓ_s is the smallest positive integer such that $s \equiv sq^{\ell_s} \pmod{n}$, and is the size of the q-cyclotomic coset. The smallest integer in C_s is called the *coset leader* of C_s. Let $\Gamma_{(n,q)}$ be the set of all the coset leaders. We have then $C_s \cap C_t = \emptyset$ for any two distinct elements s and t in $\Gamma_{(n,q)}$, and

$$\bigcup_{s \in \Gamma_{(n,q)}} C_s = \mathbb{Z}_n. \tag{2}$$

Hence, the distinct q-cyclotomic cosets modulo n partition $\mathbb{Z}_n$.

Let $m = \mathrm{ord}_n(q)$, and let α be a generator of $\mathrm{GF}(q^m)^*$. Put $\beta = \alpha^{(q^m-1)/n}$. Then β is a primitive n-th root of unity in $\mathrm{GF}(q^m)$. The minimal polynomial

$M_{\beta^s}(x)$ of β^s over $GF(q)$ is the monic polynomial of the smallest degree over $GF(q)$ with β^s as a root. It is straightforward to see that this polynomial is given by

$$M_{\beta^s}(x) = \prod_{i \in C_s} (x - \beta^i) \in GF(q)[x], \tag{3}$$

which is irreducible over $GF(q)$. It then follows from (2) that

$$x^n - 1 = \prod_{s \in \Gamma_{(n,q)}} M_{\beta^s}(x) \tag{4}$$

which is the factorization of $x^n - 1$ into irreducible factors over $GF(q)$. This canonical factorization of $x^n - 1$ over $GF(q)$ is crucial for the study of cyclic codes.

Let δ be an integer with $2 \leq \delta \leq n$ and let h be an integer. A *BCH code* over $GF(q)$ of length n and *designed distance* δ, denoted by $C_{(q,n,\delta,h)}$, is a cyclic code with generator polynomial

$$g_{(q,n,\delta,h)} = \mathrm{lcm}(M_{\beta^h}(x), M_{\beta^{h+1}}(x), \cdots, M_{\beta^{h+\delta-2}}(x)) \tag{5}$$

where the least common multiple is computed over $GF(q)$.

It may happen that $C_{(q,n,\delta_1,h)}$ and $C_{(q,n,\delta_2,h)}$ are identical for two distinct δ_1 and δ_2. The maximum designed distance of a BCH code is also called the *Bose distance*.

When $h = 1$, the code $C_{(q,n,\delta,h)}$ with the generator polynomial in (5) is called a *narrow-sense* BCH code. If $n = q^m - 1$, then $C_{(q,n,\delta,h)}$ is referred to as a *primitive* BCH code.

We have the following conclusion [3].

Theorem 4. *The automorphism group of the code $\overline{C_{(q,q^m-1,\delta,1)}}$ is 2-transitive.*

Theorem 5. *Let $\delta \geq 2$ be an integer. The codewords of each weight in the extended narrow-sense primitive BCH code $\overline{C_{(q,q^m-1,\delta,1)}}$ form a 2-design.*

Proof. The desired conclusion follows from Theorems 2 and 4.

Theorem 6. *Let $\delta \geq 2$ be an integer. The codewords of each weight in the dual code $\overline{C_{(2,2^m-1,\delta,1)}}^{\perp}$ form a 2-design.*

Proof. Since $\overline{C_{(2,2^m-1,\delta,1)}}$ is binary, we have

$$\mathrm{PAut}(C_{(2,2^m-1,\delta,1)}) = \mathrm{MAut}(C_{(2,2^m-1,\delta,1)}) = \mathrm{Aut}(C_{(2,2^m-1,\delta,1)}).$$

It is also known that $\mathrm{PAut}(C_{(2,2^m-1,\delta,1)}^{\perp}) = \mathrm{PAut}(C_{(2,2^m-1,\delta,1)})$ [10, p. 22]. It follows from Theorem 4 that the automorphism group $\mathrm{PAut}(\overline{C_{(2,2^m-1,\delta,1)}}^{\perp})$ is doubly transitive. The desired conclusion then follows from Theorem 2.

Table 1. Weight distribution of $\overline{C}_{(2,\,2^m-1,\,\delta_2,\,1)}$ for odd m.

Weight w	No. of codewords A_w
0	1
$2^{m-1} - 2^{(m-1)/2}$	$(2^m - 1)2^{m-1}$
2^{m-1}	$2(2^m - 1)(2^{m-1} + 1)$
$2^{m-1} + 2^{(m-1)/2}$	$(2^m - 1)2^{m-1}$
2^m	1

Table 2. Weight distribution of $\overline{C}_{(2,\,2^m-1,\,\delta_2)}$ for even m.

Weight w	No. of codewords A_w
0	1
$2^{m-1} - 2^{(m-2)/2}$	$(2^{m/2} - 1)2^m$
2^{m-1}	$2(2^m - 1)$
$2^{m-1} + 2^{(m-2)/2}$	$(2^{m/2} - 1)2^m$
2^m	1

Theorem 5 tells us that $\overline{C}_{(q,q^m-1,\delta,1)}$ holds 2-designs for every δ with $2 \le \delta \le q^m - 1$. However, it is very hard to determine the parameters of the 2-designs, as the weight distribution of $C_{(q,q^m-1,\delta,1)}$ and $\overline{C}_{(q,q^m-1,\delta,1)}$ are in general very difficult to settle. In the next sections, we will determine the parameters of some of the 2-designs held in $\overline{C}_{(q,q^m-1,\delta,1)}$ and $C^{\perp}_{(2,2^m-1,\delta,1)}$ for some special values of δ.

3.2   Designs held in $\overline{C}_{(2,\,2^m-1,\,\delta_2,\,1)}$ with $\delta_2 = 2^{m-1} - 1 - 2^{\lfloor(m-1)/2\rfloor}$

With the help of Theorem 5, we now describe several families of 2-designs from the narrow-sense primitive binary codes $C_{(2,\,2^m-1,\,\delta_2,\,1)}$, where $\delta_2 = 2^{m-1} - 1 - 2^{\lfloor(m-1)/2\rfloor}$.

Theorem 7. *Let $m \ge 3$ be an integer. Then for odd m, $\overline{C}_{(2,2^m-1,\delta_2,1)}$ holds 2-$(2^m, k, \lambda)$ designs with the following pairs of (k, λ):*

- $(k,\lambda) = \left(2^{m-1} - 2^{(m-1)/2},\ (2^{m-2} - 2^{(m-3)/2})(2^{m-1} - 2^{(m-1)/2} - 1)\right).$
- $(k,\lambda) = \left(2^{m-1},\ 2^{2(m-1)} - 1\right).$
- $(k,\lambda) = \left(2^{m-1} + 2^{(m-1)/2},\ (2^{m-2} + 2^{(m-3)/2})(2^{m-1} + 2^{(m-1)/2} - 1)\right).$

For even m, it holds 2-$(2^m, k, \lambda)$ designs with the following pairs of (k, λ):

- $(k,\lambda) = \left(2^{m-1} - 2^{(m-2)/2},\ (2^{m-1} - 2^{(m-2)/2})(2^{(m-2)/2} - 1)\right).$
- $(k,\lambda) = \left(2^{m-1},\ 2^{m-1} - 1\right).$
- $(k,\lambda) = \left(2^{m-1} + 2^{(m-2)/2},\ 2^{(m-2)/2}(2^{m-1} + 2^{(m-2)/2} - 1)\right).$

Proof. We present an outline of the proof. One can settle the weight distribution of $C_{(2,2^m-1,\delta_2,1)}$ with that of the code $C_{(2,2^m-1,\delta_2,0)}$ given in [8]. One can then prove that the code $\overline{C_{(2,2^m-1,\delta_2,1)}}$ has length 2^m, and dimension

$$\overline{k} = \begin{cases} 2m+1 & \text{for odd } m, \\ \frac{3m}{2}+1 & \text{for even } m. \end{cases} \tag{6}$$

With the weight distribution of $C_{(2,2^m-1,\delta_2,1)}$ settled before, one can prove that the weight distribution of $\overline{C_{(2,2^m-1,\delta_2,1)}}$ is given in Tables 1 and 2 for odd m and even m, respectively. The desired conclusions then follow from the weight distribution of the code, Theorem 5 and (1).

Theorem 6 tells us that the code $\overline{C_{(2,2^m-1,\delta_2,1)}}^{\perp}$ holds also 2-designs for both even and odd m. One can prove that the support designs of the code $\overline{C_{(2,2^m-1,\delta_2,1)}}^{\perp}$ are in fact 3-designs for odd m. We omit the detail here.

To determine the parameters of some of the 2-designs held in $\overline{C_{(2,2^m-1,\delta_2,1)}}^{\perp}$ for even m, we need to determine the weight distribution of the code for even m.

Lemma 1. *Let $m \geq 4$ be even. Then the weight distribution of $\overline{C_{(2,2^m-1,\delta_2,1)}}^{\perp}$ is given by*

$$2^{(3m+2)/2}\overline{A}_k^{\perp}$$

$$= (1+(-1)^k)\binom{2^m}{k} + \frac{1+(-1)^k}{2}(-1)^{\lfloor k/2 \rfloor}\binom{2^{m-1}}{\lfloor k/2 \rfloor}v$$

$$+u \sum_{\substack{0 \leq i \leq 2^{m-1}-2^{(m-2)/2} \\ 0 \leq j \leq 2^{m-1}+2^{(m-2)/2} \\ i+j=k}} [(-1)^i+(-1)^j]\binom{2^{m-1}-2^{(m-2)/2}}{i}\binom{2^{m-1}+2^{(m-2)/2}}{j}$$

for $0 \leq k \leq 2^m$, where

$$u = (2^{m/2}-1)2^m \text{ and } v = 2^{m+1}-2.$$

In addition, $\overline{C_{(2,2^m-1,\delta_2,1)}}^{\perp}$ has parameters $[2^m, 2^m-1-3m/2, 4]$.

Proof. With the weight distribution of $\overline{C_{(2,2^m-1,\delta_2,1)}}$ given in Table 2, one can prove the desired conclusions with the help of Theorem 3. We omit the lengthy details.

Theorem 8. *Let ≥ 4 be an even integer. Let $\overline{A}_i^{\perp}$ denote the number of codewords with weight i in $\overline{C_{(2,2^m-1,\delta_2,1)}}^{\perp}$ for all $0 \leq i \leq 2^m$. Then for every i with $\overline{A}_i^{\perp} \neq 0$, the supports of the codewords with weight i in this code form a 2-$(2^m, i, \lambda)$ design with*

$$\lambda = \frac{\overline{A}_i^{\perp}\binom{i}{2}}{\binom{2^m}{2}},$$

where these $\overline{A}_i^{\perp}$ are given in Lemma 1.

Proof. The desired conclusions follow from Theorem 6 and (1).

Corollary 1. *Let $m \geq 4$ be an even integer. Then the supports of all codewords of weight 4 in $\overline{C_{(2,2^m-1,\delta_2,1)}}^{\perp}$ give a 2-$(2^m, 4, 2^{(m-2)/2} - 1)$ design.*

Proof. By Lemma 1, we have

$$\overline{A}_4^{\perp} = \frac{2^{m-2}(2^{(m-2)/2} - 1)(2^m - 1)}{3}.$$

The desired conclusions then follow from Theorem 8.

Corollary 2. *Let $m \geq 4$ be an even integer. Then the supports of all codewords of weight 6 in $\overline{C_{(2,2^m-1,\delta_2,1)}}^{\perp}$ give a 2-$(2^m, 6, \lambda)$ design, where*

$$\lambda = \frac{(2^{m-1} - 2)(2^{(3m-4)/2} - 5 \times 2^{(m-2)/2} + 4)}{3}.$$

Proof. By Lemma 1, we have

$$\overline{A}_6^{\perp} = \frac{2^m(2^m - 1)(2^{m-2} - 1)(2^{(3m-4)/2} - 5 \times 2^{(m-2)/2} + 4)}{45}.$$

The desired conclusions then follow from Theorem 8.

Corollary 3. *Let $m \geq 4$ be an even integer. Then the supports of all codewords of weight 8 in $\overline{C_{(2,2^m-1,\delta_2,1)}}^{\perp}$ give a 2-$(2^m, 8, \lambda)$ design, where*

$$\lambda = \frac{(h^2 - 1)(32h^7 - 184h^5 + 406h^3 - 132h^2 - 308h + 213)}{45}$$

and $h = 2^{(m-2)/2}$.

Proof. By Lemma 1, we have

$$\overline{A}_8^{\perp} = \frac{h^2(h^2 - 1)(4h^2 - 1)(32h^7 - 184h^5 + 406h^3 - 132h^2 - 308h + 213)}{630},$$

where $h = 2^{(m-2)/2}$. The desired conclusions follow from Theorem 8.

3.3   Designs Held in $\overline{C_{(q,Q^m-1,\delta_2,1)}}$ for odd q

With the help of Theorem 5, we now describe several families of 2-designs from the narrow-sense primitive nonbinary codes $C_{(q,q^m-1,\delta_2,1)}$, where $\delta_2 = (q - 1)q^{m-1} - 1 - q^{\lfloor(m-1)/2\rfloor}$ and q is an odd prime.

Theorem 9. *Let $m \geq 2$ be an integer and let q be an odd prime. Then for odd m, $\overline{C_{(q,q^m-1,\delta_2,1)}}$ holds 2-(q^m, k, λ) designs with the following pairs of (k, λ):*

$$- (k, \lambda) = \left((q - 1)q^{m-1} - q^{\frac{m-1}{2}}, \frac{((q-1)q^{m-1} - q^{\frac{m-1}{2}})((q-1)q^{m-1} - q^{\frac{m-1}{2}} - 1)}{2}\right).$$

Table 3. Weight distribution of $\overline{C_{(q,\,q^m-1,\,\delta_2,\,1)}}$ for odd $m \geq 3$ and odd q.

Weight w	No. of codewords A_w
0	1
$(q-1)q^{m-1} - q^{(m-1)/2}$	$(q-1)q^m(q^m-1)/2$
$(q-1)q^{m-1}$	$(q^m+q)(q^m-1)$
$(q-1)q^{m-1} + q^{(m-1)/2}$	$(q-1)q^m(q^m-1)/2$
q^m	$q-1$

Table 4. Weight distribution of $\overline{C_{(q,\,q^m-1,\,\delta_2,\,1)}}$ for even $m \geq 2$ and odd q.

Weight w	No. of codewords A_w
0	1
$(q-1)q^{m-1} - q^{(m-2)/2}$	$(q-1)(q^{3m/2} - q^m)$
$(q-1)q^{m-1}$	$q^{m+1} - q^m$
$(q-1)q^{m-1} + (q-1)q^{(m-2)/2}$	$q^{3m/2} - q^m$
q^m	$q-1$

- $(k,\lambda) = \left((q-1)q^{m-1},\ (q^{m-1}+1)((q-1)q^{m-1}-1)\right)$.
- $(k,\lambda) = \left((q-1)q^{m-1} + q^{\frac{m-1}{2}},\ \dfrac{((q-1)q^{m-1}+q^{\frac{m-1}{2}})((q-1)q^{m-1}+q^{\frac{m-1}{2}}-1)}{2}\right)$.

For even $m \geq 2$, it holds 2-(q^m, k, λ) designs with the following pairs of (k,λ):

- $(k,\lambda) = \left((q-1)q^{m-1} - q^{\frac{m-2}{2}},\ ((q-1)q^{m-1} - q^{\frac{m-2}{2}})(q^{\frac{m}{2}} - q^{\frac{m-2}{2}} - 1)\right)$.
- $(k,\lambda) = \left((q-1)q^{m-1}, (q-1)q^{m-1} - 1\right)$.
- $(k,\lambda) = \left((q-1)(q^{m-1} + q^{\frac{m-2}{2}}),\ q^{\frac{m-2}{2}}\left((q-1)(q^{m-1} + q^{\frac{m-2}{2}}) - 1\right)\right)$.

Proof. We sketch a proof below. The details of the proof are omitted. One can determine the weight distribution of the code $C_{(q,\,q^m-1,\,\delta_2,\,1)}$ from that of the subcode $C_{(q,\,q^m-1,\,\delta_2,\,0)}$, which was described in [8]. With the derived weight distribution of $C_{(q,\,q^m-1,\,\delta_2,\,1)}$, one can prove that the code $\overline{C_{(q,\,q^m-1,\,\delta_2,\,1)}}$ has the weight distribution in Tables 3 and 4 for odd and even m, respectively.

One can then prove that in the code $\overline{C_{(q,\,q^m-1,\,\delta_2,\,1)}}$ the number of supports of all codewords with weight $k \neq 0$ is equal to $\overline{A}_k/(q-1)$ for each k, where $\overline{A}_k$ denotes the total number of codewords with weight k in $\overline{C_{(q,\,q^m-1,\,\delta_2,\,1)}}$. Then the desired conclusions follow from the weight distribution of the code, Theorem 5 and (1).

Experimental data indicates that the code $\overline{C_{(q,\,q^m-1,\,\delta_2,\,1)}}^{\perp}$ holds also 2-designs for both even and odd m. However, the Assmus-Mattson Theorem may not give a proof of the 2-design property, as $\overline{C_{(q,\,q^m-1,\,\delta_2,\,1)}}^{\perp}$ has minimum distance 4 in some cases. To settle this problem in general, we need find out the automorphism group of the code $\overline{C_{(q,q^m-1,\delta_2,1)}}^{\perp}$.

Problem 1. Determine the automorphism group $\text{Aut}(\overline{C_{(q,q^m-1,\delta_2,1)}}^{\perp})$. Prove or disprove that $\text{Aut}(\overline{C_{(q,q^m-1,\delta_2,1)}}^{\perp})$ is doubly transitive.

3.4 Designs Held in $\overline{C_{(2,2^m-1,\delta_3,1)}}$ with $\delta_3 = 2^{m-1} - 1 - 2^{\lfloor(m+1)/2\rfloor}$

With the help of Theorem 5, we now describe several families of 2-designs from the narrow-sense primitive binary code $C_{(2,2^m-1,\delta_3,1)}$, where $\delta_3 = 2^{m-1} - 1 - 2^{\lfloor(m+1)/2\rfloor}$.

Theorem 10. *Let $m \geq 4$ be an integer. Then for odd m, $\overline{C_{(2,2^m-1,\delta_3,1)}}$ holds 2-$(2^m, k, \lambda)$ designs with the following pairs of (k, λ):*

$$- \left(2^{m-1} - 2^{\frac{m+1}{2}}, (2^{m-1}-1)(2^{m-4} - 2^{\frac{m-5}{2}})(2^{m-1} - 2^{\frac{m+1}{2}} - 1)/3\right).$$
$$- \left(2^{m-1} - 2^{\frac{m-1}{2}}, (5 \cdot 2^{m-1} + 4)(2^{m-2} - 2^{\frac{m-3}{2}})(2^{m-1} - 2^{\frac{m-1}{2}} - 1)/3\right).$$
$$- \left(2^{m-1}, (2^{m-1}-1)(9 \cdot 2^{2m-4} + 3 \cdot 2^{m-3} + 1)\right).$$

Table 5. The weight distribution of $\overline{C_{(2,2^m-1,\delta_3,1)}}$ for odd m.

Weight w	No. of codewords A_w
0	1
$2^{m-1} - 2^{(m+1)/2}$	$(2^m-1)2^{m-3}(2^{m-1}-1)/3$
$2^{m-1} - 2^{(m-1)/2}$	$(2^m-1)2^{m-1}(5 \cdot 2^{m-1} + 4)/3$
2^{m-1}	$2(2^m-1)(9 \cdot 2^{2m-4} + 3 \cdot 2^{m-3} + 1)$
$2^{m-1} + 2^{(m-1)/2}$	$(2^m-1)2^{m-1}(5 \cdot 2^{m-1} + 4)/3$
$2^{m-1} + 2^{(m+1)/2}$	$(2^m-1)2^{m-3}(2^{m-1}-1)/3$
2^m	1

Table 6. The weight distribution of $\overline{C_{(2,2^m-1,\delta_3,1)}}$ for even m.

Weight w	No. of codewords A_w
0	1
$2^{m-1} - 2^{m/2}$	$(2^{m/2}-1)2^{m-2}(2^{m+1} + 2^{m/2} - 1)/3$
$2^{m-1} - 2^{(m-2)/2}$	$(2^{m/2}-1)2^m(2^m + 2^{(m+2)/2} + 4)/3$
2^{m-1}	$2(2^{m/2}-1)(2^{2m-1} + 2^{(3m-4)/2} - 2^{m-2} + 2^{m/2} + 1)$
$2^{m-1} + 2^{(m-2)/2}$	$(2^{m/2}-1)2^m(2^m + 2^{(m+2)/2} + 4)/3$
$2^{m-1} + 2^{m/2}$	$(2^{m/2}-1)2^{m-2}(2^{m+1} + 2^{m/2} - 1)/3$
2^m	1

$$- \left(2^{m-1} + 2^{\frac{m-1}{2}}, (5 \cdot 2^{m-1} + 4)(2^{m-2} + 2^{\frac{m-3}{2}})(2^{m-1} + 2^{\frac{m-1}{2}} - 1)/3\right).$$

$$- \left(2^{m-1} + 2^{\frac{m+1}{2}}, (2^{m-1} - 1)(2^{m-4} + 2^{\frac{m-5}{2}})(2^{m-1} + 2^{\frac{m+1}{2}} - 1)/3\right).$$

For even m, it holds 2-$(2^m, k, \lambda)$ designs with the following pairs of (k, λ):

$$- \left(2^{m-1} - 2^{\frac{m}{2}}, (2^{\frac{m+2}{2}} - 1)(2^{m-3} - 2^{\frac{m-4}{2}})(2^{m-1} - 2^{\frac{m}{2}} - 1)/3\right).$$

$$- \left(2^{m-1} - 2^{\frac{m-2}{2}}, (2^m + 2^{\frac{m+2}{2}} + 4)(2^{m-1} - 2^{\frac{m-2}{2}})(2^{\frac{m-2}{2}} - 1)/3\right).$$

$$- \left(2^{m-1}, ((2^{\frac{m+2}{2}} - 1)2^{m-2} + 1)(2^{m-1} - 1)\right).$$

$$- \left(2^{m-1} + 2^{\frac{m-2}{2}}, (2^m + 2^{\frac{m+2}{2}} + 4)(2^{m-1} + 2^{\frac{m-2}{2}} - 1)2^{\frac{m-2}{2}}/3\right).$$

$$- \left(2^{m-1} + 2^{\frac{m}{2}}, (2^{\frac{m+2}{2}} - 1)(2^{m-3} + 2^{\frac{m-4}{2}})(2^{m-1} + 2^{\frac{m}{2}} - 1)/3\right).$$

Proof. We give an outline of the proof. The weight distribution of $C_{(2,2^m-1,\delta_3,0)}$ was settled in [8], and can be employed to determine the weight distribution of $C_{(2,2^m-1,\delta_3,1)}$, which contains $C_{(2,2^m-1,\delta_3,0)}$ as a subcode. Employing the weight distribution of $C_{(2,2^m-1,\delta_3,1)}$ obtained, one can prove that the weight distribution of $\overline{C_{(2,2^m-1,\delta_3,1)}}$ is given in Tables 5 and 6 for odd m and even m, respectively. It then follows that $\overline{C_{(2,2^m-1,\delta_3,1)}}$ has length 2^m and dimension

$$\overline{k} = \begin{cases} 3m + 1 & \text{for odd } m, \\ \frac{5m}{2} + 1 & \text{for even } m. \end{cases} \tag{7}$$

The desired conclusions then follow from the weight distribution of the code, Theorem 5 and (1).

If m is odd, $\overline{C_{(2,2^m-1,\delta_3,1)}}^{\perp}$ holds 3-designs, which are documented in [7].

If m is even, $\overline{C_{(2,2^m-1,\delta_3,1)}}^{\perp}$ does not hold 3-designs. Below we determine the parameters of some of the 2-designs held in $\overline{C_{(2,2^m-1,\delta_3,1)}}^{\perp}$. To this end, we need the following lemma.

Lemma 2. *Let $m \geq 6$ be even. Then the weight distribution of $\overline{C_{(2,2^m-1,\delta_3,1)}}^{\perp}$ is given by*

$$2^{(5m+2)/2}\overline{A}_k^{\perp}$$
$$= \left(1 + (-1)^k\right)\binom{2^m}{k} + wE_0(k) + uE_1(k) + vE_2(k),$$

where

$$u = (2^{m/2} - 1)2^{m-2}(2^{m+1} + 2^{m/2} - 1)/3,$$
$$v = (2^{m/2} - 1)2^m(2^m + 2^{(m+2)/2} + 4)/3,$$
$$w = 2(2^{m/2} - 1)(2^{2m-1} + 2^{(3m-4)/2} - 2^{m-2} + 2^{m/2} + 1),$$

and

$$E_0(k) = \frac{1 + (-1)^k}{2} (-1)^{\lfloor k/2 \rfloor} \binom{2^{m-1}}{\lfloor k/2 \rfloor},$$

$$E_1(k) = \sum_{\substack{0 \le i \le 2^{m-1} - 2^{m/2} \\ 0 \le j \le 2^{m-1} + 2^{m/2} \\ i+j=k}} [(-1)^i + (-1)^j] \binom{2^{m-1} - 2^{m/2}}{i} \binom{2^{m-1} + 2^{m/2}}{j},$$

$$E_2(k) = \sum_{\substack{0 \le i \le 2^{m-1} - 2^{\frac{m-2}{2}} \\ 0 \le j \le 2^{m-1} + 2^{\frac{m-2}{2}} \\ i+j=k}} ((-1)^i + (-1)^j) \binom{2^{m-1} - 2^{\frac{m-2}{2}}}{i} \binom{2^{m-1} + 2^{\frac{m-2}{2}}}{j},$$

and $0 \le k \le 2^m$.

In addition, $\overline{C_{(2,2^m-1,\delta_3,1)}}^{\perp}$ has parameters $[2^m, 2^m - 1 - 5m/2, 6]$.

Proof. With the weight distribution of $\overline{C_{(2,2^m-1,\delta_3,1)}}$ given in Table 6, one can prove the desired conclusions with the help of Theorem 3. We omit the very lengthy details.

Theorem 11. *Let* $m \ge 4$ *be an even integer. Let* $\overline{A_i}^{\perp}$ *denote the number of codewords with weight* i *in* $\overline{C_{(2,2^m-1,\delta_3,1)}}^{\perp}$ *for all* $0 \le i \le 2^m$. *Then for every* i *with* $\overline{A_i}^{\perp} \ne 0$, *the supports of the codewords with weight* i *in this code form a* $2\text{-}(2^m, i, \lambda)$ *design with*

$$\lambda = \frac{\overline{A_i}^{\perp} \binom{i}{2}}{\binom{2^m}{2}},$$

where these $\overline{A_i}^{\perp}$ *are given in Lemma 2.*

Proof. The desired conclusions follows from Theorems 6 and (1).

Corollary 4. *Let* $m \ge 4$ *be an even integer. Then the supports of all codewords of weight 6 in* $\overline{C_{(2,2^m-1,\delta_3,1)}}^{\perp}$ *give a* $2\text{-}(2^m, 6, \lambda)$ *design, where*

$$\lambda = \frac{(2^{(m-2)/2} - 2)(2^{m-2} - 1)}{3}.$$

Proof. By Lemma 2, we have

$$\overline{A_6}^{\perp} = \frac{2^{m-1}(2^{(m-2)/2} - 2)(2^{m-2} - 1)(2^m - 1)}{45}.$$

The desired conclusions then follow from Theorem 11.

Table 7. The weight distribution of $C^{\perp}_{(2,\,2^m-1,\,5,\,1)}$ for even m.

Weight w	No. of codewords A_w
0	1
$2^{m-1} - 2^{m/2}$	$(2^m - 1)(2^{m-3} + 2^{(m-4)/2})/3$
$2^{m-1} - 2^{(m-2)/2}$	$(2^m - 1)(2^m + 2^{m/2})/3$
2^{m-1}	$(2^m - 1)(2^{m-2} + 1)$
$2^{m-1} + 2^{(m-2)/2}$	$(2^m - 1)(2^m - 2^{m/2})/3$
$2^{m-1} + 2^{m/2}$	$(2^m - 1)(2^{m-3} - 2^{(m-4)/2})/3$

Corollary 5. *Let $m \geq 4$ be an even integer. Then the supports of all codewords of weight 8 in $\overline{C_{(2,2^m-1,\delta_3,1)}}^{\perp}$ give a 2-$(2^m, 8, \lambda)$ design, where*

$$\lambda = \frac{(h^2 - 1)(8h^5 - 46h^3 + 50h^2 + 56h - 95)}{45}$$

and $h = 2^{(m-2)/2}$.

Proof. By Lemma 2, we have

$$\overline{A}^{\perp}_8 = \frac{h^2(h^2 - 1)(4h^2 - 1)(8h^5 - 46h^3 + 50h^2 + 56h - 95)}{630},$$

where $h = 2^{(m-2)/2}$. The desired conclusions then follow from Theorem 11.

3.5   Designs Held in $\overline{C_{(2,2^m-1,5,1)}}$ and $\overline{C_{(2,2^m-1,5,1)}}^{\perp}$ for even $m \geq 4$

In this section, we will determine the parameters of some of the 2-designs held in both $\overline{C_{(2,2^m-1,5,1)}}$ and $\overline{C_{(2,2^m-1,5,1)}}^{\perp}$ for even $m \geq 4$. Before doing this, we need to settle the weight distribution of of the two codes.

A proof of the following theorem can be found in [12]

Theorem 12. *$C^{\perp}_{(2,2^m-1,5,1)}$ has dimension $2m$, and the weight distribution of Table 7 for even m.*

Lemma 3. *Let $m \geq 4$ be even. The code $\overline{C_{(2,2^m-1,5,1)}}^{\perp}$ has length 2^m, dimension $2m + 1$ and the weight distribution in Table 8.*

Proof. The conclusion on the dimension of the code follows from Theorem 7. The desired conclusion on the weight distribution of $\overline{C_{(2,2^m-1,5,1)}}^{\perp}$ follows from the weight distribution of $C^{\perp}_{(2,2^m-1,5,1)}$ in Table 7. We omit the very lengthy details of the proof.

Theorem 13. *Let $m \geq 4$ be even. Then $\overline{C_{(2,2^m-1,5,1)}}^{\perp}$ holds 2-$(2^m, k, \lambda)$ designs with the following pairs of (k, λ):*

Table 8. Weight distribution of $\overline{C_{(2,\,2^m-1,\,5,\,1)}}^{\perp}$ for even $m \geq 4$.

Weight w	No. of codewords $\overline{A}_w^{\perp}$
0	1
$2^{m-1} - 2^{m/2}$	$(2^m - 1)2^{m-2}/3$
$2^{m-1} - 2^{(m-2)/2}$	$(2^m - 1)2^{m+1}/3$
2^{m-1}	$(2^m - 1)(2^{m-1} + 2)$
$2^{m-1} + 2^{(m-2)/2}$	$(2^m - 1)2^{m+1}/3$
$2^{m-1} + 2^{m/2}$	$(2^m - 1)2^{m-2}/3$
2^m	1

$$- \left(2^{m-1} - 2^{\frac{m}{2}}, \; (2^{m-3} - 2^{(m-4)/2})(2^{m-1} - 2^{m/2} - 1)/3\right).$$
$$- \left(2^{m-1} - 2^{\frac{m-2}{2}}, \; (2^m - 2^{m/2})(2^{m-1} - 2^{(m-2)/2} - 1)/3\right).$$
$$- \left(2^{m-1}, \; (2^{m-2} + 1)(2^{m-1} - 1)\right).$$
$$- \left(2^{m-1} + 2^{\frac{m-2}{2}}, \; (2^m + 2^{m/2})(2^{m-1} + 2^{(m-2)/2} - 1)/3\right).$$
$$- \left(2^{m-1} + 2^{\frac{m}{2}}, \; (2^{m-3} + 2^{(m-4)/2})(2^{m-1} + 2^{m/2} - 1)\right)/3.$$

Proof. The desired conclusions then follow from the weight distribution of the code in Table 8, Theorem 5 and (1).

Lemma 4. *Let $m \geq 4$ be even. Then the weight distribution of* $\overline{C_{(2,2^m-1,5,1)}}$ *is given by*

$$2^{2m+1}\overline{A}_k = \left(1 + (-1)^k\right)\binom{2^m}{k} + wE_0(k) + uE_1(k) + vE_2(k),$$

where

$$u = (2^m - 1)2^{m-2}/3,$$
$$v = (2^m - 1)2^{m+1}/3,$$
$$w = (2^m - 1)(2^{m-1} + 2),$$

and

$$E_0(k) = \frac{1 + (-1)^k}{2}(-1)^{\lfloor k/2 \rfloor}\binom{2^{m-1}}{\lfloor k/2 \rfloor},$$

$$E_1(k) = \sum_{\substack{0 \leq i \leq 2^{m-1} - 2^{m/2} \\ 0 \leq j \leq 2^{m-1} + 2^{m/2} \\ i + j = k}} ((-1)^i + (-1)^j)\binom{2^{m-1} - 2^{m/2}}{i}\binom{2^{m-1} + 2^{m/2}}{j},$$

$$E_2(k) = \sum_{\substack{0 \le i \le 2^{m-1}-2^{(m-2)/2} \\ 0 \le j \le 2^{m-1}+2^{(m-2)/2} \\ 1+j=k}} ((-1)^i + (-1)^j)\binom{2^{m-1} - 2^{\frac{m-2}{2}}}{i}\binom{2^{m-1} + 2^{\frac{m-2}{2}}}{j}),$$

and $0 \le k \le 2^m$.

In addition, $\overline{C_{(2,2^m-1,5,1)}}$ has parameters $[2^m, 2^m - 1 - 2m, 6]$.

Proof. With the weight distribution of $\overline{C_{(2,2^m-1,5,1)}}^{\perp}$ given in Table 8, one can prove the desired conclusions with the help of Theorem 3. The very lengthy details are omitted.

Theorem 14. *Let ≥ 4 be an even integer. Let $\overline{A}_i$ denote the number of codewords with weight i in $\overline{C_{(2,2^m-1,5,1)}}$ for all $0 \le i \le 2^m$. Then for every i with $\overline{A}_i \ne 0$, the supports of the codewords with weight i in this code form a 2-$(2^m, i, \lambda)$ design with*

$$\lambda = \frac{\overline{A}_i \binom{i}{2}}{\binom{2^m}{2}},$$

where these $\overline{A}_i$ are given in Lemma 4.

Proof. The desired conclusions follows from Theorem 6 and (1).

Corollary 6. *Let $m \ge 4$ be an even integer. Then the supports of all codewords of weight 6 in $\overline{C_{(2,2^m-1,5,1)}}$ give a 2-$(2^m, 6, \lambda)$ design, where*

$$\lambda = \frac{2 \times (2^{m-2} - 1)^2}{3}.$$

Proof. By Lemma 4, we have

$$\overline{A}_6 = \frac{2^m(2^m - 1)(2^{m-2} - 1)^2}{45}.$$

The desired conclusions then follow from Theorem 14.

Corollary 7. *Let $m \ge 4$ be an even integer. Then the supports of all codewords of weight 8 in $\overline{C_{(2,2^m-1,5,1)}}$ give a 2-$(2^m, 8, \lambda)$ design, where*

$$\lambda = \frac{(h^2 - 1)(16h^6 - 92h^4 + 162h^2 - 95)}{630},$$

where $h = 2^{(m-2)/2}$.

Proof. By Lemma 4, we have

$$\overline{A}_8 = \frac{h^2(h^2 - 1)(4h^2 - 1)(16h^6 - 92h^4 + 162h^2 - 95)}{630}.$$

The desired conclusions then follow from Theorem 14.

Corollary 8. *Let $m \geq 4$ be an even integer. Then the supports of all codewords of weight 10 in $\overline{C}_{(2,2^m-1,5,1)}$ give a 2-$(2^m, 10, \lambda)$ design, where*

$$\lambda = \frac{2(h^2 - 1)(16h^{10} - 160h^8 + 666h^6 - 1401h^4 + 1498h^2 - 679)}{315},$$

where $h = 2^{(m-2)/2}$.

Proof. By Lemma 4, we have

$$\overline{A}_{10} = \frac{4h^2(4h^2 - 1)(h^2 - 1)(16h^{10} - 160h^8 + 666h^6 - 1401h^4 + 1498h^2 - 679)}{14175}.$$

The desired conclusions then follow from Theorem 14.

4   Concluding Remarks

It was known that the extended code of a narrow-sense primitive code over $\mathrm{GF}(q)$ holds 2-designs. But little is known about the parameters of these 2-designs. The contribution of this extended abstract is to determine the parameters of some of the 2-designs held in the extended codes of a few families of narrow-sense primitive codes over $\mathrm{GF}(q)$. Even for the special families of narrow-sense primitive BCH codes treated in this extended abstract, we were able to find the parameters of only a small number of the 2-designs held in their extended codes. As seen before, it is a very difficult problem to determine the parameters of these 2-designs. The weight distribution of narrow-sense primitive BCH codes is known only in a few special cases. This partially explains the difficulty in determining the parameters of the 2-designs.

Another difficulty lies in the fact that many codewords of the same Hamming weight in a code $\mathrm{GF}(q)$ may have the same support when $q \geq 3$. The following problem is also very hard.

Problem 2. Let C be a linear code of length n over $\mathrm{GF}(q)$, where $q \geq 3$. Given i with $A_i \neq 0$, what is the relation between the total number of supports of all codewords of weight i in C and A_i?

Only when $i = d$ is the minimum distance of the code, the answer to this problem is known. Problem 2 is challenging, but very useful in the theory of t-designs.

We inform the reader that some of the narrow-sense primitive BCH codes dealt with in this extended abstract were employed in [7,9] for constructing 2-designs and 3-designs. But the parameters of the 2-designs in this extended abstract and those of the 2-designs in [7,9] are different. In fact, this extended abstract complements [7,9]. For all of the 2-$(2^m, k, \lambda)$ designs presented in this extended abstract, m is mostly even. For all the 2-$(2^m - 1, k, \lambda)$ designs and 2-$(2^m, k, \lambda)$ designs documented in [7,9], m must be odd. Another major difference is that all the 2-designs of this paper are obtained via the automorphism groups of the underlying codes, and their design property cannot be proved with the

Assmus-Mattson Theorem when m is even, while all the designs in [7,9] are obtained via the Assmus-Mattson Theorem.

Since this is an extended abstract, we omitted all the very lengthy proofs. We will give detailed proofs of these results elsewhere in the future.

References

1. Assmus Jr., E.F., Key, J.D.: Designs and Their Codes. Cambridge University Press, Cambridge (1992)
2. Assmus Jr., E.F., Mattson Jr., H.F.: Coding and combinatorics. SIAM Rev. **16**, 349–388 (1974)
3. Berger, T., Charpin, P.: The automorphism group of BCH codes and of some affine-invariant codes on an extension field. Des. Codes Crypt. **18**, 29–53 (1999)
4. Beth, T., Jungnickel, D., Lenz, H.: Design Theory. Cambridge University Press, Cambridge (1999)
5. Colbourn, C.J., Mathon, R.: Steiner systems. In: Colbourn, C.J., Dinitz, J. (eds.) Handbook of Combinatorial Designs, pp. 102–110. CRC Press, New York (2007)
6. Ding, C.: Codes from Difference Sets. World Scientific, Singapore (2015)
7. Ding, C.: Infinite families of t-designs from a type of five-weight codes. arXiv:1607.04813
8. Ding, C., Fan, C., Zhou, Z.: The dimension and minimum distance of two classes of primitive BCH codes. Finite Fields Appl. **45**, 237–263 (2017)
9. Ding, C., Li, C.: Infinite families of 2-designs and 3-designs from linear codes. arXiv:1607.04813
10. Huffman, W.C., Pless, V.: Fundamentals of Error-Correcting Codes. Cambridge University Press, Cambridge (2003)
11. Jungnickel, D., Tonchev, V.D.: Exponential number of quasi-symmetric SDP designs and codes meeting the Grey-Rankin bound. Des. Codes Cryptogr. **1**, 247–253 (1991)
12. Kasami, T.: Weight distributions of Bose-Chaudhuri-Hocquenghem codes. In: Bose, R.C., Dowlings, T.A. (eds.) Combinatorial Mathematics and Applications, Chapter 20. University of North Carolina Press, Chapel Hill (1969)
13. Kennedy, G.T., Pless, V.: A coding-theoretic approach to extending designs. Discrete Math. **142**, 155–168 (1995)
14. Khosrovshahi, G.B., Laue, H.: t-designs with $t \geq 3$. In: Colbourn, C.J., Dinitz, J. (eds.) Handbook of Combinatorial Designs, pp. 79–101. CRC Press, New York (2007)
15. Kim, J.-L., Pless, V.: Designs in additive codes over GF(4). Des. Codes Cryptogr. **30**, 187–199 (2003)
16. MacWilliams, F.J., Sloane, N.J.A.: The Theory of Error-Correcting Codes. North-Holland, Amsterdam (1977)
17. Pless, V.: Codes and designs-existence and uniqueness. Discrete Math. **92**, 261–274 (1991)
18. Reid, C., Rosa, A.: Steiner systems $S(2, 4)$ - a survey. Electron. J. Comb. (2010). #DS18
19. Tonchev, V.D.: Quasi-symmetric designs, codes, quadrics, and hyperplane sections. Geom. Dedicata **48**, 295–308 (1993)
20. Tonchev, V.D.: Codes and designs. In: Pless, V.S., Huffman, W.C. (eds.) Handbook of Coding Theory, Vol. II, pp. 1229–1268. Elsevier, Amsterdam (1998)

21. Tonchev, V.D.: Codes. In: Colbourn, C.J., Dinitz, J.H. (eds.) Handbook of Combinatorial Designs, 2nd edn., pp. 677–701. CRC Press, New York (2007)
22. van Lint, J.H.: Introduction to Coding Theory, 3rd edn. Springer, New York (1999)
23. Yuan, J., Carlet, C., Ding, C.: The weight distribution of a class of linear codes from perfect nonlinear functions. IEEE Trans. Inf. Theory **52**, 712–717 (2006)

A Median Nearest Neighbors LDA for Anomaly Network Detection

Zyad Elkhadir[1(✉)], Khalid Chougdali[2], and Mohammed Benattou[1]

[1] LASTID Laboratory, Ibn Tofail University, Kenitra, Morocco
zyad.elkhadir@gmail.com
[2] GEST Research Group, National School of Applied Sciences (ENSA),
Ibn Tofail University, Kenitra, Morocco
chougdali@gmail.com

Abstract. The Linear Discriminant Analysis (LDA) is a powerful linear feature reduction technique. It often produces satisfactory results under two conditions. The first one requires that the global data structure and the local data structure must be coherent. The second concerns data classes distribution nature. It should be a Gaussian distribution. Nevertheless, in pattern recognition problems, especially network anomalies detection, these conditions are not always fulfilled. In this paper, we propose an improved LDA algorithm, the median nearest neighbors LDA (median NN-LDA), which performs well without satisfying the above two conditions. Our approach can effectively get the local structure of data by working with samples that are near to the median of every data class. The further samples will be essential for preserving the global structure of every class. Extensive experiments on two well known datasets namely KDDcup99 and NSL-KDD show that the proposed approach can achieve a promising attack identification accuracy.

Keywords: LDA · median NN-LDA · Network anomaly detection · NSL-KDD · KDDcup99

1 Introduction

The linear discriminant analysis (LDA) [1] is a family of techniques whose role is dimensionality reduction and feature extraction. Fishers LDA is one of the most known LDA methods. It has been used successfully in a variety of pattern recognition problems including network anomalies detection [2–4]. The key procedure behind Fishers LDA or LDA is to employ the well-known Fisher criterion to extract a linearly independent discriminant vectors and exploit them as basis by which samples are projected into a new space. These vectors contribute in maximizing the ratio of the inter-class distance to intra-class distance in the obtained space.

In literature, many works have been proposed to ameliorate the performance and the accuracy of the classical LDA. These works can be generally divided into

© Springer International Publishing AG 2017
S. El Hajji et al. (Eds.): C2SI 2017, LNCS 10194, pp. 128–141, 2017.
DOI: 10.1007/978-3-319-55589-8_9

two categories. The first category tries to solve the small sample size (SSS) problem, which always happens when the data dimension is greater than the number of training samples. As noticed in previous contributions, to overcome the SSS problem, direct linear discriminant analysis (Direct LDA) [5] eliminates the null space of the inter-class scatter matrix as a first step. After that, it extracts the discriminant information from the null space of the intra-class scatter matrix. In the same way, Null space LDA [6] exploited the valuable discriminant vectors of the null space of the intra-class scatter matrix with the help of PCA [7]. These vectors are used rather than the eigenvectors of the classical LDA. The authors of the last method also demonstrated that the extracted vectors are equivalent to the optimal LDA discriminant vectors obtained in the original space.

In [8] we can see an exponential discriminant analysis algorithm that derive the most discriminant information which exists in the intra-class scatter matrix's null space. However, the procedures employed by the aforementioned algorithms destroy a big part of discriminant information essential for classification. Another technique to overcome the (SSS) problem is presented in [9]. It employed an optimization criterion which used a generalized singular value decomposition. This technique is operational regardless of whether the dimension of data is greater than the number of training samples. Alternatively, an ensemble learning framework was developed by Wang and Tang [10] in order to preserve the significant discriminant information by random sampling on feature vectors and training samples. In [11], three LDA approaches were proposed to solve the SSS problem: regularized discriminant analysis [12], discriminant common vectors [13], and Maximum Margin Criterion (MMC) [14]. Another famous approach to address the SSS consists in using PCA + LDA to get the discriminant features (i.e., apply PCA on data before LDA). Nevertheless, this method may lose valuable discriminant information in PCAs stage.

The second part of works deals with the incremental versions of the LDA. This kind of LDA is very useful for online learning tasks. One of their main advantages is that the feature extraction method does not need to save the entire data matrix in the memory. In [15], QR decomposition with a LDA-based incremental algorithm were proposed. In [16], the authors developed many incremental LDAs which have a common point. The algorithms have to update in every step the between-class and within-class scatter matrices. Another incremental LDA is presented in [17]. Here, the authors showcase a good mechanism to update the scatter matrices. Besides the above two kind of improvements of LDA, there are also some LDA-based algorithms such as R1 LDA [18], L1 LDA [19], Median LDA [20] and pseudo LDA [21].

Unfortunately, all these aforementioned LDA methods pay more attention to the global structure of classes. As a result, the produced discriminant vectors are often skewed. Before going through the explanation of this fact, we give an overview of class distribution types. In general, there is two kind of complementary distributions. One is local and the other is global. The first one represents a portion of samples that defines in a certain manner the real distribution nature of every class. In the other hand, the global distribution determines the class boundaries and helps us to separate as much as possible the classes. However, in reality, the last distribution is in most of cases not Gaussian and has a more complex structure. In addition, it is often incoherent with the first type of distribution. All these assumptions lead to an inaccurate discriminant vectors.

In order to address this matter, previous works [22–24] exploited local information to obtain optimal discriminant vectors. Nonetheless, in these works, it is necessary to calculate a matrix where each of it element is a distance between two data samples, in addition, we have to do an eigen decomposition of a huge matrix generated by the entire training set. For network intrusion detection field it will be a time consuming and even an infeasible task. As a result, it is difficult to implement these approaches.

In this paper, to deal with the drawback of the global LDA, we propose a kind of local LDA namely Median Nearest Neighbors LDA. The method takes into account also preserving the global structure. Our approach consists of two parts. The first part is to find a proper number of nearest neighbors to the median of every class training set. The determined nearest neighbors will be used to compute the within-class scatter matrix. In the second part, the rest of samples which are further from the median will determine the between-class scatter matrix.

The rest of this paper is organized as follows. In Sect. 2, we outline the classical LDA. Section 3 presents in details the proposed approach. Section 4 introduces the two well known network datasets KDDcup99 and NSL-KDD. In Sect. 5 we give the experimental results and illustrate the effectiveness of the algorithm and compare it to some of the above LDA approaches. Finally, Sect. 6 offers our conclusions.

2   Linear Discriminant Analysis

The conventional LDA aims to reduce dimensionality while keeping the maximum of class-discriminatory information. This operation is realized by projecting original data onto a lower dimensional space with taking into account maximizing separation of different classes on the one hand, and minimizing dispersion of samples of the same class on the other hand. Mathematically speaking, suppose we have a data matrix $X = [x_1, \ldots, x_n] \in \mathbb{R}^{d \times n}$ composed of n samples, our purpose is to find a linear transformation $G \in \mathbb{R}^{d \times l}$ that transforms each vector x_i to a new vector x_i^l in the reduced l-dimensional space as follows:

$$x_i^l = G^T x_i \in \mathbb{R}^l (l < d)$$

The data matrix X can be rewritten as $X = [X_1, \ldots, X_k]$ such that k is the number of classes and $X_i \in \mathbb{R}^{d \times n_i}$ represents samples of the ith class, n_i is the sample size of the ith class and $\sum_{i=1}^{k} n_i = n$. LDA operates on three important matrices namely within-class, between-class and total-scatter matrices which are defined as follows:

$$S_w = (1/n) \sum_{i=1}^{k} \sum_{x \in X_i} (x - c_i)(x - c_i)^T \tag{1}$$

$$S_b = (1/n) \sum_{i=1}^{k} n_i(c_i - c)(c_i - c)^T \tag{2}$$

$$S_t = (1/n) \sum_{i=1}^{n} (x_i - c)(x_i - c)^T \tag{3}$$

c_i is the mean of the ith class, and c is the general mean. It can be proved that $S_t = S_w + S_b$ [1]. It follows from (1) and (2) that:

$$trace(S_w) = (1/n) \sum_{i=1}^{k} \sum_{x \in X_i} ||x - c_i||^2 \tag{4}$$

$$trace(S_b) = (1/n) \sum_{i=1}^{k} n_i ||c_i - c||^2 \tag{5}$$

The trace of S_w gives us an idea on how every sample is close to its class mean. The trace of S_b shows us how each class is far from the global mean. In the dimensionality reduced space transformed by G, the three scatter matrices become:

$$S_w^l = G^T S_w G$$
$$S_b^l = G^T S_b G$$
$$S_t^l = G^T S_t G$$

The optimal projection matrix can be gained by maximizing the following objective function:

$$G = \arg \max \frac{trace(S_b)}{trace(S_w)} \tag{6}$$

When S_w is invertible, the solutions to (6) can be obtained by performing the following generalized eigenvalue decomposition:

$$S_w^{-1} S_b g_i = \lambda_i g_i \tag{7}$$

where $G = [g_1, \ldots, g_l]$.

Setting aside the famous (SSS) problem, LDA suffers from another matter. It uses the global structure information of the total training samples to determine the linear discriminant vectors. In general, the use of these vectors to extract features from the samples may lead to erroneous classification. The potential reason behind this phenomenon seems to be that the global distribution of the data does not represent the real distribution nature of every class. In other words, the global distribution is not always consistent with the local distribution. Moreover, the non Gaussian nature of data might cause a nonlinear boundaries between the classes. So it becomes difficult to use global linear discriminant vectors to separate the data.

3 The Proposed Method

To overcome the aforementioned LDA drawbacks, we propose to exploit the local distribution of every class. To do that we were based on the concept of median. In probability theory and statistics, the median is defined as a sample that separates the higher half of a probability distribution from the lower half. It is the middle value in a distribution, above and below which lie an equal number of samples. From this assumption, we observe that the samples which are close to the median represent the central distribution of every class and match logically with the local distribution. In the other hand we can assimilate the further samples to the global distribution, since they exist naturally in the boundaries of the class and facilitate the separation of classes. With this concept we dissociate the two distributions. Therefore, we resolve the matter of distribution's consistency.

Our approach (median NN-LDA) also performs well even if the data is not Gaussian or has nonlinear boundaries. Since it can extract the global structures of the data through determining the samples which are far from the median, the method can obtain a number of local linear discriminant vectors which approximate the nonlinear boundary between the classes.

In mathematical terms, X_i will be divided into X_i^w and X_i^b.

Let $X_i^w = [x_1, \ldots, x_p] \in \mathbb{R}^{d \times p}$ represents the p median nearest neighbors of every class.

Let $X_i^b = [x_{p+1}, \ldots, x_{n_i}] \in \mathbb{R}^{d \times (n_i - p)}$ contains the $n_i - p$ samples which are far from the median of every class.

The local distribution X_i^w will be exploited by the new within class scatter matrix S_w', since it measures the intra-class compactness. In the other hand, the global distribution represented by X_i^b is required to compute the new between-class scatter matrix S_b' and more specifically the general mean c.

Then the Eqs. (1) and (2) will be rewritten as follow:

$$S_w' = (1/p) \sum_{i=1}^{k} \sum_{x \in X_i^w} (x - c_i^w)(x - c_i^w)^T \tag{8}$$

$$S_b' = (1/p) \sum_{i=1}^{k} (c_i^w - c)(c_i^w - c)^T \tag{9}$$

Where c_i^w is the mean of X_i^w, c_i^b is the mean of X_i^b and $c = \frac{1}{k} \sum_{i=1}^{k} (c_i^b)$ is the general mean.

As a consequence, Eqs. (4) and (5) will be replaced by:

$$trace(S_w') = (1/p) \sum_{i=1}^{k} \sum_{x \in X_i^w} ||x - c_i^w||^2 \tag{10}$$

$$trace(S_b') = (1/p) \sum_{i=1}^{k} n_i ||c_i^w - c||^2 \tag{11}$$

We obtain the discriminant vectors by maximizing the following objective function:

$$G' = \arg \max \frac{trace(S'_b)}{trace(S'_w)} \tag{12}$$

The solution can be reached by performing:

$$(S'_w)^{-1} S'_b g'_i = \lambda'_i g'_i \tag{13}$$

where $G' = [g'_1, \ldots, g'_l]$.

In order to deal with the singularity problem, we propose to apply an intermediate dimensionality reduction stage, such as principal component analysis (PCA) [7] to reduce the data dimensionality before applying median NN-LDA.

4   The Simulated Databases and Its Transformation

4.1   KDDcup99

The KDDcup99 [25] intrusion detection datasets relies on the 1998 DARPA initiative, which offers to researchers in intrusion detection field a benchmark where to evaluate various approaches. This dataset is composed of many connections.

A connection is a sequence of TCP packets which begins and ends at some well defined times. In this laps of time, a data flows from a source IP address to a target IP address under a defined protocol.

Every connection is composed of 41 features and it is labeled as normal or malicious. If the connection is malicious, it falls into one of four categories:

1. Probing: surveillance and other probing, e.g., port scanning;
2. U2R: unauthorized access to local superuser (root) privileges, e.g., various buffer overflow attacks;
3. DOS: denial-of-service, e.g. syn flooding;
4. R2L: unauthorized access from a remote machine, e.g. password guessing.

We have worked with "kddcup.data_10_percent" as training dataset and "corrected" as testing dataset. The training set contains 494,021 records which is divided as follow: 97,280 are normal connection records, the rest corresponds to attacks. In the other side, the test set contains 311,029 records composed of 60,593 normal connections. It is important to note that:

1. the test data probability distribution is not like that of the training data;
2. the test data contains some new kind of attacks which are dispersed as follow: 4 U2R attack types, 4 DOS attack, 7 R2L attack and 2 Probing attacks. All these attacks do not belong to the training dataset, a fact that makes the IDS's work more challenging.

4.2 NSL-KDD

NSL-KDD [26] is a new version of KDDcup99 dataset. This dataset has some advantages over the old one and has addressed some of it critical problems. Here are the important ones:

1. Duplicate records from the training set are removed.
2. Redundant records from the test set are eliminated to improve the intrusion detection performance.
3. Each difficulty level group contains a number of records which is inversely proportional to the percentage of records in the original KDD data set. As a consequence, we will have a more precise evaluation of different machine learning techniques.
4. It is possible to exploit the complete dataset without selecting a random small portion of data because the number of records in the train and test sets are acceptable. Consequently, evaluation results of different research works will be consistent and comparable.

4.3 Transformation Process

In order to successfully apply the approach on the datasets, as a crucial step, we have converted all the discrete attributes values of the datasets to continuous values. To accomplish that, we applied the following procedure: every discrete attribute i which takes k different values will be represented as k coordinates composed of ones and zeros. For example, we know that the protocol type attribute has three values tcp, udp or icmp. According to the procedure, all these values will be transformed to the corresponding coordinates $(1, 0, 0)$, $(0, 1, 0)$ or $(0, 0, 1)$.

5   Experiments and Discussion

In this section, in order to demonstrate the effectiveness of the proposed method, we conduct a series of experiments with KDDcup99 and NSL-KDD. Meanwhile, we also compare median NN-LDA performance with LDA, direct LDA, null space LDA, R1 LDA, pseudo LDA in an all-round way.

We can employ the following measures to evaluate these methods:

$$DR = \frac{TP}{TP + FN} \times 100 \tag{14}$$

$$FPR = \frac{FP}{FP + TN} \times 100 \tag{15}$$

In network security jargon, (DR) refers to Detection Rate and (FPR) is False Positive Rate. True positives (TP) are attacks correctly predicted. False negatives (FN) represent intrusions classified as normal instances, false positive (FP) refer to normal instances wrongly classified, and true negatives (TN) are normal instances classified as normal. Therefore, the most performant feature extraction method, is the one which produces a high DR and a low FPR.

In our experiments, we varied the size of training samples and kept test dataset intact with the following composition (100 normal data, 100 DOS data, 50 U2R data, 100 R2L data, and 100 PROBE). To reduce the variation of the detection rate (DR), we adopt the mean of twenty runs. Since our aim is to evaluate the efficacy of feature extraction method, we use a simple classifier, the nearest neighbor classifier.

The first experiment consists in defining the adequate number of samples p which represent the local structure of every class. In theory, it is difficult to do that. The most suitable p is affected by several factors such as the total number

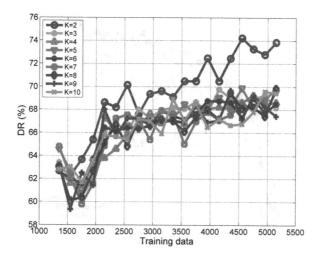

Fig. 1. Detection rate of different K for KDDcup99

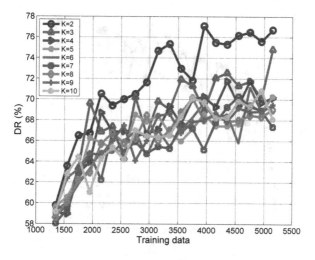

Fig. 2. Detection rate of different K for NSL-KDD

of training samples, the number of total classes, the distribution of the samples. Therefore, the value of p often needs to be empirically determined. For instance, we consider p as $\frac{n_i}{K}$ and we varied K from 2 to 10. Figures 1 and 2 show us that $p = \frac{n_i}{2}$ is the value which obtains the highest average detection rate (DR) for KDDcup99 and NSL-KDD. Consequently, we set p to this value in the next experiments.

In the second experiment we compare our proposed method to the following algorithms: LDA, median LDA, null space LDA, Direct LDA and pseudo LDA. To avoid the (SSS) problem, PCA is used as the first stage of the LDA,

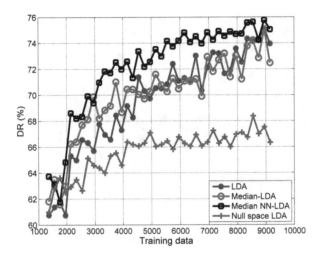

Fig. 3. Training data vs. detection rate for KDDcup99

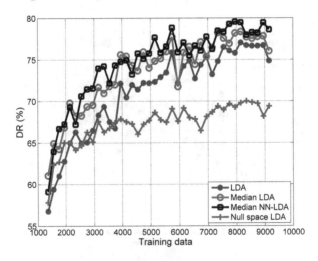

Fig. 4. Training data vs. detection rate for NSL-KDD

median LDA and median NN-LDA algorithms. Hence, these algorithms can also be viewed as the PCA + LDA, PCA + median LDA, PCA + median NN-LDA. We have chosen 3 principal components in the first stage of these methods. In the second stage we have chosen 3top features. The rest of LDA algorithms exploit the 4 top discriminant vectors. Having said that, we increased the number of training data and we visualized it influence on DR and FPR of every method.

Figures 3, 4, 5 and 6 illustrate the results we found when we compare our approach to LDA, median LDA and null space LDA for the two datasets. According to the first two figures, we observe that our approach takes the lead in attacks

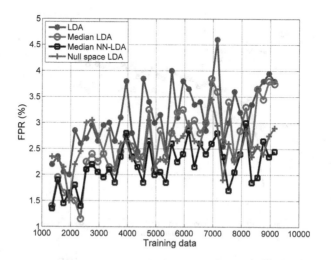

Fig. 5. Training data vs. FPR for KDDcup99

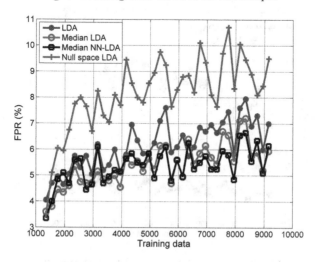

Fig. 6. Training data vs. FPR for NSL-KDD

detection as the training data grows up. The reason behind this phenomenon seems to be that more we have training samples, the easier the local structure around every class median can be captured. In addition, when we increase the number of training samples, the boundaries of every class become more structured and separable. This truth helps as much as possible in preserving the global distribution. The rest of figures depict the relationship between training samples and FPR. It is clear that median NN-LDA produces the lowest false positive rate compared to the other methods. This fact proves the high ability of our approach to recognize the normal network instances regardless of training samples size.

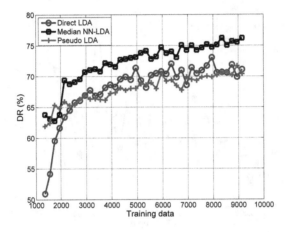

Fig. 7. Training data vs. detection rate for KDDcup99

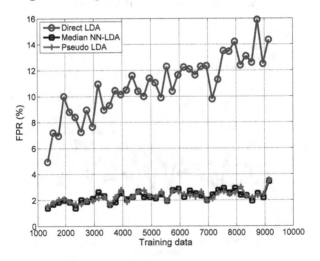

Fig. 8. Training data vs. FPR for KDDcup99

To further evaluate the performance of our approach, we compare it to other LDA methods such as Direct LDA and pseudo LDA. Figures 7, 8, 9 and 10 expose the obtained results while using KDDcup99 and NSL-KDD. As we have done in the previous experiments, we varied the number of training samples from 1350 to 9150 and illustrate DR and FPR behaviors.

As regards the first dataset, we observe from Fig. 7 that median NN-LDA overcomes the two approaches once the size of training data is superior than 2000. In the other hand, Fig. 8 shows that Pseudo LDA and the proposed approach give the fewest number of false positives.

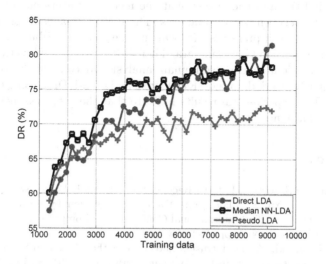

Fig. 9. Training data vs. detection rate for NSL-KDD

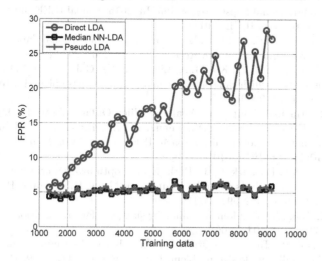

Fig. 10. Training data vs. FPR for NSL-KDD

In case we use NSL-KDD, it is shown from Fig. 9 that in term of DR, median NN-LDA surpasses Direct LDA and Pseudo LDA when the training dataset size is less than 8000. Once this value is exceeded, Direct LDA starts to compete with median NN-LDA. Concerning FPR, Fig. 10 asserts that our approach still gives satisfactory results.

6   Conclusion

In this paper, a novel feature extraction method called median NN-LDA is proposed. In this LDA approach we exploit the median of every class to compute the within and between scatter matrices. There are two advantages of median NN-LDA, one is that it preserves the local and the global distributions, the other is it insensitivity to non Gaussian data. Therefore, the proposed method is more robust than traditional linear discriminant analysis. We conduct the experiments on two popular Network data sets (KDDcup99 and NSL-KDD), using many LDA approaches. The experimental results indicate that the proposed method has a promising performance.

References

1. Fukunaga, R.: Statistical Pattern Recognition. Academic Press, New York (1990)
2. Thapngam, T., Yu, S., Zhou, W.: DDoS discrimination by linear discriminant analysis (LDA). In: 2012 International Conference on Computing, Networking and Communications (ICNC), pp. 532–536. IEEE (2012)
3. An, W., Liang, M.: A new intrusion detection method based on SVM with minimum within-class scatter. Secur. Commun. Netw. **6**(9), 1064–1074 (2013)
4. Subba, B., Biswas, S., Karmakar, S.: Intrusion detection systems using linear discriminant analysis and logistic regression. In: 2015 Annual IEEE India Conference (INDICON), pp. 1–6. IEEE (2015)
5. Yu, H., Yang, J.: A direct LDA algorithm for high-dimensional data with application to face recognition. Pattern Recogn. **34**(10), 2067–2070 (2001)
6. Chen, L.F., Liao, H.Y.M., Ko, M.T., Lin, J.C., Yu, G.J.: A new LDA-based face recognition system which can solve the small sample size problem. Pattern Recogn. **33**(10), 1713–1726 (2000)
7. Jolliffe, I.: Principal Component Analysis. Wiley Online Library (2002)
8. Zhang, T., Fang, B., Tang, Y.Y., Shang, Z., Xu, B.: Generalized discriminant analysis: a matrix exponential approach. IEEE Trans. Syst. Man Cybern. Part B Cybern. **40**(1), 186–197 (2010)
9. Ye, J., Janardan, R., Park, C.H., Park, H.: An optimization criterion for generalized discriminant analysis on under sampled problems. IEEE Trans. Pattern Anal. Mach. Intell. **26**(8), 982–994 (2004)
10. Wang, X., Tang, X.: Random sampling for subspace face recognition. Int. J. Comput. Vis. **70**(1), 91–104 (2006)
11. Liu, J., Chen, S., Tan, X.: A study on three linear discriminant analysis based methods in small sample size problem. Pattern Recogn. **41**(1), 102–116 (2008)
12. Dai, D.Q., Yuen, P.C.: Regularized discriminant analysis and its application to face recognition. Pattern Recogn. **36**(3), 845–847 (2003)

13. Cevikalp, H., Neamtu, M., Wilkes, M., Barkana, A.: Discriminative common vectors for face recognition. IEEE Trans. Pattern Anal. Mach. Intell. **27**(1), 4–13 (2005)
14. Li, H., Jiang, T., Zhang, K.: Efficient and robust feature extraction by maximum margin criterion. IEEE Trans. Neural Netw. **17**(1), 157–165 (2006)
15. Ye, J., Li, Q., Xiong, H., Park, H., Janardan, R., Kumar, V.: IDR/QR: an incremental dimension reduction algorithm via QR decomposition. IEEE Trans. Knowl. Data Eng. **17**(9), 1208–1222 (2005)
16. Pang, S., Ozawa, S., Kasabov, N.: Incremental linear discriminant analysis for classification of data streams. IEEE Trans. Syst. Man Cybern. Part B Cybern. **35**(5), 905–914 (2005)
17. Kim, T.K., Wong, S.F., Stenger, B., Kittler, J., Cipolla, R.: Incremental linear discriminant analysis using sufficient spanning set approximations. In: IEEE Conference on Computer Vision and Pattern Recognition, CVPR 2007, pp. 1–8. IEEE (2007)
18. Li, X., Hu, W., Wang, H., Zhang, Z.: Linear discriminant analysis using rotational invariant L1 norm. Neurocomputing **73**(13), 2571–2579 (2010)
19. Wang, H., Lu, X., Hu, Z., Zheng, W.: Fisher discriminant analysis with L1-norm. IEEE Trans. Cybern. **44**(6), 828–842 (2014)
20. Yang, J., Zhang, D., Yang, J.Y.: Median LDA: a robust feature extraction method for face recognition. In: IEEE International Conference on Systems, Man and Cybernetics, SMC 2006, vol. 5, pp. 4208–4213. IEEE (2006)
21. Golub, G.H., Van Loan, C.F.: Matrix Computations. Johns Hopkins Studies in the Mathematical Sciences. Hopkins University Press, Baltimore (1996)
22. Sugiyama, M., Idé, T., Nakajima, S., Sese, J.: Semi-supervised local fisher discriminant analysis for dimensionality reduction. Mach. Learn. **78**(1–2), 35–61 (2010)
23. Chen, H.T., Chang, H.W., Liu, T.L.: Local discriminant embedding and its variants. In: IEEE Computer Society Conference on Computer Vision and Pattern Recognition, CVPR 2005, vol. 2, pp. 846–853. IEEE (2005)
24. Wang, H., Chen, S., Hu, Z., Zheng, W.: Locality-preserved maximum information projection. IEEE Trans. Neural Netw. **19**(4), 571–585 (2008)
25. http://kdd.ics.uci.edu/databases/kddcup99/
26. http://nsl.cs.unb.ca/NSL-KDD/

Linearly Homomorphic Authenticated Encryption with Provable Correctness and Public Verifiability

Patrick Struck, Lucas Schabhüser[(✉)], Denise Demirel, and Johannes Buchmann

Technische Universität Darmstadt, Darmstadt, Germany
patrick.struck@stud.tu-darmstadt.de,
{lschabhueser,ddemirel,buchmann}@cdc.informatik.tu-darmstadt.de

Abstract. In this work the first linearly homomorphic authenticated encryption scheme with public verifiability and provable correctness, called LEPCoV, is presented. It improves the initial proposal by avoiding false negatives during the verification algorithm. This work provides a detailed description of LEPCoV, a comparison with the original scheme, a security and correctness proof, and a performance analysis showing that all algorithms run in reasonable time for parameters that are currently considered secure. The scheme presented here allows a user to outsource computations on encrypted data to the cloud, such that any third party can verify the correctness of the computations without having access to the original data. This makes this work an important contribution to cloud computing and applications where operations on sensitive data have to be performed, such as statistics on medical records and tallying of electronically cast votes.

Keywords: Authenticated encryption · Public verifiability · Cloud computing

1 Introduction

In this work the first "**L**inearly homomorphic authenticated **E**ncryption with **P**rovable **C**orrectness and public **V**erifiability" (LEPCoV) scheme is presented. It improves Catalano et al.'s instantiated scheme [12] by avoiding false negatives during the verification algorithm.

Outsourcing data and computations to the cloud has become an increasingly important aspect of IT. These new techniques provide a higher level of efficiency and flexibility and are therefore very valuable for private and commercial users. However, they also pose new risks for data security. Thus, secure outsourcing is a highly relevant research field. Cloud technologies must ensure that no malicious party gets access to the outsourced data and that no unauthorized modifications can be performed, i.e. the solutions must provide confidentiality and integrity.

© Springer International Publishing AG 2017
S. El Hajji et al. (Eds.): C2SI 2017, LNCS 10194, pp. 142–160, 2017.
DOI: 10.1007/978-3-319-55589-8_10

Both security goals can be provided by encrypting and, respectively, signing the data before outsourcing it to the cloud.

To allow for computations on the outsourced data, encryption and signature schemes with homomorphic properties were developed. However, so far most works focused on improving either of these schemes. Thus, Catalano et al. [12] developed a framework called "linearly homomorphic authenticated encryption with public verifiability" (LAEPuV) that allows to combine both primitives into one unified solution. They show that their framework can be instantiated with the Paillier cryptosystem and any linearly homomorphic signature scheme supporting the same message space. Furthermore, they provide a concrete instantiation using a variant of the linearly homomorphic signature scheme by Catalano et al. [11]. Since their primitive is linearly homomorphic, operations can be performed directly on the signed encrypted data and the correctness of the outcome can be verified. However, their concrete instantiation leads to false negatives, i.e. there are many functions for which the verification algorithm rejects correct computations on honestly generated ciphers. Thus, their solution does not provide correctness for all functions to be evaluated. Note that this affects the proposed instantiation rather than the generic construction. Furthermore, so far no work has tested the efficiency of their solution in practice.

Our Contribution. In this paper we propose an instantiation for LAEPuV, called LEPCoV, based on [12] that does not lead to false negatives. Besides a detailed description of LEPCoV, we also present a comparison with the scheme proposed by Catalano et al. highlighting our improvements. Furthermore, we prove that our solution is secure and ensures correctness when evaluating functions over authenticated encrypted data. Another shortcoming of the work by Catalano et al. is that an efficiency evaluation is missing. Measuring the runtime of an instantiation is important before putting it into practice. Thus, we run a performance analysis for different security parameters and dataset sizes. The tests show that our algorithms run in reasonable time for parameters that are currently considered secure. In addition, further efficiency improvements are possible and highlighted at the end of this work.

Structure. Our work is structured as follows. After providing the relevant definitions and the framework for LAEPuV in Sect. 2, we describe the instantiated scheme by Catalano et al. [12] in Sect. 3. Based on this, in Sect. 4 we point out the shortcomings of the scheme particularly with respect to correctness. Following this, in Sect. 5 we show how the correctness of the original solution can be improved, present our revised scheme LEPCoV, and prove its security and correctness. Finally, in Sect. 6 we demonstrate the practical use of our instantiation by providing the average runtimes of the algorithms based on our implementation and conclude in Sect. 7 with a summary of our contribution and possible future work.

1.1 Related Work

There are several homomorphic encryption schemes, like Paillier [19], ElGamal [14], and Benaloh [6], which allow computations on messages by performing a corresponding computation on the ciphers. Anyhow, none of these schemes address authenticity nor integrity of the data encrypted.

A general definition of homomorphic signature schemes is given by Johnson et al. [17], as a redefined version of the concept by Desmedt [13]. Linearly homomorphic signature schemes have been defined by Boneh et al. [7]. Based on this, other works [2–4,9–11,15,16], which provide either frameworks or realizations, have been proposed. However, these schemes keep neither the input data nor the output data confidential.

Authenticated encryption schemes aim at providing both privacy and authenticity. An and Bellare, for instance, introduced in [1] a new paradigm called *encryption with redundancy* achieving both security goals by adding some redundant information to the data to be encrypted. Later, Bellare and Namprempre [5] defined the term *authenticated encryption* together with corresponding security aspects. However, both works consider symmetric encryption and do not provide a solution for asymmetric encryption. Thus, closer to the setting described in this paper is the term *homomorphic authenticated encryption* defined by Joo and Yun [18]. While their scheme allows more functionalities, it is neither practical nor does it provide public verifiability. Thus, Catalano et al. [12] proposed a framework and an instantiation for a linearly homomorphic authenticated encryption scheme providing public verifiability. In this work we further improve their instantiation by providing provable correctness and a higher level of efficiency.

2 Notation and Preliminaries

In this section we provide the notation and preliminaries needed for our construction. We also give an intuition to the setup proposed by Catalano et al. [12] followed by the definition for Linearly Homomorphic Authenticated Encryption with Public Verifiability (LAEPuV). Afterwards, we present the hardness assumptions on which the security of the instantiation proposed by Catalano et al. and correspondingly our solution is based on.

2.1 Notation

Throughout this work we write $[k] = \{1, 2, ..., k\}$ for the natural number less or equal than k. For two integers $a, b \in \mathbb{Z}$, we write $\lfloor \frac{a}{b} \rfloor$ for the integer division of a and b, $a \mid b$ if a is a factor of b, and $a \nmid b$ if a is not a factor of b.

For a set S we write $s \xleftarrow{\$} S$ to indicate that s is chosen uniformly at random from S. We use $\mathcal{H}$ to describe a family of collision resistant hash functions which images can be interpreted as elements of $\mathbb{Z}_{N_E^2}$, where $N_E = p \cdot q$ for two primes p, q of equal size.

The i-th unit vector of $\mathbb{Z}^k$ is denoted by e_i. We denote functions as vectors of coefficients, i.e. $f = (f_1, ..., f_k)$. Note that for $f = e_i$ function evaluation $f(m_1, \ldots, m_k)$ returns m_i.

2.2 Setup

Catalano et al. [12] introduced a cryptographic primitive called LAEPuV that allows a user Alice to outsource encrypted data and computations on this data to the cloud. For this to be secure the cloud must keep the data received confidential and provide measures that allow verifying the integrity of the computation results. Optimally, the results are publicly verifiable enabling third parties such as external auditors to perform these checks.

To ensure confidentiality Alice could encrypt her data using a homomorphic encryption scheme. Due to its homomorphic properties, functions can be evaluated over the messages by evaluating corresponding functions over the ciphers. This allows Alice to outsource the computations to a cloud such that it neither learns the input nor the result. However, Alice has to trust that the cloud evaluates the functions correctly.

To ensure integrity of the result Alice could sign her data using a homomorphic signature scheme before outsourcing it to the cloud. This allows Alice to delegate computations such that Alice, or any third party on behalf of Alice, can verify the correctness of the computations. However, without using an encryption scheme to encrypt the data, the cloud would learn the input and the output of the computations. Thus, both schemes must be combined. More precisely, Alice encrypts her data, signs the ciphers, and asks the cloud to evaluate the function over the ciphers. When Alice receives the resulting cipher along with its (homomorphically computed) signature from the cloud, she can verify the computation using the signature and obtain the message by decrypting the cipher.

A naive combination of these primitives requires that the cipher space of the encryption scheme and the message space of the homomorphic signature scheme are equal. The message space of the Paillier cryptosystem is $\mathbb{Z}_N$, where $N = pq$ for two primes p, q of equal size while the corresponding cipher space is $\mathbb{Z}_{N^2}$. This leads to a performance problem as the homomorphic signature scheme has to support a significantly larger message space than the Paillier cryptosystem. Thus, Catalano et al. [12] proposed a method which allows combining the Paillier cryptosystem with a homomorphic signature scheme in a more efficient manner. Instead of signing the ciphers, the scheme masks the ciphers and signs the decrypted masked ciphers which have the same size as the original messages. The framework for the combination of both schemes will be presented in the next subsection. Details regarding the instantiated scheme by Catalano et al. follow in Sect. 3.

2.3 LAEPuV

Catalano et al. [12] introduced a cryptographic primitive called linearly homomorphic authenticated encryption with public verifiability (LAEPuV). These schemes allow Alice to outsource encrypted data to the cloud such that the cloud can do computations for Alice which are publicly verifiable. Below, we formally define LAEPuV schemes.

Definition 1 (Linearly Homomorphic Authenticated Encryption with Public Verifiability (LAEPuV) [12]). *A linearly homomorphic authenticated encryption with public verifiability (LAEPuV) scheme is a tuple of five PPT algorithms* $\mathcal{L} = (\mathsf{AKeyGen}, \mathsf{AEncrypt}, \mathsf{AVerify}, \mathsf{ADecrypt}, \mathsf{AEval})$:

$\mathsf{AKeyGen}(1^\kappa, k)$: *The input is a security parameter* κ *and the maximum number* k *of encrypted messages in each dataset. The output is a key pair* $(\mathsf{sk}, \mathsf{pk})$, *where* sk *is the secret key for decrypting and signing and* pk *is the public key used for verification and evaluation. The message space* $\mathcal{M}$, *the cipher space* $\mathcal{C}$, *and dataset identifier space* $\mathcal{D}$ *are implicitly defined by the public key* pk.

$\mathsf{AEncrypt}(\mathsf{sk}, \tau, i, m)$: *The input is a secret key* sk, *a dataset identifier* τ, *an index* $i \in [k]$, *and a message* m. *The output is a cipher* c.

$\mathsf{AVerify}(\mathsf{pk}, \tau, c, \boldsymbol{f})$: *The input is a public key* pk, *a dataset identifier* τ, *a cipher* c, *and a linear function* $\boldsymbol{f}$. *The output is either* 1, *i.e. the cipher is valid, or* 0, *i.e. the cipher is invalid.*

$\mathsf{ADecrypt}(\mathsf{sk}, \tau, c, \boldsymbol{f})$: *The input is a secret key* sk, *a dataset identifier* τ, *a cipher* c, *and a linear function* $\boldsymbol{f}$. *The output is a message* m *if* c *is valid and* $\perp$ *if* c *is invalid, respectively.*

$\mathsf{AEval}(\mathsf{pk}, \tau, \boldsymbol{f}, \{c_i\}_{i \in [k]})$: *The input is a public key* pk, *a dataset identifier* τ, *a linear function* $\boldsymbol{f}$, *and* k *ciphers* $\{c_i\}_{i \in [k]}$. *The output is a cipher* c.

In the following we provide the definitions for both the security and the correctness of linearly homomorphic authenticated encryption with public verifiability (LAEPuV) schemes.

Definition 2. *We call a linearly homomorphic authenticated encryption with public verifiability scheme* $\mathcal{L} = (\mathsf{AKeyGen}, \mathsf{AEncrypt}, \mathsf{AVerify}, \mathsf{ADecrypt}, \mathsf{AEval})$ *LH-IND-CCA secure, if the advantage of an adversary in the LH-IND-CCA game [12] is negligible in the security parameter* κ.

Definition 3. *We call a linearly homomorphic authenticated encryption with public verifiability scheme* $\mathcal{L} = (\mathsf{AKeyGen}, \mathsf{AEncrypt}, \mathsf{AVerify}, \mathsf{ADecrypt}, \mathsf{AEval})$ *correct, if for any key pair* $(\mathsf{sk}, \mathsf{pk}) \leftarrow \mathsf{AKeyGen}(1^\kappa, k)$ *the three conditions below are satisfied.*

Condition 1. *For any message* $m \in \mathcal{M}$, *any dataset identifier* $\tau \in \mathcal{D}$, *and any index* $i \in [k]$ *it holds that*

$$\mathsf{ADecrypt}(\mathsf{sk}, \tau, \mathsf{AEncrypt}(\mathsf{sk}, \tau, i, m), \boldsymbol{e}_i) = m.$$

Condition 2. *For any cipher* $c \in \mathcal{C}$, *any dataset identifier* $\tau \in \mathcal{D}$, *and any linear function* $\boldsymbol{f} = (f_1, ..., f_k) \in \mathbb{Z}_{N_E}^k$ *it holds that*

$$\mathsf{AVerify}(\mathsf{pk}, \tau, c, \boldsymbol{f}) = 1 \Leftrightarrow \exists m \in \mathcal{M} : \mathsf{ADecrypt}(\mathsf{sk}, \tau, c, \boldsymbol{f}) = m.$$

Condition 3. *For any dataset identifier $\tau \in \mathcal{D}$, any messages $m_1, ..., m_k \in \mathcal{M}$ with corresponding ciphers $c_1, ..., c_k \in \mathcal{C}$ such that $c_i \leftarrow \mathsf{AEncrypt}(\mathsf{sk}, \tau, i, m_i)$ for $i \in [k]$, and any linear function $\boldsymbol{f} = (f_1, ..., f_k) \in \mathbb{Z}_{N_E}^k$ it holds that*

$$\mathsf{ADecrypt}(\mathsf{sk}, \tau, \mathsf{AEval}(\mathsf{pk}, \tau, \boldsymbol{f}, \{c_i\}_{i \in [k]}), \boldsymbol{f}) = \boldsymbol{f}(m_1, ..., m_k).$$

2.4 Hardness Assumptions

Below we define the hardness assumptions needed for [12] and our construction, i.e. the decisional composite residuosity assumption (DCRA) and the strong RSA assumption.

Definition 4 (Decisional Composite Residuosity Assumption [12]). *We say the decisional composite residuosity assumption (DCRA) holds if there exists no PPT $\mathcal{A}$ that can distinguish between a random element from $\mathbb{Z}_{N^2}^*$ and one from the set $\{z^N : z \in \mathbb{Z}_{N^2}^*\}$ (i.e. the set of N-th residues modulo N^2), when N is the product of two random primes proper size.*

Definition 5 (Strong RSA Assumption [11]). *Let N be a random RSA modulus of length κ, where $\kappa \in \mathbb{N}$ is the security parameter, and z be a random element in $\mathbb{Z}_N$. Then we say that the strong RSA assumption holds if for any PPT adversary $\mathcal{A}$ it holds that*

$$Pr[(y, e) \leftarrow \mathcal{A}(N, z) : y^e = z \mod N \wedge e \neq 1] \leq \mathsf{negl}(\kappa).$$

3 LAEPuV Scheme CMP14 by Catalano et al.

Catalano et al. [12] proposed the first linearly homomorphic authenticated encryption with public verifiability scheme, henceforth referred to as CMP14. The scheme is based on the Paillier cryptosystem [19] and a variant of the linearly homomorphic signature scheme by Catalano et al. [11].

Instead of simply signing the cipher, the idea of the scheme is as follows. It encrypts the message contained in a dataset using the Paillier cryptosystem. The resulting cipher is masked by multiplying it with the hash of the dataset identifier concatenated with the index of the message within the dataset. This masked cipher is decrypted and the resulting message is signed using the linearly homomorphic signature scheme. Hereby, the message space of the linearly homomorphic signature scheme can be of size $\mathbb{Z}_{N_E}$, where $\mathbb{Z}_{N_E}$ is the message space of the Paillier cryptosystem, instead of the larger cipher space $\mathbb{Z}_{N_E^2}$. We describe the scheme below.

$\mathsf{AKeyGen}(1^\kappa, k)$: On input a security parameter κ and an integer k, the algorithm samples four (safe) primes p_E, q_E, p_S and q_S of size $\kappa/2$ such that for $N_E = p_E q_E$ and $N_S = p_S q_S$ it holds that $\varphi(N_S) = (p_S - 1)(q_S - 1)$ and N_E are coprime, i.e. $\gcd(N_E, \varphi(N_S)) = 1$. Subsequently, it samples an element $g \in \mathbb{Z}_{N_E^2}^*$ of order N_E and $k + 2$ elements $g_0, g_1, h_1, ..., h_k$ uniformly at random from $\mathbb{Z}_{N_S}^*$. Then, it chooses an (efficiently computable) injective function H_p which maps arbitrary strings to prime numbers of length $l < \kappa/2$ and a hash function $H \leftarrow \mathcal{H}$. The algorithm returns the key pair $(\mathsf{sk}, \mathsf{pk})$, where $\mathsf{sk} = (p_E, q_E, p_S, q_S)$ and $\mathsf{pk} = (N_E, g, N_S, g_0, g_1, h_1, ..., h_k, H, H_p)$.

$\mathsf{AEncrypt}(\mathsf{sk}, \tau, i, m)$: On input a secret key sk, a dataset identifier τ, an index $i \in [k]$, and a message m, the algorithm computes the Paillier encryption $C \leftarrow g^m \beta^{N_E} \mod N_E^2$ of m, where $\beta \xleftarrow{\$} \mathbb{Z}_{N_E}^*$, and the masking $R \leftarrow H(\tau||i)$. It computes a tuple $(a, b) \in \mathbb{Z}_{N_E} \times \mathbb{Z}_{N_E}^*$ such that $g^a b^{N_E} = CR \mod N_E^2$ by invoking the following steps [19]:

- Obtain a by decrypting CR using the Paillier cryptosystem [19].
- Compute $c_* \leftarrow CRg^{-a} \mod N_E$.
- Set $b \leftarrow c_*^{N_E^{-1} \mod \lambda} \mod N_E$, where $\lambda = \mathsf{lcm}(p_E - 1, q_E - 1)$.

Then, it obtains the prime $e \leftarrow H_p(\tau)$, chooses a random element $s \in \mathbb{Z}_{eN_E}$, and computes x such that

$$x^{eN_E} = g_0^s h_i g_1^a \mod N_S$$

Finally, it returns the cipher $c = (C, a, b, e, s, \tau, x)$.

$\mathsf{AVerify}(\mathsf{pk}, \tau, c, \boldsymbol{f})$: On input a public key pk, a dataset identifier τ, a cipher $c = (C, a, b, e, s, \tau, x)$, and a linear function $\boldsymbol{f} = (f_1, ..., f_k)$, the algorithm computes $e \leftarrow H_p(\tau)$, $\boldsymbol{f}' = \frac{\boldsymbol{f} - (\boldsymbol{f} \mod eN_E)}{eN_E}$, and $\hat{x} = \frac{x}{\prod_{i=1}^{k} h_i^{f_i'}}$. It checks if

$$a, s \in \mathbb{Z}_{eN_E} \tag{1}$$

$$\hat{x}^{eN_E} = g_0^s \prod_{i=1}^{k} h_i^{f_i} g_1^a \mod N_S \tag{2}$$

$$g^a b^{N_E} = C \prod_{i=1}^{k} H(\tau||i)^{f_i} \mod N_E^2 \tag{3}$$

If all checks pass, the algorithm returns 1, i.e. c is a valid cipher. Otherwise, it returns 0, i.e. c is an invalid cipher.

$\mathsf{ADecrypt}(\mathsf{sk}, \tau, c, \boldsymbol{f})$: On input a secret key sk, a dataset identifier τ, a cipher $c = (C, a, b, e, s, \tau, x)$, and a linear function $\boldsymbol{f} = (f_1, ..., f_k)$, the algorithm runs $\mathsf{AVerify}(\mathsf{pk}, \tau, c, \boldsymbol{f})$ to check if c is a valid cipher, i.e. whether $\mathsf{AVerify}(\mathsf{pk}, \tau, c, \boldsymbol{f}) = 1$. If true, the algorithm returns the message m obtained by decrypting C using the Paillier cryptosystem. Otherwise, it returns $\perp$.

$\mathsf{AEval}(\mathsf{pk}, \tau, \boldsymbol{f}, \{c_i\}_{i \in [k]})$: On input a public key pk, a dataset identifier τ, a linear function $\boldsymbol{f} = (f_1, ..., f_k)$, and k ciphers $c_i = (C_i, a_i, b_i, e_i, s_i, \tau_i, x_i)$, the algorithm first checks if there exists $i \in [k]$ such that $\tau \neq \tau_i$ or $H_p(\tau) \neq e_i$. If true, the algorithm aborts. Otherwise, the algorithm computes $e \leftarrow H_p(\tau)$ and

$$C \leftarrow \prod_{i=1}^{k} C_i^{f_i} \mod N_E^2 \qquad\qquad a \leftarrow \sum_{i=1}^{k} f_i a_i \mod N_E$$

$$b \leftarrow \prod_{i=1}^{k} b_i^{f_i} \mod N_E^2 \qquad\qquad s \leftarrow \sum_{i=1}^{k} f_i s_i \mod e N_E$$

$$s' \leftarrow \left(\sum_{i=1}^{k} f_i s_i - s \right) / (e N_E) \qquad x = \frac{\prod_{i=1}^{k} x_i^{f_i}}{g_0^{s'}} \mod N_S$$

Then, it returns the cipher $c = (C, a, b, e, s, \tau, x)$.

4   Shortcomings of CMP14

In this section we describe the shortcomings of CMP14. First, we show that there is a restriction regarding the functions which can be evaluated as most functions lead to false negatives during the verification algorithm, i.e. the verification algorithm rejects ciphers although they were generated honestly and correctly. It follows that CMP14 is not correct according to Definition 3 since these functions violate Condition 3. Following this, we describe a practical issue regarding the injective function H_p which makes the encryption infeasible in some datasets.

4.1   Restricted Function Evaluation

On a high level, CMP14 rejects honestly generated ciphers if the value a is reduced modulo N_E during AEval. In other words for a function $f = (f_1, ..., f_k)$ and ciphers $c_i = (C_i, a_i, b_i, e, s_i, \tau, x_i)$, where $f(a_1, ..., a_k) = \sum_{i=1}^{k} f_i a_i \geq N_E$, the verification of the cipher $c \leftarrow \mathsf{AEval}(\mathsf{pk}, \tau, f, \{c_i\}_{i \in [k]})$ fails. However, there are a few exceptions for which the verification of the cipher does not fail, namely the functions $f = (f_1, ..., f_k)$ for which $f(a_1, ..., a_k)$ is a multiple of the order of g_1, i.e. $f(a_1, ..., a_k) = q \cdot \mathrm{ord}(g_1)$, where $q \in \mathbb{N}$. Note that this is unlikely to happen, especially if using safe primes while generating keys.

We emphasize that Alice has no control over the values a_i, because they are obtained by decrypting the masked cipher CR, where C is the cipher of the message and $R \leftarrow H(\tau || i)$ is the masking. It follows that Alice can not simply adjust the functions f to ensure that $f(a_1, ..., a_k) < N_E$.

To show this more formally, we first provide a lemma which specifies the type of functions which leads to the modulo operation during AEval and show an inequality that holds for this type of functions. Then, we show that for this type of functions the verification algorithm of CMP14 fails even though the ciphers were computed honestly and correctly, which violates Condition 3.

Lemma 1. *Let* $(\mathsf{sk}, \mathsf{pk}) \leftarrow \mathsf{AKeyGen}(1^\kappa, k)$, *where* $\mathsf{sk} = (p_E, q_E, p_S, q_S)$ *and* $\mathsf{pk} = (N_E, g, N_S, g_0, g_1, h_1, ..., h_k, H, H_p)$ *be a key pair and* $a_1, ..., a_k \in \mathbb{Z}_{N_E}$ *be the* *decrypted masked (Paillier) ciphers of ciphers $c_1, ..., c_k$. Then, any linear function $\boldsymbol{f} = (f_1, ..., f_k)$ with $\mathsf{ord}(g_1) \nmid q$, where $q = \left\lfloor \frac{\boldsymbol{f}(a_1, ..., a_k)}{N_E} \right\rfloor = \left\lfloor \frac{\sum_{i=1}^{k} f_i a_i}{N_E} \right\rfloor \in \mathbb{N}$, leads to a modulo operation during AEval, i.e. it holds that*

$$g_1^{\sum_{i=1}^{k} f_i a_i} \neq g_1^{\sum_{i=1}^{k} f_i a_i \mod N_E} \mod N_S.$$

Proof. In order to prove the statement, it suffices to show that the exponents modulo the order of g_1 are not equal. Note that $\gcd(\mathsf{ord}(g_1), N_E) = 1$. This follows directly from the fact that, during $\mathsf{AKeyGen}$, N_E and N_S are generated such that $\gcd(N_E, \varphi(N_S)) = 1$ and $\mathsf{ord}(g_1) \mid \varphi(N_S)$.

Write $\boldsymbol{f}(a_1, ..., a_k) = \sum_{i=1}^{k} f_i a_i = q N_E + r$, where $r \in \{0, ..., N_E - 1\}$ and $q \in \mathbb{N}$ such that $\mathsf{ord}(g_1) \nmid q$. It holds that

$$
\begin{aligned}
\sum_{i=1}^{k} f_i a_i &= q N_E + r && \mod \mathsf{ord}(g_1) \\
&\neq r && \mod \mathsf{ord}(g_1) \\
&= q N_E + r \mod N_E && \mod \mathsf{ord}(g_1) \\
&= \sum_{i=1}^{k} f_i a_i \mod N_E && \mod \mathsf{ord}(g_1)
\end{aligned}
$$

Hence, $g_1^{\sum_{i=1}^{k} f_i a_i} \neq g_1^{\sum_{i=1}^{k} f_i a_i \mod N_E} \mod N_S$.   □

Proposition 1. *Let $(\mathsf{sk}, \mathsf{pk}) \leftarrow \mathsf{AKeyGen}(1^\kappa, k)$ be an honestly generated key pair, $c_i = (C_i, a_i, b_i, e, s_i, \tau, x_i) \leftarrow \mathsf{AEncrypt}(\mathsf{sk}, \tau, i, m_i)$ be ciphers of messages $m_i \in \mathbb{Z}_{N_E}$ for $i \in [k]$, where τ is an arbitrary dataset identifier. For any linear function $\boldsymbol{f} = (f_1, ..., f_k)$, where $\boldsymbol{f}(a_1, ..., a_k)$ leads to a modulo operation during AEval as defined in Lemma 1, it holds that*

$$\mathsf{ADecrypt}(\mathsf{sk}, \tau, \mathsf{AEval}(\mathsf{pk}, \tau, \boldsymbol{f}, \{c_i\}_{i \in [k]}), \boldsymbol{f}) \neq \boldsymbol{f}(m_1, ..., m_k)$$

which violates Condition 3 of Definition 3.

Proof. Let $(\mathsf{sk}, \mathsf{pk}) \leftarrow \mathsf{AKeyGen}(1^\kappa, k)$ be an honestly generated key pair, τ be an arbitrary dataset identifier, $m_1, ..., m_k \in \mathbb{Z}_{N_E}$ be arbitrary messages, and $c_i = (C_i, a_i, b_i, e, s_i, \tau, x_i) \leftarrow \mathsf{AEncrypt}(\mathsf{sk}, \tau, i, m_i)$ be the resulting ciphers. Let $\boldsymbol{f} = (f_1, ..., f_k)$ be a linear function such that $\boldsymbol{f}(a_1, ..., a_k)$ satisfies Lemma 1 and $c \leftarrow \mathsf{AEval}(\mathsf{pk}, \tau, \boldsymbol{f}, \{c_i\}_{i \in [k]})$ be the cipher that is obtained by evaluating the function $\boldsymbol{f}$ over the ciphers c_i. It holds that

$$\hat{x}^{eN_E} = \left(\frac{x}{\prod_{i=1}^{k} h_i^{f_i'}}\right)^{eN_E}$$

$$= \left(\frac{\prod_{i=1}^{k} x_i^{f_i}}{g_0^{s'} \prod_{i=1}^{k} h_i^{f_i'}}\right)^{eN_E}$$

$$= \frac{\prod_{i=1}^{k} (g_0^{s_i} h_i g_1^{a_i})^{f_i}}{\left(g_0^{\frac{\sum_{i=1}^{k} f_i s_i - s}{eN_E}} \prod_{i=1}^{k} h_i^{\frac{f_i - (f_i \mod eN_E)}{eN_E}}\right)^{eN_E}}$$

$$= \frac{g_0^{\sum_{i=1}^{k} f_i s_i} \prod_{i=1}^{k} (h_i^{f_i}) g_1^{\sum_{i=1}^{k} f_i a_i}}{g_0^{\sum_{i=1}^{k} f_i s_i - s} \prod_{i=1}^{k} h_i^{f_i - (f_i \mod eN_E)}}$$

$$= g_0^{s} \prod_{i=1}^{k} h_i^{f_i \mod eN_E} g_1^{\sum_{i=1}^{k} f_i a_i}$$

$$\overset{\text{Lemma 1}}{\neq} g_0^{s} \prod_{i=1}^{k} h_i^{f_i} g_1^{\sum_{i=1}^{k} f_i a_i} \mod N_E$$

$$= g_0^{s} \prod_{i=1}^{k} h_i^{f_i} g_1^{a}$$

This yields $\mathsf{AVerify}(\mathsf{pk}, \tau, \mathsf{AEval}(\mathsf{pk}, \tau, \boldsymbol{f}, \{c_i\}_{i\in[k]}), \boldsymbol{f}) = 0$, hence, it holds that $\mathsf{ADecrypt}(\mathsf{sk}, \tau, \mathsf{AEval}(\mathsf{pk}, \tau, \boldsymbol{f}, \{c_i\}_{i\in[k]}), \boldsymbol{f}) = \perp \neq \boldsymbol{f}(m_1, ..., m_k)$ which violates Condition 3. □

Proposition 1 proves that CMP14 is not correct according to Definition 3 as there occur false negatives. However, it does not state whether this shortcoming affects the practical use of the scheme, i.e. whether Alice can prevent false negatives by choosing the messages and functions carefully.

Hence, we implemented and tested CMP14. The results show that CMP14 is impractical since, regardless of the security parameter κ and the dataset size k, even small functions, e.g. adding two messages, mainly lead to false negatives. Below we provide an example which illustrates this.

Table 1 shows a 16 bit key pair (the relevant values) and four messages m_i along with their random encryption values β_i and the maskings R_i, which allow to compute the values a_i. While the addition of m_4 and either m_2 or m_3 works, the addition of m_1 and any other message m_i as well as the addition of m_2 and m_3 lead to a false negative. We stress that, regardless of the actual function values, combining three or four of these messages, i.e. at most one function value is 0, always yields a false negative. We further emphasize that the values in Table 1 are generated arbitrarily and not specially constructed for this shortcoming.

Table 1. Example values for false negatives.

Key pair				
Secret key sk		Public key pk		
$p_E = 151$ $q_E = 149$		$N_E = 22499$	$N_E^2 = 506205001$	$g = 457224679$
Encryption values				
m	β	R	m	β	R
$m_1 = 17$	$\beta_1 = 14296$	$R_1 = 64489750$	$m_3 = 19$	$\beta_3 = 1576$	$R_3 = 157182719$
$m_2 = 4$	$\beta_2 = 17791$	$R_2 = 18170490$	$m_4 = 92$	$\beta_4 = 6190$	$R_4 = 365721887$

4.2   Infeasible Encryption in Some Datasets

CMP14 also suffers from a minor practical issue regarding the function H_p which binds a unique prime to each dataset. There is no check whether the prime and the order of the group $\mathbb{Z}_{N_S}$ are coprime. If that is not the case, computing the signature value x during AEncrypt is equivalent to breaking the RSA assumption, which is assumed to be infeasible.

More formally, let e be a prime such that $\gcd(eN_E, \varphi(N_S)) \neq 1$. Under the strong RSA assumption (see Definition 5) one can not compute x such that $x^{eN_E} = g_0^s h_i g_1^a \mod N_S$ in polynomial time. Hence, for the dataset identified by τ, where $H_p(\tau) = e$, AEncrypt can not be executed efficiently.

5   Our Improved Scheme LEPCoV

In this section we describe our improved scheme LEPCoV based on CMP14. We start with a high-level description of the changes followed by a detailed description of the scheme. Finally, we show that our scheme is both secure and correct according to Definitions 2 and 3, respectively.

5.1   High-Level Description of Our Changes

First, we simplify the verification of ciphers. We require that all functions to be evaluated are described as vectors of coefficients where each coefficient is a value smaller than N_E. Note that this restriction still allows to express all linear functions. More precisely, let $m \in \mathbb{Z}_{N_E}$ be a message, $\beta \in \mathbb{Z}_{N_E}^*$ be a random encryption value, and $C = g^m \beta^{N_E} \in \mathbb{Z}_{N_E^2}$ be a Paillier cipher of this message. For any integer f, it holds that

$$\text{Decrypt}(C^f) = fm \mod N_E$$
$$= (f \mod N_E)m \mod N_E$$
$$= \text{Decrypt}(C^{f \mod N_E})$$

where $\text{Decrypt}(C)$ is the Paillier decryption of C. This allows us to simplify the verification algorithm as the values $\boldsymbol{f}'$ and $\hat{x}$ are no longer necessary.

To address the shortcoming that efficient encryption is not feasible in some datasets, as described in Sect. 4.2, we do not use the function H_p. Instead, Alice generates the prime for each dataset by herself and binds the prime to the dataset by signing the dataset identifier and the prime using a signature scheme $\mathcal{S} = (\mathsf{KeyGen}, \mathsf{Sign}, \mathsf{Verify})$. Hence, for each dataset, Alice can generate a unique prime e such that $\gcd(eN_E, \varphi(N_S)) = 1$, which guarantees that Alice can encrypt messages in this dataset.

The other core problem of CMP14 is the evaluation of functions for which the value a is reduced during AEval, as descried in Sect. 4.1. Note that due to $\gcd(N_E, \varphi(N_S)) = 1$, one can not generate g_1 of order N_E to trivially fix this shortcoming. Thus, to avoid this problem, we change the computation of x during AEval. We stress that, in CMP14, the problem if a is reduced during AEval does not occur if s is reduced during AEval, because x is multiplied with the inverse of $g_0^{s'}$. Hence, similar to s' we compute a new value a'. Note however that simply multiplying x also with the inverse of $g_1^{a'}$ does not suffice as a and s are not elements within the same group. Instead, we multiply x with the inverse of $g_1^{a'e^{-1}}$, where e^{-1} is the inverse element of e modulo $\varphi(N_S)$. Since the efficient computation of e^{-1} requires the factorization of N_S, Alice has to compute and publish e^{-1}.

Based on the changes described above, a cipher $c = (C, a, b, e, e^{-1}, \sigma_e, s, \tau, x)$ of a message m in LEPCoV, contains the Paillier encryption C of m, the decrypted masked cipher a along with its random encryption value b, the prime e and its inverse element e^{-1}, the signature σ_e of $\tau || e$, the random signature value s, the dataset identifier τ, and the signature x of a. Since a LAEPuV scheme does not require the complete dataset to verify a cipher, we have to address that Alice might not store the ciphers locally. Note that the values e, e^{-1}, σ_e are the same for each cipher within the same dataset. Thus, it is sufficient to assume that Alice keeps record of these values, i.e. she has access to a list L which contains tuples of dataset identifiers τ and the values e, e^{-1}, σ_e. If Alice runs AEncrypt with dataset identifier τ the first time, she computes e, e^{-1}, σ_e and stores $(\tau, e, e^{-1}, \sigma_e)$ in the list L. Otherwise, Alice takes the values from the list L. We emphasize that the list L allows Alice to generate a unique prime for each dataset.

5.2   Description of the Scheme

Below we provide a detailed description of LEPCoV and highlight the differences compared to CMP14. In the description, $\mathcal{S} = (\mathsf{KeyGen}, \mathsf{Sign}, \mathsf{Verify})$ describes a signature scheme used to bind primes to datasets.

AKeyGen($1^\kappa, k$): On input a security parameter κ and an integer k, the algorithm samples the four (safe) primes p_E, q_E, p_S, q_S, the group elements $g_0, g_1, h_1, ..., h_k \in \mathbb{Z}_{N_S}^*$ and $g \in \mathbb{Z}_{N_E^2}^*$ of order N_E, and the hash function $H \in \mathcal{H}$ as described for the original approach. In addition, it runs KeyGen(1^κ) to obtain a key pair $(\mathsf{sk}_S, \mathsf{pk}_S)$ of $\mathcal{S}$ and returns the key pair $(\mathsf{sk}, \mathsf{pk})$, where $\mathsf{sk} = (p_E, q_E, p_S, q_S, \mathsf{sk}_S)$ and $\mathsf{pk} = (N_E, g, N_S, g_0, g_1, h_1, ..., h_k, H, \mathsf{pk}_S)$ along with an empty list L.

AEncrypt(sk, τ, i, m): On input a secret key sk, a dataset identifier τ, an index $i \in [k]$, and a message m, the algorithm computes R, the Paillier encryption C of the message m, and (a, b) as described for CMP14. In addition, if τ is used the first time, it chooses a not yet used prime e of length $l < \kappa/2$ such that $\gcd(eN_E, \varphi(N_S)) = 1$, computes its inverse $e^{-1} \mod \varphi(N_S)$ and its signature $\sigma_e \leftarrow \mathsf{Sign}(\mathsf{sk}_S, \tau||e)$, and stores $(\tau, e, e^{-1}, \sigma_e)$ in the list L.

Otherwise, it takes $(\tau, e, e^{-1}, \sigma_e)$ from the list L. Then, it chooses $s \xleftarrow{\$} \mathbb{Z}_{eN_E}$, computes the value x such that $x^{eN_E} = g_0^s h_i g_1^a \mod N_S$, and returns the cipher $c = (C, a, b, e, e^{-1}, \sigma_e, s, \tau, x)$.

AVerify(pk, $\tau, c, \boldsymbol{f}$): On input a public key pk, a dataset identifier τ, a cipher $c = (C, a, b, e, e^{-1}, \sigma_e, s, \tau, x)$, and a linear function $\boldsymbol{f} = (f_1, .., f_k)$, the algorithm checks if

$$\mathsf{Verify}(\mathsf{pk}_S, \tau||e, \sigma_e) = 1$$
$$a, s \in \mathbb{Z}_{eN_E}$$

$$x^{eN_E} = g_0^s \prod_{i=1}^{k} h_i^{f_i} g_1^a \mod N_S$$

$$g^a b^{N_E} = C \prod_{i=1}^{k} H(\tau||i)^{f_i} \mod N_E^2$$

If all four checks pass, the algorithm returns 1, i.e. c is a valid cipher. Otherwise, it returns 0, i.e. c is an invalid cipher.

ADecrypt(sk, $\tau, c, \boldsymbol{f}$): On input a secret key sk, a dataset identifier τ, a cipher $c = (C, a, b, e, e^{-1}, \sigma_e, s, \tau, x)$, and a linear function $\boldsymbol{f} = (f_1, ..., f_k)$, the algorithm runs AVerify(pk, $\tau, c, \boldsymbol{f}$) to check if c is a valid cipher. If true, the algorithm returns the message m obtained by decrypting C using the Paillier cryptosystem. Otherwise, it returns $\bot$.

AEval(pk, $\tau, \boldsymbol{f}, \{c_i\}_{i \in [k]}$): On input a public key pk, a dataset identifier τ, a linear function $\boldsymbol{f}$, and k ciphers $c_i = (C_i, a_i, b_i, e_i, e_i^{-1}, \sigma_{e_i}, s_i, \tau_i, x_i)$, the algorithm checks if there exists an index $l \in [k]$ such that $\tau \neq \tau_l$, like in CMP14, or $\mathsf{Verify}(\mathsf{pk}_S, \tau||e_l, \sigma_{e_l}) = 0$. Furthermore, the algorithm checks if there are two indexes $i \neq j \in [k]$ such that $e_i \neq e_j$. If one of the checks is true, the algorithm aborts. Otherwise, the algorithm sets $e = e_1$, $e^{-1} = e_1^{-1}$, $\sigma_e = \sigma_{e_1}$, computes C, a, b, s, and s' like in the original approach, and

$$a' \leftarrow \left(\sum_{i=1}^{k} f_i a_i - a \right) / N_E \qquad x = \frac{\prod_{i=1}^{k} x_i^{f_i}}{g_0^{s'} g_1^{a'e^{-1}}} \mod N_S$$

Then, it returns the cipher $c = (C, a, b, e, e^{-1}, \sigma_e, s, \tau, x)$.

5.3   Security

The security of our improved scheme LEPCoV, according to Definition 2, is given in the theorem below.

Theorem 1. *The linearly homomorphic authenticated encryption with public verifiability scheme LEPCoV, described above, is secure according to Definition 2.*

Proof. For lack of space, we only sketch the proof. In CMP14, the injective function H_p ensures that each dataset is associated with a unique prime e. In LEPCoV, these primes are generated by Alice, hence, she can generate a unique prime for each dataset. The signature scheme $S = (\text{KeyGen}, \text{Sign}, \text{Verify})$ is used to bind primes to datasets, thus, the security of S guarantees that only the prime numbers chosen by Alice are accepted.

In case of the original scheme, an adversary $\mathcal{A}$ has to compute the eN_E-th root to forge a signature, which, under the strong RSA assumption (see Definition 5), is infeasible for the parameters chosen. In LEPCoV, Alice publishes the inverse of e, hence, the adversary has only to compute the N_E-th root in order to forge a signature. However, under the strong RSA assumption, this remains infeasible for the parameters chosen.

Based on these changes, the following statement by Catalano et al. [12] applies to the linearly homomorphic signature scheme used in LEPCoV: *The signature scheme is an unforgeable signature scheme under chosen message attacks according to the definition by Boneh and Freeman [8], if the strong RSA assumption (see Definition 5) holds [12, Theorem 31].*

Based on this, the following statement proves the security of LEPCoV: *In the random oracle model, if (1) the DCRA (see Definition 4) holds, (2) H is a random oracle and (3) the linearly homomorphic signature scheme over $\mathbb{Z}_{N_E}$ is unforgeable (under chosen message attacks), the scheme LEPCoV, described in Sect. 5, is LH-IND-CCA secure [12, Theorem 6].*   □

5.4   Correctness

The correctness of LEPCoV, described above, follows from the following theorem which is proven below.

Theorem 2. *The linearly homomorphic authenticated encryption with public verifiability scheme LEPCoV, described above, is correct according to Definition 3.*

Proof. In the following, we show that each condition described in Definition 3 holds. Throughout this proof, let $(\text{sk}, \text{pk}) \leftarrow \text{AKeyGen}(1^\kappa, k)$ be a key pair, where $\text{sk} = (p_E, q_E, p_S, q_S, \text{sk}_S)$ and $\text{pk} = (N_E, g, N_S, g_0, g_1, h_1, ..., h_k, H, \text{pk}_S)$.

Condition 1: Let $m \in \mathbb{Z}_{N_E}$ be an arbitrary message, τ be an arbitrary dataset identifier, $i \in [k]$, $c = (C, a, b, e, e^{-1}, \sigma_e, s, \tau, x) \leftarrow \text{AEncrypt}(\text{sk}, \tau, i, m)$ be the encryption of m, and $f = e_i$.

By construction we have $a, s \in \mathbb{Z}_{eN_E}$ and $\mathsf{Verify}(\mathsf{pk}_S, \tau || e, \sigma_e) = 1$. It holds that

$$x^{eN_E} = g_0^s h_i g_1^a = g_0^s h_i \prod_{\substack{j=1 \\ j \neq i}}^k h_j^0 g_1^a = g_0^s \prod_{j=1}^k h_j^{f_j} g_1^a$$

and

$$g^a b^{N_E} = CR = CH(\tau || i) = CH(\tau || i) \prod_{\substack{j=1 \\ j \neq i}}^k H(\tau || j)^0 = C \prod_{j=1}^k H(\tau || j)^{f_j}$$

which yields $\mathsf{AVerify}(\mathsf{pk}, \tau, \mathsf{AEncrypt}(\mathsf{sk}, \tau, i, m), \boldsymbol{f}) = 1$. Thus, $\mathsf{ADecrypt}$ returns the Paillier decryption of C, i.e. $\mathsf{ADecrypt}(\mathsf{sk}, \tau, \mathsf{AEncrypt}(\mathsf{sk}, \tau, i, m), \boldsymbol{e}_i) = m$.

Condition 2: We prove the equivalence by showing that both implications are satisfied.

$\Leftarrow$: Let $m \in \mathcal{M} = \mathbb{Z}_{N_E}$ be a message, $\boldsymbol{f} = (f_1, ..., f_k)$ be a linear function with $f_i < N_E$ for $i \in [k]$, and c be a cipher such that $\mathsf{ADecrypt}(\mathsf{sk}, \tau, c, \boldsymbol{f}) = m$. The fact that $\mathsf{ADecrypt}(\mathsf{sk}, \tau, c, \boldsymbol{f}) \neq \perp$ directly leads to $\mathsf{AVerify}(\mathsf{pk}, \tau, c, \boldsymbol{f}) = 1$.

$\Rightarrow$: Let $c = (C, a, b, e, e^{-1}, \sigma_e, s, \tau, x) \in \mathcal{C}$ be a cipher, τ be a dataset identifier, and $\boldsymbol{f}$ be a linear function such that $\mathsf{AVerify}(\mathsf{pk}, \tau, c, \boldsymbol{f}) = 1$. Since $\mathrm{ord}(g) = N_E$, this guarantees that the Paillier decryption of C yields $m \in \mathcal{M} = \mathbb{Z}_{N_E}$. Thus,

$$\exists m \in \mathcal{M} : \mathsf{ADecrypt}(\mathsf{sk}, \tau, c, \boldsymbol{f}) = m.$$

Condition 3: Let τ be an arbitrary dataset identifier, $m_1, ..., m_k \in \mathbb{Z}_{N_E}$ be messages, and $c_i = (C_i, a_i, b_i, e, e^{-1}, \sigma_e, s_i, \tau, x_i) \leftarrow \mathsf{AEncrypt}(\mathsf{sk}, \tau, i, m_i)$ be the cipher obtain by encrypting the message m_i for $i \in [k]$.

Let $\boldsymbol{f} = (f_1, ..., f_k)$ be a linear function such that $f_i < N_E$ for all $i \in [k]$ and $c = (C, a, b, e, e^{-1}, \sigma_e, s, \tau, x) \leftarrow \mathsf{AEval}(\mathsf{pk}, \tau, \boldsymbol{f}, \{c_i\}_{i \in [k]})$ be the cipher obtained by evaluating the function $\boldsymbol{f}$ over the ciphers c_i.

By construction it holds that $\mathsf{Verify}(\mathsf{pk}_S, \tau || e, \sigma_e) = 1$. During AEval, s and a are reduced modulo eN_E and N_E, respectively. Thus, $s, a \in \mathbb{Z}_{eN_E}$. In order to show that $\mathsf{AVerify}(\mathsf{pk}, \tau, c, \boldsymbol{f}) = 1$, it remains to show that

$$x^{eN_E} = g_0^s \prod_{i=1}^k h_i^{f_i} g_1^a \mod N_S \tag{4}$$

$$g^a b^{N_E} = C \prod_{i=1}^k H(\tau || i)^{f_i} \mod N_E^2 \tag{5}$$

For Eq. (4) we have

$$x^{eN_E} = \frac{(\prod_{i=1}^{k} x_i^{f_i})^{eN_E}}{(g_0^{s'} g_1^{a'e^{-1}})^{eN_E}} = \frac{\prod_{i=1}^{k}(g_0^{s_i} h_i g_1^{a_i})^{f_i}}{(g_0^{s'} g_1^{a'e^{-1}})^{eN_E}}$$

$$= \frac{g_0^{\sum_{i=1}^{k} f_i s_i} \prod_{i=1}^{k} h_i^{f_i} g_1^{\sum_{i=1}^{k} f_i a_i}}{\left(g_0^{(\sum_{i=1}^{k} f_i s_i - s)/(eN_E)} g_1^{((\sum_{i=1}^{k} f_i a_i - a)/(N_E))e^{-1}}\right)^{eN_E}}$$

$$= \frac{g_0^{\sum_{i=1}^{k} f_i s_i} \prod_{i=1}^{k} h_i^{f_i} g_1^{\sum_{i=1}^{k} f_i a_i}}{g_0^{\sum_{i=1}^{k} f_i s_i - s} \left(g_1^{(\sum_{i=1}^{k} f_i a_i - a)/(eN_E)}\right)^{eN_E}}$$

$$= \frac{g_0^{\sum_{i=1}^{k} f_i s_i} \prod_{i=1}^{k} h_i^{f_i} g_1^{\sum_{i=1}^{k} f_i a_i}}{g_0^{\sum_{i=1}^{k} f_i s_i - s} g_1^{\sum_{i=1}^{k} f_i a_i - a}} = g_0^s \prod_{i=1}^{k} h_i^{f_i} g_1^a$$

For Eq. (5) we obtain

$$C \prod_{i=1}^{k} H(\tau||i)^{f_i} = \prod_{i=1}^{k} C_i^{f_i} \prod_{i=1}^{k} H(\tau||i)^{f_i} = \prod_{i=1}^{k} (C_i H(\tau||i))^{f_i}$$

$$= \prod_{i=1}^{k} (g^{a_i} b_i^{N_E})^{f_i} = g^{\sum_{i=1}^{k} f_i a_i} \prod_{i=1}^{k} b_i^{f_i N_E} = g^a b^{N_E}$$

Thus, it holds that $\mathsf{AVerify}(\mathsf{pk}, \tau, c, \boldsymbol{f}) = 1$. Finally, we have

$$C = \prod_{i=1}^{k} C_i^{f_i} = \prod_{i=1}^{k} (g^{m_i} \beta_i^{N_E})^{f_i} = g^{\sum_{i=1}^{k} f_i m_i} \prod_{i=1}^{k} \beta_i^{f_i N_E}$$

hence Paillier decryption yields $\sum_{i=1}^{k} f_i m_i = \boldsymbol{f}(m_1, ..., m_k)$, which leads to

$$\mathsf{ADecrypt}(\mathsf{sk}, \tau, \mathsf{AEval}(\mathsf{pk}, \tau, \boldsymbol{f}, \{c_i\}_{i \in [k]}), \boldsymbol{f}) = \boldsymbol{f}(m_1, ..., m_k)$$

Thus, LEPCoV satisfies Conditions 1–3 which proves the statement.   □

We stress that g is an element of order N_E. Thus, the verification check in Eq. (5) does not fail if a is reduced during AEval. Also keep in mind that due to $\gcd(N_E, \varphi(N_S)) = 1$, one can not generate g_1 of order N_E to trivially fix the shortcoming of CMP14 described in Sect. 4.1.

6   Implementation

We implemented LEPCoV in Java and measured the average runtimes of the algorithms on an Intel® Core M-5Y71 CPU @ 1.20 GHz with 8 GB RAM. We run our experiments for different security parameters $\kappa \in \{1024, 2048, 3072, 4096\}$ and dataset sizes $k \in \{50, 100, 500\}$. Note that AVerify is not considered in the experiments as its runtime is similar to ADecrypt.

Table 2. Average runtimes (in ms) of AKeyGen, AEncrypt, ADecrypt, and AEval for different security parameters κ and dataset sizes k.

	$\kappa = 1024$ bits			$\kappa = 2048$ bits		
	$k = 50$	$k = 100$	$k = 500$	$k = 50$	$k = 100$	$k = 500$
AKeyGen	285	299	346	2501	2686	2862
AEncrypt	65	62	69	502	537	560
ADecrypt	86	109	403	571	724	1925
AEval	57	112	1042	297	854	19995
	$\kappa = 3072$ bits			$\kappa = 4096$ bits		
	$k = 50$	$k = 100$	$k = 500$	$k = 50$	$k = 100$	$k = 500$
AKeyGen	10038	9994	10190	25279	25755	26040
AEncrypt	1804	1787	1791	4029	4040	4078
ADecrypt	1945	2290	9679	4199	4645	16810
AEval	737	1953	43211	1255	3250	67233

Table 2 summarizes the average runtimes of AKeyGen, AEncrypt, ADecrypt, and AEval. It shows that AKeyGen and AEval are, as expected, the most expensive algorithms followed by ADecrypt and AEncrypt. Note that AKeyGen is only performed once and AEval is outsourced to the cloud. Thus, Alice only has two run the two less expensive algorithms AEncrypt and ADecrypt. The runtime of AEncrypt depends only on the security parameter κ. Therefore, the constant and relatively cheap costs of the encryption allow executing it on a device with less computation power. The runtime of ADecrypt (and AVerify) depends, besides the security parameter κ, also on the dataset size k and can be executed on a more powerful device. For a security parameter of $\kappa = 2048$ bits, which is currently assumed secure, and dataset size $k \leq 100$, AEncrypt and ADecrypt take less than a second. Note that for growing dataset size k, ADecrypt becomes faster than AEval. It follows that the size of the datasets processed must be taken into account when considering this scheme for an application. However, further efficiency improvements, e.g. using the Chinese remainder theorem to speed up the Paillier cryptosystem, and implementation-based optimizations, like parallelized code, are still possible.

Summarized the tests show that for parameters that are currently considered secure all algorithms run in a reasonable amount of time.

7   Conclusion

In this paper we proposed the first provable correct linearly homomorphic authenticated encryption with public verifiability (LAEPuV) scheme LEPCoV that is based on the CMP14 scheme by Catalano et al. [12]. We showed to what extent our scheme improves the original approach, proved our scheme secure, and showed that all algorithms run in reasonable time for currently recommended security parameters.

For future work we plan to further improve the efficiency of our implementation by implementing additional optimizations. Furthermore, we aim at constructing homomorphic authenticated encryption schemes with public verifiability for a wider class of supported functions.

Acknowledgments. This work has been co-funded by the DFG as part of project "Long-Term Secure Archiving" within the CRC 1119 CROSSING. In addition, it has received funding from the European Union's Horizon 2020 research and innovation program under Grant Agreement No. 644962.

References

1. An, J.H., Bellare, M.: Does encryption with redundancy provide authenticity? In: Pfitzmann, B. (ed.) EUROCRYPT 2001. LNCS, vol. 2045, pp. 512–528. Springer, Heidelberg (2001). doi:10.1007/3-540-44987-6_31
2. Attrapadung, N., Libert, B.: Homomorphic network coding signatures in the standard model. In: Catalano, D., Fazio, N., Gennaro, R., Nicolosi, A. (eds.) PKC 2011. LNCS, vol. 6571, pp. 17–34. Springer, Heidelberg (2011). doi:10.1007/978-3-642-19379-8_2
3. Attrapadung, N., Libert, B., Peters, T.: Computing on authenticated data: new privacy definitions and constructions. In: Wang, X., Sako, K. (eds.) ASIACRYPT 2012. LNCS, vol. 7658, pp. 367–385. Springer, Heidelberg (2012). doi:10.1007/978-3-642-34961-4_23
4. Attrapadung, N., Libert, B., Peters, T.: Efficient completely context-hiding quotable and linearly homomorphic signatures. In: Kurosawa, K., Hanaoka, G. (eds.) PKC 2013. LNCS, vol. 7778, pp. 386–404. Springer, Heidelberg (2013). doi:10.1007/978-3-642-36362-7_24
5. Bellare, M., Namprempre, C.: Authenticated encryption: relations among notions and analysis of the generic composition paradigm. J. Crypt. **21**(4), 469–491 (2008)
6. Benaloh, J.: Dense probabilistic encryption. In: Proceedings of the Workshop on Selected Areas of Cryptography, pp. 120–128 (1994)
7. Boneh, D., Freeman, D., Katz, J., Waters, B.: Signing a linear subspace: signature schemes for network coding. In: Jarecki, S., Tsudik, G. (eds.) PKC 2009. LNCS, vol. 5443, pp. 68–87. Springer, Heidelberg (2009). doi:10.1007/978-3-642-00468-1_5
8. Boneh, D., Freeman, D.M.: Homomorphic signatures for polynomial functions. In: Paterson, K.G. (ed.) EUROCRYPT 2011. LNCS, vol. 6632, pp. 149–168. Springer, Heidelberg (2011). doi:10.1007/978-3-642-20465-4_10
9. Boneh, D., Freeman, D.M.: Linearly homomorphic signatures over binary fields and new tools for lattice-based signatures. In: Catalano, D., Fazio, N., Gennaro, R., Nicolosi, A. (eds.) PKC 2011. LNCS, vol. 6571, pp. 1–16. Springer, Heidelberg (2011). doi:10.1007/978-3-642-19379-8_1
10. Catalano, D., Fiore, D., Gennaro, R., Vamvourellis, K.: Algebraic (trapdoor) one-way functions and their applications. In: Sahai, A. (ed.) TCC 2013. LNCS, vol. 7785, pp. 680–699. Springer, Heidelberg (2013). doi:10.1007/978-3-642-36594-2_38
11. Catalano, D., Fiore, D., Warinschi, B.: Efficient network coding signatures in the standard model. In: Fischlin, M., Buchmann, J., Manulis, M. (eds.) PKC 2012. LNCS, vol. 7293, pp. 680–696. Springer, Heidelberg (2012). doi:10.1007/978-3-642-30057-8_40

12. Catalano, D., Marcedone, A., Puglisi, O.: Authenticating computation on groups: new homomorphic primitives and applications. In: Sarkar, P., Iwata, T. (eds.) ASIACRYPT 2014. LNCS, vol. 8874, pp. 193–212. Springer, Heidelberg (2014). doi:10.1007/978-3-662-45608-8_11

13. Desmedt, Y.: Computer security by redefining what a computer is. In: Proceedings on the 1992–1993 Workshop on New security paradigms, pp. 160–166. ACM (1993)

14. ElGamal, T.: A public key cryptosystem and a signature scheme based on discrete logarithms. In: Blakley, G.R., Chaum, D. (eds.) CRYPTO 1984. LNCS, vol. 196, pp. 10–18. Springer, Heidelberg (1985). doi:10.1007/3-540-39568-7_2

15. Freeman, D.M.: Improved security for linearly homomorphic signatures: a generic framework. In: Fischlin, M., Buchmann, J., Manulis, M. (eds.) PKC 2012. LNCS, vol. 7293, pp. 697–714. Springer, Heidelberg (2012). doi:10.1007/978-3-642-30057-8_41

16. Gennaro, R., Katz, J., Krawczyk, H., Rabin, T.: Secure network coding over the integers. In: Nguyen, P.Q., Pointcheval, D. (eds.) PKC 2010. LNCS, vol. 6056, pp. 142–160. Springer, Heidelberg (2010). doi:10.1007/978-3-642-13013-7_9

17. Johnson, R., Molnar, D., Song, D., Wagner, D.: Homomorphic signature schemes. In: Preneel, B. (ed.) CT-RSA 2002. LNCS, vol. 2271, pp. 244–262. Springer, Heidelberg (2002). doi:10.1007/3-540-45760-7_17

18. Joo, C., Yun, A.: Homomorphic authenticated encryption secure against chosen-ciphertext attack. In: Sarkar, P., Iwata, T. (eds.) ASIACRYPT 2014. LNCS, vol. 8874, pp. 173–192. Springer, Heidelberg (2014). doi:10.1007/978-3-662-45608-8_10

19. Paillier, P.: Public-key cryptosystems based on composite degree residuosity classes. In: Stern, J. (ed.) EUROCRYPT 1999. LNCS, vol. 1592, pp. 223–238. Springer, Heidelberg (1999). doi:10.1007/3-540-48910-X_16

Constacyclic Codes over Finite Principal Ideal Rings

Aicha Batoul[1]([✉]), Kenza Guenda[1], T. Aaron Gulliver[2], and Nuh Aydin[3]

[1] Faculty of Mathematics, University of Science and Technology,
16111 Algiers, Algeria
aic.batoul@gmail.com, ken.guenda@gmail.com
[2] Department of Electrical and Computer Engineering,
University of Victoria, Victoria, BC V8W 2Y2, Canada
agullive@ece.uvic.ca
[3] Department of Mathematics and Statistics,
Kenyon College, Gambier, OH 43022, USA
aydinn@kenyon.edu

Abstract. In this paper, we study constacyclic codes over finite principal ideal rings. An isomorphism between constacyclic codes and cyclic codes over finite principal ideal rings is given. Further, an open question is partially answered by giving necessary and sufficient conditions for the existence of non-trivial cyclic self-dual codes over finite principal ideal rings. As an example of codes over a finite principal ideal ring, we study constacyclic codes over $R + vR$ where $v^2 = v$ and R is a finite chain ring.

Keywords: Codes over principal ideal rings · Self-dual codes · Cyclic codes · Constacyclic codes

1 Introduction

Although codes over rings are not new, they have attracted significant attention from the research community only since 1994 when Hammons et al. [10] established a fundamental connection between non-linear binary codes and linear codes over $\mathbb{Z}_4$.

The link between self-dual codes and unimodular lattices was given by Bannai et al. [1] and Bonnecaze et al. [5]. These results created a great deal of interest in self-dual codes over a variety of rings (see [15] and the references therein). Dougherty et al. [7,8] used the Chinese Remainder Theorem to generalize the structure of codes over principal ideal rings. They gave conditions on the existence of self-dual codes over principal ideal rings in [8]. Batoul et al. [3] gave conditions on the existence of self-dual and isodual cyclic codes over $\mathbb{F}_q + v\mathbb{F}_q$ where $v^2 = v$. In [2], conditions were given on the existence of cyclic self-dual codes over finite chain rings.

The class of constacyclic codes over rings was introduced as an extension of the class of cyclic codes over rings. Guenda and Gulliver [9] extended the

© Springer International Publishing AG 2017
S. El Hajji et al. (Eds.): C2SI 2017, LNCS 10194, pp. 161–175, 2017.
DOI: 10.1007/978-3-319-55589-8_11

structure of cyclic codes given in [6] to constacyclic codes over principal ideal rings. More recently, constacyclic codes over various commutative rings have been considered. Batoul et al. [2,4] proved that under some conditions several constacyclic codes are equivalent to cyclic codes.

In this paper, we study constacyclic codes over finite principal ideal rings. The Chinese Remainder Theorem is used to reduce the study of these codes to the study of constacyclic codes over finite chain rings. This allows us to provide conditions on the isomorphism between constacyclic codes and cyclic codes over these rings. Further, necessary and sufficient conditions are given on the existence of self-dual cyclic codes over principal ideal rings. Examples are given throughout the paper to illustrate our results, and to show that some recent results in the literature are special cases of the results presented here.

The remainder of this paper is organized as follows. In Sect. 2, we introduce some basic results on Frobenius rings, principal rings, and finite chain rings that will be useful later in the paper. The results in [2,4] are generalized to constacyclic codes over finite principal ideal rings in Sect. 3. In Sect. 4, we give necessary and sufficient conditions on the existence of self-dual cyclic codes over finite principal ideal rings. As an example of codes over a finite principal ideal ring, in Sect. 5 we study constacyclic codes over $R + vR$ where $v^2 = v$ and R is a finite chain ring. Finally, some conclusions are given in Sect. 6.

2   Preliminaries

We assume that all rings are commutative and with identity. For unexplained terminology and more details we refer the reader to [13]. Let R be a finite ring. A code C is a subset of R^n and a linear code over R is an R-submodule of R^n, in which case the code is said to have length n. We attach the standard inner product to the ambient space, i.e. $[u, v] = \sum u_i v_i$. The dual code of C is defined by

$$C^\perp = \{u \in R^n \mid [u, v] = 0 \text{ for all } v \in C\}. \tag{1}$$

We say that a code is self-orthogonal if $C \subseteq C^\perp$, and self-dual if $C = C^\perp$. The Hamming weight of a vector from R^n is the number of nonzero coordinates in the vector. The minimum weight of a code is the smallest Hamming weight of all nonzero codewords in the code. A code $C \subset R^n$ is called a free code if C is a free R-module, that is if C is isomorphic to the R-module R^k for some k.

A finite commutative ring with identity is a principal ideal ring if each proper ideal $I \subset R$ is principal. A code C over a finite principal ideal ring R and its dual satisfy the properties

$$|C||C^\perp| = |R|^n \text{ and } (C^\perp)^\perp = C. \tag{2}$$

2.1   Finite Chain Rings

In this subsection, we summarize some results from [6,14]. A finite chain ring is a finite, commutative, local, principal ideal ring R with unity $1 \neq 0$ whose ideals are ordered by inclusion. Let $\mathfrak{m} = \langle \gamma \rangle$ be the maximal ideal of the finite chain ring R. Then γ is nilpotent with nilpotency index some integer e.

The nilradical of R is $\langle \gamma \rangle$, so all elements of $\langle \gamma \rangle$ are nilpotent. Hence the elements of $R \setminus \langle \gamma \rangle$ are units. Since $\langle \gamma \rangle$ is a maximal ideal, the residue ring $\frac{R}{\langle \gamma \rangle}$ is a field which we denote by K. The natural surjective ring morphism is given by $(-)$ as follows

$$- : R \longrightarrow K$$
$$a \longmapsto \bar{a} = a \bmod \gamma \qquad (3)$$

The set R^* denotes the multiplicative group of units in R.

We define the characteristic of the finite chain ring as the prime number p which is the characteristic of the residue field K of R. Note that this is not the usual definition of the characteristic of a ring.

Let n be a positive integer and q a prime power. We denote by $ord_n(q)$ the multiplicative order of q modulo n, which is the smallest integer nonzero l such that $q^l \equiv 1 \bmod n$.

2.2 Finite Principal Ideal Rings

In this subsection, we recall some of the basic facts about finite principal ideal rings. The smallest $e \geq 1$ such that $I^e = I^{e+1} = \cdots$ in the chain $I \supset I^2 \supset I^3 \supset \cdots$ is called the index of stability of I. If I is nilpotent, then the smallest $e \geq 1$ such that $I^e = 0$ is called the index of nilpotency of I and is the same as the index of stability of I. Note that if R is local with maximal ideal M then we necessarily have $M^e = M^{e+1} = \cdots = 0$. This is not the case for non local rings.

Let $\mathfrak{m}_1, \mathfrak{m}_2, \ldots, \mathfrak{m}_k$ be the maximal ideals of a finite principal ideal ring R with $e_1, \ldots, e_k$ the corresponding indices of stability. Then the ideals $\mathfrak{m}_1^{e_1}, \mathfrak{m}_2^{e_2}, \ldots, \mathfrak{m}_k^{e_k}$ are relatively prime and satisfy

$$\prod_{i=1}^{k} \mathfrak{m}_i^{e_i} = \bigcap_{i=1}^{k} \mathfrak{m}_i^{e_i} = \{0\}.$$

From the ring version of the Chinese Remainder Theorem, the canonical ring homomorphism

$$\Psi : R \longrightarrow \prod_{i=1}^{k} R/\mathfrak{m}_i^{e_i},$$

defined by $x \longmapsto (x + \mathfrak{m}_1^{e_1}, \ldots, x + \mathfrak{m}_k^{e_k})$ is an isomorphism. Denote the local rings $R/\mathfrak{m}_i^{e_i}$ by R_i, $i = 1, \ldots, k$. The maximal ideal of R_i has nilpotency index e_i.

For a code $C \subset R^n$ over R and the maximal ideal $\mathfrak{m}_i$ of R, the $\mathfrak{m}_i$-projection of C is defined by $C_i = \Psi_i(C)$, where $\Psi_i : R^n \longrightarrow R_i^n$ is the canonical map. We extend the map Ψ to R^n as follows

$$\Psi : R^n \longrightarrow \prod_{i=1}^{k} R_i^n,$$

defined by $\Psi(u) = (\Psi_1(u), \ldots, \Psi_k(u))$ for $u \in R^n$.

Using the canonical map defined above the code defined by

$$C = \{\Psi^{-1}(u_1, \ldots, u_k); \ u_i \in C_i, \ i = 1, \ldots, k\}$$
$$= \{u \in R^n; \ \Psi_i(u) \in C_i, \ i = 1, \ldots, k\}$$

is called the Chinese Remainder Theorem product of the codes C_i and denoted by $C = CRT(C_1, \ldots, C_k)$. As a special case, we have for any finite principal ideal ring R

$$R^n = CRT(R_1^n, R_2^n, \ldots, R_k^n),$$

where the R_i are finite chain rings, and $R = \prod_{i=1}^{k} R_i$ or $R = \oplus_{i=1}^{k} R_i$, which is called the canonical decomposition of the finite principal ideal ring or the Chinese Remainder Theorem product of the local components R_i, $1 \leq i \leq k$.

3   Constacyclic Codes over Finite Principal Ideal Rings

This section considers codes over finite commutative rings which are finite principal ideal rings. Let R be a commutative ring with unity. For a given unit $\lambda \in R$, a code C is said to be constacyclic, or more specifically, λ-constacyclic, if $(\lambda c_{n-1}, c_0, c_1, \ldots, c_{n-2}) \in C$ whenever $(c_0, c_1, \ldots, c_{n-1}) \in C$. As special cases, cyclic and negacyclic codes correspond to $\lambda = 1$ and -1, respectively. The main goal of this section is to prove the existence of an isomorphism between constacyclic codes and cyclic codes over finite principal ideal rings. We first recall some results given in [2].

3.1   Constacyclic Codes over Finite Chain Rings

We begin with the following definition.

Definition 1. *Let R be a finite chain ring with residue field $\mathbb{F}_q$. A polynomial $f(x) \in R[x]$ is called basic irreducible if $\overline{f(x)}$ is irreducible in $\overline{R}[x] = \mathbb{F}_q[x]$.*
Two polynomials $f(x)$ and $g(x)$ in $R[x]$ are called coprime if

$$R[x] = \langle f(x) \rangle + \langle g(x) \rangle.$$

Let λ be a unit in a finite chain ring R. If a polynomial $f(x)$ divides $x^n - \lambda$, e.g. $x^n - \lambda = f(x)g(x)$, we refer to $g(x) = \frac{x^n - \lambda}{f(x)}$ as $\hat{f}(x)$.

Theorem 1 *([9, Theorem 4.14]).   Let R be a finite chain ring and C be a λ-constacyclic code over $R[x]$ of length n such that $(n, p) = 1$, where p is the characteristic of $\overline{R}$. Then there exists a unique set of pairwise coprime polynomials $F_0, \ldots, F_e$ in $R[x]$ such that $F_0 \cdots F_e = x^n - \lambda$ and $C = \langle \hat{F}_1, \gamma\hat{F}_2, \ldots, \gamma^{e-1}\hat{F}_e \rangle$, where $\hat{F}_j = \frac{x^n - \lambda}{F_j}$ for $0 < j \leq e$. Moreover, we have that*

$$|C| = |K|^{\sum_{j=0}^{e-1}(e-j) \deg F_{j+1}}, \tag{4}$$

where $R[x]/\langle x^n - \lambda \rangle$ is a principal ideal ring.

In some cases, a constacyclic code is equivalent to a cyclic code as given in the following corollary.

Corollary 1 *([4, Corollary 3.5]). Let R be a finite chain ring and λ, δ units in R such that $\lambda = \delta^n$. A subset I in $R[x]$ is an ideal in $R[x]/\langle x^n - 1 \rangle$ if and only if $\mu(I)$ is an ideal in $R[x]/\langle x^n - \lambda \rangle$.*

3.2   Constacyclic Codes over Finite Principal Ideal Rings

In this section, we generalize the above results to finite principal ideal rings. We fist give some results that will be useful later.

Lemma 1. *Let R be a finite principal ideal ring with canonical decomposition*

$$R = CRT(R_1, R_2, \ldots, R_k)$$

Then any unit $\lambda \in R^*$ is equal to $CRT(\lambda_1, \lambda_2, \ldots, \lambda_k)$, where $\lambda_i \in R_i^*$.

Proof. The proof is a direct consequence of the decomposition $R^* = CRT(R_1^*, R_2^*, \ldots, R_k^*)$. $\qquad\qquad\square$

Lemma 2. *Let R be a finite principal ideal ring and $\prod_{i=1}^{k} R_i$ its direct decomposition, i.e. $R = CRT(R_1, R_2, \ldots, R_k)$. R has units λ and δ such that $\lambda = \delta^n$ if and only if each finite chain ring R_i has units λ_i and δ_i such that $\lambda_i = \delta_i^n$.*

Proof. If there exist units $\lambda_i, \delta_i \in R_i$ such that $\lambda_i = \delta_i^n$ for $1 \leq i \leq k$, then $\lambda = CRT(\lambda_1, \lambda_2, \ldots, \lambda_k)$ and $\delta = CRT(\delta_1, \delta_2, \ldots, \delta_k)$ satisfy $\lambda = \delta^n$. From Lemma 1 we have that δ and λ are units in R. Conversely if R has units λ and δ such that $\lambda = \delta^n$, then from Lemma 1 $\lambda_i = \Psi_i(\lambda) = \Psi_i(\delta^n) = \Psi_i(\delta)^n = \delta_i^n$, hence the result follows. $\qquad\qquad\square$

Theorem 2. *Let R be a finite principal ideal ring, $\prod_{i=1}^{k} R_i$ its direct decomposition, and λ be a unit in R such that $\lambda = CRT(\lambda_1, \lambda_2, \ldots, \lambda_k)$ with $\lambda_i \in R_i^*$. Further, let $C = CRT(C_1, C_2, \ldots, C_k)$ be a code over R of length n with component codes C_i of length n over R_i, $1 \leq i \leq k$. Then C is λ-constacyclic code over R if and only if each C_i is a λ_i-constacyclic code over R_i.*

Proof. For $i \in \{1, \ldots, k\}$, let $\mathbb{F}_{q_i}$ be the residue field of R_i. Define the following ring homomorphism

$$\phi_i : R[x]/\langle x^n - \lambda \rangle \longrightarrow R_i[x]/\langle x^n - \lambda_i \rangle$$
$$a_0 + a_1 x + \cdots a_{n-1} x^{n-1} \longmapsto \psi_i(a_0) + \psi_i(a_1) x + \cdots + \psi_i(a_{n-1}) x^{n-1}$$

so then

$$\phi : R[x]/\langle x^n - \lambda \rangle \longrightarrow R_1[x]/\langle x^n - \lambda_1 \rangle \times R_2[x]/\langle x^n - \lambda_2 \rangle \times \cdots \times R_k[x]/\langle x^n - \lambda_k \rangle,$$

where

$$\phi(f(x)) = (\phi_1(f(x)), \phi_2(f(x)), \cdots, \phi_k(f(x))).$$

If I is an ideal of $R[x]/\langle x^n - \lambda \rangle$, then $\phi_i(I)$ is an ideal of $R_i[x]/\langle x^n - \lambda_i \rangle$. Conversely, for ideals I_i in $R_i[x]/\langle x^n - \lambda_i \rangle$ define

$$\phi^{-1}(I_1, I_2, \ldots, I_k) = I = CRT(I_1, I_2, \ldots, I_k),$$

which is an ideal in $R[x]/\langle x^n - \lambda \rangle$. Associating the λ-constacyclic codes with their corresponding ideals, we have that

$$CRT(C_1, C_2, \ldots, C_k),$$

is a λ-constacyclic code over R if and only if each C_i is a λ_i-constacyclic code over R_i. $\square$

Corollary 2. *With the above assumptions $R[x]/\langle x^n - \lambda \rangle$ is a principal ideal ring if and only if $R_i[x]/\langle x^n - \lambda_i \rangle$ is a principal ideal ring for all $1 \leq i \leq k$.*

Proof. Let C be the λ-constacyclic code of length n over R generated by $f(x) \in R[x]/\langle x^n - \lambda \rangle$. Since $C = CRT(C_1, C_2, \ldots, C_k)$, from Theorem 2 C_i is generated by $\phi_i(f(x))$ which is a polynomial in $R_i[x]/\langle x^n - \lambda_i \rangle$, so C_i is principal. Conversely, let C_i be a cyclic code of length n over R_i generated by $f_i(x) \in R_i[x]/\langle x^n - \lambda_i \rangle$, and let $f(x) \in R[x]/\langle x^n - \lambda \rangle$ be such that $f(x) = \phi^{-1}(f_1(x), f_2(x), \cdots, f_k(x))$. Since ϕ is a ring isomorphism, $f(x)$ is unique. If D the cyclic code generated by $f(x)$, then $D = CRT(C_1, C_2, \ldots, C_k)$. By the Chinese Remainder Theorem $CRT(C_1, C_2, \ldots, C_k)$ is unique, and thus $C = D$. $\square$

Example 1. *Let $\mathbb{F}_p$ be the finite field of order p and $R = \mathbb{F}_p[x]/\langle v^2 - v \rangle = \mathbb{F}_p + v\mathbb{F}_p$. Since $\langle v \rangle$ and $\langle 1 - v \rangle$ are the only maximal ideals of index of stability 1, then $R = \mathbb{F}_p/\langle v \rangle \oplus \mathbb{F}_p/\langle 1 - v \rangle \simeq \mathbb{F}_p \times \mathbb{F}_p$ is the direct decomposition of R. Note that any element c of R^n can be expressed as $c = a + vb = v(a + b) + (1 - v)a$ where $a, b \in \mathbb{F}_p^n$. Now let*

$$\begin{aligned} \psi : R^n &\longrightarrow \mathbb{F}_p^n \times \mathbb{F}_p^n \\ a + bv &\mapsto \psi(a + bv) = (\psi_1(a + bv), \psi_2(a + bv)) = (a + b, a), \end{aligned} \tag{5}$$

be the canonical R-module isomorphism. For $i = 1, 2$, let C_i be a code over $\mathbb{F}_p$ of length n and let

$$C = CRT(C_1, C_2) = \psi^{-1}(C_1 \times C_2) = \{\psi^{-1}(\mathbf{v}_1, \mathbf{v}_2) \mid \mathbf{v}_1 \in C_1, \mathbf{v}_2 \in C_2\}.$$

Then C is the Chinese product of codes C_1 and C_2. By Theorem 2, C is a λ-constacyclic code over R if and only if each C_i is a λ_i-constacyclic code over $\mathbb{F}_p$ with $\lambda = CRT(\lambda_1, \lambda_2)$. Let $\lambda = 1 - 2v = -v + (1 - v)$ so that $\lambda = CRT(-1, 1)$. Then any $(1 - 2v)$-constacyclic code C over R has the form $C = CRT(C_1, C_2)$ where C_1 is a negacyclic code over $\mathbb{F}_p$ and C_2 is a cyclic code over $\mathbb{F}_p$.

These codes have also been studied in [3, 16].

    We now generalize the results given above for finite chain rings to finite principal ideals rings.

Proposition 1. *Let n be a positive integer, and $\lambda = CRT(\lambda_1, \ldots, \lambda_k)$ and $\delta = CRT(\delta_1, \ldots, \delta_k)$ be units such that $\lambda = \delta^n$. Then the map μ defined as*

$$\mu : R[x]/\langle x^n - 1 \rangle \longrightarrow R[x]/\langle x^n - \lambda \rangle$$
$$c(x) \qquad \mapsto \mu(c(x)) = (c(\delta_1^{-1}x), \ldots, c(\delta_k^{-1}x)), \qquad (6)$$

is a ring isomorphism.

Proof. Since $R[x]/\langle x^n - 1 \rangle \simeq \Pi_{i=1}^k R_i[x]/\langle x^n - 1 \rangle$, by Lemma 2 we deduce that $\lambda = \delta^n \iff \lambda_i = \delta_i^n, \forall i \in \{1, \ldots, k\}$. Then by Corollary 1, $\Pi_{i=1}^k R_i[x]/\langle x^n - 1 \rangle \simeq \Pi_{i=1}^k R_i[x]/\langle x^n - \lambda_i \rangle \ \forall i \in \{1, \ldots, k\}$, and so $R[x]/\langle x^n - \lambda \rangle \simeq \Pi_{i=1}^k R_i[x]/\langle x^n - \lambda_i \rangle$. $\square$

If $(n, q_i) = 1$ for all $i \in \{1, \ldots, k\}$ with $\mathbb{F}_{q_i}$ the residue field of the finite chain ring R_i, then $R_i[x]/\langle x^n - \lambda_i \rangle$ is a principal ideal ring. Therefore the ideals in $R[x]/\langle x^n - \lambda \rangle$ are principal ideals, so the following result is a straightforward consequence of Corollary 1.

Corollary 3. *Let R be a finite principal ideal ring and λ, δ be units in R such that $\lambda = \delta^n$. A subset I in $R[x]$ is an ideal in $R[x]/\langle x^n - 1 \rangle$ if and only if $\mu(I)$ is an ideal in $R[x]/\langle x^n - \lambda \rangle$. Equivalently, the subset is a cyclic code C of length n over R if and only if $\mu(C)$ is a λ-constacyclic code of length n over R.*

Example 2. *Let $R = \mathbb{F}_p[x]/\langle v^2 - v \rangle \simeq \mathbb{F}_p + v\mathbb{F}_p$ and n be an odd integer. From Proposition 1, any $(1 - 2v)$-constacyclic code over R is isomorphic to a cyclic code over R.*

These codes have also been studied in [3, 16].

4 Self-dual Cyclic Codes over Finite Principal Ideal Rings

Since any finite principal ideal ring is a direct product of finite chain rings, we start by giving some results on the latter.

4.1 Cyclic Self-dual Codes over Finite Chain Rings

Here we consider cyclic self-dual codes over finite chain rings. For a polynomial $f(x)$ of degree r, let $f^*(x)$ denote its reciprocal polynomial $x^r f(x^{-1})$. The following lemma is easy to obtain.

Lemma 3. *Let $f(x)$ and $g(x)$ be polynomials in $R[x]$ with $\deg f(x) \geq \deg g(x)$ and with constants terms are units. Then the following holds.*

(i) $[f(x)g(x)]^* = f(x)^* g(x)^*$.
(i) $[f(x) + g(x)]^* = f^*(x) + x^{\deg f - \deg g} g^*(x)$.
(ii) *If $f(x)$ is monic, then $\overline{f^*(x)} = \overline{f(x)}^*$.*

The following theorem gives the structure of the dual of a cyclic code over a finite chain ring.

Theorem 3. *([6, Theorem 3.8]). Let R be a finite chain ring with characteristic p, maximal ideal γ, and index of nilpotency e. Let n be an integer such that $(p, n) = 1$ and $f_1 f_2 \ldots f_l$ be the representation of $x^n - 1$ as a product of basic irreducible pairwise coprime polynomials in $R[x]$. If C is a cyclic code of length n over R, then $C^\perp = \langle \hat{F}_0^*, \gamma \hat{F}_e^*, \ldots, \gamma^{e-1} \hat{F}_2^* \rangle$ where $F_0, F_1, \ldots, F_{e-1}$ are pairwise coprime polynomials which are divisors of $x^n - 1$ as given in Theorem 1.*

Theorem 4. *([6, Theorem 4.3]). Let R be a finite chain ring with even index of nilpotency e and maximal ideal γ. Then there exists a non-trivial self-dual cyclic code over R if and only if there exists a basic irreducible factor $f(x) \in R[x]$ of $x^n - 1$ such that $f(x)$ and $f^*(x)$ are not associate.*

The following theorem was first given first by Kanwar and López-Permouth [11] and later by Dinh and López-Permouth [6], but with an incorrect proof, so it is given here with a new proof. In [6,11], it was stated that all cyclotomic coset modulo n must be non-reversible to have $(p^r)^i \equiv -1 \bmod n$ for a positive integer i. However, only C_1, the cyclotomic coset containing 1, needs to be non-reversible. Denote by C_i the cyclotomic coset $\bmod\, n$ that contains i. First, we give a lemma that will be used in proving the theorem.

Lemma 4.

If C_1 is reversible then C_j is reversible $\forall j \in \mathbb{Z}_n$.

Proof. If C_1 is reversible, then there exists a k, $1 \le k \le ord_n(q)$, such that $q^k \equiv -1 \bmod n$. This means that $jq^k \equiv -j \bmod n$, and hence $C_j = C_{-j}$.   □

Theorem 5. *Let R be a finite chain ring with maximal ideal γ, even index of nilpotency e, and residue field K where $|R| = p^{er}$ and $|K| = p^r$. Then non-trivial cyclic self-dual codes of length n over R exist if and only if $(p^r)^i \not\equiv -1 \bmod n$ for all positive integers i.*

Proof. Let $f(x)$ be a monic basic irreducible polynomial which divides $x^n - 1$. Then $\overline{f(x)}$ is a minimal irreducible polynomial over $K = \mathbb{F}_{p^r}[x]$. Hence there exists a cyclotomic cosest C_u associated with $\overline{f(x)}$, and therefore $\overline{f(x)} = \prod_{i \in C_u}(x - \alpha^i)$, where α is a primitive nth root of unity. The reciprocal polynomial of $\overline{f(x)}$ is the polynomial $\overline{f(x)}^* = (\prod_{i \in C_u}(x - \alpha^i))^* = x^r \prod_{i \in C_u}(x^{-1} - \alpha^i) = \prod_{i \in C_{n-u}}(x - \alpha^i)$. By Lemma 3, we have that $\overline{f^*(x)} = \overline{f(x)}^*$. Then by Theorems 3 and 4, a non-trivial cyclic self-dual code exists if and only if there is a basic irreducible polynomial $f(x)$ which is a factor of $x^n - 1$ such that $f(x)$ and $f^*(x)$ are not associate. We show that this can occur if and only if $(p^r)^i \not\equiv -1 \bmod n$ for all positive integers i.

Let $\bar{f}(x) \in \mathbb{F}_{p^r}[x]$ be irreducible and $f(x)|(x^n - 1)$. Then $\bar{f}(x) = \prod_{i \in C_u}(x - \alpha^i)$ where C_u is the cyclotomic coset of n that contains u (and u is the smallest element in its class), and α is a primitive n-th root of unity. Now if $(p^r)^i \not\equiv -1 \bmod n$ for all positive integers i, then $C_1 \neq C_{-1}$. Hence $f(x) \neq f^*(x)$ where $\bar{f}(x) = \prod_{i \in C_1}(x - \alpha^i)$, and the code $(f(x)g(x), \gamma^{\frac{e}{2}} f(x)f^*(x))$ is a non-trivial self-dual code where $f(x)f^*(x)g(x) = x^n - 1$. Conversely, if a non-trivial cyclic self-dual code exists then by Theorem 4 there exists a factor $f(x)|(x^n - 1)$ with $f(x) \neq f^*(x)$. Hence $C_u \neq C_{-u}$, and then by Lemma 4 $C_1 \neq C_{-1}$ where $\bar{f}(x) = \prod_{i \in C_u}(x - \alpha^i)$. Therefore $(p^r)^i \not\equiv -1 \bmod n$ for all positive integers i, because otherwise $C_u = C_{-u}$ for all cyclotomic coset, and then $f(x) = f^*(x)$ for any $f(x)|(x^n - 1)$. $\qquad\square$

Lemma 5. *Let n and s be positive integers and q a prime power. Then the following holds.*

(i) *If $q^s \equiv -1 \bmod n$, then $ord_n(q)$ is even.*
(ii) *If n is prime, then $ord_n(q)$ is even if and only if there exists an i such that $q^i \equiv -1 \bmod n$.*

Proof. Part (i) is easy to verify. For Part (ii), assume that $ord_n(q) = 2w$ is even, so that $q^{2w} \equiv 1 \bmod n$, and hence $n|(q^w - 1)(q^w + 1)$. Since n is prime and cannot divide $q^w - 1$ (because of the order), we have $q^w = -1 \bmod n$. The converse follows from Part (i). $\qquad\square$

The following result answers the question posed in [6, p. 1734] by providing a simple criterion for the existence of cyclic self-dual codes.

Theorem 6. *Let R be a finite chain ring with maximal ideal γ, even index of nilpotency e, and $|R| = p^{er}$ where $|K| = p^r$. If n is an odd prime power, then there exists a non-trivial cyclic self-dual code of length n over R if and only if $ord_n(p^r)$ is odd.*

Proof. If there are no non-trivial self-dual codes, then by Theorem 5 there exists an integer i such that $(p^r)^i \equiv -1 \bmod n$. Then by Part (i) of Lemma 5, we have that $ord_n(p^r)$ is even.

Conversely, assume that there exists a non-trivial cyclic self-dual code. Then from Theorem 5 there is no integer i such that $p^{ri} \equiv -1 \bmod n$. We need to show that in this case $ord_n(p^r)$ is odd. For this, consider the following cases.

(i) If n is an odd prime, then by Part (ii) of Lemma 5, we have that $ord_n(p^r)$ is odd.

(ii) For $n = q^\alpha$, assume that $ord_{q^\alpha}(p^r)$ is even. We first prove the implication

$$ord_{q^\alpha}(p^r) \text{ is even} \Rightarrow ord_q(p^r) \text{ is even.}$$

If $ord_{q^\alpha}(p^r)$ is even and $ord_q(p^r)$ is odd, then there exist odd $i > 0$ such that $p^{ri} \equiv 1 \bmod q \Leftrightarrow p^{ri} = 1 + kq$. Hence $p^{riq^{\alpha-1}} = (1 + kq)^{q^{\alpha-1}} \equiv 1 \bmod q^\alpha$, because $(1 + kq)^{q^{\alpha-1}} \equiv 1 + kq^\alpha \bmod q^{(\alpha+1)}$, and so

$$p^{riq^{\alpha-1}} \equiv 1 \bmod q^\alpha. \qquad (7)$$

If i is odd and $q^{\alpha-1}$ is odd, then $ord_{q^\alpha}(p^r)$ is odd (because $ord_{q^\alpha}(p^r)|iq^{\alpha-1}$), which is a contradiction. Hence $ord_q(p^r)$ is even, so there exists some integer j such that $0 < j < ord_q(p^r)$, and $p^{rj} \equiv -1 \bmod q$. Then from (7), we have that $p^{rjq^{\alpha-1}} \equiv -1 \bmod q^\alpha$. This gives that the cyclotomic coset C_1 is reversible, which by Theorem 5 is impossible. $\qquad\square$

Remark 1. *If n is not a prime power then the condition $ord_n(p^r)$ being odd is a sufficient condition for the existence of a self-dual code of length n over R.*

For the remainder of the paper, the notation $q = \square \bmod n$ means that q is a quadratic residue modulo n.

Corollary 4 *[2, Corollary 4.8]. Let R be a finite chain ring with maximal ideal γ, even index of nilpotency e, and residue field K such that $|K| = p^r$. Then if $p_1 \ldots p_s$ is the prime factorization of an odd integer n such that $p^r = \square \bmod p_i$ and $p_i \equiv -1 \bmod 4$ for $1 \leq i \leq s$, then there exists a non-trivial cyclic self-dual code of length n over R.*

Corollary 5 *[2, Corollary 4.9]. With the previous notation, if n is an odd prime such that $n \equiv -1 \bmod 4$, then there exists a cyclic self-dual code if and only if $p = \square \bmod n$.*

For a cyclic code of length n with $(n, p) = 1$, we have the following results for the free codes. We will generalizes theses results to the finite principal ideal ring in the next Section.

Theorem 7 *([9, Theorem 4.20]). Let C be a cyclic code of length n over a finite chain ring R with characteristic p such that $(p, n) = 1$. Then C is a free cyclic code with rank k if and only if there is a polynomial $f(x) \in R[x]$ such that $f(x)|(x^n - 1)$ and $f(x)$ generates C. In this case, we have $k = n - \deg(f(x))$.*

Theorem 8 *([2, Theorem 4.13]). Let R be a finite chain ring with maximal ideal $\langle \gamma \rangle$, index of nilpotency e, and characteristic p. Then if p is odd and $(p, n) = 1$, there is no free cyclic self-dual code of length n over R.*

4.2 Self-dual Cyclic Codes over Finite Principal Ideal Rings

Let R be a finite principal ideal ring and $(R_i)_{i=1}^k$ be a direct decomposition of R. Further, let $\Psi : R^n \to \prod_{i=1}^k R_i^n$ be the canonical R-module isomorphism. For $i = 1, \ldots, k$, let C_i be a code over R_i of length n and

$$C = CRT(C_1, C_2, \ldots, C_k) = \Psi^{-1}(C_1 \times \cdots \times C_k) = \{\Psi^{-1}(\mathbf{v}_1, \mathbf{v}_2, \ldots, \mathbf{v}_k); \mathbf{v}_i \in C_i\}.$$

Theorem 9. *With the above notation, We have the following.*

(i) C is a cyclic code if and only if each C_i is a cyclic code.
(ii) $C_1, C_2, \ldots, C_k$ are self-dual codes if and only if C is a self-dual code.

Proof. Part (i) is a particular case of Theorem 2, and Part (ii) follows from the identity

$$CRT(C_1, C_2, \ldots, C_k)^\perp = CRT(C_1^\perp, C_2^\perp, \ldots, C_k^\perp).$$

$\square$

The results of Theorem 9 allow us to generalize some results on finite chain rings to finite principal ideal rings, as given by the next result:

Theorem 10. *Let $R \simeq \prod_{i=1}^{k} R/\mathfrak{m}_i^{t_i} = \prod_{i=1}^{k} R_i$, be a finite principal ideal ring, $\mathbb{F}_{q_i}$ the residue field of R_i, $1 \leq i \leq k$, n be an odd prime power, and C a cyclic code of length n over R. Then C is a self-dual code if and only if $ord_n(q_i)$ is odd for $1 \leq i \leq k$.*

Proof. Let n a power of an odd prime such that $(n, q_i) = 1$ and $C = CRT(C_1, C_2, \ldots, C_k)$ be a cyclic self-dual code over R. Then by Theorem 9, C_i is a cyclic self-dual code over R_i for all $1 \leq i \leq k$, and by Theorem 6 $ord_n(q_i)$ is odd. On the other hand, if $ord_n(q_i)$ is odd then there exists a cyclic self-dual code C_i over R_i for all $1 \leq i \leq k$. Then by Theorem 2, the cyclic code $C = CRT(C_1, C_2, \ldots, C_k)$ is a self-dual cyclic code over R. $\qquad\square$

We now generalize Corollary 4 to finite principal ideal rings.

Corollary 6. *Let $R \simeq \prod_{i=1}^{k} R/\mathfrak{m}_i^{t_i} = \prod_{i=1}^{k} R_i$ be a finite principal ideal ring, $\mathbb{F}_{q_i}$ the residue field of R_i, and n an integer such that $(n, q_i) = 1$ for $1 \leq i \leq k$. Then if $p_1 \ldots p_s$ is the prime factorization of an odd integer n such that $q_i = \square \bmod p_j$ and $p_j \equiv -1 \bmod 4$ for $1 \leq j \leq s$, then there exists a non-trivial cyclic self-dual code of length n over R.*

Proof. Let $n = p_1 \ldots p_s$ be such that $q_i = \square \bmod p_j$, and $p_j \equiv -1 \bmod 4$ for $1 \leq j \leq s$. By Corollary 4, there exists a non-trivial cyclic self-dual code C_i over R_i. Then by Theorem 2, the cyclic code $C = CRT(C_1, C_2, \ldots, C_k)$ is a self-dual code over R. $\qquad\square$

In the following, we generalize Corollary 5 to finite principal ideal rings.

Corollary 7. *With the previous notation, if n is an odd prime such that $n \equiv -1 \bmod 4$, then there exists a cyclic self-dual code over R if and only if $p_j = \square \bmod n$, where $q_j = p_j^r$.*

Proof. Let n be an odd prime such that $n \equiv -1 \bmod 4$. If $p_j = \square \bmod n$, then by Corollary 5 there exist a self-dual cyclic code C_j of length n over R_j. Thus, by Theorem 2 the cyclic code $C = CRT(C_1, C_2, \ldots, C_k)$ is a self-dual cyclic code over R. $\qquad\square$

In the following, we generalize Theorem 8 to finite principal ideal rings.

Theorem 11. *Let $R \simeq \prod_{i=1}^{k} R/\mathfrak{m}_i^{t_i} = \prod_{i=1}^{k} R_i$ be a finite principal ideal ring, $\mathbb{F}_{q_i}$ the residue field of R_i, n an integer such that $(n, q_i) = 1$ for $1 \leq i \leq k$, and $C = CRT(C_1, C_2, \ldots, C_k)$ a cyclic code over R. If there exists $i \in \{1, \ldots, k\}$ such that q_i is odd and C_i is free, then C is not self-dual.*

Proof. Let $C = CRT(C_1, C_2, \ldots, C_k)$ be a cyclic code of length n over R such that $(n, q_i) = 1$ for $1 \leq i \leq k$. Then by Theorem 8, if q_i is odd and C_i is free, C_i cannot be self-dual, so by Theorem 2 C cannot be a self-dual cyclic code of length n over R. $\qquad\square$

4.3 Cyclic Codes over Finite Principal Ideal Rings with Odd Index of Stability

In this section, we prove that there are no simple root cyclic self-dual codes over finite chain rings when the nilpotency index of the generator of the maximal ideal is odd. This result is generalized to finite principal ideal rings when the index of stability of the generator of one of the maximal ideals is odd.

Theorem 12. *Let R be a finite chain ring where $\langle \gamma \rangle$ is the maximal ideal with nilpotency index e. If e is odd and q is a prime power, then there are no nontrivial self-dual cyclic codes of length n over R such that $(n, q) = 1$.*

Proof. If $q = 2^k$, then $(n, q) = 1$ and so n must be odd. Let C be a nontrivial cyclic code of length n over R so that there exist monic and coprime polynomials $F_0, F_1, \ldots, F_{e-1}, F_e$ such that $x^n - 1 = F_0 F_1 \ldots F_{e-1} F_e$ and $C = \langle \hat{F}_1, \gamma \hat{F}_2, \ldots, \gamma^{e-1} \hat{F}_e \rangle$. If C is self-dual, then from [6, Proposition 4.1] F_i is associate with F_j for $i, j \in \{0, 1, \ldots e\}$ and $i + j \equiv 1 \bmod (e+1)$. Then $F_i = \epsilon F_j^*$ for all $i, j \in \{0, \ldots, e\}$, $i + j \equiv 1 \bmod (e+1)$, and ϵ a unit in R. Then $F_i \neq F_j^*$ since e is odd and so it cannot be that $i + i \equiv e + 2$, and therefore

$$x^n - 1 = F_0 F_0^* F_2 F_2^* F_3 F_3^* \ldots F_{\frac{e+1}{2}} F_{\frac{e+1}{2}}^*.$$

Thus none of the F_i are self-reciprocal. The polynomial $(x - 1)$ is a factor of $x^n - 1$, so there is an $0 \leq i_0 \leq e$ such that $F_{i_0} = (x - 1)g(x)$ for some polynomial $g(x)$. Hence

$$F_{i_0}^* = (x - 1)^* g(x)^* = (x - 1)g(x)^* = F_{1 - i_0 \bmod (1+e)},$$

which is impossible since the F_i are coprime for all $0 \leq i \leq e$, and $x^n - 1$ has no repeated roots since $(n, q) = 1$.   □

Theorem 13. *Let $R \simeq \prod_{i=1}^{k} R/\mathfrak{m}_i^{t_i}$ be a finite principal ideal ring, and C be a cyclic code over R. Then if one of the t_i is odd, C cannot be a self-dual code.*

Proof. By Theorem 2, C is cyclic and self-dual if and only if C_i, $1 \leq i \leq k$, is also cyclic and self-dual. However from Theorem 12, if there exists an i such that t_i is odd, then C_i cannot be self-dual.   □

5 Constacyclic Codes over $R + vR$

Let R be a finite commutative chain ring with maximal ideal $\langle \gamma \rangle$, nilpotency index e and residue field $\mathbb{F}_q$. Further, let $R + vR = \{a + vb : a, b \in R\}$ with $v^2 = v$. This ring is an example of a finite commutative principal ideal ring, and has two coprime ideals $\langle v \rangle = \{av : a \in R\}$ and $\langle 1 - v \rangle = \{a(1 - v) : a \in R\}$ with index of stability 1. Both $R_1 = R/\langle v \rangle$ and $R_2 = R/\langle 1 - v \rangle$ are isomorphic to R. By the Chinese Remainder Theorem, we have that $R + vR \simeq R_1 \times R_2 \simeq \langle v \rangle \oplus \langle 1 - v \rangle$. The motivation for considering this ring as a specific example is that the elements v

and $1-v$ are nilpotent elements such that $v+1-v = 1$, so that by [7, Proposition 2.4], any submodule N of a module M over $R + vR$ is a direct decomposition of $N_1 \oplus N_2$ where $N_1 = vN$ and $N_2 = (1-v)N$. In particular, for a positive integer n, $(R + vR)^n = v(R + vR)^n \oplus (1-v)(R+vR)^n$. Since $R + vR \simeq \langle v \rangle \oplus \langle 1-v \rangle$, let $x_i \in R + vR$ be such that $x_i = a_iv + b_i(1-v)$, $a_i, b_i \in R$. Then

$$x = (x_1, x_2, \ldots, x_n) = (a_1v+b_1(1-v), a_2v+b_2(1-v), \ldots, a_nv+b_n(1-v)) \in (R+vR)^n,$$

and

$$x = v(a_1, a_2, \ldots, a_n) + (1-v)(b_1, b_2, \ldots, b_n) \in vR^n \oplus (1-v)R^n,$$

so that $(R + vR)^n = vR^n \oplus (1-v)R^n$.

Let C be a code of length n over $R+vR$. Since C is a submodule of $(R+vR)^n$ over $R + vR$ such that

$$C = CRT(C_1, C_2) = \Psi^{-1}(C_1, C_2) = \{\Psi^{-1}(\mathbf{v}_1, \mathbf{v}_2) \mid \mathbf{v}_1 \in C_1 \ \mathbf{v}_2 \in C_2\}, \quad (8)$$

where C_1 and C_2 are codes of length n over R and the idempotent elements v and $1-v$. Since $1 + 1 - v = 1$, then $C = vC \oplus (1-v)C \simeq C_1 \times C_2$, which shows that $vC \simeq vC_1$ and $(1-v)C \simeq (1-v)C_2$. The next result is then a particular case of Theorem 2.

Theorem 14. *Let $\lambda = CRT(\lambda_1, \lambda_2) = \lambda_1v + \lambda_2(1-v)$ be a unit in $R + vR$ such that λ_1, λ_2 are units in R. Further, let C be a linear code of length n over $R + vR$. Then C is a λ-constacyclic code over $R + vR$ if and only if C_1 is a λ_1-constacyclic code of length n over R and C_2 is a λ_2-constacyclic code of length n over R.*

Example 3. *Let $\lambda = 1 - 2v = -v + (1-v)$ so that $\lambda = CRT(-1, 1)$. By Theorem 2, any $(1-2v)$-constacyclic code C over $R+vR$ is the Chinese product of a negacyclic code C_1 over R and a cyclic code C_2 over R such that $C = CRT(C_1, C_2)$.*

These codes have also been studied in [12].

In the following we give the structure of the generator polynomial of a constacyclic code over $R + vR$.

Theorem 15. *Let R be a finite commutative chain ring with maximal ideal $\langle \gamma \rangle$, nilpotency index e, and residue field $\mathbb{F}_q$. Further, let n be a positive integer such that $(n, q) = 1$, and $\lambda = \lambda_1v + \lambda_2(1-v)$ be a unit in $R + vR$ such that λ_1, λ_2 are units in R. If $C = CRT(C_1, C_2)$ is a λ-constacyclic code of length n over $R + vR$, then there are polynomials $f_1(x), f_2(x) \in R[x]$ such that $C = \langle vf_1(x), (1-v)f_2(x) \rangle$, where $C_1 = \langle f_1(x) \rangle \subseteq R[x]/\langle x^n - \lambda_1 \rangle$ and $C_2 = \langle f_2(x) \rangle \subseteq R[x]/\langle x^n - \lambda_2 \rangle$.*

Proof. Since $(n, q) = 1$, by Theorem 1 $R[x]/\langle x^n - \lambda_1 \rangle$ and $R[x]/\langle x^n - \lambda_2 \rangle$ are both principal ideal rings, so there exist polynomials $f_1(x), f_2(x) \in R[x]$ such that $C_1 = \langle f_1(x) \rangle \subseteq R[x]/\langle x^n - \lambda_1 \rangle$ and $C_2 = \langle f_2(x) \rangle \subseteq R[x]/\langle x^n - \lambda_2 \rangle$. For any $c(x) \in C$ there exist polynomials $c_1(x), c_2(x) \in R[x]$ such that $c(x) = vc_1(x) + (1-v)c_2(x)$ with $c_1(x) \in C_1$ and $c_2(x) \in C_2$. Then there are polynomials $k_1(x), k_2(x) \in R[x]$ such that

$$c_1(x) = k_1(x)f_1(x) \bmod (x^n - \lambda_1),$$
$$c_2(x) = k_2(x)f_2(x) \bmod (x^n - \lambda_2).$$

Thus, there are $r_1(x), r_2(x) \in R[x]$ such that $c_1(x) = k_1(x)f_1(x) + r_1(x)(x^n - \lambda_1)$ and $c_2(x) = k_2(x)f_2(x) + r_2(x)(x^n - \lambda_2)$.

Since $v(x^n - \lambda) = v(x^n - \lambda_1)$ and $(1-v)(x^n - \lambda) = (1-v)(x^n - \lambda_2)$, we have that

$$
\begin{aligned}
c(x) &= vc_1(x) + (1-v)c_2(x) \\
&= v(k_1(x)f_1(x) + r_1(x)(x^n - \lambda_1)) + (1-v)(k_2(x)f_2(x) + r_2(x)(x^n - \lambda_2)) \\
&= vk_1(x)f_1(x) + (1-v)k_2(x)f_2(x) + (vr_1(x) + (1-v)r_2(x))(x^n - \lambda).
\end{aligned}
$$

Hence, $c(x) = vk_1(x)f_1(x) + (1-v)k_2(x)f_2(x) \bmod (x^n - \lambda)$, so $c(x)$ is in $\langle vf_1(x), (1-v)f_2(x) \rangle \subseteq (R + vR)/\langle x^n - \lambda \rangle$. On the other hand, for any $d(x) \in \langle vf_1(x), (1-v)f_2(x) \rangle \subseteq (R + vR)/\langle x^n - \lambda \rangle$, there are polynomials $k_1(x), k_2(x) \in (R + vR)[x]$ such that

$$d(x) = k_1(x)f_1(x)v + k_2(x)f_2(x)(1-v) \bmod (x^n - \lambda).$$

Then there are $r_1(x), r_2(x) \in R[x]$ such that $vk_1(x) = vr_1(x)$ and $(1-v)k_2(x) = (1-v)r_2(x)$. Let $r(x) = vr_1(x) + (1-v)r_2(x)$ and

$$
\begin{aligned}
d(x) &= vd_1(x) + (1-v)d_2(x) \\
&= vf_1(x)r_1(x) + (1-v)f_2(x)r_2(x) + r(x)(x^n - \lambda),
\end{aligned}
$$

so that
$$
\begin{aligned}
vd_1(x) &= v(f_1(x)r_1(x) + r_1(x)(x^n - \lambda_1)), \\
(1-v)d_2(x) &= (1-v)(f_2(x)r_2(x) + r_2(x)(x^n - \lambda_2)).
\end{aligned}
$$

This means that $d_1(x) \in \langle f_1(x) \rangle \subseteq R[x]/\langle x^n - \lambda_1 \rangle$ and $d_2(x) \in \langle f_2(x) \rangle \subseteq R[x]/\langle x^n - \lambda_2 \rangle$. Hence $d_1(x) \in C_1$, $d_2(x) \in C_2$, and $d(x) \in C$, and therefore $\langle vf_1(x), (1-v)f_2(x) \rangle \subseteq C$, so that $C = \langle vf_1(x), (1-v)f_2(x) \rangle$. $\qquad \square$

Theorem 16. *With the above assumptions, let C be a λ-constacyclic over $R + vR$. Then there is a polynomial $f(x) \in (R + vR)[x]$ such that $C = \langle f(x) \rangle$.*

Proof. By Theorem 15, there are polynomials $f_1(x)$ and $f_2(x)$ over $R + vR$ such that $C = \langle vf_1(x), (1-v)f_2(x) \rangle$. Let $f(x) = vf_1(x) + (1-v)f_2(x)$, so then $\langle f(x) \rangle \subseteq C$. We have that

$$
\begin{aligned}
vf(x) &= vf_1(x) \\
(1-v)f(x) &= (1-v)f_2(x),
\end{aligned}
$$

and hence $C = \langle f(x) \rangle$. $\qquad \square$

6 Conclusion

In this paper, an isomorphism between constacyclic codes and cyclic codes over finite principal ideal rings was established. In addition, necessary and sufficient conditions were given for the existence of cyclic self-dual codes over finite principal ideals rings.

References

1. Bannai, E., Dougherty, S.T., Harada, M., Oura, M.: Type II codes, even unimodular lattices and invariant rings. IEEE Trans. Inform. Theory **45**(4), 1194–1205 (1999)
2. Batoul, A., Guenda, K., Gulliver, T.A.: On self-dual cyclic codes over finite chain rings. Des. Codes Cryptogr. **70**(3), 347–358 (2014)
3. Batoul, A., Guenda, K., Kaya, A., Yildiz, B.: Cyclic isodual and formally self-dual codes over $\mathbb{F}_q + v\mathbb{F}_q$. EJPAM **8**(1), 64–80 (2015)
4. Batoul, A., Guenda, K., Gulliver, T.A.: Some constacyclic codes over finite chain rings. Adv. Math Commun. **10**(4), 683–694 (2016)
5. Bonnecaze, A., Solé, P., Calderbank, A.R.: Quaternary quadratic residue codes and unimodular lattices. IEEE Trans. Inform. Theory **41**(2), 366–377 (1995)
6. Dinh, H., López-Permouth, S.R.: Cyclic and negacyclic codes over finite chain rings. IEEE Trans. Inform. Theory **50**(8), 1728–1744 (2004)
7. Dougherty, S.T., Kim, J.-L., Kulosman, H.: MDS codes over finite principal ideal rings. Des. Codes Cryptogr. **50**(1), 77–92 (2009)
8. Dougherty, S.T., Harada, M., Solé, P.: Self-dual codes over rings and the Chinese remainder theorem. Hokkaido Math. J. **28**, 253–283 (1999)
9. Guenda, K., Gulliver, T.A.: MDS and self-dual codes over rings. Finite Fields Appl. **18**(6), 1061–1075 (2011)
10. Hammons, A.R., Kumar, P.V., Calderbank, A.R., Sloane, N.J.A., Solé, P.: The Z_4 linearity of Kerdock, Preparata, Goethals and related codes. IEEE Trans. Inform. Theory **40**(2), 301–319 (1994)
11. Kanwar, P., López-Permouth, S.R.: Cyclc codes over the integers modulo p^m. Finite Fields Appl. **3**(4), 334–352 (1997)
12. Liao, D., Tang, Y.: A class of constacyclic codes over $R + vR$ and its Gray image. Int. J. Commun. Netw. Syst. Sci. **5**, 222–227 (2012)
13. McDonald, B.R.: Finite Rings with Identity. Marcel Dekker, New York (1974)
14. Norton, G.H., Sălăgean, A.: On the structure of linear and cyclic codes over a finite chain ring. Appl. Algebra Eng. Comm. Comput. **10**(6), 489–506 (2000)
15. Rains, E., Sloane, N.J.A.: Self-dual codes. In: Pless, V.S., Human, W.C. (eds.) Handbook of Coding Theory, pp. 177–294. Elsevier, Amsterdam (1998)
16. Zhu, S.X., Wang, L.: A class of constacyclic codes over $\mathbb{F}_p + v\mathbb{F}_p$ and its Gray image. Disc. Math. **311**(9), 2677–2682 (2011)

On Isodual Cyclic Codes over Finite Chain Rings

Aicha Batoul[1]([⊠]), Kenza Guenda[1], T. Aaron Gulliver[2], and Nuh Aydin[3]

[1] Faculty of Mathematics, University of Science and Technology,
16111 Algiers, Algeria
aic.batoul@gmail.com, ken.guenda@gmail.com
[2] Department of Electrical and Computer Engineering,
University of Victoria, Victoria, BC V8W 2Y2, Canada
agullive@ece.uvic.ca
[3] Department of Mathematics and Statistics,
Kenyon College, Gambier, OH 43022, USA
aydinn@kenyon.edu

Abstract. In this work, cyclic isodual codes over finite chain rings are investigated. These codes are monomially equivalent to their duals. Existence results for cyclic isodual codes are given based on the generator polynomials, the field characteristic, and the length. Several constructions of isodual and self-dual codes are also presented.

Keywords: Isodual codes · Self-dual codes · Cyclic codes · Finite chain rings · Codes over rings

1 Introduction

A code which is equivalent to its dual is called an isodual code. Several kinds of equivalence exist [12], but here we only consider monomial equivalence, which is the most important. For some parameters, one can prove that there are no cyclic self-dual codes over finite chain rings [2,6], whereas isodual codes can exist. Isodual codes are important because they are related to lattices. Recently, isodual cyclic codes over finite fields were constructed from duadic codes in [3]. The purpose of this paper is to extend the concept of duadic codes to finite chain rings and to extend the construction of isodual codes [3,4] to finite chain rings. Note that duadic codes over $\mathbb{Z}_4$ were presented by Langevin et al. [13], over $\mathbb{F}_2 + u\mathbb{F}_2$ by Ling et al. [15], over $\mathbb{F}_q + v\mathbb{F}_q$ by Batoul et al. [4] and over $\mathbb{Z}_{2k}$ by Bachoc et al. [1], but the general concept of duadic codes as well as the existence of isodual cyclic codes over general finite rings has not yet been examined.

The remainder of this paper is organized as follows. In Sect. 2, some preliminary results are presented. In Sect. 3, the structure of cyclic codes of length $2^a m$ over finite chain rings is given. In Sect. 4, conditions are given on the existence of isodual cyclic codes over finite chain rings, and several constructions are presented. In Sect. 5, using the lifts of duadic codes over the residue field, we give some constructions of isodual cyclic codes over finite chain rings.

© Springer International Publishing AG 2017
S. El Hajji et al. (Eds.): C2SI 2017, LNCS 10194, pp. 176–194, 2017.
DOI: 10.1007/978-3-319-55589-8_12

2 Preliminaries

In this section, we summarize some necessary results from [7,10,14]. A finite chain ring R is a finite commutative ring with identity $1 \neq 0$ and maximal principal ideal generated by an element $\gamma \in R$. Then γ is nilpotent with nilpotency index some integer e. So ideals of R form the following chain

$$\langle 0 \rangle = \langle \gamma^e \rangle \subsetneq \langle \gamma^{e-1} \rangle \subsetneq \ldots \subsetneq \langle \gamma \rangle \subsetneq R.$$

The residue field of R is $\frac{R}{\langle \gamma \rangle}$ which is denote by K.

The natural surjective ring morphism $(-)$ is given by

$$\begin{aligned} - &: R \longrightarrow K \\ b &\longmapsto \bar{b} = b \bmod \gamma. \end{aligned} \tag{1}$$

Recall that a block code C of length n is called a linear code over a finite chain ring R if it is a submodule of R^n. Here, all codes are assumed to be linear. $C \subset R^n$ is called a free code if C is a free R-module, that is if C is isomorphic to the R-module R^k for some positif integer k. C is said to be cyclic if

$$(c_{n-1}, c_0, \ldots, c_{n-2}) \in C, \text{ whenever } (c_0, c_1, \ldots, c_{n-1}) \in C.$$

We follow the usual convention of representing vectors as polynomials. And with this representation, it is well known that every cyclic code is view as an ideal of the finite ring $R[x]/(x^n - 1)$. Then C is a free cyclic code with rank k over R with characteristic p such that $(p, n) = 1$ if and only if there is a polynomial $f(x)$ in $R[x]$ which divides $(x^n - 1)$ and generates C.

We attach the standard inner product to R^n

$$[v, w] = \sum v_i w_i; \text{ for } v = (v_0, v_1, \ldots, v_{n-1}), w = (w_0, w_1, \ldots, w_{n-1}) \in R^n.$$

The dual code $C^\perp$ of C is defined as

$$C^\perp = \{v \in R^n \mid [v, w] = 0 \text{ for all } w \in C\}. \tag{2}$$

If $C \subseteq C^\perp$, the code is said to be self-orthogonal and if $C = C^\perp$, the code is self-dual.

In this paper, the notation $q = \square \bmod n$ means that q is a quadratic residue modulo n. For a prime power q and integer n such that $\gcd(q, n) = 1$, we denote by $ord_n(q)$ the multiplicative order of q modulo n. This is the smallest integer l such that $q^l \equiv 1 \bmod n$.

The function μ_b defined on $\mathbb{Z}_n = \{0, 1, \ldots, n - 1\}$ by $\mu_b(i) \equiv ib \bmod n$ is a permutation of the coordinate positions $\{0, 1, 2, \ldots, n - 1\}$ and is called a multiplier. Multipliers also act on polynomials in $R[x]$ and this gives the following ring automorphism.

$$\begin{aligned} \mu_b &: R[x]/(x^n - 1) \longrightarrow R[x]/(x^n - 1) \\ f(x) &\longmapsto \mu_b(f(x)) = f(x^b). \end{aligned} \tag{3}$$

Suppose that $f(x) = a_0 + a_1x + \ldots + a_rx^r$ is a polynomial of degree r with $f(0) = a_0$ a unit in R. Then the monic reciprocal polynomial of $f(x)$ is

$$f^*(x) = f(0)^{-1}x^r f(x^{-1}) = f(0)^{-1}x^r(\mu_{-1}(f(x))) = a_0^{-1}(a_r + a_{r-1}x + \ldots + a_0x^r).$$

If a polynomial is equal to its reciprocal, then it is called a self-reciprocal polynomial over R.

The following lemma is easily deduced.

Lemma 1. *Let $f(x)$ and $g(x)$ be two polynomials in $R[x]$ with $\deg f(x) \geq \deg g(x)$ and with constants terms are units. Then the following holds.*

(i) $[f(x)g(x)]^* = f(x)^*g(x)^*$.
(ii) $[f(x) + g(x)]^* = f(x)^* + x^{\deg f - \deg g}g(x)^*$.
(iii) *If $f(x)$ is monic, then $\overline{f(x)^*} = \overline{f(x)}^*$.*

The following theorem gives the structure of a cyclic code (not necessarily free) and its dual over a finite chain ring.

Theorem 1 *([7]).* *Let R be a finite chain ring with maximal ideal γ and index of nilpotency e. Let C be a cyclic code over R of length n such that $(n, p) = 1$, where p is the characteristic of $\overline{R}$. Then there exists a unique family of pairwise coprime polynomials $F_i, 0 \leq i \leq e$ in $R[x]$ such that $F_0 \ldots F_e = x^n - 1$,*

$$C = \langle \hat{F}_1, \gamma\hat{F}_2, \ldots, \gamma^{e-1}\hat{F}_e \rangle \text{ and } C^\perp = \langle \hat{F}_0^*, \gamma\hat{F}_e^*, \ldots, \gamma^{e-1}\hat{F}_2^* \rangle,$$

where $\hat{F}_j = \frac{x^n - 1}{F_j}$ for $0 < j \leq e$. Moreover, we have that the ring $R[x]/(x^n - 1)$ is a principal ideal ring.

2.1   Isometries and Monomial Maps

Let $R^* = R\backslash\langle\gamma\rangle$. A monomial transformation over R^n is an R-linear homomorphism τ such that there exist units $\lambda_1, \ldots, \lambda_n$ in R^*, and a permutation $\sigma \in S_n$ such that for all $(x_1, x_2, \ldots, x_n) \in R^n$, we have $\tau(x_1, \ldots, x_n) = (\lambda_1 x_{\sigma(1)}, \lambda_2 x_{\sigma(2)}, \ldots, \lambda_n x_{\sigma(n)})$. Two linear codes C and C' of length n are called monomially equivalent if there exists a monomial transformation over R^n such that $\tau(C) = C'$. Wood [16] proved that there exists a monomial permutation between two codes over a finite chain ring if and only if there exists a linear Hamming isometry.

Several weights can be defined over rings. A weight on a code C over a finite chain ring is called homogeneous if it satisfies the following conditions:

(i) $\forall x \in C, \forall u \in R^* : w(x) = w(ux)$, and
(ii) there exists a constant $\xi = \xi(w) \in \mathbb{R}$ such that

$$\sum_{x \in U} w(x) = \xi|U|, and$$

where U is any subcode of C.

Honold and Nechaev [11] proved that for codes over a finite chain ring there exists a homogeneous weight. A linear morphism $f : R \longmapsto R$ is called a homogeneous isometry if it is a linear homomorphism which preserves the homogeneous weight.

Lemma 2 *([8]). Let R be a finite chain ring, C a linear code over R and $\phi :$ $C \longmapsto R^n$ an embedding. Then the following are equivalent:*

(i) ϕ is a homogeneous isometry
(ii) C and $\phi(C)$ are equivalent.

Here whenever two codes are said to be equivalent it is meant that they are monomially equivalent.

2.2   Galois Extensions of Finite Chain Rings

Let R be a finite chain ring with residue field $\mathbb{F}_q$ where $\mathbb{F}_{q^s}$ is the splitting field of $x^n - 1$ over $\mathbb{F}_q$ with $s = ord_n(q)$. Further, let $f(x) \in \mathbb{F}_q[x]$ be a primitive polynomial of degree s. Then since $(q^s - 1, q) = 1$, there exists a unique basic irreducible polynomial $g(x) \in R[x]$ such that $\bar{g}(x) = f(x)$.

Consider the Galois extension of R denoted by $S \simeq \frac{R[x]}{(g(x))}$. Since g is irreducible and square free, S is separable and local. Then from [9, Theorem 4.2] S has a primitive element ξ which is a root of $g(x)$ such that $\bar{\xi} = \alpha$ is a root of $f(x)$ in $\mathbb{F}_{q^s}$. Let $u \in S$ then

$$u = a_0 + a_1\xi + a_2\xi^2 + \cdots + a_{s-1}\xi^{s-1}$$

where ξ is a root of $g(x)$ such that $\bar{\xi} = \alpha$ is a root of $f(x)$ in $\mathbb{F}_{q^s}$. Thus any element $u \in S$ can be written as

$$u = a_0 + a_1\xi + a_2\xi^2 + \cdots + a_{s-1}\xi^{s-1}$$

where $a_i \in R$. The map

$$\begin{aligned} \sigma : S &\longrightarrow S \\ \xi &\longmapsto \sigma(\xi) = \xi^q, \end{aligned} \tag{4}$$

is a generator of $G_R(S)$, the Galois group of S over R, which is isomorphic to $G_{\mathbb{F}_q}(\mathbb{F}_{q^s})$, the Galois group of $\mathbb{F}_{q^s}$ over $\mathbb{F}_q$. Since $G_{\mathbb{F}_q}(\mathbb{F}_{q^s})$ is a cyclic group, the elements of R are fixed by σ and all its powers. Let $\beta = \xi^{\frac{q^s-1}{n}}$, hence the Galois extension S of R contains a primitive n-th root of unity. Denote $\bar{\beta} = \alpha$ so α is a primitive n-th root of unity in $\mathbb{F}_{q^s}$.

Lemma 3. *With the above assumptions, let $p(x) = \Pi_{i \in T}(x - \alpha^i)$ be a monic divisor of $x^n - 1$ in $\mathbb{F}_q[x]$, where T is the defining set of the cyclic code $\langle p(x) \rangle$. Then there is a unique monic factor $q(x)$ of $x^n - 1$ in $R[x]$ such that*

$$q(x) = \Pi_{i \in T}(x - \beta^i),$$

and $\bar{q}(x) = p(x)$

Proof. Let $q(x)$ be the unique monic Hensel lift of $p(x)$ which is a divisor of $x^n - 1$ in $\mathbb{F}_q[x]$, and define

$$\tilde{q}(x) = \Pi_{i \in T}(x - \beta^i), \ i \in \mathbb{Z}_n.$$

From (4) we have that $\sigma(\tilde{q}(x)) = \tilde{q}(x)$, so $\tilde{q}(x)$ has coefficients from R. Further $\overline{\tilde{q}(x)} = \overline{\Pi_{i \in T}(x - \beta^i)} = \Pi_{i \in T}(x - \alpha^i) = p(x) = \overline{q}(x)$ and since $q(x)$ is unique, we have that

$$q(x) = \Pi_{i \in T}(x - \beta^i). \qquad \square$$

3   Cyclic Codes of Length $2^a m$ over R

Let R be a finite chain ring with residue field $\mathbb{F}_q$ such that q is an odd prime power, and m an odd integer such that $(m, q) = 1$. In the following we give the structure of cyclic codes of length $2^a m$ where $a \geq 1$.

We begin with the following lemma.

Lemma 4. *There exists a primitive 2^a-th root of the unity α in R^* if and only if $q \equiv 1 \bmod 2^a$. Further, $x^{2^a} - 1 = \prod_{k=1}^{2^a}(x - \alpha^k)$ in $R[x]$.*

Proof. Since q is an odd prime power, by [5, Proposition 4.2], there exists a primitive 2^a-th root of the unity in R^* if and only if there exists a primitive 2^a-th root of unity in $\mathbb{F}_q$. If there exists a primitive 2^a-th root of unity α in $\mathbb{F}_q^*$, then $\alpha^{2^a} = 1$, so that 2^a divides $q - 1$. Conversely, if 2^a divides $q - 1$ then there exists an integer k such that $q = k2^a + 1$. If α is a primitive element of $\mathbb{F}_q^*$, then

$$1 = \alpha^{q-1} = (\alpha^k)^{2^a} \text{ and } ord(\alpha^k) = \frac{ord(\alpha)}{(k, ord(\alpha))} = \frac{q-1}{(k, q-1)} = \frac{k2^a}{(k, k2^a)} = 2^a.$$

Let α be a primitive 2^a-th root of the unity in R^*. Since $(2^a, q) = 1$, it must be that $\overline{\alpha}$ is a primitive 2^a-th root of unity in $\mathbb{F}_q^*$ so that $x^{2^a} - 1 = \prod_{k=1}^{2^a}(x - \overline{\alpha}^k)$ in $\mathbb{F}_q[x]$. By Lemma 4, the monic polynomial $x^{2^a} - 1$ factors uniquely as a product of monic basic irreducible pairwise coprime polynomials over R. Furthermore, there is a one-to-one correspondence between the set of basic irreducible polynomial divisors of $x^{2^a} - 1$ in $R[x]$ and the set of irreducible divisors of $\overline{x^{2^a} - 1}$ in $\mathbb{F}_q[x]$. If $x^{2^a} - 1 = \prod_{k=1}^{2^a}(x - a_k)$, then $\overline{(x - a_k)} = (x - \overline{a_k}) = (x - (\overline{\alpha})^k)$. Since $\overline{(x - \alpha^k)} = (x - \overline{\alpha}^k) = (x - (\overline{\alpha})^k)$, from the unique decomposition of $x^{2^a} - 1$ in $R[x]$, the result follows. $\qquad \square$

Lemma 5

(1) *If there exists a primitive 2^a-th root of unity α in R^*, then α^{2^i} is a primitive 2^{a-i}-th root of unity in R^*; for all $i \leq a$.*

(2) *Let α be a primitive 2^a-th root of the unity in R^*. Then α^m is also a primitive 2^a-th root of the unity in R^*.*

(3) *If $a \geq 2$, then $\prod_{k=1}^{2^a} \alpha^k = 1$.*

Proof. By [5, Proposition 4.2], there exists a primitive 2^a-th root of the unity in R^* if and only if there exists a primitive 2^a-th root of unity in $\mathbb{F}_q$ and using [3, Lemma 3.2] we have: For (1) Let i, $i \le a$. In the cyclic group $\mathbb{F}_q^*$, then $ord(\overline{\alpha}^{2^i}) = \frac{ord(\overline{\alpha})}{(2^i, ord(\overline{\alpha}))} = \frac{2^a}{(2^i, 2^a)} = \frac{2^a}{2^i} = 2^{a-i}$.

For part (2), since $(2^a, m) = 1$, $ord(\overline{\alpha^m}) = ord((\overline{\alpha})^m) = \frac{ord(\overline{\alpha})}{(m, ord(\overline{\alpha}))} = \frac{2^a}{(m, 2^a)} = 2^a$.

For part (3), since $(x^{2^a} - 1) = \prod_{k=1}^{2^a}(x - \alpha^k)$, $\prod_{k=1}^{2^a} \alpha^k = (-1)^{2^a - 1}$. $\qquad\square$

Lemma 6. *There exist a unique monic basic irreducible pairwise coprime factors $g_i(x)$, $i \in \{1, 2, \ldots, r\}$ of $x^m - 1$ such that*

$$x^m - 1 = (x - 1) \prod_{i=1}^{r} g_i(x), \tag{5}$$

in $R[x]$.

Proof. Let m be an integer such that $(p, m) = 1$. Since $\mathbb{F}_q[x]$ is a unique factorization domain (UFD), there exist unique monic irreducible pairwise coprime factors $f_i(x)$, $i \in \{1, 2, \ldots, r\}$ which satisfy $x^m - 1 = (x - 1) \prod_{i=1}^{r} f_i(x)$ over $\mathbb{F}_q$. Thus, there exist unique monic pairwise coprime polynomials $x - a$, $g_i(x)$, $i \in \{1, 2, \ldots, r\}$ which are factors of $x^m - 1$ in $R[x]$ such that $\overline{x - a} = x - \overline{a} = x - 1$, $\overline{g_i(x)} = f_i(x)$ for all $i \in \{1, 2, \ldots, r\}$. This gives that

$$x^m - 1 = (x - a) \prod_{i=1}^{r} g_i(x), \tag{6}$$

in $R[x]$. Substituting $x = 1$ into (6), we obtain

$$(1 - a) \prod_{i=1}^{r} g_i(1) = 0.$$

Since $(m, q) = 1$, $x^m - 1$ has simple roots. Then $\overline{g_i}(1) = f_i(1) \ne 0$ for all $i \in \{1, 2, \ldots, r\}$. This means that for all $i \in \{1, 2, \ldots, r\}$ $g_i(1)$ are invertible elements of R. Therefore $a = 1$ and

$$x^m - 1 = (x - 1) \prod_{i=1}^{r} g_i(x),$$

in $R[x]$. $\qquad\square$

3.1   Free Cyclic Codes of Length $2^a m$ over R

For thereafter, R is a finite chain ring with residue field $\mathbb{F}_q$, $q = p^t$ be an odd prime power and $n = 2^a m$ a positive integer such that m is an odd integer, $a \ge 1$ and $(m, p) = 1$ such that $q \equiv 1 \bmod 2^a$.

Before giving the structure of free cyclic codes of length $2^a m$ over R, we need the following proposition.

Proposition 1. *If R^* contains a primitive 2^a-root of unity and $x - 1$, $g_i(x)$, $1 \le i \le r$, are the monic basic irreducible pairwise coprime factors of $x^m - 1$ in $R[x]$, then*

$$x^{2^a m} - 1 = (x^{2^a} - 1) \prod_{i=1}^{r} g_i(\alpha^{-k}x).$$

Proof. Assume that $x^m - 1 = (x - 1) \prod_{i=1}^{r} g_i(x)$ (so that $g_0(x) = (x - 1)$). Since $(m, p) = 1$, by [10, Theorem 4.3] and Lemma 6 this is the unique factorization of $x^m - 1$ into monic basic irreducible pairwise coprime polynomials over R. Let $\alpha \in R^*$ be a primitive 2^a-th root of unity and let $1 \le k \le 2^a$. Then

$$
\begin{aligned}
(\alpha^{-k}x)^m - 1 &= (\alpha^{-k}x - 1) \prod_{i=1}^{r} g_i(\alpha^{-k}x) \\
(\alpha^{-k})^m(x^m - (\alpha^k)^m) &= \alpha^{-k}(x - \alpha^k) \prod_{i=1}^{r} g_i(\alpha^{-k}x) \\
(x^m - \alpha^{km}) &= \alpha^{k(m-1)}(x - \alpha^k) \prod_{i=1}^{r} g_i(\alpha^{-k}x) \\
(x^m - (\alpha^m)^k) &= \alpha^{k(m-1)}(x - \alpha^k) \prod_{i=1}^{r} g_i(\alpha^{-k}x),
\end{aligned}
$$

and by Lemma 4 α^m is also a primitive 2^a-th root of unity in R^*. We have that

$$
\begin{aligned}
\prod_{k=1}^{2^a}(x^m - (\alpha^m)^k) &= \prod_{k=1}^{2^a} \alpha^{k(m-1)}(x - \alpha^k) \prod_{i=1}^{r} g_i(\alpha^{-k}x) \\
&= \prod_{k=1}^{2^a} \alpha^{k(m-1)} \prod_{k=1}^{2^a}(x - \alpha^k) \prod_{k=1}^{2^a} \prod_{i=1}^{r} g_i(\alpha^{-k}x) \\
&= \prod_{k=1}^{2^a} \frac{\alpha^{km}}{\alpha^k} \prod_{k=1}^{2^a}(x - \alpha^k) \prod_{k=1}^{2^a} \prod_{i=1}^{r} g_i(\alpha^{-k}x) \\
&= (x^{2^a} - 1) \prod_{k=1}^{2^a} \prod_{i=1}^{r} g_i(\alpha^{-k}x).
\end{aligned}
$$

Since $(x^{2^a m} - 1) = ((x^m)^{2^a} - (\alpha^m)^{2^a}) = \prod_{k=1}^{2^a}(x^m - \alpha^{km})$, the result follows. $\square$

We now give the structure of free cyclic codes of length $2^a m$ over R.

Corollary 1. *If R^* contains a primitive 2^a-root of unity α and $(x - 1), g_i(x)$, $1 \le i \le r$ are the monic basic irreducible factors of $x^m - 1$ in $R[x]$, then a free cyclic code C of length $n = 2^a m$ is generated by $\prod_{k=1}^{2^a}((x - \alpha^k)^{l_k} \prod_{i=1}^{r} g_i^{j_i}(\alpha^{-k}x))$ with $0 \le l_k, j_i \le 1$.*

Proof. By [10, Theorem 4.16], any free cyclic code of length $2^a m$ is generated by a divisor of $x^{2^a m} - 1$, and by Proposition 1 we have that

$$(x^{2^a m} - 1) = \prod_{k=1}^{2^a}((x - \alpha^k) \prod_{i=1}^{r} g_i(\alpha^{-k}x)),$$

and the result follows. $\square$

Next, the structure of cyclic codes (not necessarily free) of length $2^a m$ over R are examined.

Theorem 2. *Let C be a code of length $2^a m$ over R. Then C is a cyclic code of length $2^a m$ over R if and only if $C \simeq \bigoplus_{1 \le i \le 2^a} C_i$, where C_i are cyclic codes of length m over R.*

Proof. Since $q \equiv 1 \bmod 2^a$, by Lemma 4 there exists a primitive 2^a-th root of unity $\alpha \in R^*$ such that $\alpha^{2^a} = 1$. By Lemma 4, α^m is also a primitive 2^a-th root of unity in R^* (note that m is odd), and thus

$$(x^{2^a m} - 1) = \prod_{i=1}^{2^a} (x^m - (\alpha^m)^i).$$

Since $(2^a m, p) = 1$, there are no repeated roots so the polynomials $(x^m - \alpha^i)$, $i \in \{1, \ldots, 2^a\}$ are coprime. Then by the Chinese remainder Theorem we have the following ring isomorphism

$$\frac{R[x]}{(x^{2^a m} - 1)} \simeq \prod_{i=1}^{2^a} \frac{R[x]}{(x^m - (\alpha^i)^m)}.$$

From Theorem 4.3 in [5], we have that $\frac{R[x]}{(x^m - \alpha^i)} \simeq \frac{R[x]}{(x^m - 1)}$, $\forall i \in \{1, \ldots, 2^a\}$, so then

$$\frac{R[x]}{(x^{2^a m} - 1)} \simeq \prod_{i=1}^{2^a} \frac{R[x]}{(x^m - 1)}.$$

Thus any ideal I of $\frac{R[x]}{(x^{2^a m} - 1)}$ is equivalent to a direct sum of 2^a ideals I_i of $\frac{R[x]}{(x^m - 1)}$. Therefore, a cyclic code over R is a direct sum of 2^a cyclic codes of length m over R. □

4   The Existence of Cyclic Isodual Codes over Finite Chain Rings

In this section, conditions are given on the existence of cyclic isodual codes over finite chain rings. Explicit constructions of monomial isodual cyclic free codes for odd characteristics are also presented. We begin with the following result.

Theorem 3. *Let C be a cyclic code of length n over R generated by the polynomial $g(x)$, and λ a unit in R such that $\lambda^n = 1$. Then the following holds:*

(i) C is equivalent to the cyclic code generated by $g^*(x)$.
(ii) C is equivalent to the cyclic code generated by $g(\lambda x)$.

Proof. Let μ_{-1} defined as in (3) it is a weight preserving linear transformation for codes over finite chain rings. Let $c(x) = c_0 + c_1 x + c_2 x^2 + \ldots + c_k x^k \in C$, then $\mu_{-1}(c(x)) = c(x^{-1}) = x^{n-k}(c_k + c_{k-1}x + c_{k-2}x^2 + \ldots + c_0 x^k)$. This shows that the multiplier μ_{-1} is weight preserving, so from [16] C and $\mu_{-1}(C)$ are monomially equivalent codes. Let $g(x)$ and $g'(x)$ be the generator polynomials of C and $\mu_{-1}(C)$, respectively. Since μ_{-1} is a ring automorphism, C and $\mu_{-1}(C)$ have the same dimension, so the polynomials $g(x)$ and $g'(x)$ have the same degree.

From the definition of the reciprocal polynomial of $g(x)$, $g^*(x) \in \mu_{-1}(C)$ so that $g'(x)$ divides $g^*(x)$. For $g(0) \in R^*$, $g^*(x)$ and $g(x)$ have the same degree so that $g^*(x)$ and $g'(x)$ also have the same degree, and thus generate the same cyclic code. Therefore the free cyclic code generated by $g(x)$ is equivalent to the cyclic code generated by $g^*(x)$. Let

$$\phi : \frac{R[x]}{(x^n-1)} \longrightarrow \frac{R[x]}{(x^n-1)}$$
$$f(x) \longmapsto \phi(f(x)) = f(\lambda x).$$

For polynomials $f(x)$, $g(x) \in R[x]$ we have that $f(x) \equiv g(x) \bmod (x^n - 1)$ if and only if there exists a polynomial $h(x) \in R[x]$ such that

$$f(x) - g(x) = h(x)(x^n - 1).$$

Thus it must be that

$$\begin{aligned} f(\lambda x) - g(\lambda x) &= h(\lambda x)((\lambda x)^n - 1) \\ &= h(\lambda x)((\lambda)^n x^n - 1) \\ &= h(\lambda x)(x^n - 1), \end{aligned}$$

which is true if and only if $f(\lambda x) \equiv g(\lambda x) \bmod (x^n - 1)$. Thus for $f(x), g(x) \in R[x]/(x^n - 1)$

$$\phi(f(x)) = \phi(g(x)),$$

if and only if

$$g(x) = f(x),$$

where ϕ is well defined and one-to-one. It is obvious that ϕ is onto, and it is easy to verify that ϕ is a ring homomorphism. Therefore ϕ is a ring isomorphism. If $C = \langle g(x) \rangle$, then $\phi(C) = \langle g(\lambda x) \rangle$. Furthermore, ϕ is a weight preserving linear transformation for codes over finite chain rings. Let $c(x) = c_0 + c_1 x + c_2 x^2 + \ldots + c_k x^k \in C$. Since $c_i = 0$ is equivalent to $\lambda^i c_i = 0$ ($\lambda^i \neq 0$), we have that $\phi(c(x)) = c_0 + \lambda c_1 x + \lambda^2 c_2 x^2 + \ldots + \lambda^k c_k x^k$. Then the Hamming weights $wt(c(x))$ and $wt(\phi(c(x)))$ are equal, so from [16], C and $\phi(C)$ are monomially equivalent codes.   □

Remark 1. *With the same assumptions as in Theorem 3:*

- *C is equivalent to the cyclic code generated by $g^*(\lambda x)$.*
- *C is equivalent to the cyclic code generated by $(g(\lambda x))^*$.*

In the following we take q an odd prime power such that $q \equiv 1 \bmod 2^a$, with $a \geq 1$ an integer and m an odd integer with $(m, q) = 1$. We obtain these next results:

Theorem 4. *Let $f(x)$ be a polynomial such that $x^m - 1 = (x - 1)f(x)$. The free cyclic codes of length $2^a m$ generated by*

$$(x^{2^{a-1}} - 1) \prod_{k=0}^{2^{a-1}-1} f(\alpha^{-2k-1}x),$$

and

$$(x^{2^{a-1}} + 1) \prod_{k=1}^{2^{a-1}} f(\alpha^{-2k}x),$$

are isodual codes of length $2^a m$.

Proof. By Lemma 4, if $q \equiv 1 \mod 2^a$, there exists a primitive 2^a-th root of unity $\alpha \in R^*$ such that $\alpha^{2^a} = 1$. Suppose that $x^m - 1 = (x-1)f(x)$, then

$$(x^{2^a m} - 1) = (x^{2^a} - 1) \prod_{k=1}^{2^a} f(\alpha^{-k}x).$$

Further, we have $(x^{2^a} - 1) = (x^{2^{a-1}} - 1)(x^{2^{a-1}} + 1)$, so that

$$(x^{2^a m} - 1) = (x^{2^{a-1}} - 1)(x^{2^{a-1}} + 1) \prod_{k=1}^{2^a} f(\alpha^{-k}x)$$

$$(x^{2^a m} - 1) = (x^{2^{a-1}} - 1)(x^{2^{a-1}} + 1) \prod_{k=1}^{2^{a-1}} f(\alpha^{-2k}x) \prod_{k=0}^{2^{a-1}-1} f(\alpha^{-2k-1}x).$$

Let

$$g(x) = (x^{2^{a-1}} - 1) \prod_{k=0}^{2^{a-1}-1} f(\alpha^{-2k-1}x),$$

so that

$$h(x) = (x^{2^{a-1}} + 1) \prod_{k=1}^{2^{a-1}} f(\alpha^{-2k}x)$$

and $h^*(x) = g^*(\alpha x)$ By Theorem 3(i), C is equivalent to the cyclic code generated by $g^*(x)$, and by Theorem 3(ii), the cyclic code generated by $g^*(x)$ is equivalent to the cyclic code generated by $g^*(\alpha x) = h^*(x)$. As the latter code is $C^\perp$, C is isodual. Then the cyclic code generated by $g(x)$ is isodual. The same result is obtained for

$$g(x) = (x^{2^{a-1}} + 1) \prod_{k=1}^{2^{a-1}} f(\alpha^{-2k}x).$$

□

Example 1. *Let $R = \mathbb{Z}_9$, $a = 1$ and $m = 5$ so that $n = 10$. The polynomial $f(x)$ in Theorem 4 is $f(x) = x^4 + x^3 + x^2 + x + 1$.*
   Both of the polynomials

$$g_1(x) = x^5 + 7x^4 + 2x^3 + 7x^2 + 2x + 8$$

and

$$g_2(x) = x^5 + 2x^4 + 2x^3 + 2x^2 + 2x + 1$$

generate isodual codes. Both codes have minimum weight 4.

Theorem 5. *Let $g_1(x)$ and $g_2(x)$ polynomials in $R[x]$ such that $x^m - 1 = (x - 1)g_1(x)g_2(x)$. The free cyclic codes of length $2^a m$ generated by*

$$(x^{2^{a-1}} - 1) \prod_{k=1}^{2^{a-1}} g_i(\alpha^{-2k}x) \prod_{k=0}^{2^{a-1}-1} g_j(\alpha^{-2k-1}x),$$

and

$$(x^{2^{a-1}} + 1) \prod_{k=1}^{2^{a-1}} g_i(\alpha^{-2k}x) \prod_{k=0}^{2^{a-1}-1} g_j(\alpha^{-2k-1}x),$$

$i, j \in \{1, 2\}, i \neq j$, *are isodual codes of length $2^a m$ over R where $\alpha \in R^*$ a primitive 2^a-th root of unity.*

Proof. By Lemma 4, since $q \equiv 1 \bmod 2^a$, there exists a primitive 2^a-th root of unity $\alpha \in R^*$ such that $\alpha^{2^a} = 1$. Suppose that $x^m - 1 = (x - 1)g_1(x)g_2(x)$, then

$$(x^{2^a m} - 1) = (x^{2^a} - 1) \prod_{k=1}^{2^a} g_1(\alpha^{-k}x)g_2(\alpha^{-k}x).$$

Since $(x^{2^a} - 1) = (x^{2^{a-1}} - 1)(x^{2^{a-1}} + 1)$, we have

$$(x^{2^a m} - 1) = (x^{2^{a-1}} - 1)(x^{2^{a-1}} + 1) \prod_{k=1}^{2^a} g_1(\alpha^{-k}x)g_2(\alpha^{-k}x),$$

$$(x^{2^a m} - 1) = (x^{2^{a-1}} - 1)(x^{2^{a-1}} + 1) \prod_{k=1}^{2^{a-1}} g_1(\alpha^{-2k}x)g_2(\alpha^{-2k}x) \prod_{k=0}^{2^{a-1}-1} g_1(\alpha^{-2k-1}x)g_2(\alpha^{-2k-1}x).$$

Let

$$g(x) = (x^{2^{a-1}} - 1) \prod_{k=1}^{2^{a-1}} g_i(\alpha^{-2k}x) \prod_{k=0}^{2^{a-1}-1} g_j(\alpha^{-2k-1}x), \quad i \neq j,$$

then the free cyclic code generated by $g(x)$ is isodual. We then have

$$h(x) = (x^{2^{a-1}} + 1) \prod_{k=0}^{2^{a-1}-1} g_i(\alpha^{-2k-1}x) \prod_{k=1}^{2^{a-1}} g_j(\alpha^{-2k}x),$$

and $h^*(x) = g^*(\alpha x)$ from Theorem 3, so the cyclic code $\langle g(x) \rangle$ is isodual. The same result is obtained for

$$g(x) = (x^{2^{a-1}} + 1) \prod_{k=1}^{2^{a-1}} g_i(\alpha^{-2k}x) \prod_{k=0}^{2^{a-1}-1} g_j(\alpha^{-2k-1}x), \quad i \neq j.$$

$\square$

Example 2. Let $R = \mathbb{Z}_9$, $a = 1, m = 11$, so that $n = 22$. The polynomials that appear in the factorization of $x^{11} - 1$ over $\mathbb{Z}_9$ are $g_1 = x^5 + 3x^4 + 8x^3 + x^2 + 2x + 8$ and $g_2 = x^5 + 7x^4 + 8x^3 + x^2 + 6x + 8$. There are four possible isodual codes of length 22. Two of these codes are given by the following generator polynomials $g_1(x) = 8x^{11} + 5x^{10} + x^9 + 5x^7 + 4x^6 + 3x^5 + 6x^4 + 8x^3 + 6x + 8$ and $g_2(x) = 8x^{11} + 3x^{10} + x^8 + 6x^7 + 6x^6 + 4x^5 + 4x^4 + 8x^2 + 5x + 1$. The minimum Hamming weight of each code is 7.

Remark 2. If

$$g(x) = (x^{2^{a-1}} - 1) \prod_{k=1}^{2^a - 1} g_i(\alpha^{-2k}x) \prod_{k=1}^{2^a - 1} g_j(\alpha^{-2k}x), \ i \neq j,$$

then $g(x) = (x^{2^{a-1}m} - 1) = (x^{\frac{n}{2}} - 1)$, and the free cyclic code generated by $g(x)$ is isodual.

In the following we give construction of isodual codes as a direct sum of isoduals codes. Before we need the following lemma.

Lemma 7. Let C_1 and C_2 be linear codes of lengths n_1 and n_2, respectively over R, and define the direct sum as $C_1 \oplus C_2 = \{(c_1|c_2), \ c_1 \in C_1, \ c_2 \in C_2\}$. Then the following holds

(i) $(C_1 \oplus C_2)^\perp = C_1^\perp \oplus C_2^\perp$.
(ii) If C_1 and C_2 are isodual codes with minimum weights d_1 and d_2, respectively, then $C_1 \oplus C_2$ is an isodual code of length $n_1 + n_2$ with minimum weight $\min(d_1, d_2)$.

Proof. It is easy to verify the inclusion $C_1^\perp \oplus C_2^\perp \subseteq (C_1 \oplus C_2)^\perp$. From [16], we have $|C_1^\perp \oplus C_2^\perp| = |(C_1 \oplus C_2)^\perp|$, and part (i) follows.
For part (ii), let τ_1 and τ_2 be monomial permutations such that $\tau_1(C_1) = C_1^\perp$ and $\tau_2(C_2) = C_2^\perp$. Let $\tau_1(C_1) \oplus \tau_1(C_1) = C'$ then $C' = \sigma(C_1 \oplus C_2)$, with $\sigma(i) = \tau_1(i)$ for $1 \leq i \leq n$ and $\sigma(i) = \tau_2(i)$ for $n + 1 \leq i \leq 2n$. Then $\tau_1(C_1) \oplus \tau_2(C_2) = C_1^\perp \oplus C_2^\perp$. Since $(C_1 \oplus C_2)^\perp = C_1^\perp \oplus C_2^\perp$, $C_1 \oplus C_2$ is equivalent to $(C_1 \oplus C_2)^\perp$. The minimum weight follows from the direct sum of linear codes. □

The direct sum of two cyclic codes over R in Lemma 7 may not be a cyclic code. The following theorem gives conditions for this direct sum to be a cyclic code over R.

Theorem 6. Let R be a finite chain ring with residue field $\mathbb{F}_q$, q an odd prime power and m an odd integer such that $(m, q) = 1$. Let C_i, $1 \leq i \leq 2^a$ ($a \geq 1$ an integer), be cyclic isodual codes over R of length m. We then have

(i) $C_i \oplus C_j$, $\forall i, j, 1 \leq i, j \leq 2^a$, are cyclic isodual codes of length $2m$ over R.
(ii) If $q \equiv 1 \mod 2^a, (a \geq 2)$ then the direct sum $\bigoplus_{i \in \{1, \ldots, 2^a\}} C_i$ is a cyclic isodual code of length $2^a m$ over R.

Proof. The results follow from Lemma 7. □

5   Isodual Cyclic Codes over Finite Chain Rings from Duadic Codes

The previous section gave conditions on the existence of isodual cyclic codes over finite chain rings and constructions for these codes. However, a more straightforward means of finding these codes is desirable. Further, determining codes with good minimum distance is important. We recall some results about duadic codes which will be used in this section. Of course isodual codes cannot be duadic since their length is even. Let q be a prime power and let m be a positive odd integer such that $(m, q) = 1$. Then for $0 \leq i < m$, the q-cyclotomic coset of i (mod m) is defined as

$$Cl(i) = \{iq^l \pmod{m} | l \in \mathbb{N}\}.$$

Let α be a primitive m-th root of unity in an extension field of $\mathbb{F}_q$, and C be a cyclic code over $\mathbb{F}_q$ of length m generated by a polynomial $f(x)$. C is uniquely determined by its defining set $T = \{0 \leq i < m \mid f(\alpha^i) = 0\}$. Hence the defining set of a cyclic code over $\mathbb{F}_q$ is the union of some q-cyclotomic cosets.

Let S_1 and S_2 be unions of cyclotomic cosets modulo m such that $S_1 \cap S_2 = \emptyset$, $S_1 \cup S_2 = \mathbb{Z}_m \backslash \{0\}$, and $\mu_a S_i \mod n = S_{(i+1) \mod 2}$. Then the triple μ_a, S_1, S_2 is called a splitting modulo m. The odd-like duadic codes D_1 and D_2 are the cyclic codes over $\mathbb{F}_q$ with defining sets S_1 and S_2 and generator polynomials $f_1(x) = \Pi_{i \in S_1}(x - \alpha^i)$ and $f_2(x) = \Pi_{i \in S_2}(x - \alpha^i)$, respectively. The even-like duadic codes C_1 and C_2 are the cyclic codes over $\mathbb{F}_q$ with defining sets $\{0\} \cup S_1$ and $\{0\} \cup S_2$, respectively.

5.1   Lifts of Duadic Codes over Finite Chain Rings

In this section R is a finite chain ring with maximal ideal $\langle \gamma \rangle$, nilpotency index e, and residue field $\mathbb{F}_q$, $q = p^t$.

Lemma 8. *Let n be an odd integer such that $(p, n) = 1$ and $q \equiv \square \mod n$. Then there exists a pair of monic factors of $x^n - 1$, $g_i(x)$, $i \in \{1, 2\}$, such that*

$$x^n - 1 = (x - 1)g_1(x)g_2(x),$$

in $R[x]$.

Proof. Let n be an odd integer such that $(p, n) = 1$ and $q \equiv \square \mod n$. Then there exists a pair of odd-like duadic codes over $\mathbb{F}_q$ generated by $f_1(x)$ and $f_2(x)$, respectively, with $x^n - 1 = (x-1)f_1(x)f_2(x)$ over $\mathbb{F}_q$. Since $x - 1$, $f_1(x)$ and $f_2(x)$ are monic coprime factors of $x^n - 1$ over $\mathbb{F}_q$, there exist, by Hensel's Lemma, unique monic pairwise coprime polynomials $x - a$, $g_1(x)$, $g_2(x)$ which are factors of $x^n - 1$ in $R[x]$ such that $x - \bar{a} = x - 1$, $\overline{g_1(x)} = f_1(x)$ and $\overline{g_2(x)} = f_2(x)$. This gives

$$x^n - 1 = (x - a)g_1(x)g_2(x), \tag{7}$$

in $R[x]$. Substituting $x = 1$ into (7) we obtain

$$(1 - a)g_1(1)g_2(1) = 0.$$

Since $(n, q) = 1$, $x^n - 1$ has simple roots. Then $\overline{g_1}(1) = f_1(1) \neq 0$ and $\overline{g_2}(1) = f_2(1) \neq 0$. This gives that $g_1(1)$ and $g_2(1)$ are both invertible elements of R. Therefore $a = 1$ and

$$x^n - 1 = (x - 1)g_1(x)g_2(x),$$

in $R[x]$.   □

Let n be an odd integer such that $(p, n) = 1$ and $q \equiv \square \mod n$. Let g_i, $i \in \{1, 2\}$, be the lifted polynomials of f_i, where the f_i are generator polynomials of the duadic codes over $\mathbb{F}_q$. Then we define the following cyclic codes over R by:

Definition 1. *Let the free cyclic codes over R defined by*

$$D'_1 = \langle g_1(x) \rangle, D'_2 = \langle g_2(x) \rangle, C'_1 = \langle (x - 1)g_1(x) \rangle, \ and \ C'_2 = \langle (x - 1)g_2(x) \rangle, \quad (8)$$

and if e is even let the non free cyclic codes over R defined by

$$E_1 = \langle (x - 1)g_1(x), \gamma^{\frac{e}{2}} g_1(x)g_2(x) \rangle, \ and \ E_2 = \langle (x - 1)g_2(x), \gamma^{\frac{e}{2}} g_1(x)g_2(x) \rangle. \quad (9)$$

In the following we give some proprieties of duadic codes given by the Definition 1.

Proposition 2. *Let D'_i, and C'_i $i \in \{1, 2\}$ be the codes given by Definition 1*

(i) If the splitting is given by μ_{-1}, then $D'^{\perp}_1 = C'_1$ and $D'^{\perp}_2 = C'_2$.
(ii) If the splitting is not given by μ_{-1}, then $D'^{\perp}_1 = C'_2$ and $D'^{\perp}_2 = C'_1$.

Proof. We know that if $g(x)$ is a generator polynomial of a free cyclic code C of length n over R, then the dual code $C^{\perp}$ of C is the free cyclic code whose generator polynomial is $h^*(x)$ where $h^*(x)$ is the monic reciprocal polynomial of $h(x) = (x^n - 1)/g(x)$. By Lemma 1 and [Lemma 5.1 [3]] we have the result.   □

Proposition 3. *With the assumptions given in Definition 1 $D'_1 = \langle g_1(x) \rangle$ and $D'_2 = \langle g_2(x) \rangle$ are equivalent cyclic codes over R.*

Proof. Let $f_1(x) = \Pi_{i \in S_1}(x - \alpha^i)$ where α is a primitive n-th root of unity in $\mathbb{F}_q$. By Lemma 3, there exists $\beta \in S$, where S is a Galois extension of R, such that $\bar{\beta} = \alpha$ and $g_1(x) = \Pi_{i \in S_1}(x - \beta^i)$. Then $\mu_b(g_1(x)) = \Pi_{i \in S_1}(x - \beta^{bi}) = \epsilon \Pi_{j \in S_2}(x - \beta^j)$ where ϵ is a unit in R since the splitting is given by μ_b. Thus from [16], D'_1 and D'_2 are monomially equivalent cyclic codes over R.   □

Lemma 9. *Let G be a generator matrix of C'_1 (resp. C'_2). Then the following hold:*

(i)

$$\begin{pmatrix} 1\,1\ldots 1 \\ G \end{pmatrix} \quad (10)$$

is a generator matrix of D'_1 (resp. D'_2).

(ii)

$$\left(\begin{array}{c} G \\ \gamma^{\frac{e}{2}} \ \gamma^{\frac{e}{2}} \ \ldots \ \gamma^{\frac{e}{2}} \end{array}\right) \tag{11}$$

is a generator matrix of E_1 (resp. E_2).

Proof. For part (i), we know that D_1' and C_1' are cyclic codes of length n over R with generator polynomials $g_1(x)$ and $(x-1)g_1(x)$, respectively. Since $(x-1)$, $g_1(x)$ and $g_2(x)$ are pairwise coprime over R, there are polynomials $a(x)$ and $b(x)$ in $R[x]$ such that

$$a(x)g_2(x)g_1(x) + b(x)(x-1)g_1(x) = g_1(x).$$

Therefore,

$$a(x)(x^{n-1} + x^{n-2} + \cdots + x + 1) + b(x)(x-1)g_1(x) = g_1(x),$$

so (10) is a generator matrix of D_1'. A similar result holds for D_2', with G a generator matrix of C_2'.

For part (ii), we first prove that $\langle \gamma^{\frac{e}{2}} \rangle \not\subseteq C_1'$, where C_1' is the cyclic code of length n generated by $(x-1)g_1(x)$ over R. The codeword $\gamma^{\frac{e}{2}}(1^n)$ can be expressed as the polynomial $\gamma^{\frac{e}{2}} + \gamma^{\frac{e}{2}}x + \gamma^{\frac{e}{2}}x^2 + \cdots + \gamma^{\frac{e}{2}}x^{n-1}$. Substituting $x = 1$ into this polynomial, we obtain $n\gamma^{\frac{e}{2}} \neq 0$ since the characteristic of R is prime to n. Therefore $\gamma^{\frac{e}{2}} + \gamma^{\frac{e}{2}}x + \gamma^{\frac{e}{2}}x^2 + \cdots + \gamma^{\frac{e}{2}}x^{n-1}$ is not a multiple of $x - 1$, so that $\langle \gamma^{\frac{e}{2}} \rangle \not\subseteq C_1'$. It follows that E_1 has generator matrix (11), where G is a generator matrix of C_1'. A similar result holds for E_2, with G a generator matrix of C_2'. $\square$

Remark 3. *Since D_1' and D_2' are monomially equivalent codes, from Proposition 3, E_1 and E_2 are also monomially equivalent cyclic codes.*

Theorem 7. *With the previous notation the following hold:*

(i) If the splitting is given by μ_{-1}, then E_1, E_2 and $E_1 \oplus E_2$ are self-dual.
(ii) If the splitting is left invariant by μ_{-1}, then E_1 and E_2 are isodual cyclic codes over R.

Proof. Let $f_i, i \in \{1,2\}$, be the generator polynomials of the odd-like duadic codes over $\mathbb{F}_q$ of length n. Then we have $x^n - 1 = (x - 1)f_1(x)f_2(x)$ over $\mathbb{F}_q$. If the splitting is given by μ_{-1} then $f_1^*(x) = \epsilon f_2(x)$ and $f_2^*(x) = \epsilon f_1(x)$. Hence by Lemma 1 their lifts have the same properties so that

$$g_1^*(x) = \alpha g_2(x) \text{ and } g_2^*(x) = \alpha g_1(x),$$

with α a unit in R such that $\overline{\alpha} = \epsilon$. Then for

$$E_1 = \langle (x - 1)g_1(x), \gamma^{\frac{e}{2}}g_1(x)g_2(x) \rangle,$$

by Theorem 1 we have that

$$E_1^{\perp} = \langle (x - 1)^* g_2^*(x), \gamma^{\frac{e}{2}} g_1^*(x)g_2^*(x) \rangle = \langle (x - 1)g_1(x), \gamma^{\frac{e}{2}}g_1(x)g_2(x) \rangle.$$

This means E_1 is self dual. A similar proof shows that E_2 is also self dual over R. Since $(E_1 \oplus E_2)^\perp = E_1^\perp \oplus E_2^\perp = E_1 \oplus E_2$, $E_1 \oplus E_2$ is self-dual.

If the splitting is not given by μ_{-1}, then $f_1^*(x) = \epsilon f_1(x)$ and $f_2^*(x) = \epsilon f_2(x)$. Hence by Lemma 1 their lifts have the same properties, so that $g_1^*(x) = \alpha g_1(x)$ and $g_2^*(x) = \beta g_2(x)$, where α and β are units in R. Then for

$$E_1 = \langle (x-1)g_1(x), \gamma^{\frac{e}{2}} g_1(x) g_2(x) \rangle,$$

by Theorem 1 we have that

$$E_1^\perp = \langle (x-1)^* g_2^*(x), \gamma^{\frac{e}{2}} g_1^*(x) g_2^*(x) \rangle = \langle (x-1)g_2(x), \gamma^{\frac{e}{2}} g_1(x) g_2(x) \rangle = E_2.$$

Then E_1 and E_2 are duals of each other over R. Since they are monomially equivalent, they are isodual cyclic codes over R. $\qquad\square$

Example 3. *For $n = 11 \equiv -1 \bmod 4$ and $q = 3 \equiv \square \bmod 11$, there exists a pair of odd-like duadic codes over $\mathbb{F}_3$ generated by $f_1(x)$ and $f_2(x)$, respectively. Let $g_1(x)$ and $g_2(x)$ be the corresponding Hensel lifts over $\mathbb{Z}_9$. We have the factorization*

$$x^{11} - 1 = (x-1)(x^5 + 3x^4 + 8x^3 + x^2 + 2x - 1)(x^5 - 2x^4 - x^3 + x^2 - 3x - 1),$$

over $\mathbb{Z}_9$, so for $g_1(x) = x^5 + 3x^4 + 8x^3 + x^2 + 2x - 1$, we have $g_1^*(x) = -(x^5 - 2x^4 - x^3 + x^2 - 3x - 1) = -g_2(x)$. Therefore

$$C = \langle (x-1)g_i(x), 3g_i(x)g_j^*(x) \rangle,$$

is a self-dual code.

Example 4. *For $n = 31 \equiv -1 \bmod 4$ and $q = 2 \equiv \square \bmod 31$, there exists a pair of odd-like duadic codes over $\mathbb{F}_2$ generated by $f_1(x)$ and $f_2(x)$, respectively. Let $g_1(x)$ and $g_2(x)$ be the corresponding Hensel lifts over $\mathbb{Z}_4$. We have the factorization*

$$\begin{aligned}
x^{31} - 1 = {}&(x-1)(x^5 + 3x^2 + 2x + 3)(x^5 + 2x^4 + 3x^3 + x^2 + 3x + 3)\\
&(x^5 + 3x^4 + x^2 + 3x + 3)(x^5 + 2x^4 + x^3 + 3)\\
&(x^5 + x^4 + 3x^3 + x + 3)(x^5 + x^4 + 3x^3 + x^2 + 2x + 3).
\end{aligned}$$

over $\mathbb{Z}_4$, so for $g_1(x) = (x^5 + 3x^2 + 2x + 3)(x^5 + 2x^4 + 3x^3 + x^2 + 3x + 3)(x^5 + 3x^4 + x^2 + 3x + 3)$, we have $g_1^*(x) = -(x^5 + 2x^4 + x^3 + 3)(x^5 + x^4 + 3x^3 + x + 3)(x^5 + x^4 + 3x^3 + x^2 + 2x + 3) = -g_2(x)$ Therefore

$$C = \langle (x-1)g_i(x), 2g_i(x)g_j^*(x) \rangle,$$

is a self-dual code.

5.2 Construction of Free Isodual Cyclic Codes over Finite Chain Rings Using Lifts of Duadic Codes

Let n be an integer such that $(n, q) = 1$ so that $R[x]/(x^n - 1)$ is a principal ideal ring. The free cyclic codes over R are generated by factors of $x^n - 1$ [10], hence from Theorems 3 and 4 we obtain the following Theorem.

Theorem 8. *Let R be a finite chain ring with residue field $\mathbb{F}_q$, and suppose there exists a pair of odd-like Duadic codes $D_i = \langle f_i(x) \rangle$, $i = 1, 2$, of length m. Further, let $g_i(x) \in R[x]$ be the Hensel lift of $f_i(x) \in \mathbb{F}_q[x]$. We then have the following:*

(i) *The cyclic codes C_{ij} and C'_{ij} over R generated by*

$$(x^{2^{a-1}} - 1) \prod_{k=1}^{2^{a-1}} g_i(\alpha^{-2k}x) \prod_{k=0}^{2^{a-1}-1} g_j(\alpha^{-2k-1}x),$$

and

$$(x^{2^{a-1}} + 1) \prod_{k=1}^{2^{a-1}} g_i(\alpha^{-2k}x) \prod_{k=0}^{2^{a-1}-1} g_j(\alpha^{-2k-1}x),$$

$i, j \in \{1, 2\}, i \neq j$, *respectively, where $\alpha \in R^*$ is a primitive 2^a-th root of unity, are isodual codes of length $2^a m$.*

(ii) *If the splitting is given by μ_{-1}, then the cyclic codes C_{ii} and C'_{ii} over R generated by*

$$(x^{2^{a-1}} - 1) \prod_{k=1}^{2^a} g_i(\alpha^{-k}x),$$

and

$$(x^{2^{a-1}} + 1) \prod_{k=1}^{2^a} g_i(\alpha^{-k}x),$$

respectively, where $\alpha \in R^*$ is a primitive 2^a-th root of unity, are isodual codes of length $2^a m$.

(iii) *If the splitting is not given by μ_{-1}, then the dual of the cyclic code generated by*

$$(x^{2^{a-1}} - 1) \prod_{k=1}^{2^a} g_i(\alpha^{-k}x),$$

is equivalent to the cyclic code generated by

$$(x^{2^{a-1}} + 1) \prod_{k=1}^{2^a} g_j(\alpha^{-k}x).$$

Proof. For part (i), we use Theorem 5.

Let $C_{ii} = \langle g_{ii}(x) \rangle = \langle (x^{2^{a-1}} - 1) \prod_{k=1}^{2^a} g_i(\alpha^{-k}x) \rangle$. If the splitting is given by μ_{-1} then $f_1^*(x) = \epsilon f_2(x)$ and $f_2^*(x) = \epsilon f_1(x)$. Then by Lemma 1, $g_1^*(x) = \beta g_2(x)$ and $g_2^*(x) = \alpha g_1(x)$, so

$$C_{ii}^{\perp} = \langle h_{ii}^*(x) \rangle = \langle (x^{2^{a-1}} + 1) \prod_{k=1}^{2^a} g_i(\alpha^{-k}x)^* \rangle = \langle (x^{2^{a-1}} - 1) \prod_{k=1}^{2^a} g_i(\alpha^{-k}x) \rangle = \langle \beta g_{ii}(\alpha x) \rangle,$$

where α and β are units in R. Therefore, $C_{ii} \simeq C_{ii}^{\perp}$. The proof for the codes generated by $g_{ii}(x) = (x^{2^{a-1}} + 1) \prod_{k=1}^{2^a} g_i(\alpha^{-k}x)$ is similar.

If the splitting is not given by μ_{-1}, by Lemma 1 $g_1^*(x) = \beta g_1(x)$ and $g_2^*(x) = \alpha g_2(x)$. Then

$$C_{ii}^{\perp} = \langle h_{ii}^*(x) \rangle = \langle (x^{2^{a-1}} + 1) \prod_{k=1}^{2^a} g_i(\alpha^{-k}x)^* \rangle = \langle (x^{2^{a-1}} - 1) \prod_{k=1}^{2^a} g_i(\alpha^{-k}x) \rangle = \langle \beta g_{jj}(\alpha x) \rangle,$$

where α and β are units in R. Therefore $C_{ii} \simeq C_{jj}^{\perp}$. The proof for the codes generated by $g_{ii}(x) = (x^{2^{a-1}} + 1) \prod_{k=1}^{2^a} g_i(\alpha^{-k}x)$ is similar. $\square$

Example 5. *For $R = \mathbb{Z}_{25}$, $q = 5$ and $m = 11$, $5 \equiv 16 \mod 11$, so there exist duadic codes generated by f_i, $1 \leq i \leq 2$. Since $11 \equiv -1 \mod 4$, all splittings are given by μ_{-1} and we have*

$$(x^{11} - 1) = (x - 1)(x^5 + 17x^4 + 24x^3 + x^2 + 16x + 24)(x^5 + 9x^4 + 24x^3 + x^2 + 8x + 24)$$
$$= (x - 1)g_1(x)g_2(x)$$

$\langle (x - 1)g_i(x)g_j(-x) \rangle$, $1 \leq i, j \leq 2$, $i \neq j$, *is an isodual cyclic code of length 22 with minimum distance 8.*

$\langle (x + 1)g_i(x)g_i(-x) \rangle$, $1 \leq i \leq 2$, *is an isodual cyclic code with minimum distance 6.*

Example 6. *For $R = \mathbb{Z}_{289}$, $q = 17 \equiv 1 \mod 2^4$, $\alpha^{2^4} = 20^{2^4} = 1$, $m = 19$ and $17 \equiv 36 \mod 19$. $(x^{19} - 1) = (x - 1)f_1(x)f_2(x)$ where $f_1(x) = (x^9 + 208x^8 + 287x^7 + 83x^6 + 210x^5 + 205x^4 + 80x^3 + 2x^2 + 207x + 288)$ and $f_2(x) = (x^9 + 82x^8 + 287x^7 + 209x^6 + 84x^5 + 79x^4 + 206x^3 + 2x^2 + 81x + 288)$*

Since $19 \equiv -1 \mod 4$, all the splittings are given by μ_{-1}

$$(x^8 - 1) \prod_{k=1}^{16} g_i(20^{-k}x), \quad and \ (x^8 + 1) \prod_{k=1}^{16} g_i(20^{-k}x), 1 \leq i \leq 2$$

$1 \leq i \leq 2$, *generate isodual codes of length $2^4 19$, over R.*

References

1. Bachoc, C., Gulliver, T.A., Harada, M.: Isodual codes over Z_{2k} and isodual lattices. J. Algebra. Combin. **12**, 223–240 (2000)
2. Batoul, A., Guenda, K., Gulliver, T.A.: On self-dual cyclic codes over finite chain rings. Des. Codes Cryptogr. **70**(3), 347–358 (2014)
3. Batoul, A., Guenda, K., Gulliver, T.A.: Repeated-root isodual cyclic codes over finite fields. In: El Hajji, S., Nitaj, A., Carlet, C., Souidi, E.M. (eds.) C2SI 2015. LNCS, vol. 9084, pp. 119–132. Springer, Cham (2015). doi:10.1007/978-3-319-18681-8_10
4. Batoul, A., Guenda, K., Kaya, A., Yildiz, B.: Cyclic isodual and formally self-dual codes over $\mathbb{F}_q + v\mathbb{F}_q$. EJPAM **8**(1), 64–80 (2015)
5. Batoul, A., Guenda, K., Gulliver, T.A.: Some constacyclic codes over finite chain rings. Adv. Math Commun. **10**(4), 683–694 (2016)
6. Batoul, A., Guenda, K., Gulliver, T.A.: Constacyclic codes over finite principal ideal rings. In: C2SI (2017, accepted)
7. Dinh, H., López-Permouth, S.R.: Cyclic and negacyclic codes over finite chain rings. IEEE Trans. Inform. Theory **50**, 1728–1744 (2004)
8. Greferath, M., Schmidt, S.E.: Finite-ring combinatorics and Macwilliams' equivalence theorem. J. Combin. Theory A **92**(1), 17–28 (2000)
9. Ganske, G., McDonald, B.R.: Finite local rings. Rocky Mountain J. Math. **3**(4), 521–540 (1973)
10. Guenda, K., Gulliver, T.A.: MDS and self-dual codes over rings. Finite Fields Appl. **18**(6), 1061–1075 (2012)
11. Honold, T., Nechaev, A.: Weighted modules and representation of codes. Tech. Univ. Menchen, Fak. Math Report, Beitrage Zur Geometrie and Algebra, 36 (1998)
12. Jia, Y., Ling, S., Xing, C.: On self-dual cyclic codes over finite fields. IEEE Trans. Inform. Theory **57**(4), 2243–2251 (2011)
13. Langevin, P.: Duadic $\mathbb{Z}_4$-codes. Finite Fields Appl. **6**, 309–326 (2000)
14. Norton, G.H., Sălăgean, A.: On the structure of linear and cyclic codes over a finite chain ring. Finite Fields Appl. **10**, 489–506 (2000)
15. Ling, S., Solé, P.: Duadic codes over $F_2 + uF_2$. AAECC, 365–379 (2001). doi:10.1007/s002000100079
16. Wood, J.A.: Extension theorems for linear codes over finite rings. In: Mora, T., Mattson, H. (eds.) AAECC 1997. LNCS, vol. 1255, pp. 329–340. Springer, Heidelberg (1997). doi:10.1007/3-540-63163-1_26

Revisiting the Efficient Key Generation of ZHFE

Yasuhiko Ikematsu$^{(\boxtimes)}$, Dung H. Duong, Albrecht Petzoldt,
and Tsuyoshi Takagi

Institute of Mathematics for Industry, Kyushu University,
744 Motooka, Nishi-ku, Fukuoka 819-0395, Japan
{y-ikematsu,duong,petzoldt,takagi}@imi.kyushu-u.ac.jp

Abstract. ZHFE, proposed by Porras et al. at PQCrypto'14, is one of
the very few existing multivariate encryption schemes and a very promising candidate for post-quantum cryptosystems. The only one drawback
is its slow key generation. At PQCrypto'16, Baena et al. proposed an
algorithm to construct the private ZHFE keys, which is much faster
than the original algorithm, but still inefficient for practical parameters. Recently, Zhang and Tan proposed another private key generation
algorithm, which is very fast but not necessarily able to generate all the
private ZHFE keys. In this paper we propose a new efficient algorithm for
the private key generation of the ZHFE scheme. Our algorithm reduces
the complexity from $O(n^{2\omega+1})$ by Baena et al. to $O(n^{\omega+3})$, where n is
the number of variables and $2 < \omega < 3$ is a linear algebra constant.
We also estimate the number of possible keys generated by all existing
private key generation algorithms for ZHFE. Our algorithm generates as
many private ZHFE keys as the original and Baena et al.'s ones. This
makes our algorithm be the best appropriate for the ZHFE scheme.

Keywords: Post quantum cryptography · Multivariate cryptography ·
Encryption schemes · ZHFE

1 Introduction

In 1997, P. Shor [21] gave polynomial time quantum algorithms to factor large
integers and to solve discrete logarithms. Thus, as soon as large-scale quantum
computer are built, almost all public key cryptosystems currently used in practice
such as RSA, DSA and ECC will become insecure. Post-Quantum Cryptography
(PQC) stands for the study of cryptosystems that have the potential to resist
such quantum computer attacks [1].

Recently, PQC has taken a lot of attention and become more and more important in the cryptographic research community, including also some authorities
such as the American National Security Agency (NSA), who recommended gov-
ernmental organizations to switch their security infrastructures from schemes
such as RSA and ECC [9] to post quantum cryptosystems, and the National
Institute of Standards and Technology (NIST), which is preparing to develop
standards for these schemes [14]. Among all possible candidates for PQC, multivariate public key cryptography (MPKC) [7] is one of the main candidates for the

© Springer International Publishing AG 2017
S. El Hajji et al. (Eds.): C2SI 2017, LNCS 10194, pp. 195–212, 2017.
DOI: 10.1007/978-3-319-55589-8_13

Table 1. Complexity of key generation algorithms for ZHFE scheme. Here n is the number of variables, D is the degree chosen for efficient decryption.

Algorithm	Complexity	$q = 2$	$q = 3$	$q = 5$	$q = 7$
Original [20]	$\mathcal{O}(n^{3\omega})$	100%	100%	100%	100%
Baena et al. [2]	$\mathcal{O}(n^{2\omega+1})$	99.9%	99.9%	99.9%	99.9%
Ours	$\mathcal{O}(n^{\omega+3})$	99.5%	99.9%	99.9%	99.9%
Zhang-Tan [24]	$\mathcal{O}(\log_q D)$	28.9%	56.0%	76.0%	83.7%

standardization. Multivariate schemes are in general very fast and require only modest computational resources, which makes them attractive for the use on low cost devices like smart cards and RFID chips [3,5]. In the area of digital signatures, there exists a large number of practical multivariate schemes [8,11,18]. The great difficulty for MPKC is encryption.

The C^* scheme introduced by Matsumoto and Imai [13], hence the name MI scheme, was considered to be the first encryption scheme. After MI was broken by Patarin [15], many encryption schemes have been proposed but then efficiently broken. Notably, Patarin invented the Hidden Field Equation cryptosystem (HFE) [16] which replaces the central map of the MI scheme by a low degree univariate polynomial. However, using low degree polynomials in the central map makes HFE be broken [6,12]. In order to thwart the attack, Porras et al. [19,20] cleverly proposed at PQCrypto'14 an interesting encryption scheme called ZHFE, which uses two high degree HFE polynomials in the central map, but a chosen low degree D polynomial for efficient decryption; see Sect. 2.2 for more details.

The ZHFE scheme [19,20] is one of the few existing multivariate encryption schemes at the moment, among ABC [22], SRP [25] and EFC [23]. However, what makes ZHFE important and attractive is its efficiency and thorough security analysis, see [17,20]. One drawback of ZHFE is its super slow key generation process, which involves solving large linear systems; the original method [20] for generating the private key needs to solve a linear system of about n^3 variables, resulting in a complexity of $\mathcal{O}(n^{3\omega})$, where $2 < \omega < 3$ is a linear algebra constant. At PQCrypto'16, Baena et al. [2] proposed an improved algorithm which reduces the complexity of this step to $\mathcal{O}(n^{2\omega+1})$. Their idea is to re-arrange the HFE polynomials (see Proposition 1). As a result, the matrix associated to the large linear system forms a shape close to a block diagonal matrix. For practical parameters, this algorithm is much faster than the original one but still inefficient. Recently, Zhang and Tan [24] proposed an algorithm which requires very little computation; their algorithm reduces the complexity to $\mathcal{O}(\log_q D)$ which makes their algorithm very fast; here D is the degree of the secret polynomial (see Sect. 2.2 for more details). However, their algorithm is based on the invertibility condition of some linear map, which is not necessarily fulfilled, and this prevents their algorithm from generating all the private ZHFE keys; see Sect. 2.3 for more details. Therefore, their structured key generation algorithm may possibly weaken the security of the scheme.

Our contribution. In this paper, we propose a new private key generation algorithm of the ZHFE scheme. The complexity of our algorithm is $\mathcal{O}(n^{\omega+3})$ which improves the one by Baena et al. [2]; for example, for 96-bit security parameters ($q = 7, D = 105, n = 55$) and 111-bit security parameters ($q = 17, D = 595, n = 55$), our algorithm is around 15 and 256 times faster than that of Baena et al. [2] respectively (see our implementation results in Table 3). Moreover, our algorithm generates as many private ZHFE keys as that of Baena et al. [2]. Our method is as follows: we first analyze again the algebraic structure of the central map in ZHFE scheme, following the route of Baena et al. [2]. At some stage, instead of working in the base field, we lift our problem to the extension field and use the properties of the extension field to construct an algorithm which is simpler and more efficient than that of Baena et al. [2]; see Sect. 3 for more details.

We also estimate the number of private ZHFE keys that all existing algorithms generate in Table 2. Zhang and Tan's algorithm [24] generates only those private ZHFE keys, for which the corank of a given linear map $\mathcal{L}$ is 0. As Table 2 shows, this condition is, in the case of $q = 2$, fulfilled by only 28.9% of all possible keys, which means that the algorithm of [24] generates only a small part of the keys. In contrast to this, our algorithm generates nearly 100% of the keys, since it can deal with linear maps $\mathcal{L}$ of corank < 3. This, together with its efficiency, makes our algorithm to be the most appropriate private key generation algorithm of the ZHFE scheme.

Organization. Our paper is organized as follows: we briefly recall the ZHFE scheme and the various private key generation of Porras et al. [20], Baena et al. [2] and Zhang and Tan [24] in Sect. 2. Our algorithm is explicitly introduced and analyzed in Sect. 3. In Sect. 4 we present a MAGMA implementation of our algorithm and compare it with Baena's algorithm with respect to running time and memory consumption. Finally, we conclude our paper in Sect. 5.

2 The ZHFE Scheme and Its Key Generation Algorithms

In this section, we briefly recall the basic concepts of multivariate encryption schemes and the ZHFE scheme [20]. We also recall the key generation process in the ZHFE scheme and the improved algorithms by Baena et al. [2] and Zhang and Tan [24].

2.1 Multivariate Public Key Cryptography

The basic objects of multivariate public key cryptography are systems of multivariate quadratic polynomials over a finite field $\mathbb{F}$. The security of multivariate schemes is based on the *MQ-Problem* which asks for a solution of a given system of multivariate quadratic polynomials over the field $\mathbb{F}$. The MQ-Problem is proven to be NP-Hard even for quadratic polynomials over the field GF(2) [10]. To build a public key cryptosystem on the basis of the MQ-Problem, one starts with an easily invertible quadratic map $\mathcal{F}: \mathbb{F}^n \to \mathbb{F}^m$ (*central map*). To hide the

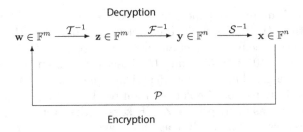

Fig. 1. General workflow of multivariate encryption schemes

structure of $\mathcal{F}$ in the public key, one composes it with two invertible affine (or linear) maps $\mathcal{T} : \mathbb{F}^m \to \mathbb{F}^m$ and $\mathcal{S} : \mathbb{F}^n \to \mathbb{F}^n$. The *public key* is therefore given by $\mathcal{P} = \mathcal{T} \circ \mathcal{F} \circ \mathcal{S}$. The *private key* consists of $\mathcal{T}, \mathcal{F}$ and $\mathcal{S}$. In this paper we consider multivariate encryption schemes. For these schemes, we require $n \leq m$.

Encryption: to encrypt a message $\mathbf{x} \in \mathbb{F}^n$, one simply computes $\mathbf{w} = \mathcal{P}(\mathbf{x})$ from the public key.

Decryption: to decrypt a given ciphertext $\mathbf{w} \in \mathbb{F}^m$, one computes recursively $\mathbf{z} = \mathcal{T}^{-1}(\mathbf{w}), \mathbf{y} = \mathcal{F}^{-1}(\mathbf{z})$ and $\mathbf{x} = \mathcal{S}^{-1}(\mathbf{y})$. Here $\mathbf{y}$ is the preimage of $\mathbf{z}$ under the easy to invert central map $\mathcal{F}$. The condition $n \leq m$ guarantees that this pre image and therefore the recovered plaintext will be unique.

Figure 1 shows a graphical illustration of the encryption and decryption process of multivariate schemes.

2.2   The ZHFE Encryption Scheme

Let $\mathbb{F}$ be a finite field with q elements and $\mathbb{K}$ a degree n extension of $\mathbb{F}$. Let $\phi : \mathbb{K} \to \mathbb{F}^n$ be an $\mathbb{F}$-isomorphism between $\mathbb{K}$ and the vector space $\mathbb{F}^n$. Consider two HFE polynomials F_1 and F_2:

$$F_1 = \sum a_{i,j} X^{q^i + q^j} + \sum a_i' X^{q^i} + a'', \quad F_2 = \sum b_{i,j} X^{q^i + q^j} + \sum b_i' X^{q^i} + b'', \quad (1)$$

where the coefficients of F_1 and F_2 are undetermined. Next randomly choose $4n$ scalars $\alpha_1, ..., \alpha_{2n}, \beta_1, ..., \beta_{2n}$ of $\mathbb{K}$. Define four linear polynomials:

$$L_{00}(X) = \sum_{i=1}^{n} \alpha_i X^{q^{i-1}}, \quad L_{01}(X) = \sum_{i=1}^{n} \alpha_{n+i} X^{q^{i-1}},$$

$$L_{10}(X) = \sum_{i=1}^{n} \beta_i X^{q^{i-1}}, \quad L_{11}(X) = \sum_{i=1}^{n} \beta_{n+i} X^{q^{i-1}}. \quad (2)$$

We construct the following polynomial with q-Hamming weight three:

$$\Psi(X) := X \left(L_{00}(F_1) + L_{01}(F_2) \right) + X^q \left(L_{10}(F_1) + L_{11}(F_2) \right). \quad (3)$$

Fix a positive integer D. This D must be chosen such that each univariate polynomial equation over $\mathbb{K}$ of degree less than or equal to D can be solved efficiently by Berlekamp's algorithm. In order to generate a ZHFE key, we have to determine the coefficients of F_1, F_2 such that

$$\deg \Psi(X) \leq D.$$

In this paper, we propose an efficient algorithm to choose such coefficients of F_1, F_2; cf. Sect. 3. Once such coefficients are given, the ZHFE scheme [19, 20] is constructed as follows. Randomly choose invertible affine transformations $\mathcal{S}$ and $\mathcal{T}$ on $\mathbb{F}^n$ (resp. $\mathbb{F}^{2n}$). Then the public key $\mathcal{P} : \mathbb{F}^n \to \mathbb{F}^{2n}$ is given by

$$\mathcal{P} = \mathcal{T} \circ (\phi \times \phi) \circ (F_1, F_2) \circ \phi^{-1} \circ \mathcal{S}.$$

This is a $2n$-tuple of quadratic polynomials over $\mathbb{F}$ in n variables.

Public Key: The field $\mathbb{F}$ and the map $\mathcal{P}$.

Private Key: $\alpha_1, \ldots, \alpha_{2n}, \beta_1, \ldots, \beta_{2n}, F_1, F_2, \Psi, \mathcal{S}$ and $\mathcal{T}$.

Encryption: For a plaintext message $x \in \mathbb{F}^n$ with redundant information, the ciphertext is $w = \mathcal{P}(x) \in \mathbb{F}^{2n}$.

Decryption: For a given ciphertext $w \in \mathbb{F}^{2n}$, we first compute $(W_0, W_1) = (\phi^{-1} \times \phi^{-1})(\mathcal{T}^{-1}(w)) \in \mathbb{K} \times \mathbb{K}$. Next we consider the equation of degree $\max\{D, q\}$:

$$\Psi(X) - X(L_{00}(W_0) + L_{01}(W_1)) - X^q(L_{10}(W_0) + L_{11}(W_1)) = 0.$$

We can solve this equation efficiently by our choice of D. For each solution X_0 of this equation, we compute $x_0 = \mathcal{S}^{-1} \circ \phi(X_0)$. Then we can find the plaintext among the resulting x_0 thanks to the added redundant information.

2.3   Algorithms for the Private Key Generation of ZHFE Scheme

As seen above, the central part of the private key generation of ZHFE scheme is the computation of suitable coefficients of F_1 and F_2 and of Ψ for given $\alpha_1, \ldots, \alpha_{2n}, \beta_1, \ldots, \beta_{2n}$. In this section we introduce the known algorithms for this step.

The Original Algorithm. In the original papers [19, 20], F_1 and F_2 were computed by solving a large linear system over the small field $\mathbb{F}$ obtained from vanishing coefficients of Ψ. The size of this linear system is about n^3. Thus the complexity of this private key generation is $\mathcal{O}(n^{3\omega})$, where $2 < \omega < 3$ is a linear algebra constant. In fact, this algorithm is very inefficient for practical parameters (See [2, Table 3, Old method]).

Table 2. Ratio of linear map $\mathcal{L}$ in Eq. (4) over all possible linear maps $\mathrm{M}_{2n}(\mathbb{F})$ with respect to the corank of $\mathcal{L}$

q	n	corank $\mathcal{L} = 0$	corank $\mathcal{L} \leq 1$	corank $\mathcal{L} \leq 2$	corank $\mathcal{L} \leq 3$
$q = 2$	$n \geq 35$	28.9%	86.6%	99.5%	99.9%
$q = 3$	$n \geq 35$	56.0%	98.0%	99.9 %	99.9 %
$q = 5$	$n \geq 35$	76.0 %	99.8 %	99.9 %	99.9 %
$q = 7$	$n \geq 35$	83.7%	99.9 %	99.9 %	99.9%

Baena et al.'s Algorithm. At PQCrypto'16, Baena et al. [2] proposed a new improved algorithm for the private key generation of ZHFE scheme. Their idea is to re-arrange the HFE polynomials (see Proposition 1). As a result, the matrix associated to the large linear system forms a shape close to a block diagonal matrix. Then the complexity of this algorithm is $\mathcal{O}(n^{2\omega+1})$. This algorithm is much faster than the original one, but still inefficient for practical parameters. We obtain our algorithm by improving this one. Thus we will explain this algorithm in our language in Sect. 3.1.

Zhang and Tan's Algorithm. Recently, Zhang and Tan proposed [24] the algorithm that constructs the central map (F_1, F_2) and Ψ so that $\Psi := XF_1 + X^q F_2$ has degree D at most. Thus the algorithm requires only very little computation. In fact, the complexity is $\mathcal{O}(\log_q D)$. But this algorithm does not necessarily give all private ZHFE keys. Strictly speaking, if we define a linear map $\mathcal{L}$ over $\mathbb{F}$ on $\mathbb{K}^2$ by

$$\mathcal{L} : \mathbb{K}^2 \ni (X, Y) \mapsto (L_{00}(X) + L_{01}(Y), L_{10}(X) + L_{11}(Y)) \in \mathbb{K}^2, \qquad (4)$$

then this algorithm can generate all private keys with $\mathcal{L}$ nonsingular. $\mathcal{L}$ can be represented as a matrix in $M_{2n}(\mathbb{F})$ due to $\mathbb{K} \cong \mathbb{F}^n$. We stress that the corank of $\mathcal{L}$ is crucial for the efficient construction of private ZHFE keys, where the corank of $\mathcal{L}$ is $2n -$ Rank $\mathcal{L}$. In particular, if $\mathcal{L}$ is singular (corank of $\mathcal{L} \geq 1$), the private keys is not necessarily generated by this algorithm. To be more precise, assume that we have found polynomials F_1, F_2 such that $\Psi = XF_1 + X^q F_2$ is of degree less than or equal to D. In order to find another $\alpha'_1, \ldots, \alpha'_{2n}, \beta'_1, \ldots, \beta'_{2n} \in \mathbb{K}$ such that the corresponding polynomial

$$\Psi' = X \left(L'_{00}(F_1) + L'_{01}(F_2) \right) + X^q \left(L'_{10}(F_1) + L'_{11}(F_2) \right)$$

is of degree less than or equal to given D, then one needs to solve about n^2 equations in $4n$ variables $\alpha'_1, \ldots, \alpha'_{2n}, \beta'_1, \ldots, \beta'_{2n}$. For recommended parameters $(q = 7, D = 105, n = 55)$ one has a system of 3016 equations in 220 variables, which has at most one solution. Hence the linear map $\mathcal{L}'$ corresponding to the later private keys has corank 0. Hence, Zhang and Tan's algorithm does not work for corank $\mathcal{L} \geq 1$. The ratio of $\mathcal{L}$ with respect to the corank of $\mathcal{L}$ is given by Table 2 if the linear map $\mathcal{L}$ is randomly distributed in $M_{2n}(\mathbb{F})$ and n is large enough, for example $n \geq 35$.

3   Our New Key Generation Algorithm for ZHFE

In this section, we propose our new private key generation algorithm of ZHFE scheme. Here, we assume that n is odd, say $n = 2l + 1$, and $q > 2$. The reason why we assume n odd will be explained in Remark 1.

3.1   Baena et al.'s Algorithm

Since our new algorithm is obtained by improving Baena et al.'s one [2], we explain it here in our language.

Let F be an HFE polynomial. If F is a linear combination of $X^{q^{i-1}+q^{j-1}}$, $(1 \leq i, j \leq n)$ over $\mathbb{K}$, then it is called a quadratic HFE polynomial. For $1 \leq d \leq l+1$ and $1 \leq i \leq n$, set $X_{d,i} := X^{q^{i-1}+q^{i-1+d-1}}$.

Proposition 1. ([2, Sect. 3.1]). *Every quadratic HFE polynomial F can be uniquely written as*

$$F = \sum_{1 \leq d \leq l+1} \sum_{1 \leq i \leq n} a_{d,i} X_{d,i}, \quad (a_{d,i} \in \mathbb{K}).$$

In Proposition 1, we call $a_{d,i}$ *the (d,i)-coefficient* of F, and write $F_{d,i} = a_{d,i}$.

We represent the two quadratic HFE polynomials F_1, F_2 of Eq. (1) according to Proposition 1 as follows:

$$F_1 = \sum_{1 \leq d \leq l+1} \sum_{1 \leq i \leq n} a_{d,i} X_{d,i}, \quad F_2 = \sum_{1 \leq d \leq l+1} \sum_{1 \leq i \leq n} b_{d,i} X_{d,i}. \tag{5}$$

Here the coefficients are to be determined. Randomly choose $4n$ scalars $\alpha_1, ..., \alpha_{2n}, \beta_1, ..., \beta_{2n}$ of $\mathbb{K}$ and set

$$\bar{F}_1 := L_{00}(F_1) + L_{01}(F_2), \quad \bar{F}_2 := L_{10}(F_1) + L_{11}(F_2), \tag{6}$$

where the L_{ij} is defined as in Eq. (2). Thus

$$\Psi = X\bar{F}_1 + X^q \bar{F}_2. \tag{7}$$

Our goal is to determine the coefficients $a_{d,i}, b_{d,i}$ of Eq. (5) such that $\deg \Psi \leq D$.

First we compute the (d,i)-coefficients $\bar{F}_{1,d,i}, \bar{F}_{2,d,i}$ of the two quadratic HFE polynomials $\bar{F}_1, \bar{F}_2$. For n scalars $z_1, z_2, \ldots, z_n \in \mathbb{K}$, we define an $n \times n$ matrix by

$$L_1(z_1, z_2, \ldots, z_n) := (z_{j-i+1}^{q^{i-1}})_{i,j} = \begin{pmatrix} z_1 & z_2 & z_3 & \cdots & z_n \\ z_n^q & z_1^q & z_2^q & \cdots & z_{n-1}^q \\ z_{n-1}^{q^2} & z_n^{q^2} & z_1^{q^2} & \cdots & z_{n-2}^{q^2} \\ \vdots & \vdots & \vdots & \ddots & \vdots \\ z_2^{q^{n-1}} & z_3^{q^{n-1}} & z_4^{q^{n-1}} & \cdots & z_1^{q^{n-1}} \end{pmatrix}. \tag{8}$$

Here $j - i + 1$ is calculated modulo n. By using this notation, we can represent the n-tuple $(\bar{F}_{i,d,1}, \bar{F}_{i,d,2}, \ldots, \bar{F}_{i,d,n})$ as follows:

Lemma 1 ([2, Corollary 1]). (i) *For any d, we have*

$$(\bar{F}_{1,d,1}, \bar{F}_{1,d,2}, ..., \bar{F}_{1,d,n}) = (\alpha_1, ..., \alpha_n) \cdot L_1(a_{d,1}, ..., a_{d,n})$$
$$+ (\alpha_{n+1}, ..., \alpha_{2n}) \cdot L_1(b_{d,1}, ..., b_{d,n}). \tag{9}$$

(ii) *For any* d, *we have*

$$
\begin{aligned}
(\bar{F}_{2,d,1}, \bar{F}_{2,d,2}, ..., \bar{F}_{2,d,n}) &= (\beta_1, ..., \beta_n) \cdot L_1(a_{d,1}, ..., a_{d,n}) \\
&\quad + (\beta_{n+1}, ..., \beta_{2n}) \cdot L_1(b_{d,1}, ..., b_{d,n}).
\end{aligned}
\tag{10}
$$

Lemma 2 ([2, Lemma 1]).

$$
X \cdot X_{d,i} = X^q \cdot X_{d',i'} \iff
\begin{cases}
d = d' = l+1, i = 2, i' = l+2, \\
d' = d-1, i = i' = n+3-d, (2 \le d \le l+1), \\
d' = d+1, i = 2, i' = 1, (1 \le d \le l).
\end{cases}
\tag{11}
$$

If $X \cdot X_{d,i} = X^q \cdot X_{d',i'}$ then we write $(d, i) \rightsquigarrow (d', i')$. By this lemma, we can describe the conditions for F_1, F_2 so that $\deg \Psi \le D$.

Corollary 1. *If the coefficients* $a_{d,i}, b_{d,i}$ *of* F_1, F_2 *satisfy the following three conditions, then we have* $\deg \Psi \le D$.

  (i) $\bar{F}_{1,d,i} = -\bar{F}_{2,d',i'}$ *for any* $(d,i) \rightsquigarrow (d',i')$ *such that* $\deg X \cdot X_{d,i} > D$.
  (ii) $\bar{F}_{1,d,i} = 0$ *if* (d,i) *is not in Lemma 2 and satisfies* $\deg X \cdot X_{d,i} > D$.
  (iii) $\bar{F}_{2,d',i'} = 0$ *if* (d',i') *is not in Lemma 2 and satisfies* $\deg X^q \cdot X_{d',i'} > D$.

Proof. By Lemmas 1 and 2, it is easy to compute the coefficients of degree $> D$ in Ψ. Then the conditions of F_1, F_2 so that $\deg \Psi \le D$ are equivalent to (i),(ii) and (iii).

Note that $\deg(X \cdot X_{d,i}) = 1 + q^{i-1} + q^{(i-1+d-1 \bmod n)}$. Also $\deg(X^q \cdot X_{d',i'}) = q + q^{i'-1} + q^{(i'-1+d'-1 \bmod n)}$.

Finally, it follows from Lemma 1 and Corollary 1 that:

Theorem 1. *Randomly choose* $4n$ *scalars* $\alpha_1, ..., \beta_{2n}$ *of* $\mathbb{K}$. *Also we take any scalars* $c_{j,d,i} \in \mathbb{K}$, $(1 \le j \le 2, 1 \le d \le l+1, 1 \le i \le n)$ *with the assumptions* (i),(ii),(iii) *in Corollary 1. If* $a_{d,i}$ *and* $b_{d,i}$ *are solutions of equations*

$$
\begin{aligned}
(c_{1,d,1}, c_{1,d,2}, \ldots, c_{1,d,n}) &= (\alpha_1, \ldots, \alpha_n) \cdot L_1(a_{d,1}, \ldots, a_{d,n}) \\
&\quad + (\alpha_{n+1}, \ldots, \alpha_{2n}) \cdot L_1(b_{d,1}, \ldots, b_{d,n}) \qquad (A_d), \\
(c_{2,d,1}, c_{2,d,2}, \ldots, c_{2,d,n}) &= (\beta_1, \ldots, \beta_n) \cdot L_1(a_{d,1}, \ldots, a_{d,n}) \\
&\quad + (\beta_{n+1}, \ldots, \beta_{2n}) \cdot L_1(b_{d,1}, \ldots, b_{d,n}) \qquad (B_d),
\end{aligned}
$$

for any $1 \le d \le l+1$, *then* F_1, F_2 *satisfy that* $\deg \Psi \le D$. *Also we have*

$$
\Psi = \sum_{1 \le d \le l+1} \left(\sum_{1 \le i \le n} c_{1,d,i} X \cdot X_{d,i} + \sum_{1 \le i \le n} c_{2,d,i} X^q \cdot X_{d,i} \right).
$$

The equations $(A_d), (B_d)$ in Theorem 1 are not linear systems in $a_{d,i}, b_{d,i}$. They can be reduced to linear systems over the small field $\mathbb{F}$. Baena et al. [2] obtained F_1, F_2 and Ψ by solving such linear systems over the small field $\mathbb{F}$. Our strategy in obtaining F_1, F_2 and Ψ is to lift equations $(A_d), (B_d)$ to linear systems over the big field $\mathbb{K}$; cf. Sect. 3.2.

3.2 Main Idea

Here we explain the main idea of the proposed algorithm for efficient private key generation of ZHFE scheme.

For any $1 \leq d \leq l+1$, set

$$x_{d,i} := a_{d,n+2-i}^{q^{i-1}}, \quad y_{d,i} := b_{d,n+2-i}^{q^{i-1}}.$$

Then

$$a_{d,i} = x_{d,n+2-i}^{q^{i-1}}, \quad b_{d,i} = y_{d,n+2-i}^{q^{i-1}}.$$

Also we have

$$L_1(a_{d,1}, a_{d,2}, \ldots, a_{d,n}) = \begin{pmatrix} x_{d,1} & x_{d,n}^q & x_{d,n-1}^{q^2} & \cdots & x_{d,2}^{q^{n-1}} \\ x_{d,2} & x_{d,1}^q & x_{d,n}^{q^2} & \cdots & x_{d,3}^{q^{n-1}} \\ x_{d,3} & x_{d,2}^q & x_{d,1}^{q^2} & \cdots & x_{d,4}^{q^{n-1}} \\ \vdots & \vdots & \vdots & \ddots & \vdots \\ x_{d,n} & x_{d,n-1}^q & x_{d,n-2}^{q^2} & \cdots & x_{d,1}^{q^{n-1}} \end{pmatrix}.$$

By using these, the equation (A_d) is equivalent to the following (A_d'):

$$(c_{1,d,1}, c_{1,d,2}, \ldots, c_{1,d,n}) = (\alpha_1, \ldots, \alpha_n, \alpha_{n+1}, \ldots, \alpha_{2n}) \begin{pmatrix} x_{d,1} & x_{d,n}^q & \cdots & x_{d,2}^{q^{n-1}} \\ x_{d,2} & x_{d,1}^q & \cdots & x_{d,3}^{q^{n-1}} \\ \vdots & \vdots & \ddots & \vdots \\ x_{d,n} & x_{d,n-1}^q & \cdots & x_{d,1}^{q^{n-1}} \\ y_{d,1} & y_{d,n}^q & \cdots & y_{d,2}^{q^{n-1}} \\ y_{d,2} & y_{d,1}^q & \cdots & y_{d,3}^{q^{n-1}} \\ \vdots & \vdots & \ddots & \vdots \\ y_{d,n} & y_{d,n-1}^q & \cdots & y_{d,1}^{q^{n-1}} \end{pmatrix},$$

which is equivalent to the following equation (A_d''):

$$(c_{1,d,1}, c_{1,d,2}^{q^{n-1}}, \ldots, c_{1,d,n}^q) = (x_{d,1}, \ldots, x_{d,n}, y_{d,1}, \ldots, y_{d,n}) \begin{pmatrix} \alpha_1 & \alpha_2^{q^{n-1}} & \cdots & \alpha_n^q \\ \alpha_2 & \alpha_3^{q^{n-1}} & \cdots & \alpha_1^q \\ \alpha_3 & \alpha_4^{q^{n-1}} & \cdots & \alpha_2^q \\ \vdots & \vdots & \ddots & \vdots \\ \alpha_n & \alpha_1^{q^{n-1}} & \cdots & \alpha_{n-1}^q \\ \alpha_{n+1} & \alpha_{n+2}^{q^{n-1}} & \cdots & \alpha_{2n}^q \\ \alpha_{n+2} & \alpha_{n+3}^{q^{n-1}} & \cdots & \alpha_{n+1}^q \\ \vdots & \vdots & \ddots & \vdots \\ \alpha_{2n} & \alpha_{n+1}^{q^{n-1}} & \cdots & \alpha_{2n-1}^q \end{pmatrix}.$$

Remark 1. If we assume that n is even, then we can not obtain a linear system as above. In fact, if n is even, then in the case $d = n/2 + 1$, $x_{d,i}^{q^{j-1}}$ and $x_{d,i}^{q^{j-1+n/2}}$ appear on each j-column in the matrix in (A'_d). Thus we can not have a linear system as the linear system $(A''_{n/2+1})$. That is the reason why we consider n odd in this paper.

Similarly, the equation (B_d) is equivalent to the following equation (B''_d):

$$
(c_{2,d,1}, c_{2,d,2}^{q^{n-1}}, \ldots, c_{2,d,n}^q) = (x_{d,1}, \ldots, x_{d,n}, y_{d,1}, \ldots, y_{d,n})
\begin{pmatrix}
\beta_1 & \beta_2^{q^{n-1}} & \cdots & \beta_n^q \\
\beta_2 & \beta_3^{q^{n-1}} & \cdots & \beta_1^q \\
\beta_3 & \beta_4^{q^{n-1}} & \cdots & \beta_2^q \\
\vdots & \vdots & \ddots & \vdots \\
\beta_n & \beta_{n+1}^{q^{n-1}} & \cdots & \beta_{n-1}^q \\
\beta_{n+1} & \beta_{n+2}^{q^{n-1}} & \cdots & \beta_{2n}^q \\
\beta_{n+2} & \beta_{n+3}^{q^{n-1}} & \cdots & \beta_{n+1}^q \\
\vdots & \vdots & \ddots & \vdots \\
\beta_{2n} & \beta_{n+1}^{q^{n-1}} & \cdots & \beta_{2n-1}^q
\end{pmatrix}.
$$

For n scalars $z_1, z_2, \ldots, z_n$ of $\mathbb{K}$, define an $n \times n$ matrix by

$$
L_2 \begin{pmatrix} z_1 \\ z_2 \\ \vdots \\ z_n \end{pmatrix} := (z_i^{q^{n-j+1}})_{i,j} =
\begin{pmatrix}
z_1 & z_2^{q^{n-1}} & z_3^{q^{n-2}} & \cdots & z_n^q \\
z_2 & z_3^{q^{n-1}} & z_4^{q^{n-2}} & \cdots & z_1^q \\
z_3 & z_4^{q^{n-1}} & z_5^{q^{n-2}} & \cdots & z_2^q \\
\vdots & \vdots & \vdots & \ddots & \vdots \\
z_n & z_1^{q^{n-1}} & z_2^{q^{n-2}} & \cdots & z_{n-1}^q
\end{pmatrix}.
$$

By using this notation, set

$$
L := \begin{pmatrix}
L_2 \begin{pmatrix} \alpha_1 \\ \alpha_2 \\ \vdots \\ \alpha_n \end{pmatrix} & L_2 \begin{pmatrix} \beta_1 \\ \beta_2 \\ \vdots \\ \beta_n \end{pmatrix} \\
L_2 \begin{pmatrix} \alpha_{n+1} \\ \alpha_{n+2} \\ \vdots \\ \alpha_{2n} \end{pmatrix} & L_2 \begin{pmatrix} \beta_{n+1} \\ \beta_{n+2} \\ \vdots \\ \beta_{2n} \end{pmatrix}
\end{pmatrix} \in M_{2n}(K).
\tag{12}
$$

Remark 2. It is easy to prove that $\operatorname{Rank} L = \operatorname{Rank} \mathcal{L}$, where $\mathcal{L}$ is defined in Sect. 2.3.

Now we can restate Theorem 1 by using this L as follows:

Theorem 2. *Randomly choose 4n scalars $\alpha_1, \ldots \beta_{2n}$ of $\mathbb{K}$. Also we take any scalars $c_{j,d,i} \in \mathbb{K}$ $(1 \leq j \leq 2, 1 \leq d \leq l+1, 1 \leq i \leq n)$ with the assumptions (i),(ii),(iii) in Corollary 1. Let $x_{d,i}$ and $y_{d,i}$ be solutions of the linear system*

$$(c_{1,d,1}, c_{1,d,2}^{q^{n-1}}, \ldots, c_{1,d,n}^{q}, c_{2,d,1}, c_{2,d,2}^{q^{n-1}}, \ldots, c_{2,d,n}^{q}) = (x_{d,1}, \ldots, x_{d,n}, y_{d,1}, \ldots, y_{d,n}) \cdot L \quad (\star)$$

for any $1 \leq d \leq l+1$. If we set

$$F_1 = \sum_{1 \leq d \leq l+1} \sum_{1 \leq i \leq n} x_{d,n+2-i}^{q^{i-1}} X_{d,i}, \quad F_2 = \sum_{1 \leq d \leq l+1} \sum_{1 \leq i \leq n} y_{d,n+2-i}^{q^{i-1}} X_{d,i},$$

then F_1, F_2 satisfy $\deg \Psi \leq D$. Also we have

$$\Psi = \sum_{1 \leq d \leq l+1} \left(\sum_{1 \leq i \leq n} c_{1,d,i} X \cdot X_{d,i} + \sum_{1 \leq i \leq n} c_{2,d,i} X^q \cdot X_{d,i} \right).$$

Proof. Solving (A_d) is equivalent to solving (A_d''). Similarly, solving (B_d) is equivalent to solving (B_d''). Also solving (A_d'') and (B_d'') is equivalent to solving $(\star)$. Thus we have the theorem.

Thus we can reduce the equations $(A_d), (B_d)$ in Theorem 1 to the linear system $(\star)$ over the big field $\mathbb{K}$.

3.3   Our Proposed Algorithm

Here, we explain an algorithm to solve the linear systems in Theorem 2. This is our new algorithm to generate F_1, F_2 and Ψ; see Algorithm 2 in Appendix A for overview of our algorithm in this section.

Set

$$c_{d,i} := c_{1,d,i}, \quad c_{d,n+i} := c_{2,d,i} \quad \text{for } d, i.$$

Take a sequence $1 \leq i_1 < i_2 < \cdots < i_{m-1} < i_m \leq 2n$, where $1 \leq m \leq 2n$. We denote by $L[i_1, i_2, \ldots, i_m]$ the $2n \times m$ matrix that is obtained by leaving each i_j-column of L. Similarly, we define

$$(c_1, c_2, \ldots, c_n, c_{n+1}, c_{n+2}, \ldots, c_{2n})[i_1, i_2, \ldots, i_m] := (c_{i_1}, c_{i_2}, \ldots, c_{i_m}).$$

Now, we explain our algorithm that gives solutions of the linear systems $(\star)$ for well chosen scalars $c_{d,i}$.

$\underline{d = l+1}$

$$S_{l+1}' := \{i \mid 1 \leq i \leq n, \deg X \cdot X_{l+1,i} \leq D\} \cup \{n+i' \mid 1 \leq i' \leq n, \deg X^q \cdot X_{l+1,i'} \leq D\},$$

$$S_{l+1} := \{1, \ldots, 2n\} \setminus (S_{l+1}' \cup \{l+3, n+1\}).$$

Randomly choose a scalar z in $\mathbb{K}$. For any $i \in S_{l+1}$, set

$$c_{l+1,i} := \begin{cases} z & \text{if } i = 2 \text{ and } 2 \in S_{l+1}, \\ -z & \text{if } i = n+l+2 \text{ and } 2 \in S_{l+1}, \\ 0 & \text{otherwise.} \end{cases}$$

Then we consider the following linear system:

$$(c_{l+1,1}, c_{l+1,2}^{q^{n-1}}, \ldots, c_{l+1,n}^q, c_{l+1,n+1}, c_{l+1,n+2}^{q^{n-1}}, \ldots, c_{l+1,2n}^q)[S_{l+1}]$$
$$= (x_{l+1,1}, \ldots, x_{l+1,n}, y_{l+1,1}, \ldots, y_{l+1,n}) \cdot L[S_{l+1}].$$

Note that since the scalars $c_{l+1,i}$ ($i \notin S_{l+1}$) do not occur in this system, this system is well-defined. After we find a solution $(x_{l+1,1}, \ldots, x_{l+1,n}, y_{l+1,1}, \ldots, y_{l+1,n})$ of this system, the other scalars $c_{l+1,i}$, ($i \notin S_{l+1}$) are given by the formula

$$(c_{l+1,1}, c_{l+1,2}^{q^{n-1}}, \ldots, c_{l+1,n}^q, c_{l+1,n+1}, c_{l+1,n+2}^{q^{n-1}}, \ldots, c_{l+1,2n}^q)$$
$$= (x_{l+1,1}, \ldots, x_{l+1,n}, y_{l+1,1}, \ldots, y_{l+1,n}) \cdot L.$$

$1 < d < l+1$

$$S_d' := \{i \mid 1 \le i \le n, \ \deg X \cdot X_{d,i} \le D\} \cup \{n+i' \mid 1 \le i' \le n, \ \deg X^q \cdot X_{d,i'} \le D\},$$

$$S_d := \{1, \ldots, 2n\} \smallsetminus (S_d' \cup \{(n+2-d \mod n)+1, n+1\}).$$

For any $i \in S_d$, we set

$$c_{d,i} := \begin{cases} -c_{d+1,n+1} & \text{if } i = 2 \text{ and } 2 \in S_d, \\ -c_{d+1,n+2-d} & \text{if } i = 2n+2-d \text{ and } 2n+2-d \in S_d, \\ 0 & \text{otherwise.} \end{cases}$$

Then we consider the following linear system:

$$(c_{d,1}, c_{d,2}^{q^{n-1}}, \ldots, c_{d,n}^q, c_{d,n+1}, c_{d,n+2}^{q^{n-1}}, \ldots, c_{d,2n}^q)[S_d]$$
$$= (x_{d,1}, \ldots, x_{d,n}, y_{d,1}, \ldots, y_{d,n}) \cdot L[S_d].$$

After we find a solution $(x_{d,1}, \ldots, x_{d,n}, y_{d,1}, \ldots, y_{d,n})$ of this system, the other scalars $c_{d,i}$, ($i \notin S_d$) are given by the formula

$$(c_{d,1}, c_{d,2}^{q^{n-1}}, \ldots, c_{d,n}^q, c_{d,n+1}, c_{d,n+2}^{q^{n-1}}, \ldots, c_{d,2n}^q)$$
$$= (x_{d,1}, \ldots, x_{d,n}, y_{d,1}, \ldots, y_{d,n}) \cdot L.$$

$d = 1$

$$S_1' := \{i \mid 1 \le i \le n, \ \deg X \cdot X_{1,i} \le D\} \cup \{n+i' \mid 1 \le i' \le n, \ \deg X^q \cdot X_{1,i'} \le D\},$$

$$S_1 := \{1, \ldots, 2n\} \smallsetminus S_1'.$$

For any $i \in S_1$, we set

$$c_{1,i} := \begin{cases} -c_{2,n+1} & \text{if } i = 2 \text{ and } 2 \in S_1, \\ -c_{2,1} & \text{if } i = n+1 \text{ and } n+1 \in S_1, \\ 0 & \text{otherwise.} \end{cases}$$

Then we consider the following linear system:

$$(c_{1,1}, c_{1,2}^{q^{n-1}}, \ldots, c_{1,n}^{q}, c_{1,n+1}, c_{1,n+2}^{q^{n-1}}, \ldots, c_{1,2n}^{q})[S_1]$$
$$= (x_{1,1}, \ldots, x_{1,n}, y_{1,1}, \ldots, y_{1,n}) \cdot L[S_1].$$

After we find a solution $(x_{1,1}, \ldots, x_{1,n}, y_{1,1}, \ldots, y_{1,n})$ of this system, the other scalars $c_{1,i}$, $(i \notin S_1)$ are given by the formula

$$(c_{1,1}, c_{1,2}^{q^{n-1}}, \ldots, c_{1,n}^{q}, c_{1,n+1}, c_{1,n+2}^{q^{n-1}}, \ldots, c_{1,2n}^{q})$$
$$= (x_{1,1}, \ldots, x_{1,n}, y_{1,1}, \ldots, y_{1,n}) \cdot L.$$

Finally, we have quadratic HFE polynomials F_1, F_2 and Ψ such that $\deg \Psi \leq D$:

$$F_1 = \sum_{1 \leq d \leq l+1} \sum_{1 \leq i \leq n} x_{d,n+2-i}^{q^{i-1}} X_{d,i}, \quad F_2 = \sum_{1 \leq d \leq l+1} \sum_{1 \leq i \leq n} y_{d,n+2-i}^{q^{i-1}} X_{d,i},$$

$$\Psi = \sum_{1 \leq d \leq l+1} \left(\sum_{i \in S'_d, i \leq n} c_{d,i} X \cdot X_{d,i} + \sum_{i \in S'_d, i > n} c_{d,i} X^q \cdot X_{d,i-n} \right).$$

Remark 3. If corank $L \leq 2$, then each $L[S_d]$ has the full rank. Thus all the above linear systems have solutions. Therefore, if corank $L \leq 2$, then our algorithm terminates. Also if corank $L \geq 3$, then our algorithm failed in our experiments. But Table 2 implies that the class of L with corank ≥ 3 is very small in total. Thus we may take L to be corank $L \leq 2$. Notice that Baena et al.'s algorithm [2] succeeds for corank $L \leq 5$. For L with higher corank, their algorithm also works, but produces $\Psi = 0$ making the corresponding ZHFE scheme insure under linearization attack. However, it suffices to only consider linear maps L of corank less than 3 for their majority, cf. Table 2.

Algorithm 1. Generating Matrix L of Corank r (Section 4.2)

Input : a field $\mathbb{F}$ with q elements, integers n and r, $\mathbb{K}$ the extension field of degree n over $\mathbb{F}$, an $\mathbb{F}$-basis $(\theta_1, \ldots, \theta_n)$ of $\mathbb{K}$
Output: $\alpha_1, \ldots, \alpha_{2n}, \beta_1, \ldots, \beta_{2n}, L$ with corank r (See (12) for L)

$M \leftarrow (\theta_i^{q^{j-1}})_{1 \leq i, j \leq n}$;
$A, B \leftarrow Random(GL_{2n}(\mathbb{F}))$;
$L' \leftarrow \begin{pmatrix} M^{-1} & \\ & M^{-1} \end{pmatrix} \cdot A \cdot \begin{pmatrix} 1_{2n-r} & \\ & 0_r \end{pmatrix} \cdot B \cdot \begin{pmatrix} M & \\ & M \end{pmatrix}$, $2n \times 2n$ matrix;
$\alpha_i \leftarrow L'_{i,1}$: the $(i,1)$-entry of L', $(1 \leq i \leq 2n)$;
$\beta_i \leftarrow L'_{i,n+1}$ $(1 \leq i \leq 2n)$;
$L \leftarrow (L'_1, L'_n, L'_{n-1}, \ldots, L'_2, L'_{n+1}, L'_{2n}, L'_{2n-1}, \ldots, L'_{n+2})$;
where L'_i is the i-th column of L'

---

4   Complexity and Implementation Results

In this section, we give the complexity and implementation results for our private key generation algorithm of ZHFE scheme.

4.1   The Complexity of the Proposed Algorithm

We can easily prove the complexity of our proposed algorithm (Algorithm 2) discussed in Sect. 3.3 in the following theorem.

Theorem 3. *The complexity of our algorithm in Sect. 3.3 is given by $\mathcal{O}(n^{\omega+3})$.*

Proof. In our algorithm proposed in Sect. 3.3, we obtain a private ZHFE key by solving $l + 1$ linear systems over the big field $\mathbb{K}$. Here each linear system has at most $2n$ variables and at most $2n$ equations. Thus the complexity is

$$(l+1) \times (2n)^{\omega} \times (\log q^n)^2 = \mathcal{O}(n^{\omega+3}). \qquad \square$$

Thus our algorithm improves the original algorithm of $\mathcal{O}(n^{3\omega})$ and Baena et al.'s algorithm of $\mathcal{O}(n^{2\omega+1})$ (See Table 1).

4.2   Our Experiments

In order to perform the experiments of generating the private ZHFE keys, we need to decide the matrix $L \in M_{2n}(\mathbb{K})$ in Eq. (12), where L is generated by $\alpha_1, \ldots, \alpha_{2n}, \beta_1, \ldots, \beta_{2n}$ in Eq. (2). Note that our proposed algorithm works only for matrices L with corank $0, 1$ and 2 (cf. Remark 3), and thus we have to investigate how to generate such a matrix. In the following we describe an algorithm for generating the matrix L of any corank $0 \leq r \leq 2n$. For the matrix $\begin{pmatrix} 1_{2n-r} & \\ & 0_r \end{pmatrix}$ in $M_{2n}(\mathbb{K})$ of corank r, we multiply random invertible matrices $A, B \in GL_{2n}(\mathbb{F})$ and matrices $\begin{pmatrix} M^{-1} & \\ & M^{-1} \end{pmatrix}, \begin{pmatrix} M & \\ & M \end{pmatrix}$ from both sides. The resulting matrix of corank r implies $\alpha_1, \ldots, \alpha_{2n}, \beta_1, \ldots, \beta_{2n}$ used for the private ZHFE keys. The explicit algorithm is described in Algorithm 1.

On the other hand, as can be seen in Table 2, in most cases the corank L is 0 or 1 for randomly chosen $\alpha_1, \ldots, \alpha_{2n}, \beta_1, \ldots, \beta_{2n}$. Note that the corank of L is equal to the corank $\mathcal{L}$ in Eq. (4) (cf. Remark 2). Therefore if we generate the matrix L by Algorithm 1 with $r \leq 2$, then we can generate almost all instances L for the key generation algorithms of ZHFE scheme.

4.3   Comparison of Timings

The implementation results and the comparison between our algorithm and Baena et al.'s algorithm [2] are presented in Table 3. All the experiments in this section were performed using Magma V2.20-10 [4] with a processor Intel(R) Core(TM) i5-4300U CPU @ 1.90 GHz, running Windows 7 Professional SP1.

Table 3. The comparison of timings between Baena et al.'s algorithm [2] and our algorithm.

			Our algorithm		Baena et al.'s	
q	D	n	CPU time [s]	Max Memory [MB]	CPU time [s]	Max Memory [MB]
7	105	15	0.09	10	0.59	11
7	105	31	3.47	11	22.27	43
7	**105**	**55**	**39.19**	**18**	**607.06**	**338**
17	105	15	0.13	9	3.06	14
17	105	31	3.91	11	348.57	81
17	**595**	**55**	**62.91**	**22**	**15350.79**	**683**

Notice that according to the estimation of Zhang and Tan [24], the parameters $(q = 7, n = 55, D = 105)$, which is recommended in the original paper [20], and $(q = 17, n = 55, D = 595)$ are for 96-bit and 111-bit security level respectively. In Table 3 we present the timings of our experiments using these parameters and in addition we run experiments under other parameters $n = 15, 31$.

Our algorithm in Table 3 presents timing to generate a private ZHFE key, that is, $\alpha_1, \ldots, \alpha_{2n}$, $\beta_1, \ldots, \beta_{2n}$, L, F_1, F_2 and Ψ. Here, we used Algorithm 1 to generate $\alpha_1, \ldots, \alpha_{2n}, \beta_1, \ldots, \beta_{2n}$ and L with corank $L \leq 2$. For example, for the recommended parameters $(q = 7, D = 105, n = 55)$ at 96-bit security, our algorithm takes 39.19 s and the max memory is 18 mega bytes.[1] For comparison, we also present timing to generate such a private ZHFE key by Baena et al.'s algorithm. For the recommended parameters $(q = 7, D = 105, n = 55)$, our algorithm is around 15 times faster than that of Baena et al. [2].

5   Conclusion

In this paper, we proposed a new efficient algorithm for generating private keys of the ZHFE scheme [20]. Our algorithm has complexity $O(n^{\omega+3})$ which improves the original [19] and Baena's [2] algorithm whose complexities are $O(n^{3\omega})$ and $O(n^{2\omega+1})$ respectively. Here n is the number of variables and $2 < \omega < 3$ is a linear algebra constant. Our algorithm is in practice very fast compared to that of Baena et al.: for recommended parameter $(q = 7, n = 55, D = 105)$ at 96-bit security, our algorithm is around 15 times faster than that of Baena et al.; cf. Table 3. Moreover, in contrast to Zhang and Tan's algorithm [24], our algorithm generates as many private ZHFE keys as the previous ones [2,20], as estimated in Table 2. Although our algorithm works for linear maps L with corank $L \leq 2$ (cf. Remark 3), it already generates around 99% private keys in total (cf. Table 2). This makes our algorithm to be the most appropriate for generating private ZHFE keys.

---

[1] Here we used the Magma's command `GetMaximumMemoryUsage` to measure max memory. Note also that we used the Magma's command `Solution` to solve linear systems in the algorithm.

Acknowledgments. This work was supported by CREST, JST. The second author also acknowledges the Japanese Society for the Promotion of Science (JSPS) for financial support under grant KAKENHI 16K17644.

A   Our Algorithm in Sect. 3.3

The expression $f \overset{R}{\leftarrow} W$ denotes that f is an element chosen uniformly at random from the set W.

Algorithm 2. Our Proposed Algorithm (Section 3.3)

Input : $\mathbb{F}$: field with q elements, $n = 2l + 1$: odd integer, $\mathbb{K}$: extension field of degree n,
       L: the $2n \times 2n$ matrix chosen by Algorithm 1, D: interger
Output: F_1, F_2, Ψ: private key

$c_d \leftarrow (0, 0, \ldots, 0)$, length $2n$, $\ 1 \le d \le l+1$;
if $2 \in S_{l+1}$ **then**
   | $c_{l+1,2} \leftarrow Random(\mathbb{K})$;
   $\lfloor$ $c_{l+1,n+l+2} \leftarrow -c_{l+1,2}$;

$c'_{l+1} \leftarrow (c_{l+1,1}, c_{l+1,2}^{q^{n-1}}, c_{l+1,3}^{q^{n-2}}, \ldots, c_{l+1,n}^{q}, c_{l+1,n+1}, c_{l+1,n+2}^{q^{n-1}}, \ldots, c_{l+1,2n}^{q})$;
$f \overset{R}{\leftarrow} W := \{f \in \mathbb{K}^{2n} \mid f \cdot L[S_{l+1}] = c'_{l+1}[S_{l+1}]\}$;
$g \leftarrow f \cdot L$;
$c_{l+1} \leftarrow (g_1, g_2^{q}, \ldots, g_n^{q^{n-1}}, g_{n+1}, g_{n+2}^{q}, \ldots, g_{2n}^{q^{n-1}})$;
$x_{l+1} \leftarrow (f_1, \ldots, f_n)$, $y_{l+1} \leftarrow (f_{n+1}, \ldots, f_{2n})$;
$d \leftarrow l$;
while $d > 1$ **do**
   if $2 \in S_d$ **then**
      $\lfloor$ $c_{d,2} \leftarrow -c_{d+1,n+1}$;
   if $2n + 2 - d \in S_d$ **then**
      $\lfloor$ $c_{d,2n+2-d} \leftarrow -c_{d+1,n+2-d}$;
   $c'_d \leftarrow (c_{d,1}, c_{d,2}^{q^{n-1}}, c_{d,3}^{q^{n-2}}, \ldots, c_{d,n}^{q}, c_{d,n+1}, c_{d,n+2}^{q^{n-1}}, \ldots, c_{d,2n}^{q})$;
   $f \overset{R}{\leftarrow} W := \{f \in \mathbb{K}^{2n} \mid f \cdot L[S_d] = c'_d[S_d]\}$;
   $g \leftarrow f \cdot L$;
   $c_d \leftarrow (g_1, g_2^{q}, \ldots, g_n^{q^{n-1}}, g_{n+1}, g_{n+2}^{q}, \ldots, g_{2n}^{q^{n-1}})$;
   $x_d \leftarrow (f_1, \ldots, f_n)$, $y_d \leftarrow (f_{n+1}, \ldots, f_{2n})$;
   $\lfloor$ $d \leftarrow d - 1$;

if $2 \in S_1$ **then**
   $\lfloor$ $c_{1,2} \leftarrow -c_{2,n+1}$;
if $n + 1 \in S_1$ **then**
   $\lfloor$ $c_{1,n+1} \leftarrow -c_{2,1}$;
$c'_1 \leftarrow (c_{1,1}, c_{1,2}^{q^{n-1}}, c_{1,3}^{q^{n-2}}, \ldots, c_{1,n}^{q}, c_{1,n+1}, c_{1,n+2}^{q^{n-1}}, \ldots, c_{1,2n}^{q})$;
$f \overset{R}{\leftarrow} W := \{f \in \mathbb{K}^{2n} \mid f \cdot L[S_1] = c'_1[S_1]\}$;
$g \leftarrow f \cdot L$;
$c_1 \leftarrow (g_1, g_2^{q}, \ldots, g_n^{q^{n-1}}, g_{n+1}, g_{n+2}^{q}, \ldots, g_{2n}^{q^{n-1}})$;
$x_1 \leftarrow (f_1, \ldots, f_n)$, $y_1 \leftarrow (f_{n+1}, \ldots, f_{2n})$;
$F_1 \leftarrow \sum_{1 \le d \le l+1} \sum_{1 \le i \le n} x_{d,n+2-i}^{q^{i-1}} X_{d,i}$;
$F_2 \leftarrow \sum_{1 \le d \le l+1} \sum_{1 \le i \le n} y_{d,n+2-i}^{q^{i-1}} X_{d,i}$;
$\Psi \leftarrow \sum_{1 \le d \le l+1} \left(\sum_{i \in S'_d, i \le n} c_{d,i} X \cdot X_{d,i} + \sum_{i \in S'_d, i > n} c_{d,i} X^{q} \cdot X_{d,i-n} \right)$;

References

1. Bernstein, D.J., Buchmann, J., Dahmen, E.: Post-Quantum Cryptography. Springer, Heidelberg (2009)
2. Baena, J.B., Cabarcas, D., Escudero, D.E., Porras-Barrera, J., Verbel, J.A.: Efficient ZHFE key generation. In: Takagi, T. (ed.) PQCrypto 2016. LNCS, vol. 9606, pp. 213–232. Springer, Cham (2016). doi:10.1007/978-3-319-29360-8_14
3. Bogdanov, A., Eisenbarth, T., Rupp, A., Wolf, C.: Time-area optimized public-key engines: $\mathcal{MQ}$-cryptosystems as replacement for elliptic curves? In: Oswald, E., Rohatgi, P. (eds.) CHES 2008. LNCS, vol. 5154, pp. 45–61. Springer, Heidelberg (2008). doi:10.1007/978-3-540-85053-3_4
4. Bosma, W., Cannon, J., Playoust, C.: The Magma algebra system I: the user language. J. Symbolic Comput. **24**(3–4), 235–265 (1997)
5. Chen, A.I.-T., Chen, M.-S., Chen, T.-R., Cheng, C.-M., Ding, J., Kuo, E.L.-H., Lee, F.Y.-S., Yang, B.-Y.: SSE implementation of multivariate PKCs on modern x86 CPUs. In: Clavier, C., Gaj, K. (eds.) CHES 2009. LNCS, vol. 5747, pp. 33–48. Springer, Heidelberg (2009). doi:10.1007/978-3-642-04138-9_3
6. Courtois, N.T.: The security of hidden field equations (HFE). In: Naccache, D. (ed.) CT-RSA 2001. LNCS, vol. 2020, pp. 266–281. Springer, Heidelberg (2001). doi:10.1007/3-540-45353-9_20
7. Ding, J., Gower, J.E., Schmidt, D.S.: Multivariate Public Key Cryptosystems. Springer, Heidelberg (2006)
8. Ding, J., Schmidt, D.: Rainbow, a new multivariable polynomial signature scheme. In: Ioannidis, J., Keromytis, A., Yung, M. (eds.) ACNS 2005. LNCS, vol. 3531, pp. 164–175. Springer, Heidelberg (2005). doi:10.1007/11496137_12
9. Goodin, D.: NSA preps quantum-resistant algorithms to head off crypto-apocalypse. http://arstechnica.com/security/2015/08/nsa-preps-quantum-resistant-al-gorithms-to-head-off-crypto-apocolypse/
10. Garey, M.R., Johnson, D.S.: Computers and Intractability: A Guide to the Theory of NP-Completeness. W.H Freeman and Company, New York (1979)
11. Kipnis, A., Patarin, J., Goubin, L.: Unbalanced oil and vinegar signature schemes. In: Stern, J. (ed.) EUROCRYPT 1999. LNCS, vol. 1592, pp. 206–222. Springer, Heidelberg (1999). doi:10.1007/3-540-48910-X_15
12. Kipnis, A., Shamir, A.: Cryptanalysis of the HFE public key cryptosystem by relinearization. In: Wiener, M. (ed.) CRYPTO 1999. LNCS, vol. 1666, pp. 19–30. Springer, Heidelberg (1999). doi:10.1007/3-540-48405-1_2
13. Matsumoto, T., Imai, H.: Public quadratic polynomial-tuples for efficient signature-verification and message-encryption. In: Barstow, D., Brauer, W., Brinch Hansen, P., Gries, D., Luckham, D., Moler, C., Pnueli, A., Seegmüller, G., Stoer, J., Wirth, N., Günther, C.G. (eds.) EUROCRYPT 1988. LNCS, vol. 330, pp. 419–453. Springer, Heidelberg (1988). doi:10.1007/3-540-45961-8_39
14. National Institute of Standards and Technology: Report on Post Quantum Cryptography, NISTIR draft 8105. http://csrc.nist.gov/publications/drafts/nistir-8105/nistir_8105_draft.pdf
15. Patarin, J.: Cryptanalysis of the Matsumoto and Imai public key scheme of eurocrypt 88. In: Coppersmith, D. (ed.) CRYPTO 1995. LNCS, vol. 963, pp. 248–261. Springer, Heidelberg (1995). doi:10.1007/3-540-44750-4_20
16. Patarin, J.: Hidden fields equations (HFE) and isomorphisms of polynomials (IP): two new families of asymmetric algorithms. In: Maurer, U. (ed.) EUROCRYPT 1996. LNCS, vol. 1070, pp. 33–48. Springer, Heidelberg (1996). doi:10.1007/3-540-68339-9_4

17. Perlner, R., Smith-Tone, D.: Security analysis and key modification for ZHFE. In: Takagi, T. (ed.) PQCrypto 2016. LNCS, vol. 9606, pp. 197–212. Springer, Heidelberg (2016). doi:10.1007/978-3-319-29360-8_13

18. Petzoldt, A., Chen, M.-S., Yang, B.-Y., Tao, C., Ding, J.: Design principles for HFEv- based multivariate signature schemes. In: Iwata, T., Cheon, J.H. (eds.) ASIACRYPT 2015. LNCS, vol. 9452, pp. 311–334. Springer, Heidelberg (2015). doi:10.1007/978-3-662-48797-6_14

19. Porras, J., Baena, J., Ding, J.: New candidates for multivariate trapdoor functions. Cryptology ePrint Archive, Report 2014/387 (2014)

20. Porras, J., Baena, J., Ding, J.: ZHFE, a new multivariate public key encryption scheme. In: Mosca, M. (ed.) PQCrypto 2014. LNCS, vol. 8772, pp. 229–245. Springer, Heidelberg (2014). doi:10.1007/978-3-319-11659-4_14

21. Shor, P.: Polynomial-time algorithms for prime factorization and discrete logarithms on a quantum computer. SIAM J. Comput. **26**(5), 1484–1509 (1997)

22. Tao, C., Diene, A., Tang, S., Ding, J.: Simple matrix scheme for encryption. In: Gaborit, P. (ed.) PQCrypto 2013. LNCS, vol. 7932, pp. 231–242. Springer, Heidelberg (2013). doi:10.1007/978-3-642-38616-9_16

23. Szepieniec, A., Ding, J., Preneel, B.: Extension field cancellation: a new central trapdoor for multivariate quadratic systems. In: Takagi, T. (ed.) PQCrypto 2016. LNCS, vol. 9606, pp. 182–196. Springer, Cham (2016). doi:10.1007/978-3-319-29360-8_12

24. Zhang, W., Tan, C.H.: On the Security and key generation of the ZHFE encryption scheme. In: Ogawa, K., Yoshioka, K. (eds.) IWSEC 2016. LNCS, vol. 9836, pp. 289–304. Springer, Heidelberg (2016). doi:10.1007/978-3-319-44524-3_17

25. Yasuda, T., Sakurai, K.: A multivariate encryption scheme with rainbow. In: Qing, S., Okamoto, E., Kim, K., Liu, D. (eds.) ICICS 2015. LNCS, vol. 9543, pp. 236–251. Springer, Heidelberg (2016). doi:10.1007/978-3-319-29814-6_19

The Weight Distribution for an Extended Family of Reducible Cyclic Codes

Gerardo Vega[1(✉)] and Jesús E. Cuén-Ramos[2]

[1] Dirección General de Cómputo y de Tecnologías de Información y Comunicación, Universidad Nacional Autónoma de México, 04510 Ciudad de México, Mexico
gerardov@unam.mx
[2] Posgrado en Ciencias Matemáticas, Universidad Nacional Autónoma de México, 20059 Ciudad de México, Mexico
elisandro@ciencias.unam.mx

Abstract. The purpose of this work is to present new advances on the weight distribution of the duals of some cyclic codes with two zeros. More specifically, our contribution improves the sufficient numerical conditions that determine the weight distribution for the class of reducible cyclic codes that were studied in [19] and in [17]. Furthermore, as will be shown later, a conclusion here will be that thanks to these previous works and the present contribution, we can determine the weight distribution for an extended family of reducible cyclic codes. More specifically, we are going to determine the weight distribution for all the elements of an extended family of reducible cyclic codes that fully covers one of the open cases suggested in [20]. In addition, as will be seen further on, through our results we obtain an alternative description for one of the families of cross-correlation functions studied in [8].

Keywords: Weight distribution · Reducible cyclic codes · Gaussian periods · Cross-correlation functions

1 Introduction

It is said that a cyclic code is reducible if its parity-check polynomial is factorizable in two or more irreducible factors. Since the number of zeros of a cyclic code is the number of non-conjugated zeros of its generator polynomial (see for example [12, p. 199]), it should become clear that a reducible cyclic code whose parity-check polynomial is factorizable in $s > 1$ irreducible factors, is nothing but a cyclic code whose dual code has s zeros (non-conjugated).

Over a number of years, several authors have dedicated their efforts to solving the problem of determining the weight distribution of families of reducible or irreducible cyclic codes (see for example [3,7,9,15,18]), and this has been so because the weight distribution determines the capabilities of error detection

Partially supported by PAPIIT-UNAM IN107515.

J.E. Cuén-Ramos—Ph.D. student.

© Springer International Publishing AG 2017
S. El Hajji et al. (Eds.): C2SI 2017, LNCS 10194, pp. 213–229, 2017.
DOI: 10.1007/978-3-319-55589-8_14

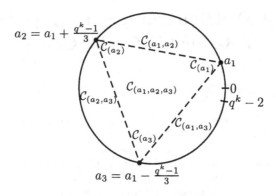

Fig. 1. All $(3, s, N)$-polygon cyclic codes with $s = 1, 2, 3$.

and correction of a given code. On the other hand, the family of cyclic codes is important because it possesses a rich algebraic structure that can be utilized in a variety of ways, particularly, in the design of very efficient coding and decoding algorithms. A recent classification of some of the families of cyclic codes that has been used by several authors (see particularly [20], and alternatively [5,6,11,21]) has shown to be a guide that helps to clarify these efforts.

In order to explain such classification, let p, t, q, k and Δ be five positive integers, such that p is a prime number, $q = p^t$, and $\Delta = (q^k - 1)/(q - 1)$. Assume that γ is a fixed primitive element of $\mathbb{F}_{q^k}$ and, for any integer a, denote by $h_a(x) \in \mathbb{F}_q[x]$ the minimal polynomial of γ^{-a}. Also, for any integers a, e and N such that $e > 1$, $e|(q^k - 1)$ and $N = \gcd(\Delta, ea)$, consider $\mathcal{I}$ to be the set of e integers given by $\mathcal{I} := \{a + \frac{q^k - 1}{e} i | 0 \le i < e\}$. Thus, for any subset $\{a_1, \ldots, a_s\} \subseteq \mathcal{I}$, with $1 \le s \le e$, we are interested in the weight distribution of the cyclic code $\mathcal{C}_{(a_1, \cdots, a_s)}$, whose parity-check polynomial is $\prod_{i=1}^{s} h_{a_i}(x)$. Throughout this work, and for simplicity, a cyclic code constructed in such a way will be called a *polygon cyclic code* that belongs to the (e, s, N) class. Alternatively, we refer to it simply as an (e, s, N)-polygon cyclic code. Since the index a_1 of any finite field element γ^{a_1}, in $\mathbb{F}_{q^k}$, is taken modulo $q^k - 1$, we now use Fig. 1 to illustrate "geometrically", all the possible polygon cyclic codes with $e = 3$, $s = 1, 2, 3$, and under the assumption that $N = \gcd(\Delta, 3a_1)$. Clearly, for a class of (e, s, N)-polygon cyclic codes, the challenge will be to determine the sufficient numerical conditions over the integers in $\mathcal{I}$, in order to give the weight distribution for some or, preferably, for all subclasses of such cyclic codes.

In this work, we are interested in polygon cyclic codes for which $e = N = 3$. In that sense, it is worth noting that the problem of determining the weight distribution for all possible $(3, 3, 3)$-polygon cyclic codes was completely solved in [20], while the class of $(3, 2, 3)$-polygon cyclic codes was recently studied in [19] and in [17]. It is important to note that, unlike [17], the study in [19] was made by considering the two possible cases: $p \equiv 1 \pmod 3$, and $p \equiv 2 \pmod 3$ (the semiprimitive case). However, it is also important to note that the study in [19] was limited to those $(3, 2, 3)$-polygon cyclic codes whose dimensions are of the form $2k$, where $3|k$ (see page 512, therein), and this, in turn, means that such study was restricted to $(3, 2, 3)$-polygon cyclic codes that satisfy the condition $9|(q^k - 1)$. In contrast to this, the study in [17] was limited to the semiprimitive case and just for these $(3, 2, 3)$-polygon cyclic codes where the condition $9|(q^k - 1)$ must not be satisfied. Therefore, as a consequence of this, it is important to stress that the subclasses of $(3, 2, 3)$-polygon cyclic codes obtained in those two works are incomplete and disjoint, despite that some of the weight distribution tables can be interchanged between them.

Thus, by considering only the semiprimitive case, one of the goals for this work is to present an improvement on the sufficient numerical conditions that determine the weight distribution of the class of $(3, 2, 3)$-polygon cyclic codes, that includes the two disjoint subclasses studied in [19] and in [17]. Clearly, this improvement will result in a more relaxed description for these class of polygon cyclic codes. A second goal is to consider this relaxed description, along with the subclass of $(3, 2, 3)$-polygon cyclic codes belonging to the $p \equiv 1 \pmod 3$ case, in order to determine the weight distribution for all the possible $(3, 2, 3)$-polygon cyclic codes, achieving in this way a solution that fully covers one of the open cases suggested in [20]. That is, we will show that by means of some of the weight distribution tables in [19] (see Tables 3 and 4 therein) and some of the

Table 1. Weight distribution

Weight	Frequency
0	1
$\frac{2(q-1)}{3dq}(q^k + 2(-1)^{kt/2}q^{k/2})$	$q^k - 1$
$\frac{2(q-1)}{3dq}(q^k - (-1)^{kt/2}q^{k/2})$	$2(q^k - 1)$
$\frac{q-1}{dq}(q^k + 2(-1)^{kt/2}q^{k/2})$	$\frac{1}{27}(q^k - 1)(q^k - 2(-1)^{kt/2}q^{k/2} - 8)$
$\frac{q-1}{dq}(q^k + (-1)^{kt/2}q^{k/2})$	$\frac{2}{9}(q^k - 1)(q^k + (-1)^{kt/2}q^{k/2} - 2)$
$\frac{q-1}{d}q^{k-1}$	$\frac{2}{9}(q^k - 1)(2q^k - (-1)^{kt/2}q^{k/2} - 1)$
$\frac{q-1}{dq}(q^k - (-1)^{kt/2}q^{k/2})$	$\frac{2}{27}(q^k - 1)(4q^k + (-1)^{kt/2}q^{k/2} - 14)$

Table 2. Weight distribution

Weight	Frequency
0	1
$\frac{2(q-1)}{3dq}(q^k - (-1)^{kt/2}q^{k/2})$	$q^k - 1$
$\frac{(q-1)}{3dq}(2q^k + (-1)^{kt/2}q^{k/2})$	$2(q^k - 1)$
$\frac{q-1}{dq}(q^k + 2(-1)^{kt/2}q^{k/2})$	$\frac{1}{27}(q^k - 1)(q^{k/2} - (-1)^{kt/2})^2$
$\frac{q-1}{dq}(q^k + (-1)^{kt/2}q^{k/2})$	$\frac{2}{9}(q^k - 1)(q^k + (-1)^{kt/2}q^{k/2} - 2)$
$\frac{q-1}{d}q^{k-1}$	$\frac{1}{9}(q^k - 1)(4q^k - 2(-1)^{kt/2}q^{k/2} - 11)$
$\frac{q-1}{dq}(q^k - (-1)^{kt/2}q^{k/2})$	$\frac{2}{27}(q^k - 1)(4q^k + (-1)^{kt/2}q^{k/2} - 5)$

tables that we are going to present here (see Tables 1 and 2), it is now possible to obtain the weight distribution of *any* $(3, 2, 3)$-polygon cyclic code over any finite field, excluding, of course, the trivial case when $q^k = 4$. Thus, our first goal is accomplished through the following:

Theorem 1. *Let p, t, q, k and Δ be as before, and suppose that $3|(p+1)$, $3|\Delta$ and $q^k \neq 4$. Let also a_1, a_2, a_3, d and n be integers such that $a_2 = a_1 + \frac{q^k-1}{3}$, $a_3 = a_1 - \frac{q^k-1}{3}$, $d = \gcd(q^k - 1, a_1, a_2, a_3)$ and $n = \frac{q^k-1}{d}$. Then $\gcd(\Delta, 3a_1) = \gcd(\Delta, 3a_2) = \gcd(\Delta, 3a_3)$. In particular, if $\gcd(\Delta, 3a_1) = 3$, then the following conditional statements are true:*

(A) If $\gcd(\Delta, a_i) = 1$, for some $i = 1, 2, 3$, then $\mathcal{C}_{(a_i)}$ is an $[n, k]$ one-weight irreducible cyclic code, whose nonzero weight is $\frac{q-1}{d}q^{k-1}$. On the other hand, if $\gcd(\Delta, a_i) = 3$, then $\mathcal{C}_{(a_i)}$ is an $[n, k]$ semiprimitive two-weight irreducible cyclic code, whose weight enumerator polynomial is

$$A(z) = 1 + \frac{2(q^k - 1)}{3} z^{\frac{q-1}{dq}(q^k - (-1)^{kt/2}q^{k/2})} + \frac{(q^k - 1)}{3} z^{\frac{q-1}{dq}(q^k + 2(-1)^{kt/2}q^{k/2})}. \tag{1}$$

(B) If $\gcd(\Delta, a_1) = \gcd(\Delta, a_2) = 1$ and $\gcd(\Delta, a_3) = 3$, then $\mathcal{C}_{(a_1,a_2)}$ is an $[n, 2k]$ cyclic code, whose weight distribution is given in Table 1, whereas $\mathcal{C}_{(a_1,a_3)}$ and $\mathcal{C}_{(a_2,a_3)}$ are $[n, 2k]$ cyclic codes with the same weight distribution given in Table 2.

(C) If $\gcd(\Delta, a_1) = \gcd(\Delta, a_2) = \gcd(\Delta, a_3)$, then $\mathcal{C}_{(a_1,a_2)}$, $\mathcal{C}_{(a_1,a_3)}$ and $\mathcal{C}_{(a_2,a_3)}$ are $[n, 2k]$ cyclic codes with the same weight distribution which is given in Table 1 if $3|a_1$, and in Table 2 otherwise.

It is interesting to note that the subclass of $(3, 2, 3)$-polygon cyclic codes described in Part *(C)*, of the previous theorem, contains completely the subclass of cyclic codes studied in [19], under the semiprimitive case. On the other hand, the cyclic codes studied in [17] are just an instance of the subclass of $(3, 2, 3)$-polygon cyclic codes described in Part *(B)*. And of course, all of this is in this way because the sufficient numerical conditions in Theorem 1 are now more relaxed than those in [19] and in [17] (see Example 1 and Remark 6 below).

This work is organized as follows: In Sect. 2 we establish the notation and the main assumption that will be considered throughout this work unless otherwise indicated (Sect. 6). We also recall, in this section, some important already known results. Section 3 is devoted to presenting some preliminary results, while in Sect. 4, we use these results in order to determine the value distribution of a particular class of exponential sums. As we will see, this value distribution will be important in order to present a formal proof of Theorem 1 in Sect. 5. In Sect. 6 we use some of the results in [19], and Theorem 1, in order to show that it is now possible to give the weight distribution of any $(3, 2, 3)$-polygon cyclic code. In Sect. 7 we present alternative description for one of the families of cross-correlation functions studied in [8]. Finally, Sect. 8 will be devoted to conclusions.

2   Definitions, Notation and Main Assumption

First of all, we set for this section and for the rest of this work, the following:

Notation. By using p, t, q, k and Δ, we will denote five positive integers such that p is a prime number, $q = p^t$ and $\Delta = (q^k - 1)/(q - 1)$. From now on, γ will denote a fixed primitive element of $\mathbb{F}_{q^k}$, and for any integer a, the polynomial $h_a(x) \in \mathbb{F}_q[x]$ will denote the minimal polynomial of γ^{-a}. For a set of l integers, $\{a_1, a_2, \cdots, a_l\}$, we will denote by $\mathcal{C}_{(a_1, \cdots, a_l)}$ the cyclic code with parity-check polynomial $\prod_{i=1}^l h_{a_i}(x)$. With the notation "$\mathrm{Tr}_{\mathbb{F}_{q^k}/\mathbb{F}_q}$", we will mean the trace mapping from $\mathbb{F}_{q^k}$ to $\mathbb{F}_q$. For any positive divisor m of $q^k - 1$ and for any integer i, we define $\mathcal{D}_i^{(m)} := \gamma^i \langle \gamma^m \rangle$, where $\langle \gamma^m \rangle$ denotes the subgroup of $\mathbb{F}_{q^k}^*$ generated by γ^m (note that $\mathcal{D}_i^{(m)} = \mathcal{D}_{i+lm}^{(m)}$, for any integer l). The m cosets $\mathcal{D}_i^{(m)}$ are called the *cyclotomic classes* of order m in $\mathbb{F}_{q^k}$. In connection with these cyclotomic classes, we recall the *cyclotomic numbers* of order m. Such cyclotomic numbers are defined by

$$(i, j)^{(m, q^k)} := |(\mathcal{D}_i^{(m)} + 1) \cap \mathcal{D}_j^{(m)}|,$$

where $(\mathcal{D}_i^{(m)} + 1) = \{x + 1 \mid x \in \mathcal{D}_i^{(m)}\}$, and $0 \leq i, j \leq m - 1$.

Let p, q, k and γ be as before; then, the canonical additive character χ, of $\mathbb{F}_{q^k}$, is defined as

$$\chi(y) := \zeta_p^{\mathrm{Tr}(y)}, \qquad \text{for all } y \in \mathbb{F}_{q^k},$$

where $\zeta_p := \exp(\frac{2\pi\sqrt{-1}}{p})$ and "Tr" is the absolute trace mapping from $\mathbb{F}_{q^k}$ to $\mathbb{F}_p$. There are several interesting properties about the canonical additive character, and, fortunately for us, these properties are perfectly well explained in Chap. 5 of [10].

Now, we set for this section and for the rest of this work, the following:

Main assumption. Through this work we assume that $3|(p+1)$ and that $3|\Delta$. Thus, since $q^k - 1 = \Delta(q - 1)$, in what follows we will reserve the Greek letter τ in order to fix $\tau = \gamma^{\frac{q^k-1}{3}}$.

Remark 1. Note that if $p \equiv -1 \pmod 3$ and $p^{kt} = q^k \equiv 1 \pmod 3$, then necessarily kt must be an even integer. Conversely, note also that if p is *semiprimitive modulo 3* (see for example [14, p. 9]), then necessarily $p \equiv -1 \pmod 3$. Lastly, note also that the finite field element τ is a primitive third root of unity satisfying $\tau^2 + \tau + 1 = 0$, and since $3|\Delta$, $\mathbb{F}_q^* \subset \mathcal{D}_0^{(3)}$.

Let χ be as before, and let i be any integer. Since $3|(q^k - 1)$, it follows that:

$$\sum_{x \in \mathbb{F}_{q^k}} \chi(\gamma^i x^3) = 1 + 3 \sum_{z \in \mathcal{D}_i^{(3)}} \chi(z).$$

We are particularly interested in the kind of exponential sums that appear in the RHS of the previous equality. These exponential sums are known as the

Gaussian periods of order 3. The following result, which is an instance of the main result in [13], gives us useful information about such Gaussian periods.

Theorem 2. *With our notation and main assumption, let η_0 and η_1 be the two integers given by:*

$$\eta_0 = \frac{-2(-1)^{kt/2}q^{\frac{k}{2}} - 1}{3}, \text{ and } \eta_1 = \frac{(-1)^{kt/2}q^{\frac{k}{2}} - 1}{3}. \tag{2}$$

Then, for any integer i, the i-th Gaussian period of order 3 is:

$$\sum_{z \in \mathcal{D}_i^{(3)}} \chi(z) = \begin{cases} \eta_0 \text{ if } i \equiv 0 \pmod{3}, \\ \eta_1 \text{ otherwise.} \end{cases}$$

Since we will be dealing with the Gaussian periods of order 3, we will also need the cyclotomic numbers of order 3. The following lemma gives us information about such cyclotomic numbers (see [1] for the general result).

Lemma 1. *Let us consider the same notation and hypotheses as in the previous theorem. Then*

$$(0,0)^{(3,q^k)} = \frac{(q^{k/2} - (-1)^{kt/2})^2}{9} - 1,$$

$$(i,0)^{(3,q^k)} = (0,i)^{(3,q^k)} = (i,i)^{(3,q^k)}$$
$$= \frac{(q^{k/2} - (-1)^{kt/2})(q^{k/2} + 2(-1)^{kt/2})}{9}, \text{ for } i = 1,2,$$

$$(1,2)^{(3,q^k)} = (2,1)^{(3,q^k)} = \frac{(q^{k/2} - (-1)^{kt/2})^2}{9}.$$

Remark 2. Note that if $3|(p+1)$, then 3 will be a common divisor of both $(q^{k/2} - (-1)^{kt/2})$ and $(q^{k/2} + 2(-1)^{kt/2})$.

3   Some Preliminary Results

The following result, that was proved in [20] (see Lemma 6 therein), gives a really smart and insightful criterion by which it is possible to determine, in a straightforward manner, the dimension of the kind of cyclic codes studied here.

Lemma 2. *With our notation, let e be any integer such that $e > 1$ and $e|(q^k - 1)$. Also let a_i, for $i = 1, \cdots, e$, be integers such that $a_{i+1} = a_1 + \frac{q^k-1}{e}i$, for $i = 1, \cdots, e-1$. If $\gcd(\Delta, ea_1) \leq q^{k/2}$ then $\deg(h_{a_i}(x)) = k$, and $h_{a_i}(x) \neq h_{a_j}(x)$, for any $1 \leq i \neq j \leq e$.*

Remark 3. We already said that under our main assumption, tk must be an even integer, therefore note that if $e = 3$ and $\gcd(\Delta, 3a_1) = 3$, then, owing to the previous lemma, all the possible pairs (q, k) will satisfy the condition $3 \leq q^{k/2}$, except for the one where we get $q^k = 4$.

Lemma 3. *With our notation and main assumption, let d be a divisor of $q - 1$. Also let i be any integer. If $\gcd(\Delta, 3d) = 3$, then*

$$\{xy \mid x \in \mathcal{D}_i^{(3d)} \text{ and } y \in \mathbb{F}_q^*\} = \frac{(q-1)}{d} * \mathcal{D}_i^{(3)},$$

where $\frac{(q-1)}{d} * \mathcal{D}_i^{(3)}$ is the multiset in which each element of $\mathcal{D}_i^{(3)}$ appears with multiplicity $\frac{(q-1)}{d}$.

Proof. Since $3 \mid \Delta$, $\mathbb{F}_q^* \subset \mathcal{D}_0^{(3)}$. But $\mathcal{D}_0^{(3d)} \subseteq \mathcal{D}_0^{(3)}$, $\gcd(\Delta, 3d) = 3$ and $(3d) \mid (q^k - 1)$, therefore the result comes from the fact that $|\mathcal{D}_i^{(3d)}| \, |\mathbb{F}_q^*| / |\mathcal{D}_i^{(3)}| = \frac{(q-1)}{d}$, and $\mathcal{D}_i^{(l)} = \gamma^i \mathcal{D}_0^{(l)}$, for any integer i and for any divisor l of $q^k - 1$. $\qquad \square$

Lemma 4. *Let a_1, a_2, a_3, and d be integers such that $a_2 = a_1 + \frac{q^k - 1}{3}$, $a_3 = a_1 - \frac{q^k - 1}{3}$ and $d = \gcd(q^k - 1, a_1, a_2, a_3)$. Then the following conditional statements are true:*

(i) *If $\gcd(\Delta, a_1) = \gcd(\Delta, a_2) = 1$, then $\gcd(q^k - 1, a_1) = \gcd(q^k - 1, a_2) = d$.*

(ii) *If $\gcd(\Delta, a_1) = \gcd(\Delta, a_2) = \gcd(\Delta, a_3)$, then there exists an element $a \in \{a_1, a_2, a_3\}$ such that $d = \gcd(q^k - 1, a) = \gcd(q^k - 1, a + \frac{q^k - 1}{3})$.*

Proof. Part *(i)*: Let $d_i = \gcd(q^k - 1, a_i)$, for $i = 1, 2$. Since $\gcd(\Delta, a_1) = \gcd(\Delta, a_2) = 1$ and $q^k - 1 = \Delta(q - 1)$, we have that $d_1 \mid (q - 1)$ and $d_2 \mid (q - 1)$. But $a_2 = a_1 + \frac{q^k - 1}{3}$, therefore $d_1 \mid a_2$ and $d_2 \mid a_1$, thus $d_1 = d_2$. Now, clearly $d \mid d_1$. On the other hand, since $d_1 \mid a_1$ and $d_1 \mid (q - 1)$, we have that $d_1 \mid a_i$, for $i = 1, 2, 3$, and therefore $d_1 \mid d$. Thus $d_1 = d$.

Part *(ii)*: Let $d_i = \gcd(q^k - 1, a_i)$, for $i = 1, 2, 3$. Without loss of generality, suppose that $d_3 = \max\{d_1, d_2, d_3\}$. Observe that $d = \gcd(\frac{q^k - 1}{3}, a_i)$, for all $i = 1, 2, 3$, and therefore either $d_i = d$ or $d_i = 3d$. Now, if $d_1 = 3d$, then $d_3 = 3d$, which implies that $3d \mid \gcd(q^k - 1, a_1, a_2, a_3)$. But this last condition is impossible, thus $d_1 = d$. In a similar way we have $d_2 = d$. The proof is now complete by taking $a = a_1$. $\qquad \square$

For our last result in this section, we want to recall that $\tau = \gamma^{\frac{q^k - 1}{3}}$.

Lemma 5. *With our notation and main assumption in mind, let b_1, b_2 and ϵ be integers such that $b_2 = b_1 + \epsilon \frac{q^k - 1}{3}$, and $\epsilon = \pm 1$. Let $d = \gcd(q^k - 1, b_1, b_2)$ and assume that $\gcd(\Delta, 3b_1) = 3$. Then the following two assertions are true:*

(i) *$d \mid (q - 1)$, and therefore if $n = \frac{q^k - 1}{d}$ then $3 \mid n$.*

(ii) *$d = \min\{\gcd(q^k - 1, b_1), \gcd(q^k - 1, b_2)\}$, and assuming $d = \gcd(q^k - 1, b_1)$, we have that $3 \mid (b_1 + b_2)$ if and only if $\gamma^\nu \tau^2 \in \mathcal{D}_0^{(3)}$, where $\nu = d\epsilon m$ and m is any integer such that $b_1 m \equiv d \pmod{q^k - 1}$.*

Proof. Part *(i)*: Since $d|b_1$ and $d|b_2$, we have $d|\frac{q^k-1}{3}$. But $\frac{q^k-1}{3} = \frac{\Delta}{3}(q-1)$ and $\gcd(\frac{\Delta}{3}, d) = 1$ (due that $d|b_1$ and $\gcd(\frac{\Delta}{3}, b_1) = 1$), therefore $d|(q-1)$.

Part *(ii)*: Let $d_i = \gcd(q^k - 1, b_i)$, with $i = 1, 2$. Clearly, either $d_1 = d_2$ or d_1 and d_2 differ by a multiple of 3, therefore $d = \min\{d_1, d_2\}$. Now, note that $3 \nmid m$, and this is so because if $3|m$, then $3d$ divides both $q^k - 1$ and $b_1 m$, which in turn implies that $3d|d$, and clearly this is impossible. Now, since $3|(q^k - 1)$ and $\nu = d\epsilon m$, we have that $\nu \equiv \epsilon b_1 m^2 \equiv \epsilon b_1 \pmod 3$. But $\tau = \gamma^{\epsilon(b_2-b_1)}$, therefore $\gamma^\nu \tau^2 \in \mathcal{D}_0^{(3)}$ if and only if $3|\epsilon(2b_2 - b_1)$, and this will happen if and only if $3|(b_1 + b_2)$. $\square$

4   The Value Distribution of a Class of Exponential Sums

Let ν be an integer. Then, for any $\alpha, \beta \in \mathbb{F}_{q^k}$, we are interested in determining the value distribution of the exponential sum of the form:

$$\mathcal{F}_\nu(\alpha, \beta) := \sum_{i=0}^{2} \sum_{z \in \mathcal{D}_0^{(3)}} \chi(z\gamma^{\nu i}(\alpha + \beta\tau^i)). \tag{3}$$

As we will see later, in order to obtain the value distribution of $\mathcal{F}_\nu(\alpha, \beta)$, it will be important to determine in which of three cyclotomic classes of order 3, the finite field element $\gamma^\nu \tau^2$ is contained. Consequently, throughout this section, we reserve the letter r to denote the integer in $\{0, 1, 2\}$ such that $\gamma^\nu \tau^2 \in \mathcal{D}_r^{(3)}$.

Another key component in order to achieve our goal in this section is to define the following ten sets:

$$\mathcal{E}_{i,j} := \{(\alpha, -\alpha\tau^{-j}) \mid \gamma^{\nu(j+1)}(\alpha - \tau\alpha) \in \mathcal{D}_i^{(3)}\}, \quad \text{for } i, j = 0, 1, 2, \text{ and}$$
$$\mathcal{G} := \{(\alpha, -\beta) \in \mathbb{F}_{q^k} \times \mathbb{F}_{q^k} \mid (\alpha - \tau^j\beta) \neq 0, \text{ for } j = 0, 1, 2\}.$$

Remark 4. By the previous definition, note that $(\alpha, -\alpha\tau^{-j}) \in \mathcal{E}_{i,j}$ if and only if $\gamma^{\nu(j+1)}(1 - \tau)\alpha \in \mathcal{D}_i^{(3)}$. In consequence we have that these ten sets are pairwise disjoint, and their cardinalities are $|\mathcal{E}_{i,j}| = |\mathcal{D}_i^{(3)}| = \frac{q^k-1}{3}$, for all $i, j = 0, 1, 2$, and $|\mathcal{G}| = q^{2k} - 1 - 9|\mathcal{E}_{0,0}| = (q^k - 1)(q^k - 2)$. Furthermore, due to Remark 1, note that if $\gamma^{\nu(j+1)}(\alpha - \tau\alpha) \in \mathcal{D}_i^{(3)}$, for some $i, j = 0, 1, 2$, then $\gamma^{\nu(j+2)}(\alpha - \tau^2\alpha) = \gamma^{\nu(j+2)}(\tau + 1)(\alpha - \tau\alpha) = -\gamma^\nu \tau^2 \gamma^{\nu(j+1)}(\alpha - \tau\alpha) \in \gamma^\nu \tau^2 \mathcal{D}_i^{(3)}$. Therefore, by considering the Gaussian periods of order 3, in Theorem 2 we have

$$\mathcal{F}_\nu(\alpha, -\alpha\tau^{-j}) = \frac{q^k - 1}{3} + \eta_i + \eta_{i+r},$$

where the subscripts of η, in the previous equation, are reduced modulo 3, and where, in addition, we are defining $\eta_2 := \eta_1$.

Now, for each $(\alpha, -\beta) \in \mathcal{G}$, we define the function $f_{\alpha,\beta} : \{0, 1, 2\} \to \{0, 1, 2\}$, given by the rule $f_{\alpha,\beta}(i) = j$ if and only if $\gamma^{\nu i}(\alpha - \tau^i\beta) \in \mathcal{D}_j^{(3)}$. With the help of these functions we induce a partition of the set $\mathcal{G}$ into the following four disjoint subsets:

$$\mathcal{S}_l := \{(\alpha, -\beta) \in \mathcal{G} \mid W_H(f_{\alpha,\beta}(0), f_{\alpha,\beta}(1), f_{\alpha,\beta}(2)) = l \}, \quad \text{for } l = 0, 1, 2, 3,$$

where $W_H(\cdot)$ stands for the usual Hamming weight function.

Remark 5. For any $\alpha, \beta \in \mathbb{F}_{q^k}$, we define $u_i := \gamma^{\nu i}(\alpha + \tau^i \beta)$, for $i = 0, 1, 2$. Note that if $(\alpha, \beta) \in \mathcal{G}$ then, owing to Theorem 2, we have

$$\mathcal{F}_\nu(\alpha, \beta) = \begin{cases} 3\eta_0 & \text{if } (\alpha, \beta) \in \mathcal{S}_0, \\ 2\eta_0 + \eta_1 & \text{if } (\alpha, \beta) \in \mathcal{S}_1, \\ \eta_0 + 2\eta_1 & \text{if } (\alpha, \beta) \in \mathcal{S}_2, \\ 3\eta_1 & \text{if } (\alpha, \beta) \in \mathcal{S}_3. \end{cases} \tag{4}$$

On the other hand, it is not difficult to see that these u_i's values satisfy: $u_0 + \gamma^{-\nu}\tau u_1 + \gamma^{-2\nu}\tau^2 u_2 = 0$. Also note that if we arbitrarily choose the values of, say, u_1 and u_2, then there must exist a unique vector $(\alpha, \beta) \in \mathbb{F}_{q^k}^2$, such that $u_1 = \gamma^\nu(\alpha + \tau\beta)$, $u_2 = \gamma^{2\nu}(\alpha + \tau^2\beta)$ and $u_0 = -(\gamma^{-\nu}\tau u_1 + \gamma^{-2\nu}\tau^2 u_2)$. Therefore, if we want to calculate, for example $|\mathcal{S}_0|$, then we can assume, without loss of generality, that u_2 can take any value in $\mathcal{D}_0^{(3)}$. This leads us to $\frac{q^k-1}{3}$ possible choices for u_2. But $\gamma^\nu \tau^2 \in \mathcal{D}_r^{(3)}$, $u_0 = -\frac{u_1}{\gamma^\nu \tau^2}(\frac{u_2}{\gamma^\nu \tau^2 u_1} + 1)$ and $-1 \in \mathcal{D}_0^{(3)}$ (in fact, recall that $\mathbb{F}_q^* \subset \mathcal{D}_0^{(3)}$), thus, in order that u_0 and u_1 also belong to $\mathcal{D}_0^{(3)}$, it is necessary that $(\frac{u_2}{\gamma^\nu \tau^2 u_1} + 1) \in \mathcal{D}_r^{(3)}$, and due to Lemma 1, the number of such instances is given by the cyclotomic number $(r, -r)^{(3,q^k)}$. Consequently, we have $|\mathcal{S}_0| = \frac{q^k-1}{3}(r, -r)^{(3,q^k)}$. In a quite similar way, one can obtain $|\mathcal{S}_1|$, $|\mathcal{S}_2|$ and $|\mathcal{S}_3|$.

Keeping in mind the previous definitions and their remarks, we now present the following:

Lemma 6. *With our current notation and main assumption, we have that*

$$|\mathcal{S}_0| = \frac{q^k - 1}{3}(r, -r)^{(3,q^k)},$$

$$|\mathcal{S}_1| = 2(q^k - 1)(0, 1)^{(3,q^k)},$$

$$|\mathcal{S}_2| = (q^k - 1)(q^k - (r, -r)^{(3,q^k)} - 4(0, 1)^{(3,q^k)} - 2),$$

$$|\mathcal{S}_3| = 2(q^k - 1)(0, 1)^{(3,q^k)} + \frac{2(q^k - 1)}{3}(r, -r)^{(3,q^k)}.$$

Furthermore, if χ denotes the canonical additive character of $\mathbb{F}_{q^k}$, and if η_0 and η_1 are as in Theorem 2, then, for any $\alpha, \beta \in \mathbb{F}_{q^k}$, we have

$$\mathcal{F}_\nu(\alpha, \beta) = \begin{cases} q^k - 1 & if \ (\alpha, \beta) = (0,0), \\ \frac{q^k-1}{3} + 2\eta_1 - 2\delta_{0,r}(\eta_1 - \eta_0) & if \ (\alpha, \beta) \in \cup_{j=0}^2 \mathcal{E}_{r,j}, \\ \frac{q^k-1}{3} + \eta_0 + \eta_1 + \delta_{0,r}(\eta_1 - \eta_0) & if \ (\alpha, \beta) \in \cup_{i \neq r} \cup_{j=0}^2 \mathcal{E}_{i,j}, \\ 3\eta_0 & if \ (\alpha, \beta) \in \mathcal{S}_0, \\ 2\eta_0 + \eta_1 & if \ (\alpha, \beta) \in \mathcal{S}_1, \\ \eta_0 + 2\eta_1 & if(\alpha, \beta) \in \mathcal{S}_2, \\ 3\eta_1 & if \ (\alpha, \beta) \in \mathcal{S}_3, \end{cases}$$

where $\delta_{i,j}$ is the Kronecker delta symbol ($\delta_{i,j}$ is equal to 1 if $i = j$, and 0 otherwise).

Proof. The first assertion comes from Remark 5. On the other hand, the second assertion comes directly from (4), Remark 4, and from the definitions of the sets $\mathcal{E}_{i,j}$ and $\mathcal{S}_l$, with $i, j = 0, 1, 2$, and $l = 0, 1, 2, 3$. □

Table 3. *Value distribution of $\mathcal{F}_\nu(\alpha, \beta)$ when $\gamma^\nu \tau^2 \in \mathcal{D}_0^{(3)}$*

Value	Frequency
$q^k - 1$	1
$\frac{q^k-1}{3} + 2\eta_0$	$q^k - 1$
$\frac{q^k-1}{3} + 2\eta_1$	$2(q^k - 1)$
$3\eta_0$	$\frac{1}{27}(q^k - 1)(q^k - 2(-1)^{kt/2}q^{k/2} - 8)$
$2\eta_0 + \eta_1$	$\frac{2}{9}(q^k - 1)(q^k + (-1)^{kt/2}q^{k/2} - 2)$
$\eta_0 + 2\eta_1$	$\frac{2}{9}(q^k - 1)(2q^k - (-1)^{kt/2}q^{k/2} - 1)$
$3\eta_1$	$\frac{2}{27}(q^k - 1)(4q^k + (-1)^{kt/2}q^{k/2} - 14)$

Considering the actual values of the cyclotomic numbers in Lemma 1, the following result is an important consequence of the previous lemma.

Corollary 1. *Consider the same hypotheses as in the previous lemma. Then the value distribution of the character sum $\mathcal{F}_\nu(\alpha, \beta)$ is given in Table 3 if $\gamma^\nu \tau^2 \in \mathcal{D}_0^{(3)}$, and in Table 4 otherwise.*

5   A Formal Proof of Theorem 1

Let b_1, b_2 and d be integers in such a way that $b_2 = b_1 \pm \frac{q^k-1}{3}$, whereas that $d = \gcd(q^k - 1, b_1, b_2)$. Then, the following result gives the sufficient numerical conditions over the integers b_1 and b_2, which guarantee that $\mathcal{C}_{(b_1, b_2)}$ is a cyclic code of length $\frac{q^k-1}{d}$ and dimension $2k$, whose weight distribution is given either in Table 1 or in Table 2.

Table 4. *Value distribution of $\mathcal{F}_\nu(\alpha, \beta)$ when $\gamma^\nu \tau^2 \notin \mathcal{D}_0^{(3)}$*

Value	Frequency
$q^k - 1$	1
$\frac{q^k-1}{3} + 2\eta_1$	$q^k - 1$
$\frac{q^k-1}{3} + \eta_0 + \eta_1$	$2(q^k - 1)$
$3\eta_0$	$\frac{1}{27}(q^k - 1)(q^{k/2} - (-1)^{kt/2})^2$
$2\eta_0 + \eta_1$	$\frac{2}{9}(q^k - 1)(q^k + (-1)^{kt/2}q^{k/2} - 2)$
$\eta_0 + 2\eta_1$	$\frac{1}{9}(q^k - 1)(4q^k - 2(-1)^{kt/2}q^{k/2} - 11)$
$3\eta_1$	$\frac{2}{27}(q^k - 1)(4q^k + (-1)^{kt/2}q^{k/2} - 5)$

Lemma 7. *With our notation and main assumption in mind, suppose also that $q^k \neq 4$. Let b_1, b_2, ϵ, d and n be integers such that $b_2 = b_1 + \epsilon\frac{q^k-1}{3}$, $\epsilon = \pm 1$, $d = \gcd(q^k - 1, b_1, b_2)$ and $n = \frac{q^k-1}{d}$. Let us assume that $\gcd(\Delta, 3b_1) = 3$. Then $\mathcal{C}_{(b_1,b_2)}$ is an $[n, 2k]$ cyclic code with the weight distribution given in Table 1 if $3|(b_1 + b_2)$, and given in Table 2 if $3 \nmid (b_1 + b_2)$.*

Proof. Clearly $\mathcal{C}_{(b_1,b_2)}$ has length n. On the other hand, because $q^k \neq 4$, and owing to Remark 3, we can conclude that $\mathcal{C}_{(b_1,b_2)}$ has dimension $\deg(h_{b_1}(x)h_{b_2}(x)) = 2k$.

Before continuing with the proof, observe that, since $\mathcal{C}_{(b_1,b_2)} = \mathcal{C}_{(b_2,b_1)}$ and $d = \min\{\gcd(q^k - 1, b_1), \gcd(q^k - 1, b_2)\}$ (see Part *(ii)* of Lemma 5), we can assume without loss of generality that $d = \gcd(q^k - 1, b_1)$.

Now, for each $\alpha, \beta \in \mathbb{F}_{q^k}$, we define $c(n, b_1, b_2, \alpha, \beta)$ as the vector of length n over $\mathbb{F}_q$, given by:

$$(\mathrm{Tr}_{\mathbb{F}_{q^k}/\mathbb{F}_q}(\alpha(\gamma^{b_1})^i + \beta(\gamma^{b_2})^i))_{i=0}^{n-1}.$$

Thanks to Delsarte's Theorem (see, for example, [2]), it is well known that

$$\mathcal{C}_{(b_1,b_2)} = \{c(n, b_1, b_2, \alpha, \beta) \mid \alpha, \beta \in \mathbb{F}_{q^k}\}.$$

Thus the Hamming weight of any codeword $c(n, b_1, b_2, \alpha, \beta) \in \mathcal{C}_{(b_1,b_2)}$ is equal to $n - Z(\alpha, \beta)$, where

$$Z(\alpha, \beta) = \#\{\, i \mid 0 \leq i < n, \text{ and } \mathrm{Tr}_{\mathbb{F}_{q^k}/\mathbb{F}_q}(\alpha\gamma^{b_1 i} + \beta\gamma^{b_2 i}) = 0 \,\}.$$

If χ' and χ are, respectively, the canonical additive characters of $\mathbb{F}_q$ and $\mathbb{F}_{q^k}$, then

$$Z(\alpha, \beta) = \frac{1}{q}\sum_{i=0}^{n-1}\sum_{y\in\mathbb{F}_q} \chi'(\mathrm{Tr}_{\mathbb{F}_{q^k}/\mathbb{F}_q}(y(\alpha\gamma^{b_1 i} + \beta\gamma^{b_2 i})))$$

$$= \frac{n}{q} + \frac{1}{q}\sum_{i=0}^{n-1}\sum_{y\in\mathbb{F}_q^*} \chi(y\gamma^{b_1 i}(\alpha + \beta\tau^{\epsilon i})),$$

where the last equality arises because $\chi(\cdot) = \chi'(\mathrm{Tr}_{\mathbb{F}_{q^k}/\mathbb{F}_q}(\cdot))$, $b_2 = b_1 + \epsilon\frac{q^k-1}{3}$ and $\tau = \gamma^{\frac{q^k-1}{3}}$. Since $d|(b_2-b_1)$, $\tau = \gamma^{\epsilon(b_2-b_1)} = \gamma^{dl}$ for some integer l. On the other hand, since $d = \gcd(q^k-1, b_1)$, $mb_1 \equiv d \pmod{q^k-1}$, for some integer m. Therefore $\tau^i = (\gamma^{b_1 i})^{ml}$, and

$$Z(\alpha, \beta) = \frac{n}{q} + \frac{1}{q}\sum_{i=0}^{n-1}\sum_{y\in\mathbb{F}_q^*}\chi(y\gamma^{di}(\alpha + \beta\tau^{\epsilon mi})).$$

But, owing to Part *(i)* of Lemma 5, we have that $3|n$, thus

$$\{\gamma^{di} \mid 0 \le i < n\} = \mathcal{D}_0^{(d)} = \mathcal{D}_0^{(3d)} \cup \mathcal{D}_d^{(3d)} \cup \mathcal{D}_{2d}^{(3d)}.$$

Therefore,

$$Z(\alpha, \beta) = \frac{n}{q} + \frac{1}{q}\sum_{i=0}^{2}\sum_{x\in\mathcal{D}_{di}^{(3d)}}\sum_{y\in\mathbb{F}_q^*}\chi(xy(\alpha + \beta\tau^{\epsilon mi})).$$

Now, $d|(q-1)$ (see again Part *(i)* of Lemma 5), and clearly $d|b_1$ and $\gcd(\frac{\Delta}{3}, b_1) = 1$, therefore $\gcd(\Delta, 3d) = 3$. Thus, after applying Lemma 3, we obtain

$$Z(\alpha, \beta) = \frac{n}{q} + \frac{q-1}{dq}\sum_{i=0}^{2}\sum_{z\in\mathcal{D}_{di}^{(3)}}\chi(z(\alpha + \beta\tau^{\epsilon mi})),$$

and because 3 does not divide neither m (see proof of Part *(ii)* of Lemma 5) nor ϵ, we can apply the variable change $i \mapsto \epsilon mi$, and thereby obtaining

$$Z(\alpha, \beta) = \frac{n}{q} + \frac{q-1}{dq}\sum_{i=0}^{2}\sum_{z\in\mathcal{D}_{d\epsilon mi}^{(3)}}\chi(z(\alpha + \beta\tau^i))$$

$$= \frac{n}{q} + \frac{q-1}{dq}\sum_{i=0}^{2}\sum_{z\in\mathcal{D}_0^{(3)}}\chi(z\gamma^{d\epsilon mi}(\alpha + \beta\tau^i))$$

$$= \frac{n}{q} + \frac{q-1}{dq}\mathcal{F}_\nu(\alpha, \beta),$$

where, for the last equality, we are considering $\nu = d\epsilon m$ and we are also using the notation for class of exponential sum defined in (3). Finally, the assertion about the weight distribution of $\mathcal{C}_{(b_1,b_2)}$ comes now from (2), Corollary 1, Part *(ii)* of Lemma 5 and from the fact that the Hamming weight of any codeword in $\mathcal{C}_{(b_1,b_2)}$ is equal to $n - Z(a, b)$.   □

We are now able to present a formal proof of Theorem 1.

Proof. For the way in which the integers a_i, $i = 1, 2, 3$, were defined, it must be clear that $\gcd(\Delta, 3a_1) = \gcd(\Delta, 3a_2) = \gcd(\Delta, 3a_3)$.

Part *(A)*: Owing to Theorem 2 of [16], we know that $\gcd(\Delta, a_i) = 1$ if and only if $\mathcal{C}_{(a_i)}$ is an $[n, k]$ one-weight irreducible cyclic code, whose nonzero weight is $\frac{q-1}{d}q^{k-1}$. On the other hand, if $\gcd(\Delta, a_i) = 3$ then, since $p \equiv -1 \pmod 3$, we have that $\mathrm{Ord}_3(p) = 2$. Therefore, owing to Theorem 7 of [16], we know that $\mathcal{C}_{(a_i)}$ is an $[n, k]$ semiprimitive two-weight irreducible cyclic, whose weight enumerator polynomial is given by (1).

Part *(B)*: Due to Part *(i)* of Lemma 4, we know that $d = \gcd(q^k - 1, a_1) = \gcd(q^k - 1, a_2)$. Now, it is easy to see that $3|(a_1 + a_2)$ and that 3 does not divide neither $a_1 + a_3$ nor $a_2 + a_3$. Thus, the two assertions follow as direct applications of Lemma 7.

Part *(C)*: If $\gcd(\Delta, a_1) = \gcd(\Delta, a_2) = \gcd(\Delta, a_3)$, then, thanks to Part *(ii)* of Lemma 4 and since $\mathcal{C}_{(a_i,a_j)} = \mathcal{C}_{(a_j,a_i)}$, for all $1 \le i \ne j \le 3$, we can assume that $d = \gcd(q^k - 1, a_1) = \gcd(q^k - 1, a_2)$. Now, it is not difficult to see that $3|(a_i + a_j)$, for $1 \le i \ne j \le 3$, if and only if $\gcd(\Delta, a_1) = 3$. Therefore, the assertion follow as direct application of Lemma 7.   □

The following are direct applications of Theorem 1.

Example 1. With our notation, let $p = q = 5$, $k = 2$ and $a_1 = 17$. Then $\Delta = 6$, $\frac{q^k-1}{3} = 8$, $a_2 = 1$, $a_3 = 9$, $d = 1$ and $n = 24$. Clearly, $3|(p+1)$ and $\gcd(\Delta, 3a_1) = 3$. Since $\gcd(\Delta, 17) = \gcd(\Delta, 1) = 1$, we can be sure that $\mathcal{C}_{(17)}$ and $\mathcal{C}_{(1)}$ are one-weight irreducible cyclic codes of length 24 and dimension 2, whose nonzero weight is 20. Meanwhile, since $\gcd(\Delta, 9) = 3$, $\mathcal{C}_{(9)}$ is a semiprimitive two-weight irreducible cyclic code of length 24 and dimension 2, whose weight enumerator polynomial is $A(z) = 1 + 8z^{12} + 16z^{24}$. On the other hand, we can also be sure that $\mathcal{C}_{(17,1)}$, $\mathcal{C}_{(17,9)}$ and $\mathcal{C}_{(1,9)}$ are cyclic codes of length 24 and dimension 4. In addition, the weight enumerator polynomial for the code $\mathcal{C}_{(17,1)}$ is $A(z) = 1 + 24z^8 + 24z^{12} + 144z^{16} + 288z^{20} + 144z^{24}$, whereas the weight enumerator polynomial for the codes $\mathcal{C}_{(17,9)}$ and $\mathcal{C}_{(1,9)}$ is $A(z) = 1 + 80z^{12} + 120z^{16} + 264z^{20} + 160z^{24}$.

Remark 6. It is interesting to note that since $3 \nmid k$, all the $(3, 2, 3)$-polygon cyclic codes, in the previous example are outside of the class of cyclic codes studied in [19]. In a similar way, since $\gcd(\Delta, \frac{q^k-1}{3} - 1) \ne 3$, all the $(3, 2, 3)$-polygon cyclic codes, in the previous example are outside of the class of cyclic codes studied in [17].

Example 2. With our notation, let $p = 2$, $q = 4$, $k = 3$ and $a_1 = 1$. Then $\Delta = \frac{q^k-1}{3} = 21$, $a_2 = 22$, $a_3 = 43$, $d = 1$ and $n = 63$. Clearly, $3|(p+1)$ and $\gcd(\Delta, 3a_1) = 3$. Since $\gcd(\Delta, a_i) = 1$, for $i = 1, 2, 3$, we can be sure that $\mathcal{C}_{(1)}$, $\mathcal{C}_{(22)}$ and $\mathcal{C}_{(43)}$ are one-weight irreducible cyclic codes of length 63 and dimension 3, whose nonzero weight is 48. On the other hand, we can also be sure that $\mathcal{C}_{(1,22)}$, $\mathcal{C}_{(1,43)}$ and $\mathcal{C}_{(22,43)}$ are cyclic codes of length 63, dimension 6 and weight enumerator polynomial $A(z) = 1 + 126z^{30} + 252z^{36} + 756z^{42} + 1827z^{48} + 1134z^{54}$.

Example 3. With our notation, let $p = 2$, $q = 4$, $k = 3$ and $a_1 = 3$. Then $\Delta = \frac{q^k-1}{3} = 21$, $a_2 = 24$, $a_3 = 45$, $d = 3$ and $n = 21$. Clearly, $3|(p+1)$

and $\gcd(\Delta, 3a_1) = 3$. Since $\gcd(\Delta, a_i) = 3$, for $i = 1, 2, 3$, we can be sure that $\mathcal{C}_{(3)}$, $\mathcal{C}_{(24)}$ and $\mathcal{C}_{(45)}$ are semiprimitive two-weight irreducible cyclic codes of length 21 and dimension 3, whose weight enumerator polynomial is $A(z) = 1 + 21z^{12} + 42z^{18}$. On the other hand, we can also be sure that $\mathcal{C}_{(3,24)}$, $\mathcal{C}_{(3,45)}$ and $\mathcal{C}_{(24,45)}$ are cyclic codes of length 21, dimension 6 and weight enumerator polynomial $A(z) = 1 + 63z^8 + 294z^{12} + 756z^{14} + 1890z^{16} + 1092z^{18}$.

6   Determining the Weight Distribution of Any $(3, 2, 3)$-Polygon Cyclic Code

It is relevant to note that the techniques employed in the study of the cyclic codes in [19] and those employed here (in the semiprimitive case) are different. And this is so because the former used some results on elliptic curves, and the case $p = 2$ was treated separately. In contrast, as we already saw, the techniques in this work are the standard ones, which basically rely on the availability of some Gaussian periods. However, beyond these differences, the important issue here is that by combining these two works it is now possible to give the weight distribution of any $(3, 2, 3)$-polygon cyclic code. We formally state this result, by means of the following:

Theorem 3. *With our notation, suppose that $q^k \neq 4$ and let a be any integer such that $\gcd(\Delta, 3a) = 3$. Then the weight distribution of $\mathcal{C}_{(a, a + \epsilon\frac{q^k-1}{3})}$, with $\epsilon = \pm 1$, can be determined either through Tables 1 and 2, or through Tables 3 and 4 in [19].*

Proof. Clearly if $p \equiv 0 \pmod 3$, then the condition $\gcd(\Delta, 3a) = 3$ cannot be met.

If $p \equiv 1 \pmod 3$, then $p^{kti} = q^i \equiv 1 \pmod 3$, for any non negative integer i. But observe that $\Delta = q^{k-1} + q^{k-2} + \cdots + 1$ and $\gcd(\Delta, 3a) = 3$, therefore $3 | k$. If $d = \gcd(q^k - 1, a, a + \epsilon\frac{q^k-1}{3})$, then, since $d | a$ and $\gcd(\frac{\Delta}{3}, a) = 1$, we have that $\gcd(\Delta, 3d) = 3$. In addition, owing to Part *(i)* of Lemma 5, $d | (q - 1)$. Let $h = \frac{q-1}{d}$ and observe that $\gcd(\Delta, 3\frac{q-1}{h}) = 3$ and $3 | \gcd(q - 1, hk)$. Now, since $\mathcal{C}_{(a, a + \epsilon\frac{q^k-1}{3})} = \mathcal{C}_{(a + \epsilon\frac{q^k-1}{3}, a)}$, it is not difficult to see that the cyclic codes $\mathcal{C}_{(a, a + \epsilon\frac{q^k-1}{3})}$ and $\mathcal{C}_{(\frac{q-1}{h}, \frac{q-1}{h} + \frac{q^k-1}{3})}$ have the same weight distribution. Thus, the whole picture here implies that we have $p \equiv 1 \pmod 3$, $3 | k$, $h | (q - 1)$, $\gcd(\Delta, 3\frac{q-1}{h}) = 3$, $3 | \gcd(q - 1, hk)$, and note that $3 | \frac{q-1}{h} \Leftrightarrow 3 | a$. Therefore, by simply checking whether or not $3 | a$, we can determine the weight distribution of $\mathcal{C}_{(\frac{q-1}{h}, \frac{q-1}{h} + \frac{q^k-1}{3})}$ through Theorem 2 in [19].

Finally, if $p \equiv -1 \pmod 3$, then, by simply checking whether or not $3 | (2a + \epsilon\frac{q^k-1}{3})$, we can determine the weight distribution of $\mathcal{C}_{(a, a + \epsilon\frac{q^k-1}{3})}$ directly from Lemma 7. $\square$

Note that in accordance with the open cases that were suggested in the Conclusion of [20], the previous result completely solves the case when $t = 2$ and $e = N = 3$ (we are now using the same notation as in [20]).

7   An Extended Family of Cross-Correlation Functions

With our notation, let χ be the canonical additive character of $\mathbb{F}_{q^k}$, then, for any integer μ with $0 \le \mu < q^k - 1$, the cross-correlation function, $C_\delta(\mu)$, of the maximum-length sequence $\{\mathrm{Tr}_{\mathbb{F}_{q^k}/\mathbb{F}_q}(\gamma^i)\}_{i=0}^\infty$, of period $q^k - 1$, and its δ-decimated version $\{\mathrm{Tr}_{\mathbb{F}_{q^k}/\mathbb{F}_q}(\gamma^{\delta i})\}_{i=0}^\infty$, of period $\frac{q^k-1}{\gcd(q^k-1,\delta)}$, is defined as:

$$C_\delta(\mu) = \sum_{x \in \mathbb{F}_{q^k}^*} \chi(\gamma^{-\mu}x - x^\delta).$$

Clearly if $\gcd(q^k - 1, \delta) = 1$, then $C_\delta(\mu)$ corresponds to a cross-correlation function between two maximal linear recurring sequences. Finding the values of the cross-correlation function between two different maximal linear recurring sequences of period $q^k - 1$, seems to be a very difficult problem [8]. However, when $\delta - 1 \equiv 0 \pmod{q-1}$, this problem is equivalent to that of the determination of the weight distribution of the reducible cyclic code $\mathcal{C}_{(1,\delta)}$ over $\mathbb{F}_q$ (see, for example, [4, Subsect. 4.3]). More specifically, suppose that $w_1, w_2, \cdots, w_N$ are the nonzero weights of $\mathcal{C}_{(1,\delta)}$, and for $1 \le i \le N$, let A_i be the number of words of weight w_i in $\mathcal{C}_{(1,\delta)}$. Thus, if $\gcd(q^k - 1, \delta) = 1$ and $\delta - 1 \equiv 0 \pmod{q-1}$, then, for each nonzero weight w_i $(1 \le i \le N)$, $C_\delta(\mu)$ will take the value $q^k - 1 - \frac{qw_i}{q-1}$, $\frac{A_i}{q^k-1}$ times if $w_i \ne (q-1)q^{k-1}$, and $\frac{A_i}{q^k-1} - 2$ times if $w_i = (q-1)q^{k-1}$. With this discussion in mind, we now present an alternative description for one of the families of cross-correlation functions studied in [8]:

Theorem 4. *With our notation, assume $p \equiv 2 \pmod 3$ and $3 \mid \Delta$. Suppose also that $q^k \ne 4$. For some $\epsilon = \pm 1$ and some $0 \le i < k$, let $\delta = \frac{1}{3}\epsilon(q^k - 1) + q^i$, and $f = \frac{1}{3}\epsilon q^{-i}(q^k - 1) \not\equiv 2 \pmod 3$. Then $C_\delta(\mu)$ corresponds to a cross-correlation function between two different maximal linear recurring sequences, and $C_\delta(\mu)$ takes on the following values. When $f \equiv 0 \pmod 3$, then:*

   (i) *-1 occurs $\frac{1}{9}(4q^k + 2(-1)^{kt/2+1}q^{k/2} - 29)$ times,*
   (ii) *$-1 + (-1)^{kt/2+1}q^{k/2}$ occurs $\frac{1}{9}(2q^k + 2(-1)^{kt/2}q^{k/2} - 4)$ times,*
   (iii) *$-1 + (-1)^{kt/2}q^{k/2}$ occurs $\frac{1}{27}(8q^k + 2(-1)^{kt/2}q^{k/2} - 10)$ times,*
   (iv) *$-1 + 2(-1)^{kt/2+1}q^{k/2}$ occurs $\frac{1}{27}(q^k + 2(-1)^{kt/2+1}q^{k/2} + 1)$ times,*
   (v) *$-1 + \frac{1}{3}(q^k + 2(-1)^{kt/2}q^{k/2})$ occurs 1 time,*
   (vi) *$-1 + \frac{1}{3}(q^k + (-1)^{kt/2+1}q^{k/2})$ occurs 2 times.*

When $f \equiv 1 \pmod 3$, then:

   (i) *-1 occurs $\frac{1}{9}(4q^k + 2(-1)^{kt/2+1}q^{k/2} - 20)$ times,*
   (ii) *$-1 + (-1)^{kt/2+1}q^{k/2}$ occurs $\frac{1}{9}(2q^k + 2(-1)^{kt/2}q^{k/2} - 4)$ times,*
   (iii) *$-1 + (-1)^{kt/2}q^{k/2}$ occurs $\frac{1}{27}(8q^k + 2(-1)^{kt/2}q^{k/2} - 28)$ times,*
   (iv) *$-1 + 2(-1)^{kt/2+1}q^{k/2}$ occurs $\frac{1}{27}(q^k + 2(-1)^{kt/2+1}q^{k/2} - 8)$ times,*
   (v) *$-1 + \frac{1}{3}(q^k + 2(-1)^{kt/2}q^{k/2})$ occurs 2 time,*
   (vi) *$-1 + \frac{1}{3}(q^k + 4(-1)^{kt/2+1}q^{k/2})$ occurs 1 times.*

Proof. Since $3 \nmid (f + 1)$, $\gcd(q^k - 1, \delta) = 1$, and therefore $C_\delta(\mu)$ corresponds to a cross-correlation function between two maximal linear recurring sequences. Furthermore, due to Lemma 2 and Remark 3, these linear recurring sequences are different (their recursion polynomials are different). On the other hand, observe that $\delta - 1 = (\epsilon \frac{\Delta}{3} + \frac{q^i - 1}{q - 1})(q - 1)$, and therefore $\delta - 1 \equiv 0 \pmod{q - 1}$. Now, since γ^{-1} and γ^{-q^i} are roots of the same minimal polynomial, we have $C_{(1,\delta)} = C_{(q^i, q^i + \epsilon \frac{q^k - 1}{3})} = C_{(q^i, \delta)}$, and clearly, $d = \gcd(q^k - 1, q^i, \delta) = 1$ and $\gcd(\Delta, 3q^i) = 3$. On the other hand, since $f \not\equiv 2 \pmod 3$, it follows that $f \equiv 1 \pmod 3$ if and only if $3 | (q^i + \delta)$. Thus, in accordance with Lemma 7, we can see now that $C_{(q^i, \delta)}$ is a reducible cyclic code of length $q^k - 1$ over $\mathbb{F}_q$, whose weight distribution is given either by Table 1 or by Table 2 (with $d = 1$ in these tables). Thus, the remaining part of the proof follows directly from our previous discussion and from such tables.   □

Remark 7. Note that in the trivial case in which $q^k = 4$ (that is, $p = q = 2$, $t = 1$ and $k = 2$), $C_\delta(\mu)$ gives the cross-correlation between the maximal linear recurring sequence $\{\mathrm{Tr}_{\mathbb{F}_4/\mathbb{F}_2}(\gamma^i)\}_{i=0}^\infty$ (with $\gamma^2 + \gamma + 1 = 0$), and itself.

Remark 8. In the particular case when $p \equiv 2 \pmod 3$, $q = p$ (that is, $t = 1$), and $k \equiv 0 \pmod 2$, observe that necessarily $3 | (p^k - 1)$, but recall that $(p^k - 1) = \Delta(q - 1)$, and because $3 \nmid (q - 1)$, we can then conclude that $3 | \Delta$. In consequence, this shows that the family of cross-correlation functions, in the previous theorem, gives an alternative description for the family of cross-correlation functions studied in Theorem 4.11 of [8].

Example 4. With the notation of Theorem 4, let $p = 2$, $q = 4$, $k = 3$, $\epsilon = 1$ and $i = 0$. Thus, $\delta = 22$ and $f = 21 \not\equiv 2 \pmod 3$. Furthermore, in accordance with Example 2, $C_{(1,22)}$ is a cyclic code over $\mathbb{F}_4$ of length 63, dimension 6 and weight enumerator polynomial $A(z) = 1 + 126z^{30} + 252z^{36} + 756z^{42} + 1827z^{48} + 1134z^{54}$. On the other hand, thanks to Theorem 4, we can be sure that $C_{22}(\mu)$ corresponds to a cross-correlation function between two maximal linear recurring sequences of period 63 over $\mathbb{F}_4$, that takes the values -1, 7, -9, 15 and 23, and whose corresponding frequencies of occurrence are 27, 12, 18, 4 and 2.

8   Conclusion

A recent topic of interest has been to obtain the weight distribution for the kind of reducible cyclic codes whose parity-check polynomials are given by products of the form $h_a(x)h_{a + \frac{q^k - 1}{3}}(x)$. A particular class of this kind of cyclic codes was the main subject of study in [19] and in [17]. In this work we presented a complement of these two works, and with it, we formally prove that one of the open cases suggested in [20] has been now completely settled. In addition, as was shown above, through our results we were able to present an alternative description for one of the families of cross-correlation functions studied in [8]. Related with this last issue, we believe that it could be interesting to find the values of the cross-correlation functions in Theorem 4, when $f \equiv 2 \pmod 3$. Of course, under this circumstance such functions will not correspond to cross-correlation functions between two maximal linear recurring sequences.

References

1. Baumert, L., Mills, W., Ward, R.: Uniform cyclotomy. J. Number Theor. **14**(1), 67–82 (1982)
2. Delsarte, P.: On subfield subcodes of Reed-Solomon codes. IEEE Trans. Inf. Theor. **21**(5), 575–576 (1975)
3. Ding, C.: The weight distribution of some irreducible cyclic codes. IEEE Trans. Inf. Theor. **55**(3), 955–960 (2009)
4. Ding, C., Li, C., Li, N., Zhou, Z.: Three-weight cyclic codes and their weight distributions. Discrete Math. **339**(2), 415–427 (2016)
5. Ding, C., Liu, Y., Ma, C., Zeng, L.: The weight distributions of the duals of cyclic codes with two zeros. IEEE Trans. Inf. Theor. **57**(12), 8000–8006 (2011)
6. Dinh, H., Li, C., Yue, Q.: Recent progress on weight distributions of cyclic codes over finite fields. J. Algebra Comb. Discrete Appl. **2**(1), 39–63 (2014)
7. Feng, K., Luo, J.: Weight distribution of some reducible cyclic codes. Finite Fields Appl. **14**(2), 390–409 (2008)
8. Helleseth, T.: Some results about the cross-correlation function between two maximal linear sequences. Discrete Math. **16**(3), 209–232 (1976)
9. Helleseth, T.: Some two-weight codes with composite parity-check polynomials. IEEE Trans. Inf. Theor. **22**, 631–632 (1976)
10. Lidl, R., Niederreiter, H.: Finite Fields. Cambridge University Press, Cambridge (1983)
11. Ma, C., Zeng, L., Liu, Y., Feng, D., Ding, C.: The weight enumerator of a class of cyclic codes. IEEE Trans. Inf. Theor. **57**(1), 397–402 (2011)
12. MacWilliams, F., Sloane, N.: The Theory of Error-Correcting Codes. North-Holland, Amsterdam (1977)
13. Moisio, M.: A note on evaluations of some exponential sums. Acta Arith. **93**, 117–119 (2000)
14. Schmidt, B., White, C.: All two-weight irreducible cyclic codes? Finite Fields Appl. **8**, 1–17 (2002)
15. Vega, G.: Two-weight cyclic codes constructed as the direct sum of two one-weight cyclic codes. Finite Fields Appl. **14**(3), 785–797 (2008)
16. Vega, G.: A critical review and some remarks about one- and two-weight irreducible cyclic codes. Finite Fields Appl. **51**(33), 1–13 (2015)
17. Vega, G.: A family of six-weight reducible cyclic codes and their weight distribution. In: El Hajji, S., Nitaj, A., Carlet, C., Souidi, E.M. (eds.) C2SI 2015. LNCS, vol. 9084, pp. 184–196. Springer, Cham (2015). doi:10.1007/978-3-319-18681-8_15
18. Wolfmann, J.: Are 2-weight projective cyclic codes irreducible? IEEE Trans. Inf. Theor. **51**(2), 733–737 (2005)
19. Xiong, M.: The weight distributions of a class of cyclic codes II. Des. Codes Crypt. **72**(3), 511–528 (2012)
20. Yang, J., Xiong, M., Ding, C., Luo, J.: Weight distribution of a class of cyclic codes with arbitrary number of zeros. IEEE Trans. Inf. Theor. **59**(9), 5985–5993 (2013)
21. Yang, J., Xiong, M., Xia, L.: Weight distributions of a class of cyclic codes with arbitrary number of nonzeros in quadratic case. Finite Fields Appl. **36**, 41–62 (2015)

A NP-Complete Problem in Coding Theory with Application to Code Based Cryptography

Thierry P. Berger[1], Cheikh Thiécoumba Gueye[2], and Jean Belo Klamti[2(✉)]

[1] XLIM-MATHIS, UMR CNRS 6172, Université de limoges,
123 av. A. Thomas, 87060 Limoges Cedex, France
`thierry.berger@unilim.fr`
[2] Faculté des Sciences et Techniques, Université Cheikh Anta Diop,
DMI, LACGAA, Dakar, Senegal
{`cheikht.gueye,jeanbelo.klamti`}`@ucad.edu.sn`

Abstract. It is easy to determine if a given code $\mathcal{C}$ is a subcode of another known code $\mathcal{D}$. For most of occurrences, it is easy to determine if two codes $\mathcal{C}$ and $\mathcal{D}$ are equivalent by permutation. In this paper, we show that determining if a code $\mathcal{C}$ is equivalent to a subcode of $\mathcal{D}$ is a NP-complete problem. We give also some arguments to show why this problem seems much harder to solve in practice than the Equivalence Punctured Code problem or the Punctured Code problem proposed by *Wieschebrink* [21]. For one application of this problem we propose an improvement of the three-pass identification scheme of Girault and discuss on its performance.

Keywords: Code-based cryptography · Equivalence Subcode · Identification scheme

1 Introduction

There are some well-known NP-complete problems in Coding Theory [1,3,5, 6,13] that are related to the difficulty of correcting errors or determining the minimum distance of a code. Most of Public Key Cryptosystems (PKC) using Coding Theory are based on these problems [12,14]. However, a crucial point of such PKC is to mask the structure of codes [4,12–14]. In this paper we study the technique of permuted subcode: we want to mask the structure of a linear code $\mathcal{D}$ of length n, dimension k and minimal distance d, having for example an efficient decoding algorithm. To perform this, we choose a random subcode $\mathcal{C}'$ of $\mathcal{D}$, and a random permutation σ of the support. The public key is $\mathcal{C} = \sigma(\mathcal{C}')$, and the secret key is constituted of σ and a decoding algorithm for D. This method is directly related to the following Decision Problem:

Definition 1 *(Equivalence Subcode(ES)).* *Given two linear codes $\mathcal{C}$ and $\mathcal{D}$ of length of n and respective dimension k' and k, $k' \leq k$, over the same finite field $\mathbb{F}_q$, is there a permutation σ of the support such $\sigma(\mathcal{C})$ be a subcode of $\mathcal{D}$?*

© Springer International Publishing AG 2017
S. El Hajji et al. (Eds.): C2SI 2017, LNCS 10194, pp. 230–237, 2017.
DOI: 10.1007/978-3-319-55589-8_15

The fact that a code C' is a subcode of $\mathcal{D}$ can be easily checked using the linear algebra. To decide if two codes C and C' are equivalent by permutation in practice, it can be done in most of cases using the Support Splitting Algorithm SSA (see [16] for more details).

In order to mask the structure of a code, C. Wieschebrink presented In [21] two new NP-complete problems in Coding Theory: The "Equivalence Punctured Code" problem (EPC) and the "Punctured Code" problem (PC). These problems can be used to mask the structure of a code. However, comparing the results of [21], it seems that the method of permuted subcode is more efficient than those derived from EPC or PC.

The paper is structured as following: in Sect. 2 we give some definitions and properties relative to coding theory, in Sect. 3 we prove that the *Equivalence Subcode problem*(ES) is NP-complete, in Sect. 4 we give an application of the *ES* precisely an improvement of *Girault's* identification scheme using *ES* with its analysis performance.

2   Coding Theory Background

Let $\mathbb{F}_q$ be a finite field ($q = p^m$, p is prime). A q-ary linear code C of length n and dimension k over $\mathbb{F}_q$ is a vectorial subspace of dimension k of the full vectorial space $\mathbb{F}_q^n$. It can be specified by a full rank matrix $\mathbf{G} \in \mathbb{F}_q^{k \times n}$ called *generator matrix* of C whose rows span the code. Namely, $C = \left\{ \boldsymbol{x}\mathbf{G} \;\; such \;\; that \;\; \boldsymbol{x} \in \mathbb{F}_q^k \right\}$. A linear code can be also defined by the right kernel of matrix $\mathbf{H}$ called *parity-check matrix* of C as follows:

$$C = \left\{ \boldsymbol{x} \in \mathbb{F}_q^n \;\; s.t. \;\; \mathbf{H}\boldsymbol{x}^T = \mathbf{0} \right\}$$

The *Hamming distance* between two codewords is the number of positions(coordinates) where they differ. The minimal distance of a code is the minimal distance of all codewords.

The *weight* of a codeword $\boldsymbol{x} \in \mathbb{F}_q^n$ denoted by $wt\,(\boldsymbol{x})$ is the number of its nonzero positions. Then the minimal *weight* of a code C is the minimal *weight* of all nonzero codewords. If a code C is linear, the minimal distance is equal to the minimal *weight* of the code.

Let C be a linear code C over an arbitrary finite field $\mathbb{F}_{q^m}$ of length n, dimension k and minimal distance d. A subcode C' of C is one of its vector subspace of dimension $k' \leq k$. Then we see that minimal distance of a subcode is great than the minimal distance of the code.

Let C be a linear code C over an arbitrary finite field $\mathbb{F}_{q^m}$ of length n, dimension k and minimal distance d, generator matrice $\mathbf{G}$ and parity cheikh matrix $\mathbf{H}$. We can construct an arbitrary subcode C' of dimension $k' \leq k$ of the code by two ways as following:

1. First by choosing arbitrary a $k' \times k$ matrix $\mathbf{S}$ of rank k' then the generator matrix $\mathbf{G}'$ of the subcode C' is given by:

$$\mathbf{G}' = \mathbf{S}\mathbf{G}$$

2. Second by extending the rows of the parity check matrix $\mathbf{H}$ of the code by adding arbitrary $k - k'$ vectors linearly independant to its rows vectors.

3   Equivalence Subcode(ES) Is NP-Complete

The main result of this paper is the fact that ES is a NP-complete problem. To prove this, we will use a reduction of the 3-Dimensional Matching problem; which is a well known NP-complete problem (cf. [8]). This proof is similar to the proof of *Petrank et al.* [15]. Contrary to the *Petrank et al.* who have used a reduction to the Graph Isomorphism problem which is not NP-complete, our reduction is a reduction to a NP-complete problem.

Definition 2 *(3-Dimensional Matching (3DM)). Let $N = \{1, ..., n\}$ and $K \subset N \times N \times N$. Does K contain a matching, i.e. a subset K' of size n such that no two elements of K' agree in any coordinate?*

Theorem 1. *The Equivalence Subcode(ES) is NP-Complete*

Clearly, ES is a NP-problem, since it is sufficient to give the permutation and to verify that the permuted code $\sigma(\mathcal{C})$ is a subcode of $\mathcal{D}$, which can be done in polynomial time.

Suppose an instance (n, K) of 3DM is given. We can assume that $|K| \geq n+1$. Set $r = |K|$. To any element $\boldsymbol{x} = (x_1, x_2, x_3)$ in K we associate the incidence vector $l(\boldsymbol{x}) = (y_1, ..., y_{3n})$ such that $y_i = 0$ for all i except $y_{x_1} = y_{n+x_2} = y_{2n+x_3} = 1$. Fixing an order on K, we can construct a $r \times 3n$ incidence matrix $\mathbf{M}$ by keeping the incidence vectors $l(x)$ of the ordered elements of K. Let $\mathbf{G}_{\mathcal{D}}$ be a $r \times (3r + 3n)$ matrix defined as follows:

$$\mathbf{G}_{\mathcal{D}} = (\mathbf{I}_r | \mathbf{I}_r | \mathbf{I}_r | \mathbf{M})$$

Let $\mathcal{D}$ be the $[3r + 3n, r]$ linear code over $\mathbb{F}_q$ generated by the matrix $\mathbf{G}_{\mathcal{D}}$. Note that a change in the order of K, i.e. a permutation of the rows of $\mathbf{M}$ corresponds to a same permutation on r elements applied simultaneously to the three $\mathbf{I}_r$ matrices. In that case, we obtain a code $\mathcal{C}'$ equivalent to $\mathcal{D}$.

Lemma 1. *The minimum distance of $\mathcal{D}$ is exactly 6. Moreover, the minimum codewords are exactly the rows of $\mathbf{G}_{\mathcal{D}}$.*

Proof. The rows of $\mathbf{G}_{\mathcal{D}}$ correspond to codewords of weight 6. Since all the rows of $\mathbf{M}$ are distinct, the weight of the sum of two rows of $\mathbf{G}_{\mathcal{D}}$ is at least 8. Finally, the weight of the sum of s distinct rows is at least $3s$, which is greater than 9 for $s \geq 3$. Now we consider the $n \times (3r + 3n)$ matrix $\mathbf{G}_{\mathcal{C}}$ defined by

$$\mathbf{G}_{\mathcal{C}} = (\mathbf{I}_n | \mathbf{0}_{n \times (r-n)} | \mathbf{I}_n | \mathbf{0}_{n \times (r-n)} | \mathbf{I}_n | \mathbf{0}_{n \times (r-n)} | \mathbf{I}_n | \mathbf{I}_n | \mathbf{I}_n)$$

where $\mathbf{0}_{n \times (r-n)}$ is the $n \times (r-n)$ null matrix. Let $\mathcal{C}$ be the linear code generated by $\mathbf{G}_{\mathcal{C}}$. Note that the codewords corresponding to the rows of $\mathbf{G}_{\mathcal{C}}$ have no common coordinates. Suppose first that the instance (n, K) satisfies 3DM. Let $K' \subset K$

be the matching. Without loss of gen- erality, we order K in such a way that the n first elements are those of K'. It is possible to permute the $3n$ last columns of $\mathbf{G}_C$ in such a way the matrix $\mathbf{G}_C$ corresponds to the n first rows of $\mathbf{G}_D$. Then the code C is equivalent by permutation to a subcode of D.

Reciprocally, suppose that C is equivalent by permutation to a subcode of D. Let σ be a permutation such that $\sigma(C) \subset D$. The image of any row of $\mathbf{G}_C$ by σ is a codeword of D of weight exactly 6. From *Lemma* 1, this element is a row of D. By this mean, we obtain n distinct rows of D with the particularity that no two rows agree on any coordinate. This leads directly to a matching K' of K.

Since all operations in this process are polynomial, we obtain a reduction of 3DM to ES, and then ES is a NP-complete problem.

4   Application of ES to the Code Based Cryptography

The zero-knowledge is a central concept in cryptographie. It allows to a prover to convince a verifer that it knows a secret without the verifer learning any information about the secret.

The code based identification scheme is due to *Harari* [11] in 1989, followed by Stern [23] in 1990 who introduced the first protocol (a five-pass identification protocol) and at the same moment *Girault* [9] proposed another identification scheme (a Three-pass identification scheme). All of those identifications schemes are non-practical or broken or severely weakened (example [9]). Then there are since some improvement of the code based identification protocols: by *Stern* [25] in 1993, *Véron* [24] in 1996, *Gaborit et al.* [10] in 2007, *Cayrel et al.* [7] in 2011, and recently in 2016 by *Sendrier et al.* [17].

Then for an application of the ES we propose an improvement of the code based identification protocol.

4.1   The Girault Identification Scheme

In this subsection we give briefly the description of *Giralt*'s three-pass identification protocol. The *Girault*'s protocol is following: let $\mathbf{H}$ be a $(n-k) \times n$ binary matrix common to all users. Each prover choose randomly and keeps a binary vector $\mathbf{e}$ of length n and of weight ω then compute the public identifer $\mathbf{s} = \mathbf{H}\mathbf{e}$. It is clear that to find the binary $\mathbf{e}$ knowing the identifer s is a NP-hard problem in coding theory called *Syndrome Decoding* problem.

Definition 3 *(Syndrome Decoding problem). Given an $m \times n$ binary matrix $\mathbf{H}$ over $\mathbb{F}_2$, a target vector $\mathbf{s} \in \mathbb{F}_2^m$ and an integer $\omega > 0$ does there exist a vector $\mathbf{e} \in \mathbb{F}_2^n$ of weight $\leq \omega$ such that $\mathbf{H}\mathbf{e} = \mathbf{s}$?*

When the prover $\mathcal{P}$ wants to convince a verifier $\mathcal{V}$ that he is the owner of $\mathbf{s}$ without revealing any additional information, then they must interact through the following scheme:

Girault Identification Scheme

Key generation: Random $[n, k]$-linear code with a $(n - k) \times n$ parity-check matrix $\mathbf{H}$.

- **Private key:** A word $e \in \mathbb{F}_{q^m}^n$ of weight ω.
- **Public key:** A public identifer $s \in \mathbb{F}_{q^m}^{n-k}$ such that $s = \mathbf{H}e$

Commitments:

- $\mathcal{P}$ chooses randomly a non singular $(n - k) \times (n - k)$ binary matrix $\mathbf{S}$ and a $n \times n$ permutation matrix $\mathbf{P}$.
- $\mathcal{P}$ compute $s' = \mathbf{S}s$ and $\mathbf{H}' = \mathbf{SHP}$
- $\mathcal{P}$ sends s' and $\mathbf{H}'$ to $\mathcal{V}$.

Challenge: $\mathcal{V}$ chooses randomly $c \in \{0, 1\}$
Response:

- If $c = 0$ then $\mathcal{P}$ answer by delivering the non singular matrix $\mathbf{S}$ and the permutation matrix $\mathbf{P}$
- If $c = 1$ then $\mathcal{P}$ replies by delivering $e' = \mathbf{P}^{-1}e$

Verification

- If $c = 0$ then $\mathcal{V}$ checks that $\mathbf{H}' = \mathbf{SHP}$ and $s' = \mathbf{S}s$
- If $c = 1$ then $\mathcal{V}$ checks that the weight of e' is equal to ω, $s' = \mathbf{H}'e'$

The *Girault*'s identification protocol is based on two problems in coding theory: *Syndrome Decoding* problem and *Equivalence code* problem over the binary field. The first one is proved NP-complete in worst case in [5] and the second was severely weakened for some variants [15, 19].

For reach a high security level in order of $1 - \dfrac{1}{2^t}$ this protocol is a multi-round and it has to be repeated t times.

4.2 Version of the Girault Identification Scheme Using ES

The improvement *Girault*'s identification protocol that we proposed is given by:

Girault Identification Scheme using ES

Key generation: Random $[n, k]$-linear code with a $(n - k) \times n$ parity-check matrix $\mathbf{H}$.

- **Private key:** A word $e \in \mathbb{F}_{q^m}^n$ of weight ω.
- **Public key:** A public identifer $s \in \mathbb{F}_{q^m}^{n-k}$ such that $s = \mathbf{H}e$

Commitments:

- $\mathcal{P}$ choose randomly an integer $\ell < k$m an arbitrary $(n-k-\ell) \times (n-k)$ matrix $\mathbf{S}$ of rank $n - k - \ell$ and a permutation matrix $\mathbf{P}$.
- $\mathcal{P}$ compute $s' = \mathbf{S}s$ and $\mathbf{H}' = \mathbf{S}\mathbf{H}\mathbf{P}$
- $\mathcal{P}$ send s' and $\mathbf{H}'$ to $\mathcal{V}$.

Challenge: $\mathcal{V}$ choose randomly $c \in \{0, 1\}$
Response:

- If $c = 0$ then $\mathcal{P}$ replies by delivering the non singular matrix $\mathbf{S}$ and the permutation matrix $\mathbf{P}$.
- If $c = 1$ then $\mathcal{P}$ replies by delivering $e' = \mathbf{P}^{-1}e$

Verification

- If $c = 0$ then $\mathcal{V}$ check that $\mathbf{H}' = \mathbf{S}\mathbf{H}\mathbf{P}$ and $s' = \mathbf{S}s$
- If $c = 1$ then $\mathcal{V}$ check that the weight of e' is equal to ω, $s' = \mathbf{H}'e'$

4.3   Performance of the Scheme

This improvement of *Girault*'s identification protocol is too a three-pass identification protocol. The proof of the *completeness*, the *soundness* and the *zero-knowlege* of this improvement protocol is similar to the proof given by *Girault* in the original version [9], with a reduced commitment $\mathbf{H}'$.

We see that the security of this improvement is based on two coding theory problems, the first is a well-known problem called *Syndorme decoding* problem and the second is the *Equivalence Subcode* problem which we introduce in this paper and prove that it is NP-complete. By reducing the size of the matrix $\mathbf{H}'$ in the commitments, we see that this technique allows us to reduce the cost of the communication and space cost at each round.

One of the important performance of this improvement is that it allows us to enhance the security of the *Girault*'s identification protocol which has been weakened [15]. Compared to the last improvement given in [17] which can be used just for random linear code over a finite field with cardinality $q \geq 5$, this new improvement of Girault's identification protocol can be used for random linear code over an arbitrary finite fields. In addition it can be used by choosing a random subcode of the initial dual code over the same arbitrary finite field.

For reach a security level $1 - 1/2^t$ this improvement protocol is too multiround and it has to be repeated t times.

5   Conclusion

In this paper we introduce a new hard problem in coding theory then we called *Equivalence Subcode* problem and we prove that it is NP-complete. We improve the *Girault*'s identification protocol based on this new problem and the *Syndrome Decoding* problem.

Acknowlegment. This work was carried out with financial support of CEA-MITIC for CBC project and financial support of the government of Senegal's Ministry of Hight Education and Research for ISPQ project.

References

1. Barg, A.: Some new NP-complete coding problems. Problemy Peredachi Informatsii **30**(3), 23–28 (1994). English translation in Probl. Inform. Trans. **30**, 209–214, July–September 1994
2. Berger, T.P.: New perspectives for code-based public key cryptography. In: Codes and Lattices in Cryptography, CLC 2006, Darmstadt (2006)
3. Berger, T.P., Cayrel, P.-L., Gaborit, P., Otmani, A.: Reducing key length of the McEliece cryptosystem. In: Preneel, B. (ed.) AFRICACRYPT 2009. LNCS, vol. 5580, pp. 77–97. Springer, Heidelberg (2009). doi:10.1007/978-3-642-02384-2_6
4. Berger, T.P., Loidreau, P.: How to mask the structure of codes for a cryptographic use. Des. Codes Crypt. **35**, 63–79 (2005)
5. Berlekamp, E., McEliece, R.J., van Tilborg, H.: On the inherent intractability of certain coding problems. IEEE Trans. Inf. Theor. **24**(3), 384–386 (1978)
6. Cayrel, P.L., Diagne, M.K., Gueye, C.T.: NP-completeness of the Coset weight problem for Quasi-dyadic codes. In: International Conference on Coding theory and Cryptography ICCC 2015, Alger, Algeria (2015)
7. Cayrel, P.-L., Véron, P., Yousfi Alaoui, S.M.: A zero-knowledge identification scheme based on the q-ary syndrome decoding problem. In: Biryukov, A., Gong, G., Stinson, D.R. (eds.) SAC 2010. LNCS, vol. 6544, pp. 171–186. Springer, Heidelberg (2011). doi:10.1007/978-3-642-19574-7_12
8. Garey, E., Johnson, D.: Computers and Intractability: A Guide to the Theory of NP-Completeness. W.H. Freeman and Company, New York (1979)
9. Girault, M.: A (non-practical) three-pass identification protocol using coding theory. In: Seberry, J., Pieprzyk, J. (eds.) AUSCRYPT 1990. LNCS, vol. 453, pp. 265–272. Springer, Heidelberg (1990). doi:10.1007/BFb0030367
10. Gaborit, P., Girault, M.: Lightweight code-based authentication and signature. In: ISIT (2007)
11. Harari, S.: A new authentication algorithm. In: Cohen, G., Wolfmann, J. (eds.) Coding Theory 1988. LNCS, vol. 388, pp. 91–105. Springer, Heidelberg (1989). doi:10.1007/BFb0019849
12. McEliece, R.J.: A public-key cryptosystem based on algebraic coding theory. Jet Propulsion Lab. DSN Progress Report, Technical report (1978)
13. Misoczki, R., Barreto, P.S.L.M.: Compact McEliece keys from Goppa codes. In: Jacobson, M.J., Rijmen, V., Safavi-Naini, R. (eds.) SAC 2009. LNCS, vol. 5867, pp. 376–392. Springer, Heidelberg (2009). doi:10.1007/978-3-642-05445-7_24
14. Niederreiter, H.: Knapsack-type cryptosystems and algebraic coding theory. Probl. Control Inf. Theor. **15**(2), 159–166 (1986)

15. Petrank, E., Roth, R.M.: Is code equivalence easy to decide? IEEE Trans. Inf. Theory **43**(5), 1602–1604 (1997)
16. Sendrier, N.: Finding the permutation between equivalent codes: the support splitting algorithm. IEEE Trans. Inf. Theor. **46**(4), 1193–1203 (2000)
17. Sendrier, N., Simos, D.E.: The hardness of code equivalence over $\mathbb{F}_q$ and its application to code-based cryptography. In: Proceeding of Post-Quantum Cryptography, 5th International Workshop PQcrupto 2013, Limoges, France (2013)
18. Sidel'nikov, V.M., Shestakov, S.O.: On cryptosystems based on generalized Reed-Solomon codes. Discrete Math. **4**(3), 57–63 (1992)
19. Sendrier, N., Simos, D.E.: How easy is code equivalence over $\mathbb{F}_q$? In: Proceedings of the 8th International Workshop on Coding and Cryptography, WCC 2013 (2013, to appear). https://www.rocq.inria.fr/secret/PUBLICATIONS/codeq3.pdf. Preprint (2012)
20. Vardy, A.: The intractability of computing the minimum distance of a code. IEEE Trans. Inf. Theor. **43**(6), 1757–1766 (1997)
21. Wieschebrink, C.: Two NP-complete problems in coding theory with an application in code based cryptography. In: Proceedings of IEEE ISIT 2006, Seattle, USA, pp. 1733–1737 (2006)
22. Wieschebrink, C.: An attack on a modified niederreiter encryption scheme. In: Yung, M., Dodis, Y., Kiayias, A., Malkin, T. (eds.) PKC 2006. LNCS, vol. 3958, pp. 14–26. Springer, Heidelberg (2006). doi:10.1007/11745853_2
23. Stern, J.: An alternative to the Fiat-Shamir protocol. In: Quisquater, J.-J., Vandewalle, J. (eds.) EUROCRYPT 1989. LNCS, vol. 434, pp. 173–180. Springer, Heidelberg (1990). doi:10.1007/3-540-46885-4_19
24. Véron, P.: Improved identification schemes based on error-correcting codes. Appl. Algebra Eng. Commun. Comput. **8**(1), 57–69 (1996)
25. Stern, J.: A new identification scheme based on syndrome decoding. In: Stinson, D.R. (ed.) CRYPTO 1993. LNCS, vol. 773, pp. 13–21. Springer, Heidelberg (1994). doi:10.1007/3-540-48329-2_2

Spectral Approach for Correlation Power Analysis

Philippe Guillot[1], Gilles Millérioux[2,3], Brandon Dravie[2,3(✉)],
and Nadia El Mrabet[4]

[1] Université Paris 8, LAGA, UMR 7539, Saint-Denis, France
`philippe.guillot@univ-paris8.fr`
[2] Université de Lorraine, CRAN, UMR 7039, ESSTIN, 2 rue Jean Lamour,
54506 Vandœuvre-lès-Nancy, France
`{gilles.millerioux,brandon.dravie}@univ-lorraine.fr`
[3] CNRS, CRAN, UMR 7039, Vandoeuvre-lès-Nancy, France
[4] Ecole des Mines de Saint Etienne, SAS, Gardanne, France
`nadia.el-mrabet@emse.fr`

Abstract. This paper provides a new approach to perform Correlation Power Analysis (CPA) attack. Power analysis attacks are side channel attacks based on power consumption measures on a device running a cryptographic algorithm with a CMOS technology based circuitry. Unlike most of CPA attacks that are based on statistical attacks, this paper proposes a new approach based on spectral analysis. The interest lies in the reduction of the attack complexity. The complexity is quasi linear in the size of the table of values of the S-box whereas it is quadratic with statistical attacks. It is shown that it can be easily extended to a so-called multidimensional attack. The attack is experimented on a AES S-box.

Keywords: Correlation Power Analysis · Spectral analysis · Fourier transform

1 Introduction

The Correlation Power Analysis (CPA) is a method that allows to recover the secret information (usually the secret key) embedded in the silicon of an electronic component [3]. It consists in measuring the power consumption while running operations that involve the secret information. This method has been introduced in 2004 by researchers of Gemplus Company (Eric Brier, Christophe Clavier and Francis Olivier) in [3]. The attack follows the work of Paul Kocher proposed in 1999 [7]. The electric circuits that perform the computation in the processor are designed from a technology known as CMOS (Complementary Metal Oxide Semiconductor) [2]. A general description of this technology is provided in [9]. The outcome of CMOS architectures lies in fast transitions but, on the other hand, they are sensitive to power consumption or electromagnetic leakage.

© Springer International Publishing AG 2017
S. El Hajji et al. (Eds.): C2SI 2017, LNCS 10194, pp. 238–253, 2017.
DOI: 10.1007/978-3-319-55589-8_16

Power analysis attacks are based on the general principle that the instantaneous power consumption of a cryptographic device depends on the data it processes and on the operation it performs [10]. During a symmetric protocol, those operations are in general processed by a nonlinear function called S-box parametrized by a secret parameter [1,5]. Usually, CPA attacks rely on statistical analysis and have a quadratic complexity when resorting, for example, to Pearson correlation coefficients computation [12,13]. In this paper we present a method that relies on spectral analysis. The complexity is quasi linear in the size of the table of values of the S-box whereas it is quadratic with statistical attacks.

The article is organized as follows. In Sect. 2, having in mind the spectral analysis, preliminaries on Fourier analysis are given. In Sect. 3, the main physical principles on which the attack is based are recalled. Then, the principle of the attack is presented for a general nonlinear Boolean function implemented as an S-box. Section 4 is devoted to a so-called multidimensional attack. In Sect. 5, the spectral-based attack is experimented on a AES S-box. Finally, some future prospects are given in the conclusion of Sect. 6.

2   Preliminaries on Spectral Analysis

Having in mind a CPA attack based on spectral considerations, preliminaries on Fourier analysis must be recalled. It is worth pointing out that the suggested attack is based on a power consumption measurement on a component that processes binary data as input. Hence, the measurement can be modeled as a real valued function over the set of binary words. This section is devoted to prerequisites on Fourier analysis of this class of functions.

Let Φ be the set of real valued functions over the set of n-dimensional binary words:

$$\Phi = \big\{ \varphi : \{0,1\}^n \to \mathbb{R} \big\}$$

For any two functions φ and ψ in Φ, let us define the scalar product of φ and ψ as:

$$\langle \varphi, \psi \rangle = \sum_{x \in \{0,1\}^n} \varphi(x)\psi(x).$$

This scalar product is a symmetric bilinear form that confers to this set the structure of a 2^n dimensional Euclidean vector space over $\mathbb{R}$.

The norm associated to this scalar product is:

$$\|\varphi\| = \sqrt{\langle \varphi, \varphi \rangle} = \sqrt{\sum_{x \in \{0,1\}^n} \varphi(x)^2}$$

The norm of a function φ in Φ is called the energy of φ.

The well known Cauchy-Schwarz inequality holds:

$$\forall \varphi, \psi \in \Phi, \ \left| \langle \varphi, \psi \rangle \right| \leq \|\varphi\| \times \|\psi\|$$

where $|\cdot|$ stands for the absolute value.

The canonical basis of the space Φ is the family of characteristic functions of singletons which are by definition, for all vectors $u \in \{0,1\}^n$, the functions denoted by δ_u and defined by:

$$\delta_u : x \mapsto \begin{cases} 1 & \text{if } x = u \\ 0 & \text{else} \end{cases}$$

This basis is clearly orthonormal according to the above scalar product.

Each function $\varphi \in \Phi$ can be expressed in this basis as:

$$\varphi = \sum_{u \in \{0,1\}^n} \varphi(u)\delta_u.$$

Another basis of the space Φ is the basis of the so-called Walsh functions.

Definition 1 (Walsh functions). *The Walsh functions are the functions of Φ defined for any $u \in \{0,1\}^n$ by:*

$$\chi_u : x \mapsto \frac{1}{\sqrt{2^n}}(-1)^{u \cdot x}$$

where $x \cdot u = u_1 x_1 + u_2 x_2 + \cdots + u_n x_n$ is the dot product over the space $\{0,1\}^n$ of n–dimensional binary words over the two elements field $\mathbb{F}_2$.

The Walsh functions are pairwise orthogonal. They are presented here with a normalization coefficient equal to $1/\sqrt{2^n}$ such that they provide an orthonormal basis as stated in the following proposition.

Proposition 1. *The family $(\chi_u)_{u \in \{0,1\}^n}$ of Walsh functions is an orthonormal basis of Φ.*

Proof. Let u and v be two n-dimensional binary vectors. One has:

$$\langle \chi_u, \chi_v \rangle = \frac{1}{2^n} \sum_{x \in \{0,1\}^n} (-1)^{x \cdot (u+v)}.$$

If $u = v$ then $u + v = 0$. The above sum has 2^n terms, all equal to 1. Then $\langle \chi_u, \chi_u \rangle = 1$.

If $u + v \neq 0$, there exists a non-zero component. Let $u_i + v_i$ be this component. Then, the sum over all the vectors x for which $x_i = 0$ is the opposite of the sum over all the vectors x for which $x_i = 1$. It results that if $u \neq v$ then $\langle \chi_u, \chi_v \rangle = 0$.

This family allows to express any function $\varphi \in \Phi$ in the basis of Walsh functions

$$\varphi = \sum_{u \in \{0,1\}^n} \langle \varphi, \chi_u \rangle \chi_u.$$

Definition 2 (Fourier Transform). *The Fourier spectrum of $\varphi \in \Phi$ is the family of coefficients $\left(\langle \varphi, \chi_u \rangle \right)_{u \in \{0,1\}^n}$, of the expression of f in the basis of Walsh functions, and the Fourier transform of f is the function of Φ defined on $\{0,1\}^n$ as:*

$$u \mapsto \widehat{\varphi}(u) = \langle \varphi, \chi_u \rangle = \frac{1}{\sqrt{2^n}} \sum_{x \in \{0,1\}^n} \varphi(x)(-1)^{u \cdot x}.$$

The Fourier transform expresses a change of basis. Thus, the transformation is linear. Moreover, it is isometric as stated in the following proposition.

Proposition 2. *For any functions φ and ψ in Φ, one has:*

$$\langle \varphi, \psi \rangle = \langle \widehat{\varphi}, \widehat{\psi} \rangle.$$

Proof.

$$\langle \widehat{\varphi}, \widehat{\psi} \rangle = \sum_{x \in \{0,1\}^n} \frac{1}{\sqrt{2^n}} \sum_{u \in \{0,1\}^n} \varphi(u)(-1)^{u \cdot x} \frac{1}{\sqrt{2^n}} \sum_{v \in \{0,1\}^n} \psi(v)(-1)^{v \cdot x}$$

By inverting the summation order, it follows that

$$\langle \widehat{\varphi}, \widehat{\psi} \rangle = \frac{1}{2^n} \sum_{u \in \{0,1\}^n} \sum_{v \in \{0,1\}^n} \varphi(u)\psi(v) \sum_{x \in \{0,1\}^n} (-1)^{(u+v) \cdot x}.$$

As this latter sum equals 2^n if $u = v$ and equals 0 elsewhere, the result holds.

As a direct consequence of this proposition, it follows that for all functions φ in Φ, the so-called Parseval equality holds and expresses the energy conservation law: $\|\varphi\| = \|\widehat{\varphi}\|$.

Proposition 3 (Effect of a translation). *For any vector a in $\{0,1\}^n$, let $\tau_a : t \longmapsto t + a$ be the translation of vector a. Let φ be a function in Φ, then for all vectors $u \in \{0,1\}^n$, one has:*

$$\widehat{\varphi \circ \tau_a}(u) = (-1)^{u \cdot a} \widehat{\varphi}(u).$$

Proof.

$$\widehat{\varphi \circ \tau_a} = \frac{1}{\sqrt{2^n}} \sum_{x \in \{0,1\}^n} \varphi(x + a)(-1)^{u \cdot x} = \frac{1}{\sqrt{2^n}} \sum_{y \in \{0,1\}^n} \varphi(y)(-1)^{u \cdot (y+a)}$$

$$= (-1)^{u \cdot a} \widehat{\varphi}(u) \tag{1}$$

Proposition 4 (Fourier Transform of a constant). *Let* $k \in \{0,1\}^n$.
The Fourier transform of the constant function $x \longmapsto k$ *is:*

$$\widehat{k} = k\sqrt{2^n}\delta_0.$$

Proof. By definition,

$$\widehat{k}(u) = \frac{1}{\sqrt{2^n}} \sum_{x \in \{0,1\}^n} k(-1)^{u \cdot x} = \frac{k}{\sqrt{2^n}} \sum_{x \in \{0,1\}^n} (-1)^{u \cdot x}.$$

The latter sum equals 2^n if $u = 0$ and 0 elsewhere, and the result holds.

3   Modelling the Attack

3.1   Physical Principles

Nowadays, digital circuits are often designed with CMOS (Complementary Metal Oxide Semiconductor) technology. The main characteristic lies in the output stage of the logical gates that involves a pair of Field Effect Transistors (FET) with opposite polarity. The transistors are combined symmetrically (push-pull architecture) such that they switch between two states: on and off. The outcome of such architecture is that in a steady state mode, the current consumption is almost null. On the other hand, when the output state of a gate changes, a parasitic capacitor discharges in the complementary parasitic capacitor. This leads to power consumption while a transition occurs. As a consequence, the following assumption, which is the core idea of CPA is well-admitted:

Assumption 1. *The power consumption of CMOS circuit is proportional to the number of logic gates that switch [2].*

From this assumption, two models of consumption can be considered: the Hamming weight model and the Hamming distance model [3]. For a given calculus, the first model assumes that the gates switch from the state 0 to the result of the calculus. It turns out that the model is well suited for software implementations. The second model assumes that the gates switch from an initial state that corresponds to the former calculus result to the state corresponding to the result of the current calculus. It turns out that this model is more suitable for hardware implementations.

3.2   Target of the Attack

Many symmetric ciphering algorithms such as DES [5] or AES [1] for instance are based on alternate stages of linear and nonlinear calculations. The nonlinear stage is most often implemented in the form of S-boxes, each of one corresponding to a nonlinear function f from $\{0,1\}^n$ to $\{0,1\}^m$, whose input is the exclusive or of a data $x \in \{0,1\}^n$ and of an unknown secret subkey $k^\star \in \{0,1\}^n$, and returns a quantity $y \in \{0,1\}^m$, that is

$$y = f(x + k^\star).$$

This being the case, the attack consists in performing ciphering operations with various inputs x chosen by the adversary, then, measuring the power consumption during the calculation and finally, trying to infer the values of the secret subkey $k^\star$. Having in mind a smart card software implementation, according to the discussion in Sect. 3.1, the first model of leakage will be hereafter considered and Assumption 1 applies. Let us notice that this is not restrictive because it is easy to adapt the proposed attack to the second model.

Thus, for a given calculus of an S-box with input x and secret subkey $k^\star$, the power consumption is proportional to the quantity $\varphi(x)$ which admits the following expression:

$$\varphi(x) = \sum_{i=1}^{m} f_i(x + k^\star) + \varepsilon(x) + C, \tag{2}$$

where:

- the first term is the model leakage, i.e. the Hamming weight of $f(x+k^\star)$, with f_i the Boolean i-th component of the function f.
- the second term is a random noise denoted by $\varepsilon(x)$.
- the third term is a constant C that corresponds to the power consumption of the system which does not depend on x.

Let g be the function $x \mapsto \sum_{i=1}^{m} f_i(x)$. Equation (2) is rewritten as:

$$\varphi(x) = g(x + k^\star) + \varepsilon(x) + C.$$

However, the subkey k assumed by the adversary may not be equal to the right secret $k^\star$ and the leakage model may not be the exact one. Hence, we must introduce an error depending in particular on k. It is denoted by $\varepsilon_k(x)$ and is defined as:

$$\varepsilon_k(x) = \varphi(x) - g(x + k) - C. \tag{3}$$

In an ideal situation, that is no noise, no mismatch, and in particular when $k = k^\star$, it should be zero for any $x \in \{0,1\}^n$. Hence, we can define the objective of the attack as finding k which minimizes the quadratic error

$$E(k)^2 = \|\varepsilon_k\|^2 = \sum_{x \in \{0,1\}^n} \varepsilon_k(x)^2.$$

In the following, it is shown that a spectral approach to solve Eq. (4) will be relevant in terms of complexity of the underlying attack.

Based on Parseval's Equality, the problem is equivalent to find k which is solution to:

$$\arg\min_k E(k)^2 = \sum_{u \in \{0,1\}^n} \widehat{\varepsilon_k}(u)^2 \tag{4}$$

By applying the Fourier transform to Eq. (3), it comes, for all vectors $u \in \{0,1\}^n$:

$$\widehat{\varepsilon_k}(u) = \widehat{\varphi}(u) - (-1)^{u \cdot k}\widehat{g}(u) - C\sqrt{2^n}\delta_0(u). \tag{5}$$

The interest of considering the Fourier transform is that discarding the value at the zero vector eliminates the last term which corresponds to the consumption that does not depend on the value of x.

Hence, we define the functions $\widehat{\varphi}^\star$ and $\widehat{g}^\star$ by zeroing the value at the zero vector, that is:

$$\widehat{\varphi}^\star(u) = \begin{cases} \widehat{\varphi}(u) & \text{if } u \neq 0 \\ 0 & \text{else} \end{cases} \quad \text{and} \quad \widehat{g}^\star(u) = \begin{cases} \widehat{g}(u) & \text{if } u \neq 0 \\ 0 & \text{else} \end{cases} \tag{6}$$

then finding k solution of Eq. (4) is equivalent to finding k solution of

$$\arg\min_k E(k)^2 = \sum_{u \in \{0,1\}^n} \left(\widehat{\varepsilon_k}^\star(u)\right)^2 \tag{7}$$

with

$$\widehat{\varepsilon_k}^\star(u) = \widehat{\varphi}^\star(u) - (-1)^{u \cdot k}\widehat{g}^\star(u).$$

Expansing the terms in $E(k)^2$ yields

$$E(k)^2 = \sum_{u \in \{0,1\}^n} \widehat{\varphi}^\star(u)^2 + \sum_{u \in \{0,1\}^n} \widehat{g}^\star(u)^2 - 2 \sum_{u \in \{0,1\}^n} \widehat{\varphi}^\star(u)\widehat{g}^\star(u)(-1)^{u \cdot k} \tag{8}$$

In the right hand side, only the last term depends on k. Hence and finally, the problem is to find k solution to

$$\arg\max_k F(k) \tag{9}$$

with

$$F(k) = \sum_{u \in \{0,1\}^n} \widehat{\varphi}^\star(u)\widehat{g}^\star(u)(-1)^{u \cdot k}. \tag{10}$$

Remark 1. The function F is, up to a factor $1/\sqrt{2^n}$, nothing but the Fourier transform of the function:

$$u \mapsto \widehat{\varphi}^\star(u)\widehat{g}^\star(u).$$

3.3   Assessing the Estimation Reliability

From the above section, we assume that the attacker can choose any $x \in \{0,1\}^n$ and knows the exact instant when the operation $f(x + k)$ is performed. To circumvent such a difficulty which arises in practice, an enhancement of the attack should be proposed. We must memorize the consumption $\varphi(x)$ during a sufficient large time window Δ_t to guarantee that the computation of $f(x + k^\star)$ will be actually performed. The resulting signal is called a trace. A trace has to be measured for all 2^n value of x. Let us denote by $\varphi_t(x)$ the consumption at time t for the input value x. For a finite number of sample times t in the time window Δ_t, an estimation k of the secret subkey $k^\star$ is computed from (9). However, if $k^\star$ is not involved in the operation performed at this time or if $k^\star$ is actually involved but k is not equal to $k^\star$, the estimation of the right subkey $k^\star$, that is the result of (9) will not be reliable. Thus, it is required to find a way of assessing this reliability.

To this end, similarly as in Subsect. 3.2, let us introduce the quantity defined as:

$$F_t(k) = \sum_{u \in \{0,1\}^n} \widehat{\varphi_t}^\star(u)\widehat{g}^\star(u)(-1)^{u \cdot k}. \tag{11}$$

where

$$\widehat{\varphi_t}^\star(u) = \begin{cases} \widehat{\varphi_t}(u) & \text{if } u \neq 0 \\ 0 & \text{else} \end{cases} \tag{12}$$

For brevity, let us denote by g_k the function $x \mapsto g(x + k)$. It is recalled that this function stands for the leakage model. Let us denote by $\widehat{g}_k$ the Fourier transform of g_k and with the same motivation as in Subsect. 3.2, let us introduce $\widehat{g}_k^\star$ as the Fourier transform of $g_k^\star$ defined as:

$$\widehat{g}_k^\star(u) = \begin{cases} \widehat{g}_k(u) & \text{if } u \neq 0 \\ 0 & \text{else} \end{cases} \tag{13}$$

which allows to disregard the zero vector. According to Proposition 3, for all n–dimensional binary vectors u, we have that $\widehat{g}_k^\star(u) = (-1)^{u \cdot k}\widehat{g}^\star(u)$. Hence, Eq. (11) can be rewritten as

$$F_t(k) = \langle \widehat{\varphi_t}^\star, \widehat{g}_k^\star \rangle \tag{14}$$

Hence, for every sample times t in Δ_t, it is aimed at finding k, that is finding the solution of

$$\arg \max_k F_t(k) \tag{15}$$

Thus, k may depend on t. The scalar product $F_t(k)$ being computed for all sample times t in Δ_t, we must detect peaks. The more the matching between the measure and the model, including the key k, the higher the peaks. Hence, the orthogonality is a way of assessing the matching. However, since the detection of the peaks requires a comparison of $F_t(k)$ for all sample times t in Δ_t, a normalization must be done. To this end, we introduce the following quantity which will be called reliability coefficient.

$$r_t(k) = \frac{F_t(k)}{\|\widehat{\varphi_t^\star}\| \cdot \|\widehat{g_k}^\star\|} = \frac{\langle \widehat{\varphi_t^\star}, \widehat{g_k}^\star \rangle}{\|\widehat{\varphi_t^\star}\| \cdot \|\widehat{g_k}^\star\|}$$

Noticing that $\|\widehat{g_k^\star}\| = \|\widehat{g}^\star\|$, the reliability coefficient turns into

$$r_t(k) = \frac{F_t(k)}{\|\widehat{\varphi_t^\star}\| \cdot \|\widehat{g}^\star\|}, \tag{16}$$

with a normalization which is independent from k. According to Cauchy-Schwarz inequality it follows that $r_t(k) \in [0, 1]$

Finally, the attack consists in the following steps:

Steps 1: for every sample times t in Δ_t, finding the solution k of (15);
Steps 2: for every solution k associated to a given sample time t in Δ_t, compute $r_t(k)$ (see Eq. (16)) and detect the peaks among all the $r_t(k)$ with $t \in \Delta_t$. Those peaks correspond to the times where the key $k^\star$ match with the solution k of Steps 1.

Remark 2. If the time where $f(x + k^\star)$ is computed is exactly known, there is no need to perform Steps 2.

Let us comment on the complexity of the attack. The Fourier Transform of the function $u \mapsto g^\star(u)$ can be computed off-line once and for all from the S-box table of the leakage model. The Fourier transform $\widehat{\varphi_t^\star}$ must be computed from experimental data for every time t in Δ_t. For every time t in Δ_t, according to Remark 1, $F_t(k)$ is computed with the Fourier transforms of the function $u \mapsto \widehat{\varphi_t^\star}(u)\widehat{g}^\star(u)$ and searching for the maximum among the $F_t(k)$. Finally, the peaks are detected by computing $r_t(k)$ (see Eq. (16)). The Fourier transforms calculation can be carried out by using a fast algorithm. Such an algorithm exists, with a complexity equal to $n\,2^n$, which is quasi linear in the size of the table of values of the related function (see for example [4]). As a result, the overall complexity of the attack is quasi linear in the size of the S-box table.

Remark 3. It is worth pointing out that the quantity in Eq. (16) is related to the usual Pearson correlation coefficient. This coefficient is widely used to evaluate relationship between data. Hence, it is a popular choice for statistical analysis when it comes to perform CPA attacks. Let us recall that the Pearson correlation coefficient of two functions φ and ψ in Φ is by definition:

$$c(\varphi, \psi) = \frac{\sum_{x \in \{0,1\}^n} (\varphi(x) - m_\varphi) \times (\psi(x) - m_\psi)}{\sqrt{\sum_{x \in \{0,1\}^n} (\varphi(x) - m_\varphi)^2} \times \sqrt{\sum_{x \in \{0,1\}^n} (\psi(x) - m_\psi)^2}}, \tag{17}$$

where m_φ and m_ψ are the means of functions φ and ψ, given by:

$$m_\varphi = \frac{1}{2^n} \sum_{x \in \{0,1\}^n} \varphi(x) \quad \text{and} \quad m_\psi = \frac{1}{2^n} \sum_{x \in \{0,1\}^n} \psi(x)$$

In other words, Pearson coefficient $c(\varphi, \psi)$ can be expressed as:

$$c(\varphi, \psi) = \frac{\langle \varphi - m_\varphi, \psi - m_\psi \rangle}{\|\varphi - m_\varphi\| \cdot \|\psi - m_\psi\|}$$

As it can be noticed that $\widehat{\varphi - m_\varphi} = \widehat{\varphi}^\star$, and as the Fourier transform is isometric, it follows that the coefficient given by Eq. (16) is nothing but the Pearson correlation coefficient of the functions φ_t and g_k. Thus, it results that the value of k given by maximizing $F(k)$ in Eq. (10) is the same value that maximizes the Pearson correlation coefficient of φ and g_k.

The interest of the spectral analysis is that it can be easily extended to a so-called multidimensional attack as explained in next section.

4 Multidimensional Attack

When computing the value of $f(x + k^\star)$ in a software device, the processor computes sequentially: first $x + k^\star$ and later the value of $f(x + k^\star)$. Thus, the unknown $k^\star$ is involved at least twice within a sufficient time window Δ_t. We call multidimensional attacks, the attacks that take into account the consumption at several instants. It can be expected that more reliable results can be obtained.

The two-dimensional attack is presented thereafter because it is the one which will be used in the experiments presented in Sect. 5 but it is straightforward to generalize it to any higher finite dimension.

Let us consider two instants t_1 and t_2. Let us denote by $f^1(x + k^\star)$ and $f^2(x + k^\star)$ the values computed respectively at times t_1 and t_2. In our case, the function f^1 is the identity since the computation $x + k^\star$ is considered and the function f^2 is the S-box implementing $f(x + k^\star)$. Let φ^1 the chip consumption at time t_1 and φ^2 at time t_2. We consider the two dimensional vector of functions $\overrightarrow{\varphi} = (\varphi^1, \varphi^2)$ in a set Φ_2 of functions $\{0,1\}^n \to \mathbb{R}^2$. Let E^1 the error value given by Eq. (8) at time t_1 and E^2 be the error value at time t_2. The most likely value of k is the one minimizing the Euclidean norm of the two-dimensional vector $\overrightarrow{E} = (E^1, E^2)$. A direct computation shows that the value that minimizes this norm is the one maximizing the value of

$$F(k) = \sum_{u \neq 0} \left(\widehat{\varphi^1}(u) \widehat{g^1}(u) + \widehat{\varphi^2}(u) \widehat{g^2}(u) \right)(-1)^{u \cdot k},$$

where g^1 and g^2 are the sum of the components of f^1 and f^2. The leakage model at time t_1 is $g_k^1(x) = g^1(x + k)$ and at time t_2 is $g_k^2(x) = g^2(x + k)$.

The scalar product to consider in Φ_2 is the following:

$$\forall \overrightarrow{\varphi} = (\varphi^1, \varphi^2), \ \overrightarrow{\psi} = (\psi^1, \psi^2) \in \Phi_2, \ \langle \overrightarrow{\varphi}, \overrightarrow{\psi} \rangle = \langle \varphi^1, \psi^1 \rangle + \langle \varphi^2, \psi^2 \rangle.$$

The norm associated to this scalar product is:

$$\forall \overrightarrow{\varphi} = (\varphi^1, \varphi^2) \in \Phi_2, \ \|\overrightarrow{\varphi}\|^2 = \|\varphi^1\|^2 + \|\varphi^2\|^2.$$

Clearly, the multidimensional attack can be combined with the estimation approach described above. Let $\overrightarrow{\varphi_t}$ be the two-dimensional power consumption vector at time t. Then $\overrightarrow{\varphi_t} = (\varphi_t^1, \varphi_t^2)$, where φ_t^1 is the consumption at time t and φ_t^2 is the consumption at time $t + t_2 - t_1$. The reliability coefficient between the power consumption $(\varphi_t^1, \varphi_t^2)$ and the model (g_k^1, g_k^2) is required to assess the quality of the estimation. This reliability coefficient for the estimated value k at time t is computed as:

$$r_t(k) = \frac{\langle (\widehat{\varphi_t^1}^{\,*}, \widehat{\varphi_t^2}^{\,*}), (\widehat{g_k^1}^{\,*}, \widehat{g_k^2}^{\,*}) \rangle}{\|(\widehat{\varphi_t^1}^{\,*}, \widehat{\varphi_t^2}^{\,*})\| \cdot \|(\widehat{g^1}^{\,*}, \widehat{g^2}^{\,*})\|},$$

where, for $i \in \{1, 2\}$, $\widehat{\varphi_t^i}^{\,*}$ and $\widehat{g_k^i}^{\,*}$ are obtained by discarding the value at the zero vector, as in (6).

By using fast Fourier transform algorithm, the multidimensional attack complexity still remains quasi linear in the size of the S-box.

5   Experimental Results

The Challenge. An ATMega 163 smart card, involving an 8-bit AVR type processor, has been programmed to process the operation $f(x_i + k_i)$, where f is the AES S-Box, the k_i's are 8-bit secret key elements previously introduced in the card, the x_i's are 8-bit parameters introduced in the card by the adversary. The operation $+$ denotes the bitwise xor of bytes. The challenge of the adversary is to retrieve the secret values k_i by measuring the smart card consumption during calculations. For this purpose, the adversary uses a test bench.

Fig. 1. Test bench.

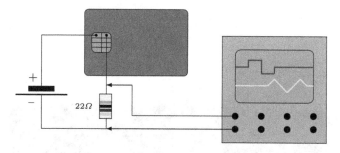

Fig. 2. Schematic of the power consumption measurement.

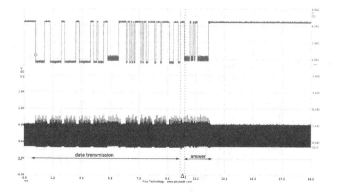

Fig. 3. Whole data exchanges (up) and the corresponding consumption signal (down).

The Test Bench. The test bench (see Fig. 1) involves the smart card, a smart card reader and an oscilloscope. The oscilloscope is a picoscope 5444b with a 200 MHz bandwidth and a sample rate of $1GS$ per second. The pins of the smart card are connected to the oscilloscope via an adapter. The chip consumption is measured through the potential drop at a resistor connected between the Vss of the card and the ground (see Fig. 2). All these devices are driven by a computer that implements the attack algorithm.

Experimental Protocol. The attack consists in recovering four secret keys k_1, k_2, k_3 and k_4 during a time window when the card processes sequentially $f(x + k_1)$, $f(x + k_2)$, $f(x + k_3)$ and $f(x + k_4)$ for 256 values of x ranging from 0 to 255.

In all subsequent figures, the upper signal is the I/O signal corresponding to the data exchanges and the lower signal is the chip consumption.

Figure 3 is an example of consumption trace and the corresponding consumption signal. The time window Δ_t during the computation of the four successive operations $f(x + k_i)$ $(i = 1, \ldots, 4)$ is the range of time when the 256 traces of consumption are memorized in the oscilloscope. The 256 trace records are each composed of 40.000 samples.

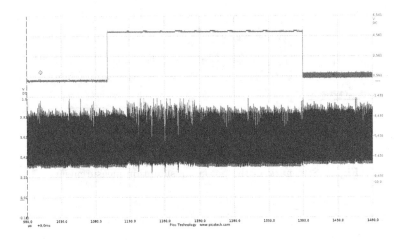

Fig. 4. Zoom on the time range Δ_t highlighting the first falling edge of the card response used for the synchronization of the traces.

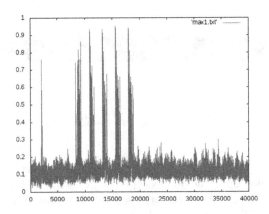

Fig. 5. Correlation peaks that correspond to scenario *(i)*.

It is crucial to synchronize properly the 256 consumption traces. For this purpose, the consumption traces are synchronised with the instant given by the first falling edge of the I/O signal that follows the calculation (Fig. 4).

For every sample t, $t \in \{1, \ldots, 40.000\}$ of each 256 traces, the values $\varphi_t(x)$, $x \in \{0, \ldots, 255\}$ are extracted. Then, for every t, $t \in \{1, \ldots, 40.000\}$, the Fourier transform of φ_t is computed. The secret key k_i is evaluated by maximizing the function F given by Eq. (10). Finally, for this value of k_i, the correlation coefficient is computed according to Eq. (16).

Experimental Results. Three distinct scenarios has been considered:

(i) an attack when the card computes the exclusive or of the secret k and the data x, that is the operation $x + k$

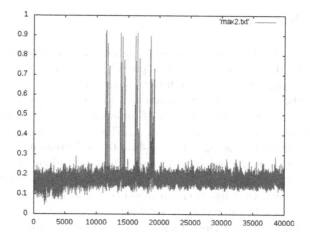

Fig. 6. Correlation peaks that correspond to scenario *(ii)*.

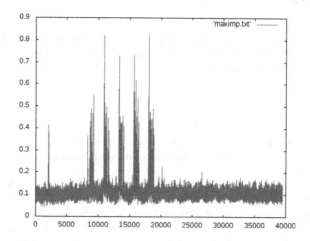

Fig. 7. Correlation peaks that correspond to scenario *(iii)*. The two-dimensional attack considers the times instants t and $t + 4.2\mu$s.

(ii) an attack when the card computes the output of the S-box, that is the operation $f(x + k)$

(iii) a two dimensional attack that combines the operation $x + k$ followed by $f(x + k)$

Figures 5, 6 and 7 show the reliability coefficient $r_t(k)$ given by Eq. (16) with respect to the time t, for the three respective situations. The attack is successful. Indeed, the correlation peaks correspond to the instants when the operations are actually performed and with the right secret key k_i. Let us notice that in Figs. 5 and 7, there are ghost peaks. However, they can be easily disregarded since they correspond to $k = 0$. One possible explanation is that the value of x is loaded in a register previously containing the zero value.

6   Conclusion

A new approach of CPA attack has been proposed. Unlike usual approaches based on statistical analysis, a spectral approach has been provided. It is based on a correlation quantity derived from the Fourier transform of the power consumption signals. The attack can be applied to any algorithms that involve S-boxes whose input is the exclusive or of data with a secret subkey. A Hamming weight leakage model has been used. The interest is the complexity of the attack is quasi linear in the size of the S-box table. Furthermore, it has been shown that it can be easily extended to a so-called multidimensional attack.

As future prospects, it remains to check how these improvements behave with counter measure that has been proposed to resist such attack [6,8,11,14]. Furthermore, it would be interesting to see whether the number of traces could be reduced. It would mean that the attack could be achieved with an approximate value of the Fourier transform. In this case, the multidimensional attack would have a further interest.

Acknowledgments. This work was supported by the Research Grants THE CASCADE ANR-13-INSE-0005-01 from the Agence Nationale de la Recherche and PEPS SISC ECHARPE 2016 from the Centre National de la Recherche Scientifique, France.

References

1. Aes, N.: Fips publication 197 - advanced encryption standard
2. Baker, R.J.: CMOS: Circuit Design, Layout, and Simulation. Wiley, Hoboken (2011)
3. Brier, E., Clavier, C., Olivier, F.: Correlation power analysis with a leakage model. In: Joye, M., Quisquater, J.-J. (eds.) CHES 2004. LNCS, vol. 3156, pp. 16–29. Springer, Heidelberg (2004). doi:10.1007/978-3-540-28632-5_2
4. Carlet, C.: Boolean functions for cryptography and error-correcting codes. In: Boolean Models and Methods in Mathematics, Computer Science, and Engineering. Cambridge Press, Cambridge (2010)
5. Des, N.: Fips publication 46–3 - data encryption standard
6. Itoh, K., Yajima, J., Takenaka, M., Torii, N.: DPA countermeasures by improving the window method. In: Kaliski, B.S., Koç, K., Paar, C. (eds.) CHES 2002. LNCS, vol. 2523, pp. 303–317. Springer, Heidelberg (2003). doi:10.1007/3-540-36400-5_23
7. Kocher, P., Jaffe, J., Jun, B.: Differential power analysis. In: Wiener, M. (ed.) CRYPTO 1999. LNCS, vol. 1666, pp. 388–397. Springer, Heidelberg (1999). doi:10.1007/3-540-48405-1_25
8. Mamiya, H., Miyaji, A., Morimoto, H.: Efficient countermeasures against RPA, DPA, and SPA. In: Joye, M., Quisquater, J.-J. (eds.) CHES 2004. LNCS, vol. 3156, pp. 343–356. Springer, Heidelberg (2004). doi:10.1007/978-3-540-28632-5_25
9. Mangard, S., Oswald, E., Popp, T.: Power Analysis Attacks - Revealing the Secrets of Smart Cards. Springer, Heidelberg (2007)
10. Messerges, T.S., Dabbish, E.A., Sloan, R.H.: Examining smart-card security under the threat of power analysis attacks. IEEE Trans. Comput. **51**(5), 541–552 (2002). doi:10.1109/TC.2002.1004593

11. Moradi, A., Poschmann, A.: Lightweight cryptography and DPA countermeasures: a survey. In: Sion, R., Curtmola, R., Dietrich, S., Kiayias, A., Miret, J.M., Sako, K., Sebé, F. (eds.) FC 2010. LNCS, vol. 6054, pp. 68–79. Springer, Heidelberg (2010). doi:10.1007/978-3-642-14992-4_7

12. Prouff, E. (ed.): Constructive Side-Channel Analysis and Secure Design - 4th International Workshop, COSADE 2013, Paris, France, 6–8 March 2013. LNCS, vol. 7864. Springer, Heidelberg (2013). doi:10.1007/978-3-642-40026-1

13. Schimmel, O., Duplys, P., Bohl, E., Hayek, J., Rosenstiel, W.: Correlation power analysis in frequency domain (2010)

14. Tillich, S., Herbst, C.: Attacking state-of-the-art software countermeasures-a case study for AES. In: Proceedings of 10th International Workshop Cryptographic Hardware and Embedded Systems - CHES 2008, Washington, D.C., USA, 10–13 August 2008, pp. 228–243 (2008). doi:10.1007/978-3-540-85053-3_15

Efficient Implementation of Hybrid Encryption from Coding Theory

Pierre-Louis Cayrel[1], Cheikh Thiecoumba Gueye[2], El Hadji Modou Mboup[2],
Ousmane Ndiaye[2], and Edoardo Persichetti[3(✉)]

[1] Laboratoire Hubert Curien, UMR CNRS 5516,
Bâtiment F 18 Rue du Professeur Benoît Lauras, 42000 Saint-Etienne, France
`pierre.louis.cayrel@univ-st-etienne.fr`
[2] Université Cheikh Anta Diop de Dakar, FST, DMI, LACGAA, Dakar, Senegal
`{cheikht.gueye,ousmane3.ndiaye}@ucad.edu.sn, domboups@yahoo.fr`
[3] Department of Mathematical Sciences,
Florida Atlantic University, Boca Raton, USA
`epersichetti@fau.edu`

Abstract. In this work we present an efficient implementation of the
Hybrid Encryption scheme based on the Niederreiter PCKS proposed by
E. Persichetti.

To achieve IND-CCA2 security (in the random oracle model), we use
an HMAC function of the message and the symmetric key, and then
apply AES128-CBC as the data encapsulation part of this hybrid scheme.
The HMAC function is based on SHA3-512. In addition, we introduce a
modification in the decapsulation algorithm, to resist a reaction attack
first proposed by Bernstein et al.

The implementation is done in C on Intel core i3 CPU and 4 GB RAM
and 64 bit OS. The code is running Debian/Linux 3.5.2, where the source
has been compiled with gcc 4.7.

Keywords: KEM-DEM · Niederreiter PKCS · Code-based cryptography · Random oracle

1 Introduction

Hybrid encryption, also known as the KEM-DEM paradigm, takes its name from
its characteristic integration of public-key and symmetric algorithms. Thanks to
its particular structure, this paradigm combines the advantages of both systems
and overcomes some disadvantages. In fact, a hybrid encryption scheme is fast
and shows no weaknesses at the key level.

A hybrid encryption scheme usually operates as follows. First of all, the
recipient generates a key pair for a public-key "key encapsulation" mechanism
(KEM). The public key is used to encrypt a (usually randomly-generated) key for
a symmetric "data encapsulation" scheme (DEM). The DEM is used to encrypt
the plaintext, and the ciphertext transmitted contains both this encryption, and
an encapsulation of the symmetric key. At the other hand, the recipient can use

© Springer International Publishing AG 2017
S. El Hajji et al. (Eds.): C2SI 2017, LNCS 10194, pp. 254–264, 2017.
DOI: 10.1007/978-3-319-55589-8_17

the private key to decapsulate the symmetric key and use that to recover the plaintext.

The paradigm was first introduced by Cramer and Shoup in [3].

In 1994, P.W. Shor [8] showed that quantum computers can break most *classical* cryptosystems, e.g. those based on the integer factorization problem or on the discrete logarithm problem. It is, therefore, crucial to develop cryptosystems that are resistant to quantum computer attacks. Cryptography based on error-correcting codes is a very promising candidate for post-quantum cryptography since code-based cryptographic schemes are usually fast and do not require special hardware, specifically no cryptographic co-processor.

Related Work

Code-based cryptography has been intensively studied since McEliece's seminal work in 1978 [5], and schemes have been implemented on several platforms, both software and hardware. Unfortunately, these systems [5,6] suffer from critical attacks that compromise their semantic security.

To solve this problem, a hybrid encryption system based on coding theory was proposed for the first time by E. Persichetti in [7]. In the scheme, a low-weight word (i.e. below the error correction capability of the considered code) is randomly generated. In the KEM, this word is used to generate the symmetric key using a key derivation function. The Niederreiter scheme is used to encrypt the random word with the recipient's public key. Then, the DEM uses this key to encrypt the plaintext. Upon receipt of the ciphertext, the recipient uses the Niederreiter private key to decode and recover the low-weight word, from which it is possible to recover the symmetric key and therefore the plaintext.

Some variants of this scheme were later proposed. The McBits scheme [1] was introduced by Bernstein et al. to provide constant-time encryption: the scheme uses a stream cipher instead of the One-Time Pad used in the original version. Recently, von Maurich et al. [10] provided a hardware implementation of another modification of the scheme based on QC-MDPC instance of Niederreiter for ARM Cortex. This variant uses AES128CBC as the symmetric scheme underlying the DEM, and SHA-256 for data integrity.

Our Contributions

In this work, we revisit Persichetti's KEM/DEM scheme [7] and present a software implementation of this new hybrid scheme. Our implementation is based on a binary Goppa code and it is done in C. As opposed to Persichetti's original work, we propose to use SHA3-512 for the integrity check and AES128-CBC instead of the One-Time Pad. Furthermore, we introduce a modification in the decapsulation algorithm: in case of a decoding failure, we apply a random permutation to the ciphertext before extracting the "mock" symmetric key via the key derivation function. This extra step allows the hybrid scheme to resist the attack given in [2].

Organization of the Paper

The paper is organized as follows: in Sect. 2, we briefly introduce the Niederreiter PKCS as well as the basic hybrid encryption structure. In Sect. 3, we present in detail the hybrid scheme with a description of the various algorithms. In Sect. 4 we present the C implementation of the new Hybrid Niederreiter scheme. Section 5 focuses on the security of the scheme, including a comparison with previous work. Finally we conclude the paper in Sect. 6.

2   Preliminaries

Notation. Throughout the paper we use the following notation:

- $\mathcal{E}^{Ne}(x)$ the encryption algorithm of the Niederreiter PKC, where x is a plaintext of length at most $\lceil \log_2 \binom{n}{t} \rceil$
- $\mathcal{D}^{Ne}(z)$ the decryption algorithm of the Niederreiter PKC, return the plaintext x corresponding to the ciphertext z.
- $\varphi : \mathbb{F}_2^l \mapsto \mathcal{W}_{n,t}$ a constant-weight encoding function, where $l = \lceil \log_2 \binom{n}{t} \rceil$ and $\mathcal{W}_{n,t}$ is the set of vectors of length n and weight t.
- $Len(x)$, for the bit-Length of x.
- $H : \mathbb{F}_2^k \mapsto \mathbb{F}_2^l$, where $l < k$ and $l = \lceil \log_2 \binom{n}{t} \rceil$, a One-Way compression function.
- $G : \mathbb{F}_2^{k_0} \mapsto \mathbb{F}_2^k$, where $k_0 < k$ and k_0 the length of the plaintext, a Cryptographically secure pseudorandom sequences generator.
- CCA2: Adaptive Chosen Ciphertext Attack Model
- KDF: Key Derivation Function

2.1   Niederreiter PKCS

This cryptosystem was introduced by H. Niederreiter in 1985 [6]. Since it makes use of the parity-check matrix rather than the generator matrix, it is often considered as a "dual" version of the McEliece cryptosystem [5].

Let H be an $r \times n$ parity-check matrix for a (n, k, t) binary Goppa code, where $r = n - k$, P an $n \times n$ random permutation matrix, S an $r \times r$ invertible matrix and γ_H a decoding algorithm for the code defined by H. Since the underlying Goppa code can only correct a certain number $t < n$ of errors, the Niederreiter scheme uses a function φ to map the message to a word of length n and weight t (detailed above). The private key is (S, H, P, γ_H), while $\widehat{H} = SHP$, the value of t and φ are made public.

Encryption: Let $x \in \mathbb{F}_2^l$ be the plaintext, then the ciphertext c is computed as $c = \widehat{H}\varphi(x)^T$.

Decryption: The recipient receives a ciphertext c and computes $\widehat{c} = S^{-1}c = HP\varphi(x)^T$. Since P is a permutation, $wt(P\varphi(x)^T) = wt(\varphi(x)^T)$, so γ_H can be used to decode it: $P\varphi(x)^T = \gamma_H(\widehat{c})$. Finally, the receiver computes $x = \varphi^{-1}(P^{-1}P(\varphi(x))^T)^T$.

3   Hybrid Encryption

A hybrid cryptosystem combines the convenience of a public-key cryptosystem with the efficiency of a symmetric-key cryptosystem. Public-key cryptosystems are convenient in that they do not require the sender and receiver to share a common secret in order to communicate securely (among other useful properties). However, they often rely on complicated mathematical computations and are thus generally much more inefficient than comparable symmetric-key cryptosystems. In many applications, the high cost of encrypting long messages in a public-key cryptosystem can be prohibitive. This is addressed by hybrid systems by using a combination of both. A hybrid cryptosystem is constructed using the two mechanisms presented below.

3.1   Key Encapsulation Mechanism (KEM)

A key encapsulation mechanism is essentially a public-key encryption scheme (PKE), with the exception that the encryption algorithm takes no input apart from the public key, and returns a pair (K, ψ_0). The string K has fixed length and it is usually obtained via a KDF, while ψ_0 is an encryption of K in the sense that $Dec_{sk}(\psi_0) = K$. Formally, a KEM consists of the following three algorithms.

- *KeyGen.* A probabilistic key generation algorithm that takes as input a security parameter 1^λ and outputs a public key pk and a private key sk.
- *Enc^{KEM}.* A probabilistic encryption algorithm that receives as input a public key pk and returns a key/ciphertext pair (K, ψ_0).
- *Dec^{KEM}.* A deterministic decryption algorithm that receives as input a private key sk and a ciphertext ψ_0 and outputs either a key K or the failure symbol $\perp$.

A KEM is required to be sound for at least all but a negligible portion of public key/private key pairs, that is, if $Enc_{pk}() = (K, \psi_0)$ then $Dec_{sk}(\psi_0) = K$ with overwhelming probability.

Definition 31. *The adaptive chosen-ciphertext attack game for a KEM proceeds as follows:*

1. *Query a key generation oracle to obtain a public key pk.*
2. *Make a sequence of calls to a decryption oracle, submitting any string ψ_0 of the proper length. The oracle will respond with $Dec_{sk}^{KEM}(\psi_0)$.*
3. *Query an encryption oracle. The oracle runs Enc_{pk}^{KEM} to generate a pair $(\widehat{K}, \widehat{\psi_0})$, then chooses a random $b \in \{0,1\}$ and replies with the"challenge" ciphertext $(K^*; \widehat{\psi_0})$ where $K^* = \widehat{K}$ if $b = 1$ or K^* is a random string of length ℓ_K otherwise.*
4. *Keep performing decryption queries. If the submitted ciphertext is ψ_0^*, the oracle will return $\perp$.*
5. *Output $b^* \in \{0, 1\}$.*

*The adversary succeeds if $b^* = b$. More precisely, we define the advantage of A
against KEM with security parameter λ as:*

$$Adv_{KEM}(A, \lambda) = \left| Pr[b^* = b] - \frac{1}{2} \right| \tag{1}$$

*We say that a KEM is secure if the advantage Adv_{KEM} of any polynomial-time
adversary A in the above CCA attack model is negligible.*

3.2   Data Encapsulation Mechanism (DEM)

The data encapsulation mechanism is a (possibly labeled) symmetric encryption
scheme (SE) that uses as a key the string K output by the KEM. In what follows
we only discuss, for simplicity, un-labeled DEMs. Formally, a DEM consists of
the following two algorithms.

- Enc^{DEM}. A deterministic encryption algorithm that receives as input a key
  K and a plaintext φ and returns a ciphertext ψ_1.
- Dec^{DEM}. A deterministic decryption algorithm that receives as input a key K
  and a ciphertext ψ_1 and outputs either a plaintext φ or the failure symbol $\perp$.

3.3   Key Derivation Function

Definition 32. *A Key Derivation Function (KDF) is a function that takes as
input a string x of arbitrary length and an integer $\ell \geq 0$ and outputs a bit string
of length ℓ (see [7]).*

Key derivation functions are also used in applications to derive keys from
secret passwords or passphrases, which typically do not have the desired properties to be used directly as cryptographic keys. In such applications, it is generally
recommended that the key derivation function be made deliberately slow so as to
frustrate brute-force attack or dictionary attack on the password or passphrase
input value.

Modern password-based key derivation functions, such as PBKDF2 (specified
in RFC 2898), use a cryptographic hash, such as SHA-2, more salt (e.g. 64 bits
and greater) and a high iteration count (often 1000 or more). NIST requires
at least 128 bits of random salt and a NIST-approved cryptographic function,
such as the SHA series or AES (functions such as MD5, for instance, are not
approved) [9]. There have been proposals to use algorithms that require large
amounts of computer memory and other computing resources to make custom
hardware attacks more difficult to mount.

Definition 33. *A Message Authentication Code (MAC) is an algorithm that
produces a short piece of information (tag) used to authenticate a message. A
MAC is defined by a function Ev that takes as input a key K of length ℓ_{MAC}
and an arbitrary string T and returns a tag to be appended to the message, that
is, a string τ of fixed length ℓ_{TAG}. A MAC that makes use of a hash function is
commonly known as HMAC.*

4   Hybrid Scheme from Coding Theory

In this section we present our variant of the scheme presented in [7]. We present the hybrid scheme as a whole rather than describing the individual mechanisms. The scheme uses the Niederreiter encryption scheme as the KEM. This KEM addresses the key generation step, as well as the first portion of the encryption and decryption algorithms. At the DEM level, the hash function SHA3-512 is used as a key derivation function (KDF) to construct the key used for symmetric encryption (AES128-CBC) as well as the integrity check (via the HMAC). The same hash function is at the base of our HMAC.

Key Generation

Following the key generation process in Sect. 2.1, choose a binary Goppa code with parity-check matrix H and special description. Return the public-key M and the private key Δ.

Encryption

Input: $M \in K_{pub}$ and the plaintext φ.
Output: the ciphertext ψ.

Choose a random word $e \in \mathcal{W}_{n,t}$. Set $H := (M|I_{n-k})$, then compute $\psi_0 := He^T$ and a key $K := KDF(e, \ell_{DEM} + \ell_{MAC})$. Parse the symmetric key as $K := (K_1||K_2)$. Compute $\psi' := Enc_{K_1}^{DEM}(\varphi)$, then set $T := \psi'$ and evaluate $\tau := Ev(K_2, T)$. Return $\psi := (\psi_0||\psi'||\tau)$.

Decryption

Input: a private key Δ and the ciphertext ψ.
Output: the plaintext φ.

Parse the ciphertext as $\psi := (\psi_0||\psi'||\tau)$, then compute $e := Decode_\Delta(\psi_0)$: if decoding succeeds compute $K := KDF(e, \ell_{DEM} + \ell_{MAC})$, else sample a random permutation σ and compute $K := KDF(\sigma(\psi_0), \ell_{DEM} + \ell_{MAC})$. Parse the symmetric key as $K := (K_1||K_2)$ and set $T = \psi'$, then compute $\tau' := Ev(K_2, T)$: if ($\tau \neq \tau'$) the verification fails and return $\bot$. Otherwise, compute $\varphi := Dec_{K_1}^{DEM}(\psi')$. Return φ.

Note that, as we anticipated, in the case of a decoding failure in the decryption phase, we added a step featuring a random permutation σ, which we apply to the ciphertext ψ_0 to compute a pseudorandom K. This simple modification allows us to resist the attack proposed by Bernstein et al. in [2], which we describe later in this paper.

5 Implementation

5.1 Description

In this part, we present an efficient implementation of the Hybrid scheme described above. As hardware platform, we used a PC with 1.80 GHz, Intel core i3 CPU and 4 GB RAM on a 64 bit OS. The code is running Debian/Linux 3.5.2, where the source has been compiled with gcc 4.7.

5.2 Overview of Functions

In this part, we describe the functions used in this implementation.

Key Generation. The key generation step features the original functions of the Niederreiter algorithm. We have a *keypair(sk, pk)* function to generate the public and private key. The keypair function outputs a Goppa polynomial that will be used to construct both keys. These two keys will be stored in the files (*pk.pem* and *sk.pem*). The symmetric key is instead computed in the encryption and decryption phase (as per KEM-DEM paradigm).

Hybrid Encryption. The encryption step is composed of Niederreiter encryption for KEM, symmetric encryption AES128-CBC for the DEM and also the SHA3-512 hash function for the KDF and HMAC. We present the list of functions used:

- *Mat-from-pk():* This function transforms the public matrix H in systematic form. This form plays a very important role on the completeness of the implementation.
- *sponge():* This function is used in the key derivation function (KDF) as well as the HMAC. This sponge is the hash function SHA3-512. We chose this function because of its speed and security.
- *AES128-CBC-encrypt-DEM():* It handles AES128-CBC symmetric encryption. This function represents the DEM part of Hybrid Encryption.
- *encrypt-Nied():* This function is used as part of Niederreiter's encryption.
- *Encrypt-HyNe():* In this function, we have implemented the encryption of the hybrid scheme. It uses the functions defined above and takes as input the plaintext, the public key and returns the ciphertext $(\psi := (\psi_0 || \psi' || \tau) :=$ Encrypt-HyNe(pk.pem, plainText)).

Hybrid Decryption. Hybrid decryption is composed of Niederreiter decryption for KEM, symmetric decryption AES128-CBC for the DEM and also the SHA3-512 hash function for the KDF and HMAC. By analogy to Hybrid encryption, decryption uses the same functions in similar decryption functions. For the implementation of the Hybrid decryption, we used the *decrypt-HyNe()* function which takes as input the private key (sk.pem) and the ciphertext (ψ) and returns the plaintext($\varphi :=$ decrypt-HyNe(sk.pem, ψ)).

6   Security and Performance

6.1   KEM-DEM Security

The KEM-DEM paradigm introduced by Cramer and Shoup in [3] provides IND-CCA2 security for the encryption scheme, provided that both components are IND-CCA2 secure (in their own sense). In [7], Persichetti details a full security proof that follows closely the paradigm; however, the proof required to introduce a modification in the scheme in order to guarantee the integrity of the simulator. In fact, since it is not possible to decide a priori whether a given word is decodable or not, it is necessary that the KEM decryption always output something. A natural suggestion was to use a pseudorandom function of the ciphertext, i.e. $KDF(\psi_0)$. Unfortunately, this choice makes the scheme vulnerable to a simple attack, which we will detail below.

Malleability. Malleability is a property of some cryptographic algorithms. An encryption algorithm is malleable if it is possible for an adversary to transform a ciphertext into another ciphertext which is decrypted to a related plaintext. That is, given an encryption of the plaintext e, it is possible to generate another ciphertext which is decrypted as $f(e)$, for a known function f, without necessarily knowing or learning e.

In the basic Niederreiter scheme, $\psi_0 := He^T$ with $wt(e) = t$, the adversary may take randomly the ith column of H denoted $H[i]$ and send to the decryption oracle $\psi_0' = \psi_0 \oplus H[i]$. The decryption will succeed if we had "1" on the ith bit of the message e. Let e_1 be a vector of length n with only the ith bit non zero. Then the oracle returns the vector e' of weight $t-1$ such that:

$$\psi_0' = He'^T = \psi_0 \oplus H[i] = He^T \oplus He_1^T = H(e \oplus e_1)^T$$

Then

$$e = e' \oplus e_1$$

The success probability of this attack is:

$$Pr = P(e[i] = 1) = \frac{\binom{t}{1}}{\binom{n}{1}} = \frac{t}{n}$$

for $n = 1024$ and $t = 50$, $Pr = 0.05$ and for $n = 2048$ and $t = 81$, $Pr = 0.04$ i.e. it succeeds at most 25 times.

It is clear then that Niederreiter's scheme is malleable.

Bernstein's Attack. Bernstein et al. in [2] managed to adapt the technique above to attack the full hybrid scheme.

Let the challenge ciphertext be (ψ_0, ψ', τ). The adversary can compute the ciphertext $(\psi_0'', \psi'', \tau'')$, where:

$- \psi_0'' = \psi_0 \oplus H[i] \oplus H[j]$, for $i \neq j$

– ψ'' is a random string of the appropriate length
– $\tau'' = Ev(K_2'', \psi'')$ where $K'' = (K_1'', K_2'') = KDF(\psi_0'', \ell_{DEM} + \ell_M AC)$

As we have seen above, $psi_0'' = H(e'')^T$ for a vector e'' that is essentially e with the two bits in positions i and j flipped. Now, if e'' has weight t or less, decoding will succeed, and the KEM will output $K = KDF(e'', m + \ell_{MAC})$. This leads to a decryption failure, since the tag will fail the verification step. On the other hand, if the weight of e'' is greater than t decoding will fail and the KEM will output exactly K''; in this case the tag passes the verification and decryption succeeds. Thus the scheme is malleable and an adversary has a way of recovering the secret vector e, and from that the plaintext m.

The attack is possible because the choice of pseudorandom function makes it too predictable for an attacker to calculate the string output by the KEM in case of a decoding failure. With our modification, instead, the KEM calculates KDF on a randomly permuted version of ψ_0, thus thwarting this simple attack.

6.2   Performance and Comparison

HyNe vs. Niederreiter PKCS. Here we present a simulation of the HyNE (Hybrid Niederreiter Encryption) algorithm versus the original Niederreiter PKCS, implemented on Intel core i3 CPU 1.80 GHz, 4 GB RAM and 64 bit OS. This comparison is based on the running time of various algorithms (key generation, data encryption and data decryption) depending on the size symmetric keys, which can be either 512, 256 or 128 bits.

The running times are listed in Table 1, for codes of length $n = 2^m$ and error correction capacity t. We measure the speed of our implementation (number of cycles of processor required to process one byte) as execution time (s)* processor frequency (Hz)/test file size (bytes).

HyNe vs. Hybrid RSA. In the same environment, we present in Table 2 the running time of a Hybrid RSA, as presented in [4], for two different exponent (i.e. RSA keys) sizes.

Discussion. In terms of comparison, we evaluate the size of the parameters and the ratio between our speeds and those obtained for Hybrid RSA [4].

As one well knows, RSA keys size are small than the ones of code-based cryptography. However, for a similar security level, the encryption is 3 time faster in HyNe than in Hybrid RSA. Nevertheless, the decryption process is still faster in the RSA.

We can also measure of our implementation by comparing it with the original Niederreiter (with no IND-CCA2 security), as reported on the first table. It is possible to see that the key generation process doesn't change, while the ratio of the full hybrid scheme to the plain scheme is respectively $(1.16, 1.19, 1.26)$ for message size $\ell_{DEM} = (128, 256, 512)$.

Table 1. Running Times of HyNe and Niederreiter PKCS

Scheme (m,t) or Scheme (m,t,ℓ_{DEM})	Cycles			Security (bit ops.)
	KeyGen	Encrypt	Decrypt	
Nied$(10,28)$	142060394	111014	3053326	60
Nied$(11,32)$	384089214	170121	3224981	88
Nied$(12,48)$	2613209726	322965	6095783	128
HyNe$(10,28,128)$	142060394	128909	3148128	60
HyNe$(11,32,128)$	384089214	176429	3575968	88
HyNe$(12,48,128)$	2613209726	349012	6196288	128
HyNe$(10,28,256)$	142060394	132245	3213051	60
HyNe$(11,32,256)$	384089214	198890	3876063	88
HyNe$(12,48,256)$	2613209726	434950	7969412	128
HyNe$(10,28,512)$	142060394	140306	3465572	60
HyNe$(11,32,512)$	384089214	201690	4993159	88
HyNe$(12,48,512)$	2613209726	492826	8860359	128

Table 2. Running Times of Hybrid RSA

Exponents (bits)	Keygen	Cycles (Enc)	Cycles (Dec)	Security (bit ops.)
1024	986400	482400	463200	80
2048	4735200	2402400	2320800	112

7   Conclusion

As a first contribution, we have presented an efficient implementation of the hybrid scheme proposed by Persichetti in [7]. The implementation is fast and practical, as shown by the comparison against an "equivalent" version of the paradigm using RSA. Moreover, we have introduced a variant that fixes the security flaw highlighted by Bernstein et al. [2]. Thanks to a simple tweak (introducing a random permutation), we are able to preserve the IND-CCA2 security of the scheme.

Acknowledgment. This work was carried out with financial support of CEA-MITIC for CBC projet and financial support from the government of Senegal's Ministry of Hight Education and Research for ISPQ Project.

References

1. Bernstein, D.J., Chou, T., Schwabe, P.: McBits: fast constant-time code-based cryptography. In: Bertoni, G., Coron, J.-S. (eds.) CHES 2013. LNCS, vol. 8086, pp. 250–272. Springer, Heidelberg (2013). doi:10.1007/978-3-642-40349-1_15
2. Bernstein, D.J., Chuengsatiansup, C., Lange, T., van Vredendaal, C.: NTRU prime (2016). http://eprint.iacr.org/2016/461

3. Cramer, R., Shoup, V.: Design and analysis of practical public-key encryption schemes secure against adaptive chosen ciphertext attack. SIAM J. Comput.**33**, 167–226 (2004). Society for Industrial and Applied Mathematics, Philadelphia

4. Alrashdan, M.T., Moghaddam, F.F., Karimi, O.: A hybrid encryption algorithm based on RSA small-e and efficient-RSA for cloud computing environments. J. Adv. Comput. Netw. **1**(3), 238–241 (2013)

5. McEliece, R.J.: A public-key cryptosystem based on algebraic coding theory. Jet Propulsion Laboratory DSN Progress Report 42–44, pp. 114–116 (1978)

6. Niederreiter, H.: Knapsack-type cryptosystems and algebraic coding theory. In: Problems of Control and Information Theory, vol. 15, pp. 159–166 (1986)

7. Persichetti, E.: Secure and anonymous hybrid encryption from coding theory. In: Gaborit, P. (ed.) PQCrypto 2013. LNCS, vol. 7932, pp. 174–187. Springer, Heidelberg (2013). doi:10.1007/978-3-642-38616-9_12

8. Shor, P.W.: Polynomial-time algorithms for prime factorization and discrete logarithms on a quantum computer. In: Proceedings of the 35th Annual Symposium on Foundations of Computer Science (1994)

9. Turan, M.S., Barker, E.B., Burr, W.E., Chen, L.: Sp 800–132, Recommendation for password-based key derivation: Part 1: storage applications. National Institute of Standards & Technology, Gaithersburg (2010)

10. Maurich, I., Heberle, L., Güneysu, T.: IND-CCA secure hybrid encryption from QC-MDPC niederreiter. In: Takagi, T. (ed.) PQCrypto 2016. LNCS, vol. 9606, pp. 1–17. Springer, Cham (2016). doi:10.1007/978-3-319-29360-8_1

On the Multi-output Filtering Model and Its Applications

Teng Wu, Yin Tan, Kalikinkar Mandal$^{(\boxtimes)}$, and Guang Gong

Department of Electrical and Computer Engineering,
University of Waterloo, Waterloo, ON N2L 3G1, Canada
{teng.wu,y24tan,kmandal,ggong}@uwaterloo.ca

Abstract. In this paper, we propose a novel technique, called multi-output filtering model, to study the non-randomness property of a cryptographic algorithm such as message authentication codes and block ciphers. A multi-output filtering model consists of a linear feedback shift register and a multi-output filtering function. Our contribution in this paper is twofold. First, we propose an attack technique under IND-CPA using the multi-output filtering model. By introducing a distinguishing function, we theoretically determine the success rate of this attack. In particular, we construct a distinguishing function based on the distribution of the linear complexity of component sequences, and apply it on studying TUAK's f_1 algorithm, AES, KASUMI, PRESENT and PRINT-cipher. We demonstrate that the success rate of the attack on KASUMI and PRESENT is non-negligible, but f_1 and AES are resistant to this attack. Second, we study the distribution of the cryptographic properties of component functions of a random primitive in the multi-output filtering model. Our experiments show some non-randomness in the distribution of algebraic degree and nonlinearity for KASUMI.

Keywords: Randomness · Distinguishing attack · TUAK · Linear complexity

1 Introduction

Let $\mathcal{C}$ be a cryptographic scheme (keyed or non-keyed) with n-bit input and m-bit output. Clearly it can be regarded as a vectorial Boolean function from $\mathbb{F}_2^n$ to $\mathbb{F}_2^m$. When $\mathcal{C}$ involves a key K, we should write $\mathcal{C}_K$ for strictness, but we prefer to use $\mathcal{C}$ for simplicity if the context is clear. In most circumstances, the cryptographic properties of $\mathcal{C}$, such as algebraic degree and nonlinearity, are difficult to be exploited due to the large values of n and m. A natural idea to overcome this difficulty is to restrict the inputs of $\mathcal{C}$ on a subspace $\mathcal{S}$ of $\mathbb{F}_2^n$. For instance, the subspace $\mathcal{S}$ can be generated by an ℓ-stage linear feedback shift register. Then we obtain a function $\mathcal{C}'$ from $\mathcal{S}$ to its image set $\mathcal{C}(\mathcal{S})$. By adapting the size of $\mathcal{S}$, we can study the cryptographic properties of $\mathcal{C}'$. If $\mathcal{C}$ has good randomness properties, it should be difficult to find a subspace $\mathcal{S}$ such that $\mathcal{C}'$ has bad randomness properties. We must mention that the above method

© Springer International Publishing AG 2017
S. El Hajji et al. (Eds.): C2SI 2017, LNCS 10194, pp. 265–281, 2017.
DOI: 10.1007/978-3-319-55589-8_18

for analyzing the cryptographic scheme $\mathcal{C}$ lies in a more general notion called *subset cryptanalysis* [19], which tries to track the statistical evolution of a certain subset of values through various operations in the cryptographic schemes. One is referred to [12] for a successful application of the subset cryptanalysis to find a 5-round collision on Keccak. More analysis of the Keccak permutation can be found in [1,6,9,11,13,22–24].

We achieve the above idea by proposing a new technique, called *multi-output filtering model*. This model aims to exploit the non-randomness property of a cryptographic algorithm $\mathcal{C}$ such as message authentication codes and block ciphers. A multi-output filtering model consists of a linear feedback shift register (LFSR) and a multi-output filtering function. The LFSR is used to generate an input subspace of $\mathcal{C}$ and $\mathcal{C}$ is used as a multi-output filtering function. This multi-output model is a generalization of the classic filtering model in stream ciphers [27] as it outputs multiple bits, instead of only one bit, for the set of inputs to $\mathcal{C}$ generated by an LFSR. Under this model, we obtain a number of component sequences and component functions. This paper is devoted to studying the randomness properties of $\mathcal{C}$ by investigating its component sequences and component functions. In this paper we restrict $\mathcal{C}$ to MACs and block ciphers, but this model can also be generalized to study other cryptographic primitives.

Recently, TUAK [31] is proposed to the 3rd Generation Partnership Project (3GPP) for providing authenticity and key derivation functionalities in mobile communications. The design of TUAK is based on the Keccak permutation [4] with 1600-bit internal state. The TUAK algorithm set contains seven different algorithms, namely f_1 to f_5 and f_1^* and f_5^*. The f_1 (or f_1^* used for resynchronisation) algorithm ensures the authenticity of messages, f_2 is used for generating responses and f_3 to f_5 and f_5^* are used as key derivation functions. The security evaluation for TUAK is essential for guaranteeing the authenticity in mobile communications. An analysis of the resistance of TUAK to many known attacks has been presented in [16]. In this paper, we restrict ourselves to the analysis of the MAC generation algorithm f_1 in the multi-output filtering model. We also consider the block ciphers PRINTcipher [18], AES [8], KASUMI [30], and PRESENT [5] in the multi-output filtering model.

Our contributions in this paper are as follows. We introduce the multi-output filtering model for analyzing non-randomness of cryptographic primitives. We consider a generic distinguishing attack framework on $\mathcal{C}$ under the indistinguishability under chosen-plaintext attack model (IND-CPA for short), which is a variant of indistinguishability of encryptions proposed by Goldwasser and Micali [14] in public-key cryptography settings. We theoretically determine the success rate of the attack. In the multi-output filtering model, we present the construction of a new type of distinguishing function based on the distributions of the linear complexity of the component sequences for IND-CPA. Applying the new distinguishing function on f_1, AES, KASUMI, PRESENT and PRINTcipher, we can distinguish the output of KASUMI and PRESENT with that of a random primitive with a non-negligible success rate. Our study shows that f_1 and AES is immune to this attack. Moreover, we study the distribution of the algebraic degree and nonlinearity of the component functions. We determine the distribution of these two properties for the component functions of a random multi-output filtering function. We perform an experiment on f_1, AES, KASUMI and PRESENT, and our experimental result shows that for KASUMI, the density of its component functions with algebraic degree less than $\ell - 2$ is greater than the random case, where ℓ is the length of the LFSR. While the degree distributions of the other primitives are similar to that of the random case.

2   Preliminaries

In this section, we present a list of notations and provide some definitions that will be used in this paper.

Notations

- $\mathbb{F}_2$: the Galois field with two elements $\{0, 1\}$;
- $\mathbb{F}_{2^n}$: a finite field with 2^n elements, defined by a primitive element α;
- $\mathbb{F}_2^n$: a vector space with 2^n elements;
- $\mathrm{NL}(f)$: the nonlinearity of a Boolean function;
- $\mathrm{LC}(\mathbf{s})$: the linear complexity of a binary sequence $\mathbf{s}$ with period N;
- $\mathcal{B}_n$: the set of all Boolean functions with n variables.

2.1   Basic Definitions on Sequences and Boolean Functions

Let $\mathbf{s} = \{s_i\}$ be a sequence generated by a linear feedback shift register (LFSR) whose recurrence relation is defined as

$$s_{\ell+i} = \sum_{j=0}^{\ell-1} c_j s_{i+j}, \quad s_i, c_i \in \mathbb{F}_2, \ i = 0, 1, \ldots \tag{1}$$

where $p(x) = \sum_{i=1}^{\ell} c_i x^i \in \mathbb{F}_2[x]$ is the characteristic polynomial of degree ℓ of the LFSR. A binary sequence $\mathbf{s}$ in Eq. (1) with period $2^\ell - 1$ generated by an LFSR is called an m-sequence. Let $\mathbf{s} = \{s_i\}$ be an m-sequence of period $2^\ell - 1$ and $f(x_0, \ldots, x_{\ell-1})$ be a Boolean function in ℓ variables. We define a sequence $\mathbf{a} = \{a_i\}$ as

$$a_i = f(s_{r_1+i}, s_{r_2+i}, \ldots, s_{r_t+i}), \ s_i, a_i \in \mathbb{F}_2, \ i \geq 0$$

where $r_1 < r_2 < \ldots < r_t < \ell$ are tap positions. Then the sequence $\mathbf{a}$ is called a *filtering sequence* and the period of $\mathbf{a}$ equals $2^\ell - 1$.

The *linear complexity* or *linear span* of a sequence is defined as the length of the shortest LFSR that generates the sequence. For an m-sequence, the linear complexity of an m-sequence is equal to the length of its LFSR [15]. The linear complexity of a nonlinear filtering sequence lies in the range of ℓ and $2^\ell - 1$ [17]. If a filtering sequence has linear complexity $2^\ell - 1$, then we say it has *optimal* linear complexity.

Definition 1. *Let f be a Boolean function from $\mathbb{F}_2^n$ to $\mathbb{F}_2$. Then f can be uniquely represented by its algebraic normal form (ANF) as $f(x) = \sum_{I \in \mathcal{P}(\{0,\ldots,n-1\})} a_I x^I$, where $a_I \in \mathbb{F}_2, x^I = \prod_{i \in I} x_i$ and $\mathcal{P}(\{0,\ldots,n-1\})$ is the power set of $\{0,\ldots,n-1\}$. The algebraic degree of f, denoted by $d(f)$, is the maximal size of I in the ANF of f such that $a_I \neq 0$.*

One of the most important properties of Boolean functions is its nonlinearity, which was proposed to measure the distance of it to all affine functions. A cryptographically strong Boolean function should have high nonlinearity to resist linear attacks [20].

Definition 2. *The Walsh transform of a Boolean function f to a point $a \in \mathbb{F}_2^n$, denoted by $W_f(a)$, is defined by $W_f(a) = \sum_{x \in \mathbb{F}_2^n} (-1)^{f(x) + a \cdot x}$ where $a \cdot x$ is the inner product of a and x. The nonlinearity of f can be defined in terms of the Walsh transform as $\mathrm{NL}(f) = 2^{n-1} - \max_{a \in \mathbb{F}_2^n} \frac{|W_f(a)|}{2}$.*

When n is an even positive integer, it is known that the maximum value of the nonlinearity of a Boolean function f is $\mathrm{NL}(f) \geq 2^{n-1} - 2^{n/2-1}$ [7]. A Boolean functions achieving this bound is called a *bent function*.

Let m and n be two positive integers. A function F, from $\mathbb{F}_2^n$ to $\mathbb{F}_2^m$, defined by $F(x) = (f_1(x), f_2(x), ..., f_m(x))$ is called a (n,m)-function, multi-output Boolean functions, or vectorial Boolean functions, where f_i's are called coordinate functions [7]. For a well-rounded treatment of sequences and Boolean functions, the reader is referred to [7, 15].

3   Multi-output Filtering Model

In this section, we provide a detailed description of the multi-output filtering model of a cryptographic primitive.

3.1   Description of the Multi-output Filtering Model

Let $\mathbf{a} = \{a_i\}_{i \geq 0}$ be a binary sequence generated by an ℓ-stage linear feedback shift register whose recurrence relation is

$$a_{\ell+i} = \sum_{j=0}^{\ell-1} c_j a_{i+j}, \ c_j \in \mathbb{F}_2, \ i \geq 0, \tag{2}$$

where $p(x) = x^\ell + \sum_{i=0}^{\ell-1} c_i x^i$ is a primitive polynomial of degree ℓ over $\mathbb{F}_2$ and $\mathrm{STATE}_j = (a_j, a_{j+1}, ..., a_{\ell-1+j})$ is called the j-th state of the LFSR. Using this LFSR, from the above sequence $\mathbf{a}$, we generate a set of messages of n bits as $\mathcal{R} = \{\mathrm{R}_j : 0 \leq j \leq 2^\ell - 2\}$ where

$$\text{ } \quad \mathrm{R}_j = (a_j, a_{j+1}, \cdots, a_{j+n-1}), \ j = 0, 1, ..., 2^\ell - 2, \tag{3}$$

where modulo $(2^\ell - 1)$ is taken over the indices of a_i's. Note that the elements in $\mathcal{R}$ are in the sequential order. We now define the multi-output filtering model on $F : \{0,1\}^k \times \{0,1\}^n \to \{0,1\}^m$. For a fixed key $K \in \{0,1\}^k$ and for each $R_j \in \{0,1\}^n$ with $0 \le j \le 2^\ell - 2$, we obtain

$$C_j = F(K, R_j) = (g_0(K, R_j), \dots, g_{m-1}(K, R_j))$$
$$\triangleq (y_{j,0}, y_{j,1}, \dots, y_{j,m-1}) \in \{0,1\}^m. \tag{4}$$

Using a matrix, we can represent the above C_j as

$$\begin{pmatrix} C_0 \\ C_1 \\ \vdots \\ C_{2^\ell-2} \end{pmatrix} = \begin{pmatrix} y_{0,0} & y_{0,1} & \cdots & y_{0,m-1} \\ y_{1,0} & y_{1,1} & \cdots & y_{1,m-1} \\ \vdots & \vdots & & \vdots \\ y_{2^\ell-2,0} & y_{2^\ell-2,1} & \cdots & y_{2^\ell-2,m-1} \end{pmatrix}. \tag{5}$$

We study the cryptographic properties of F using the matrix (5).

I. Sequence point of view: Each column in the above can be considered as a sequence of period $2^\ell - 1$ for a nonzero initial state of the LFSR. Each sequence of period $2^\ell - 1$ is called a *component sequence*. We denote the i-th component sequence by $\mathbf{s}_i$ and $\mathbf{s}_i = \{y_{0,i}, y_{1,i}, \dots, y_{2^\ell-2,i}\}$. $\mathbf{s}_i$ can also be considered as a filtering sequence with filter function g_i, $0 \le i \le m - 1$.

II. Boolean function point of view: From (4) and (5), we see the following process

$$g_i : \{\text{STATE}_j \in \mathbb{F}_2^\ell \text{ of LFSR}\} \to \{R_j \in \mathbb{F}_2^n\} \to i\text{-th comp. sequence.}$$

Thus, each component sequence can also be regarded as a Boolean function on $\mathbb{F}_2^\ell$. Note that, for a nonzero initial state, the LFSR cannot generate all-zero state, we need to query F to get the output value $F(K, 0^n)$ for all-zero input for all component Boolean functions. With a fixed K in F, using an ℓ-stage LFSR, we obtain m Boolean functions on $\mathbb{F}_2^\ell$. Mathematically, m Boolean functions $g_i : \mathbb{F}_2^\ell \to \mathbb{F}_2$ $(0 \le i \le m - 1)$ are defined as

$$g_i(K, \text{STATE}_j) = y_{j,i}, \quad (0 \le j \le 2^\ell - 2). \tag{6}$$

We call each Boolean function g_i a *component* or *coordinate function* of F.

3.2   Application to TUAK's f_1, AES, KASUMI and PRESENT

For the sake of clarity on the input assignment, we briefly explain how we apply the multi-output filtering model on TUAK's f_1, and block ciphers AES, PRESENT and KASUMI. Recall that f_1 takes K, RAND, and SQN as inputs. We fix a key K and a sequence number SQN for f_1. Table 1 shows the input assignments to the primitives in the multi-output filtering model where R_j is defined in the previous section. For details about block ciphers, see [5, 8, 18, 30].

Remark 1. For TUAK's f_1 function, in Eq. (5), recovering the last bit $y_{2^\ell-2,i}$ for each component sequence $\mathbf{s}_i$ from the previous $2^\ell - 2$ bits is equivalent to recovering $C_{2^\ell-2}$ from $\{C_0, \dots, C_{2^\ell-3}\}$. This leads to a MAC forgery attack on f_1.

Table 1. Input assignments to the primitives in the multi-output filtering model.

Primitives	$(y_{j,0}, \ldots, y_{j,m-1}) = C_j$	Block size n	Key size $|K|$	$|SQN|$
f_1	$C_j = f_1\,(K, R_j, SQN)$	128	128	48
PRINTcipher	$C_j = $ PRINTcipher (K, R_j)	48	80	–
AES_128	$C_j = $ AES_128 (K, R_j)	128	128	–
AES_256	$C_j = $ AES_256 (K, R_j)	128	256	–
PRESENT	$C_j = $ PRESENT (K, R_j)	64	80	–
KASUMI	$C_j = $ KASUMI (K, R_j)	64	80	–

4   Distinguishing Attack Model

In this section, we describe the attack model of our distinguishing attack on a message authentication code and a block cipher. The attack model is based on indistinguishability (IND) of encryptions under chosen-plaintext attack (CPA) (IND-CPA), which was first proposed by Goldwasser and Micali [14] in public-key settings. In [3], Bellare *et al.* studied the indistinguishability of encryptions under chosen-plaintext attack in the symmetric key setting. Here, we use the same attack model to distinguish MACs (or ciphertexts) in the symmetric-key setting. However, we develop a new distinguishing function based on the linear complexity of component sequences in the multi-output filtering model.

Let $F : \{0,1\}^k \times \{0,1\}^n \to \{0,1\}^m$ be a cryptographic algorithm which accepts two inputs, a key of length k and a message of length n and produces an output of length m. Assume that P_0 and P_1 are two messages of chosen by the adversary and $c_i = F(K, P_i), i = 0, 1$ for the key K. The aim of the distinguishing attack is to distinguish c_0 and c_1 for P_0 and P_1, resp. with a high probability. We denote the random oracle by $\mathcal{O}$ and the adversary by $\mathcal{A}$. The indistinguishability game [2,14] between the random oracle and the adversary is played as follows.

(1) Fixing a key K and generating the set of messages $\mathcal{R} = \{R_0, R_1, ..., R_{N-1}\}$ using an LFSR with a primitive polynomial of degree ℓ, $N = 2^\ell - 1$;
(2) The adversary $\mathcal{A}$ randomly picks up $P_0 \in \mathcal{R}$ and $P_1 \notin \mathcal{R}$ and sends both $\{P_0, P_1\}$ to $\mathcal{O}$.
(3) The random oracle picks up $P_b \xleftarrow{\$} \{P_0, P_1\}$, $b = 0$ or 1 and computes $c = F(K, P_b)$. $\mathcal{O}$ sends c to the adversary $\mathcal{A}$.
(4) Once $\mathcal{A}$ receives c as a challenge, the adversary performs a technique and decides b' and returns b' to $\mathcal{O}$ where $b' = 0$ or 1;
(5) If $b = b'$, then adversary $\mathcal{A}$ succeeds; otherwise she fails.

5   Distinguishing Attack Based on Linear Complexity

In this section, we first present a general technique to build a distinguisher of a cryptographic primitive, followed by the theoretical determination of the success probability of the distinguishing attack. In particular, we make use of the distribution of the linear complexity of component sequences of a primitive to develop a new distinguisher. Finally, we apply this technique on f_1, AES, KASUMI, and PRESENT.

5.1 A Generic Framework to Build a Distinguisher

We start this section by the following definition.

Definition 3. *Let $\mathcal{R}$ and $\mathcal{S}$ be two subsets of U, where $\mathcal{S} = U \setminus \mathcal{R}$. Let Ω be a subset of $\mathcal{R} \times \mathcal{S}$. Let $\mathcal{C}$ be a cryptographic scheme from U to some set V. For any $P_0 \in \mathcal{R}$ and $P_1 \in \mathcal{S}$, define a distinguishing function $h : \{\mathcal{C}(P_0), \mathcal{C}(P_1)\} \rightarrow \{0, 1\}$. We say that $\mathcal{C}$ is distinguishable with respect to $\mathcal{R}, \mathcal{S}, h, \Omega$ if the average probability*

$$\sum_{i \in \{0,1\}} \Pr\Big(h(c) = i \wedge c = \mathcal{C}(P_i)\Big)$$

is non-negligible compared with $1/2$, when (P_0, P_1) is randomly chosen from Ω.

Now we state the main theorem below and, due to the page limit, the proof is provided in the full paper [29].

Theorem 1. *Let the notations be the same as above. Now we define a subset $\mathcal{CS}$ of U, which is called the condition set. Let $\mathcal{S}' \subset \mathcal{S}$ and $\Omega = \mathcal{R} \times \mathcal{S}'$. For any $P_0 \in \mathcal{R}, P_1 \in \mathcal{S}'$, let us define the distinguishing function $h : \{\mathcal{C}(P_0), \mathcal{C}(P_1)\} \rightarrow \{0, 1\}$ as*

$$h(y) = \begin{cases} 0 & \text{if } y = \mathcal{C}(x) \text{ and } x \in \mathcal{CS}, \\ 1 & \text{otherwise.} \end{cases} \tag{7}$$

Define the following two probabilities

$$q_0 = \Pr\left(x_0 \in \mathcal{R} \wedge x_0 \in \mathcal{CS}\right), q_1 = \Pr\left(x_1 \in \mathcal{S}' \wedge x_1 \in \mathcal{CS}\right) \tag{8}$$

where $(x_0, x_1) \xleftarrow{\$} \Omega$. Then the average probability is

$$\sum_{i \in \{0,1\}} \Pr\left(h(c) = i \wedge c = \mathcal{C}(P_i)\right) = \frac{1 + (q_0 - q_1)}{2}. \tag{9}$$

Several remarks on Theorem 1 are as follows:

(i) An attacker will expect the probability value in (9) to be as large as possible so that she can distinguish the cryptographic scheme $\mathcal{C}$ with a high probability.

(ii) The difficulty of finding the distinguishing attack described in Theorem 1 is to find a proper condition set $\mathcal{CS}$ such that $(q_0 - q_1)$ is large.

(iii) The value of $(q_0 - q_1)$ could be negative. If the attacker uses $\overline{\mathcal{CS}}$ to replace $\mathcal{CS}$, $(q_0 - q_1)$ will be positive, and the probability will be greater than 0.5. Thus, the problem of finding a condition set such that $q_0 - q_1$ is large becomes the problem of finding the condition set such that $|q_0 - q_1|$ is large.

5.2 Distribution of the Linear Complexity of Component Sequences

We use f_1, AES, KASUMI and PRESENT as multi-output filtering functions and study the distribution of the linear complexities of their component sequences. Meidl and Niederreiter studied the expectation of the linear complexity of random binary periodic sequences in [21]. According to our experimental results, the average values of the linear complexities of the component sequences of AES, f_1, KASUMI, PRESENT are very close to the theoretical value determined in [21]. This motivates us to look at the whole distribution of the linear complexity of the component sequences instead of considering only the average value.

Testing of the Distribution of Linear Complexity. We test the distribution of the linear complexity of component sequences by choosing two (large) subsets of inputs and by comparing the distributions of the linear complexity of their component sequences. In particular, we choose one subset $\mathcal{LI}$ of the inputs generated by an ℓ-stage LFSR and the other subset $\mathcal{RI} = (\mathcal{LI} \setminus \{P_0\}) \cup \{P_1\}$, where $P_0 \xleftarrow{\$} \mathcal{LI}$ and $P_1 \xleftarrow{\$} \overline{\mathcal{LI}}$. Note that the elements in $\mathcal{LI}$ are ordered according to Eq. (3). It is clear that if the $\mathcal{C}$ has very good random property, it should not be easy to distinguish two distributions for $\mathcal{LI}$ and $\mathcal{RI}$. Our method consists of the following three steps. Now fixing a primitive $\mathcal{C}$ and an ℓ-stage LFSR:

Step 1 (Generating component sequences). We randomly choose N_{key} keys.

1. For all keys, using $\mathcal{LI}$ as the set of inputs and $\mathcal{C}$ as a multi-output filter, we obtain $m \cdot N_{key}$ component sequences. This set of component sequences is denoted by Q_1.
2. Similarly, using $\mathcal{RI}$ as the inputs, we generate another set of $m \cdot N_{key}$ component sequences, which is denoted by Q_2.

Step 2 (Computing linear complexity). We compute the linear complexities of the sequences in Q_1 and Q_2 and count the number of component sequences in Q_i with the linear complexity $2^\ell - 2$ and $2^\ell - 1$, denoted by $N_{2^\ell-1}^i$ and $N_{2^\ell-2}^i$, where $i = 1$ or 2.

Step 3 (Comparing the distributions). Now we compare two distributions by computing the slopes sl_i of the line between two points $(2^\ell - 2, N_{2^\ell-2}^i)$ and $(2^\ell - 1, N_{2^\ell-1}^i)$, where $sl_i = \frac{N_{2^\ell-1}^i - N_{2^\ell-2}^i}{(2^\ell-1)-(2^\ell-2)} = N_{2^\ell-1}^i - N_{2^\ell-2}^i$. If the difference between sl_1 and sl_2 is non-negligible, we can make use of it to build a distinguisher of $\mathcal{C}$, which is described in the next section. The worst case computational complexity for exhausting all ℓ-stage LFSRs of the above three steps is

$$\frac{\phi(2^\ell - 1)}{\ell} \times N_{key} \times 2\ell \times (2^\ell - 1) \times m, \tag{10}$$

where ϕ is the Euler phi function. Below we perform the experiment using these parameters on f_1, AES, KASUMI and PRESENT.

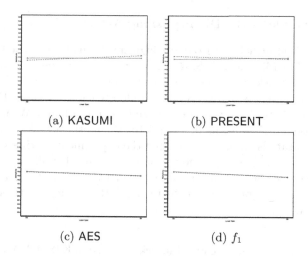

(a) KASUMI (b) PRESENT

(c) AES (d) f_1

Fig. 1. Linear complexity distribution

Distribution of f_1, AES, Kasumi and Present. In our experiment, we choose $\ell = 8$ and $N_{key} = 10^8$. By Eq. (10), the worst case complexity for the primitive f_1 is $2^{50.27}$ (some computation can be performed in a parallel way). We present the distribution of the linear complexity in which the solid (resp. dashed) line represents the distribution of sequences in Q_1 (resp. Q_2).

From Figs. 1a and b, one can observe that, for KASUMI and PRESENT, the difference of the distributions of the linear complexity for sequences in Q_1 and Q_2 is non-negligible. While Figs. 1c and d show this is not the case for AES and f_1.

5.3   The New Distinguishing Attack

We now present the details of our distinguishing attack, which is achieved through constructing a distinguishing function h based on the linear complexity distribution of the component sequences. The steps for the distinguishing attack are as follows.

1. Choosing an ℓ-stage LFSR with a primitive polynomial to generate the inputs of length n in $\mathcal{R}$ (see Eq. (3)). Constructing $\mathcal{S} = \mathbb{F}_2^n \setminus \mathcal{R}$;
2. Randomly choosing a message $P_0 \in \mathcal{R}$ and $P_1 \in \mathcal{S}$;
3. Let N_{LC} be the number of component sequences with linear complexity LC where $\ell \leq LC \leq 2^\ell - 1$;
4. Defining the condition set

$$\mathcal{CS} = \left\{ y \in \mathbb{F}_2^n \middle| \begin{array}{l} \text{using } (\mathcal{R} \setminus \{P_0\}) \cup \{y\} \text{ as the inputs of a primitive} \\ \text{in the multi-output filtering model, the slope of the} \\ \text{the points line between } (2^\ell - 2, N_{2^\ell-2}) \text{ and } (2^\ell - 1, \\ N_{2^\ell-1}) \text{ is less than } t. \end{array} \right\};$$

5. The distinguishing function h is defined in Eq. (7) using the condition set $\mathcal{CS}$; q_0, q_1 are the probability values defined in Definition 3.

5.4   An Example of the Distinguishing Attack

We now apply the attack with our distinguishing function defined in Sect. 5.3 on f_1, AES, KASUMI, and PRESENT. For simplicity, we use an 8-stage LFSR to conduct our attack. However, one can use any length LFSR. The set $\mathcal{R}$ is constructed using the 8-stage LFSR. We randomly chose 2^{10} keys. For each key, a message $P_0 \in \mathcal{R}$ and a message $P_1 \in \mathcal{S}$ are chosen randomly. We use the linear complexity based distinguishing function h to execute the attack. To test that the average success rate is stable, we repeated the experiment 20 times by choosing different groups of 2^{10} keys and found similar results for all experiments. We present the average success rate for an experiment in Table 2, where we use the upper bound of the slope t and the 8-stage LFSR the same as those in Table 6 in Appendix A.

Table 2. Average success rate of our attack on f_1, AES, KASUMI and PRESENT

Primitive	t	q_0	q_1	Avg. Succ. Rate
f_1	2	0.20398	0.194458	50.476%
AES	2	0.193848	0.20044	50.329%
KASUMI	4	0.421875	0.454103	51.612%
PRESENT	5	0.5686	0.540285	51.416%

Theorem 1 and the observations in Figs. 1a and b enable us to gain a non-negligible success rate of the attack on KASUMI and PRESENT. One can observe from the average success rate in Table 2 that the outputs of both KASUMI and PRESENT can be distinguished from a random primitive with a non-negligible probability. On the other hand, the performance of the attack on f_1 and AES is very similar to the random one.

6   Distribution of the Linear Complexity of **PRINTcipher** Under the Multi-output Filtering Model

In [19], the authors pointed out a weakness of PRINTcipher that when the input and key have some particular patterns, the output has the same pattern as the input. Since the input and the output have the same pattern, the subspace of $\mathbb{F}_2^{48}$ formed by the inputs and outputs is called *invariant subspace*. Table 3 shows one example of such patterns and the invariant subspace. Each asterisk in Table 3 represents one variant bit, which can be either 0 or 1. The bits that are fixed to 0 or 1 are called invariant bits. We use $\mathcal{I}$ to denote the invariant subspace shown in Table 3, and $\mathcal{K}$ to denote the key (composed by XOR and permute key) space shown in Table 3. The *invariant subspace attack* is addressed as follows. If the input and key are selected from $\mathcal{I}$ and $\mathcal{K}$ respectively, the output must be in $\mathcal{I}$.

Table 3. Patterns of input, key, and output

Input	00*	*10	***	***	00*	*10	***	***	00*	*10	***	***	00*	*10	***	***
XOR Key	01*	*01	***	***	01*	*01	***	***	01*	*01	***	***	01*	*01	***	***
Permute Key	0*	11	**	**	10	01	**	**	11	*0	**	**	*0	11	**	**
Output	00*	*10	***	***	00*	*10	***	***	00*	*10	***	***	00*	*10	***	***

6.1 Directly Applying Multi-output Filtering Model to PRINTcipher

We apply the multi-output filtering model to PRINTcipher with the weak keys in $\mathcal{K}$. We choose an 8-stage LFSR with primitive feedback polynomial, and initialize the LFSR with 0x01. Let $v_0, \cdots, v_7$ denote the first 8 variant bits of the input, and LFSR_i denote the i-th bit of the LFSR's internal state. The experiment is done as follows.

(1) Fix all invariant bits of the input to input patterns in Table 3.
(2) Randomly generate variant bits except for the first 8 variant bits.
(3) Run the LFSR once.
(4) Assign the internal state of the LFSR to the first 8 variant bits of the input, i.e. $v_i = \text{LFSR}_i$, for $0 \le i < 8$.
(5) Randomly generate a key, with the key pattern shown in Table 3.
(6) Compute the output of the input, and put each bit of the output into the corresponding component sequence.
(7) Repeat from Step (3) 254 times.
(8) Compute linear complexity for each component sequence.

By repeating this procedure N times, we can get the linear complexity distribution of each component sequence. Obviously, the component sequences that are composed by the invariant bits are either all 0 sequences or all 1 sequences. Thus, the linear complexity of such component sequence is either 0 or 1. The distribution of the linear complexity of each component sequence can be found in the full paper [29]. For each component sequence composed by invariant bit of the output, the linear complexity is a constant. Thus, the distribution is far away from the ideal one. Such difference shows non-randomness in PRINTcipher.

6.2 Applying Multi-output Filtering Model to PRINTcipher with Recurrent Input

A more challenging experiment is letting one invariant bit change. Denote the i-th bit of the input by b_i. The steps of this experiment are the same as above except steps (1), (2) and (4). The modified three steps are given below.

(1) Fix all invariant bits of the input to the input pattern shown in Table 3 except for b_5.
(2) Randomly generate the variant bits except for b_2, b_3, b_6, b_7, b_8, and b_9.
(4) Assign the internal state as follows.

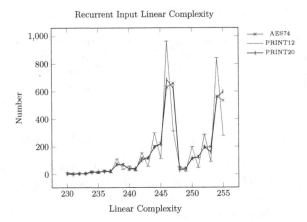

Fig. 2. Distributions of component sequence's linear complexity with recurrent inputs: AES and PRINTcipher

$$b_2 = \text{LFSR}_0, \, b_3 = \text{LFSR}_1, \, b_5 = \text{LFSR}_3, \, b_6 = \text{LFSR}_4,$$
$$b_7 = \text{LFSR}_5, \, b_8 = \text{LFSR}_6, \, b_9 = \text{LFSR}_7$$

Note that since b_4 is fixed to 1, in one period of the LFSR, each possible input occurs twice. We call this *recurrent input*.

After running the above procedure for N times, we have the linear complexity distribution of each component sequence. Since the ideal distribution of each component sequence with recurrent input is hard to compute, we run the above procedure on both PRINTcipher and AES, and use the distribution of AES as a reference. The results are shown in Fig. 2. The blue curve marked with cross represents the distribution of AES' 74th component sequence; the red curve marked with horizontal bars represents PRINTcipher's 12th component sequence, and the brown curve marked with vertical bars represents the 20th component sequence. For all AES's component sequences, the distributions of linear complexity are almost the same. However, we find the distribution of the 12th component sequence of PRINTcipher is quite different from AES. We can see that the red curve jumps dramatically.

7   Distribution of the Algebraic Degree and Nonlinearity of the Component Functions

In this section, we investigate the distribution of the algebraic degree and the nonlinearity of the component functions of f_1, AES, KASUMI, and PRESENT in the multi-output filtering model. To measure the randomness property, we first determine the distribution of the algebraic degree and the nonlinearity of component functions using a random primitive as the multi-output filter and compare this ideal distribution with that of f_1, AES, KASUMI and PRESENT obtained by performing experiments.

7.1   Algebraic Degree Distribution

The following result states the number of Boolean functions with a given algebraic degree. The first part of the result can also be found in [7].

Theorem 2. *Let f be a Boolean function on $\mathbb{F}_2^n$. Then the number of Boolean functions with algebraic degree at most d is $2^{\sum_{i=0}^{d}\binom{n}{i}}$, and the number of Boolean functions with algebraic degree exactly d is $\left(2^{\binom{n}{d}} - 1\right) \cdot 2^{\sum_{i=0}^{d-1}\binom{n}{i}}$.*

Corollary 1. *Let $\mathcal{C}$ be a random cryptographic primitive and $\mathcal{L}$ be an n-stage LFSR whose characteristic polynomial is a primitive polynomial of degree n. We use $\mathcal{C}$ as a multi-output filtering function and $\mathcal{L}$ to generate the inputs of $\mathcal{C}$. Then the probability of the component functions having degree at most d is $\frac{2^{\sum_{i=0}^{d}\binom{n}{i}}}{2^{2^n}}$. In particular, $Pr(d \leq n - 2) = \frac{1}{2^{n+1}}$.*

For a more detailed discussion about Theorem 2, we refer the reader to [29]. For f_1, AES, KASUMI and PRESENT, we perform the following test on the distribution of the algebraic degree of their component functions.

Statistical Test 1. *By Corollary 1, using an LFSR with a primitive polynomial of degree 8, the probability that the degree of the component functions is smaller than 7 is $\frac{1}{2^9} = 19.53125 \times 10^{-4}$. For f_1, AES, KASUMI and PRESENT, we apply the multi-output filtering model as in Sect. 3.1. We choose 50,000 keys for these primitives and compute the degree of the component functions. The probability of the degree is smaller than 7 is listed in the following table. From Table 4, we can see that for KASUMI, the probability $Pr(d \leq 6)$ is much higher than the one for other ciphers. To confirm this, we test another 50000 keys and found the probability is very close to it. This points out a distinguisher of KASUMI and other ciphers in Table 4.*

7.2   Nonlinearity Distribution

For a random Boolean function, we have the following result on the distribution of its nonlinearity.

Table 4. Distribution of the degree smaller than 7

Cryptographic primitive	$Pr(d \leq 6)$
Random function	19.53125×10^{-4}
f_1	19.87×10^{-4}
AES	19.77×10^{-4}
KASUMI	20.16×10^{-4}
PRESENT	19.58×10^{-4}

Theorem 3 ([7,26]). *Let c be any strictly positive real number. The density of the set*

$$\left\{ f \in \mathcal{B}_n, \ NL(f) \geq 2^{n-1} - c\sqrt{n}2^{\frac{n-1}{2}} \right\}$$

is greater than $1 - 2^{n+1-c^2 n \log_2 e}$. If $c^2 \log_2 e > 1$, then this density tends to 1 when n tends to infinity.

The reader is referred to Table 5 of [29] for the lower bound of the density of Boolean functions in $\mathcal{B}_8$. The best nonlinearity we can expect for Boolean functions in 8 variables is $2^7 - 2^3 = 120$. From Table 5 of [29], one can see that if the component functions of f_1 are random, the probability that the component Boolean functions have nonlinearity smaller than 90 is very small, which is $1 - 0.99354511316750952827725848552 4 \approx 0.00645$. In view of this, we perform the following statistical test for f_1, AES, KASUMI and PRESENT.

Statistical Test 2. *Let the LFSR and the other settings be the same as in **Statistical Test** 1. We list the distribution of the nonlinearity of the component functions of f_1 and AES in Table 5. Since only the component functions with smallest nonlinearity are important to us (as an attacker), we only list the probability that a Boolean function has nonlinearity smaller than 90 or 91. The notation $\Pr_{<W}$ denotes the probability that the nonlinearity is smaller than W. Unlike the distribution of the algebraic degree, from the above table we can not see obvious difference among these four ciphers. However, one can still see that the probability values $\Pr_{<90}$ and $\Pr_{<91}$ is still very different with the random case (although they are only the **upper bounds** of the probability).*

Table 5. The distribution of the nonlinearity of component sequences of f_1, AES, KASUMI and PRESENT

Cryptographic primitive	$\Pr_{<90}$	$\Pr_{<91}$
Random Function	0.006455	0.011597
f_1	0.000299	0.000690
AES	0.000306	0.000592
KASUMI	0.000299	0.000565
PRESENT	0.000308	0.000589

Although now we cannot derive attacks from Statistical Test 1 and Statistical Test 2, it is interesting to observe some non-randomness in the aspect of the distribution of cryptographic properties.

8   Conclusions

In this paper, we introduced the multi-output filtering model for analyzing the security of a cryptographic primitive. We proposed a general distinguish attack technique under IND-CPA and developed a new object, called a distinguishing function, to characterize the success rate of our new attack method. The importance of this new distinguishing function is demonstrated by launching attacks on KASUMI and PRESENT with non-negligible success rates. Furthermore, we studied the cryptographic properties of the component functions. By comparing the distribution of the algebraic degree and nonlinearity properties with that of a random one, we discovered that, for KASUMI, its distribution of the algebraic degree is very different, while the distributions of f_1, AES and PRESENT are not. We could not propose any immediate attack based on this observation, but it is worth to pointing it out for future research.

Acknowledgement. The authors would like to thank the reviewers of the C2SI-Carlet 2017 conference for their insightful comments to improving the quality of the paper. The authors sincerely thank Reviewer 3 for pointing out an error in Corollary 1 and also mentioning a connection between Statistical test 1 and the saturation attack.

A   Appendix

We describe the slope of the linear complexity distribution of f_1 and AES, KASUMI and PRESENT. The slope in Table 6 is the average slope over 10^8 samples. The column "Slope (L)" contains the slopes computed from the LFSR input, and the column "Slope (R)" contains the slopes computed from the random input. The last column shows the absolute value of the difference between "Slope(L)" and "Slope(R)". We can see the "Difference" of KASUMI and PRESENT are much greater than f_1 and AES.

Table 6. The slope of f_1, AES, KASUMI and PRESENT on average

Primitive	Polynomial of the LFSR	Slope (L)	Slope (R)	| Difference |
f_1	$x^8 + x^6 + x^4 + x^3 + x^2 + x^1 + 1$	-0.125	-0.124	2.210×10^{-4}
AES	$x^8 + x^7 + x^6 + x^3 + x^2 + x^1 + 1$	-0.088	-0.087	5.190×10^{-4}
KASUMI	$x^8 + x^7 + x^6 + x^5 + x^4 + x^2 + 1$	0.130	0.015	0.115
PRESENT	$x^8 + x^6 + x^5 + x^3 + 1$	-0.057	0.036	0.093

References

1. Aumasson, J.-P., Meier, W.: Zero-sum distinguishers for deduced Keccak-f and for the core functions of Luffa and Hamsi. In: Presented at the Rump Session of CHES 2009 (2009)
2. Bellare, M., Desai, A., Pointcheval, D., Rogaway, P.: Relations among notions of security for public-key encryption schemes. In: Krawczyk, H. (ed.) CRYPTO 1998. LNCS, vol. 1462, pp. 26–45. Springer, Heidelberg (1998). doi:10.1007/BFb0055718

3. Bellare, M., Desai, A., Jokipii, E., Rogaway, P.: A concrete security treatment of symmetric encryption: analysis of the DES modes of operation. In: Proceedings of the 38th Symposium on Foundations of Computer Science. IEEE (1997)

4. Bertoni, G., Daemen, J., Peeters, M., Assche, G.V.: The Keccak reference. (2011). http://keccak.noekeon.org/Keccak-reference-3.0.pdf

5. Bogdanov, A., Knudsen, L.R., Leander, G., Paar, C., Poschmann, A., Robshaw, M.J.B., Seurin, Y., Vikkelsoe, C.: PRESENT: an ultra-lightweight block cipher. In: Paillier, P., Verbauwhede, I. (eds.) CHES 2007. LNCS, vol. 4727, pp. 450–466. Springer, Heidelberg (2007). doi:10.1007/978-3-540-74735-2_31

6. Boura, C., Canteaut, A.: Zero-sum distinguishers for iterated permutations and application to KECCAK-f and Hamsi-256. In: Biryukov, A., Gong, G., Stinson, D.R. (eds.) SAC 2010. LNCS, vol. 6544, pp. 1–17. Springer, Heidelberg (2011). doi:10.1007/978-3-642-19574-7_1

7. Carlet, C.: Boolean functions for cryptography and error correcting codes. In: Crama, Y., Hammer, P.L. (eds.) The Monography Boolean Models and Methods in Mathematics, Computer Science, and Engineering, pp. 257–397. Cambridge University Press (2010)

8. Daemen, J., Rijmen, V.: The Design of Rijndael: AES - The Advanced Encryption Standard. Springer, New York (2002)

9. Daemen, J., Assche, G.: Differential propagation analysis of keccak. In: Canteaut, A. (ed.) FSE 2012. LNCS, vol. 7549, pp. 422–441. Springer, Heidelberg (2012). doi:10.1007/978-3-642-34047-5_24

10. Dinur, I., Shamir, A.: Cube attacks on tweakable black box polynomials. In: Joux, A. (ed.) EUROCRYPT 2009. LNCS, vol. 5479, pp. 278–299. Springer, Heidelberg (2009). doi:10.1007/978-3-642-01001-9_16

11. Dinur, I., Dunkelman, O., Shamir, A.: New attacks on Keccak-224 and Keccak-256. In: Canteaut, A. (ed.) FSE 2012. LNCS, vol. 7549, pp. 442–461. Springer, Heidelberg (2012). doi:10.1007/978-3-642-34047-5_25

12. Dinur, I., Dunkelman, O., Shamir, A.: Collision attacks on up to 5 rounds of SHA-3 using generalized internal differentials. Cryptology ePrint Archive, Report 2012/627. (2012). http://eprint.iacr.org/

13. Duc, A., Guo, J., Peyrin, T., Wei, L.: Unaligned rebound attack: application to Keccak. In: Canteaut, A. (ed.) FSE 2012. LNCS, vol. 7549, pp. 402–421. Springer, Heidelberg (2012). doi:10.1007/978-3-642-34047-5_23

14. Goldwasser, S., Micali, S.: Probabilistic encryption. J. Comput. Syst. Sci. **28**, 270–299 (1984)

15. Golomb, S.W., Gong, G.: Signal Design for Good Correlation - for Wireless Communication, Cryptography and Radar. Cambridge Press (2005)

16. Gong, G., Mandal, K., Tan, Y., Wu, T.: Security assessment of TUAK algorithm set. CACR Technical Report, University of Waterloo (2014)

17. Key, E.L.: An analysis of the structure and complexity of nonlinear binary sequence generators. IEEE Trans. Inf. Theory **22**, 732–736 (1976)

18. Knudsen, L., Leander, G., Poschmann, A., Robshaw, M.J.B.: PRINTCIPHER: a block cipher for IC-printing. In: Mangard, S., Standaert, F.-X. (eds.) CHES 2010. LNCS, vol. 6225, pp. 16–32. Springer, Heidelberg (2010). doi:10.1007/978-3-642-15031-9_2

19. Leander, G., Abdelraheem, M.A., AlKhzaimi, H., Zenner, E.: A cryptanalysis of PRINTCIPHER: the invariant subspace attack. In: Rogaway, P. (ed.) CRYPTO 2011. LNCS, vol. 6841, pp. 206–221. Springer, Heidelberg (2011). doi:10.1007/978-3-642-22792-9_12

20. Matsui, M.: Linear cryptanalysis method for DES cipher. In: Helleseth, T. (ed.) EUROCRYPT 1993. LNCS, vol. 765, pp. 386–397. Springer, Heidelberg (1994). doi:10.1007/3-540-48285-7_33

21. Meidl, W., Niederreiter, H.: On the expected value of the linear complexity and the k-error linear complexity of periodic sequences. IEEE Trans. Inf. Theory $48(11)$, 2817–2825 (2002)

22. Morawiecki, P., Pieprzyk, J., Srebrny, M., Straus, M.: Preimage attacks on the round-reduced Keccak with the aid of differential cryptanalysis, Cryptology ePrint Archive, Report 2013/561 (2013). http://eprint.iacr.org/

23. Morawiecki, P., Pieprzyk, J., Srebrny, M.: Rotational cryptanalysis of round-reduced KECCAK, Cryptology ePrint Archive, Report 2012/546. (2012). http://eprint.iacr.org/

24. Naya-Plasencia, M., Röck, A., Meier, W.: Practical analysis of reduced-round KEC-CAK. In: Bernstein, D.J., Chatterjee, S. (eds.) INDOCRYPT 2011. LNCS, vol. 7107, pp. 236–254. Springer, Heidelberg (2011). doi:10.1007/978-3-642-25578-6_18

25. NIST, the SHA-3 competition (2007–2012). http://csrc.nist.gov/groups/ST/hash/sha-3/index.html

26. Olejar, D., Stanek, M.: On cryptographic properties of random Boolean functions. J. Univers. Comput. Sci. $4(8)$, 705–717 (1998)

27. Rueppel, R.A.: Analysis and Design of Stream Ciphers. Springer, Berlin (1986)

28. Todo, Y.: Integral cryptanalysis on full MISTY1. In: Gennaro, R., Robshaw, M. (eds.) CRYPTO 2015. LNCS, vol. 9215, pp. 413–432. Springer, Heidelberg (2015). doi:10.1007/978-3-662-47989-6_20

29. Wu, T., Tan, Y., Mandal, K., Gong, G.: On the multi-output filtering model and its applications. CACR Technical Report, CACR 2017-01, University of Waterloo (2017). http://cacr.uwaterloo.ca/

30. 3rd generation partnership project, Technical specification group services, system aspects, 3G security, specification of the 3Gpp. confidentiality, integrity algorithms; Document 2: KASUMI specification, V. 3.1.1 (2001)

31. Specification of the TUAK algorithm set: a second example algorithm set for the 3Gpp. authentication and key generation functions $f_1, f_1^*, f_2, f_3, f_4, f_5$ and f_5^*, SP-130602, ETSI/SAGE, 13 Dec 2013. http://www.3gpp.org/ftp/tsg_sa/TSG_SA/TSGS_62/ftp-TdocsByTdoc_SP-62.htm

New Bent Functions from Permutations and Linear Translators

Sihem Mesnager[1,2,3]([✉]), Pınar Ongan[4], and Ferruh Özbudak[4,5]

[1] Department of Mathematics, University of Paris VIII, Saint-Denis, France
`smesnager@univ-paris8.fr`
[2] University of Paris XIII, LAGA, UMR 7539, CNRS, Villetaneuse, France
[3] Telecom ParisTech, Paris, France
[4] Institute of Applied Mathematics,
Middle East Technical University, Ankara, Turkey
`pinar.ongan@metu.edu.tr`
[5] Department of Mathematics, Middle East Technical University, Ankara, Turkey
`ozbudak@metu.edu.tr`

Abstract. Starting from the secondary construction originally introduced by Carlet ["On Bent and Highly Nonlinear Balanced/Resilient Functions and Their Algebraic Immunities", Applied Algebra, Algebraic Algorithms and Error-Correcting Codes, 2006], that we shall call *"Carlet's secondary construction"*, Mesnager has showed how one can construct several new primary constructions of bent functions. In particular, she has showed that three tuples of permutations over the finite field $\mathbb{F}_{2^m}$ such that the inverse of their sum equals the sum of their inverses give rise to a construction of a bent function given with its dual. It is not quite easy to find permutations satisfying such a strong condition $(\mathcal{A}_m)$. Nevertheless, Mesnager has derived several candidates of such permutations in 2015, and showed in 2016 that in the case of involutions, the problem of construction of bent functions amounts to solve arithmetical and algebraic problems over finite fields.

This paper is in the line of those previous works. We present new families of permutations satisfying $(\mathcal{A}_m)$ as well as new infinite families of permutations constructed from permutations in both lower and higher dimensions. Our results involve linear translators and give rise to new primary constructions of bent functions given with their dual. And also, we show that our new families are not in the class of Maiorana-McFarland in general.

Keywords: Boolean functions · Bent functions · Linear translators · Permutations

1 Introduction

1.1 Preliminaries

Boolean functions of n variables are binary functions over the Galois field $\mathbb{F}_{2^n}$. For even values of n, a Boolean function $f : \mathbb{F}_{2^n} \to \mathbb{F}_2$ can be represented in

© Springer International Publishing AG 2017
S. El Hajji et al. (Eds.): C2SI 2017, LNCS 10194, pp. 282–297, 2017.
DOI: 10.1007/978-3-319-55589-8_19

bivariate representation as a polynomial $f(x, y)$ in two variables $x, y \in \mathbb{F}_{2^m}$, where $m = n/2$. They are used in the pseudo-random generators of stream ciphers and S-boxes of block ciphers, and the nonlinearity of such functions represents a significant cryptographic criterion against attacks on stream and block ciphers.

The *nonlinearity* of a function $f : \mathbb{F}_{2^n} \rightarrow \mathbb{F}_2$ is the minimum Hamming distance between f and all affine functions, and it can be expressed as

$$\mathcal{N}_f := 2^{n-1} - \frac{1}{2} \max_{\omega \in \mathbb{F}_{2^n}} |\widehat{\chi_f}(\omega)|,$$

where $\widehat{\chi_f}(\omega)$ denotes the *Walsh transform* of f of an element $\omega \in \mathbb{F}_{2^n}$ defined by $\widehat{\chi_f}(\omega) := \sum_{x \in \mathbb{F}_{2^n}} (-1)^{f(x) + Tr_1^n(\omega x)}$, and $Tr_1^n(x)$ denotes the *absolute trace function* over $\mathbb{F}_2$ of an element $x \in \mathbb{F}_{2^n}$ defined by $Tr_1^n(x) := \sum_{i=0}^{n-1} x^{2^i}$. Because of the Parseval's relation $\sum_{\omega \in \mathbb{F}_{2^n}} (\widehat{\chi_f}(\omega))^2 = 2^{2n}$, $\mathcal{N}_f$ must be bounded above by $2^{n-1} - 2^{\frac{n}{2}-1}$, and this upper bound is tight for even values of n.

An n-variable Boolean function f is called *bent* if $\mathcal{N}_f = 2^{n-1} - 2^{\frac{n}{2}-1}$. So, bent functions are maximally nonlinear functions, and they exist only with even number of variables. Indeed, they are unbalanced and weak against fast algebraic attacks. The highest possible algebraic degree of a $2m$-variable bent function is m. Because of the fact that the degree is too small, even after modifications to balance them -which is sampled in [14]-, they still remain vulnerable against fast algebraic attacks.

There exists a main characterization of the bentness of a Boolean function in terms of the Walsh transform as follows: A function $f : \mathbb{F}_{2^n} \rightarrow \mathbb{F}_2$ (n even) is bent if and only if $\widehat{\chi_f}(\omega) = \pm 2^{\frac{n}{2}}$ for all $\omega \in \mathbb{F}_{2^n}$. If f is bent, then $\widehat{\chi_f}(\omega) = 2^{n/2}(-1)^{\tilde{f}(\omega)}$, for all $\omega \in \mathbb{F}_{2^n}$ defines the *dual function* $\tilde{f}$ of f, and its a known fact that bent functions always appear in pairs since their duals are also bent functions. The reader willing to see the previous results on bent functions significant for this paper may run an eye over the next subsection.

In this study, permutation polynomials and linear translators are used as the building blocks of secondary constructions of several new bent functions.

Let $f(x)$ be any polynomial in $\mathbb{F}_{2^n}[x]$. Then $f(x)$ defines a mapping $F : \mathbb{F}_{2^n} \rightarrow \mathbb{F}_{2^n}$ via $F(x) := f(x)$. This mapping F is called the *associated mapping* of the polynomial $f(x)$.

Definition 1. *A polynomial $f(x) \in \mathbb{F}_{2^n}[x]$ is called a* permutation polynomial *of $\mathbb{F}_{2^n}$ if its associated mapping is bijective.*

Definition 2 (*[6]*). *Let $m, t \in \mathbb{Z}^+$ be such that t divides m. Let f be a map from $\mathbb{F}_{2^m}$ to $\mathbb{F}_{2^t}$ and $a \in \mathbb{F}_{2^t}$. An element α of $\mathbb{F}_{2^m}$ is said to be an a-linear translator of f if and only if $f(x + \alpha u) = f(x) + au$ for every $(x, u) \in \mathbb{F}_{2^m} \times \mathbb{F}_{2^t}$.*

For a Boolean map, linear translators are not desirable. The functions with linear translators are considered to be weak for some cryptographic applications. For instance, a recent attack on hash functions proposed in [1] exploits a similar weakness of the involved mappings. A classical property of linear translators is the following.

Proposition 1. *Let $m, t \in \mathbb{Z}^+$ be such that t divides m. Let $a_1, a_2 \in \mathbb{F}_{2^t}$ and $\alpha_1, \alpha_2 \in \mathbb{F}_{2^m}$. If α_1 is an a_1-linear translator and α_2 is an a_2-linear translator of a mapping $f : \mathbb{F}_{2^m} \to \mathbb{F}_{2^t}$, then*

i. $\alpha_1 + \alpha_2$ is an $(a_1 + a_2)$-linear translator of f.
ii. $c\alpha_1$ is an (ca_1)-linear translator of f, for any $c \in \mathbb{F}_{2^t}^*$.

1.2   Previous Results and the Aim of the Paper

Bent functions were introduced by Rothaus [13] in 1976 but already studied by Dillon [4] since 1974. They have attracted a lot of research for four decades because of their own sake as interesting combinatorial objects, but also because of their relations to coding theory and applications in cryptography, sequences and other domains. A book devoted to bent functions (including variations, generalizations and applications) is [11]. A jubilee survey paper on bent functions giving an historical perspective, and making pertinent connections to designs, codes and cryptography is [3].

Despite their simple and natural definition, bent functions turned out to have a very complicated structure in general. Several constructions are known (see [3,11]) and linear translators are also used in certain constructions (see [5,6]).

Starting from a general secondary construction ([2], Theorem 3) known since 2006 that we shall call *"Carlet's secondary construction"*, Mesnager has refined recently in ([7], Theorem 4) Carlet's result and presented in [7–10] several new ways to construct new primary bent functions in bivariate representation. In this paper, instead of taking Carlet's secondary construction in its most general form, we only consider this particular case of it. This method allows to construct primary bent functions whose dual functions can be computed explicitly. This feature is important because it is often not so easy to compute the dual function of a bent function (especially in univariate representation). In particular, Mesnager has shown in [7,8] that one can construct easily such bent functions by using three permutations defined over the finite field $\mathbb{F}_{2^m}$ satisfying a condition that she denoted by $(\mathcal{A}_m)$.

Definition 3. *Let $m \in \mathbb{Z}^+$. Three pairwise distinct permutations ϕ_1, ϕ_2 and ϕ_3 of $\mathbb{F}_{2^m}$ are said to satisfy $(\mathcal{A}_m)$ if the following conditions hold:*

i. $\psi = \phi_1 + \phi_2 + \phi_3$ is a permutation of $\mathbb{F}_{2^m}$,
ii. $\psi^{-1} = \phi_1^{-1} + \phi_2^{-1} + \phi_3^{-1}$.

Mesnager has exhibited explicit families of permutations satisfying $(\mathcal{A}_m)$ in [7,8]. Furthermore, using such a method, it has been shown in [9,10,12] that constructing bent functions from involutions satisfying $(\mathcal{A}_m)$ amounts to solve arithmetical problems (using Fermat hypersurface and Lang-Weil estimations) (see [12]) and algebraic problems (for instance, based on the resolution of systems of equations over finite fields) (see [9,10]).

In this paper, the main target is to exhibit new constructions of bent functions in bivariate representation in the line of Mesnager's works mentioned above. More precisely, we are interested in pushing further the investigations initiated in the previous works and exploring new families of permutations satisfying $(\mathcal{A}_m)$. This study is motivated by the below result given in [7,8] (see also [2]) presenting a construction of bent functions from families of permutations satisfying $(\mathcal{A}_m)$.

Proposition 2 ([7, 8]). *Let $m \in \mathbb{Z}^+$. Let ϕ_1, ϕ_2 and ϕ_3 be three pairwise distinct permutations satisfying $(\mathcal{A}_m)$. Then, the Boolean function H from $\mathbb{F}_{2^m} \times \mathbb{F}_{2^m}$ to $\mathbb{F}_2$ defined by*

$$H(x,y) = Tr_1^m\big(x\phi_1(y)\big)Tr_1^m\big(x\phi_2(y)\big) + Tr_1^m\big(x\phi_1(y)\big)Tr_1^m\big(x\phi_3(y)\big) + \\ Tr_1^m\big(x\phi_2(y)\big)Tr_1^m\big(x\phi_3(y)\big), \quad (1)$$

is bent. Furthermore, its dual function $\tilde{H}$ is given by

$$\tilde{H}(x,y) = Tr_1^m\big(\phi_1^{-1}(x)y\big)Tr_1^m\big(\phi_2^{-1}(x)y\big) + Tr_1^m\big(\phi_1^{-1}(x)y\big)Tr_1^m\big(\phi_3^{-1}(x)y\big) \\ + Tr_1^m\big(\phi_2^{-1}(x)y\big)Tr_1^m\big(\phi_3^{-1}(x)y\big). \quad (2)$$

To this end, we are going to construct bent functions using linear translators of the functions of the form $f : \mathbb{F}_{2^m} \rightarrow \mathbb{F}_{2^t}$, where $t, m \in \mathbb{Z}^+$ are such that t divides m.

1.3   The Organization of the Paper

The rest of the paper is organized as follows. All our results are wrapped in Sect. 2. We start by presenting in Proposition 3 that one can construct 3-tuples of permutations satisfying $(\mathcal{A}_m)$, by selecting 3-tuples of a new family of permutations of $\mathbb{F}_{2^m}$ obtained from a permutation of $\mathbb{F}_{2^t}$ with the help of linear translators of a mapping from $\mathbb{F}_{2^m}$ to $\mathbb{F}_{2^t}$. In Propositions 4 and 5, we show that one can construct 3-tuples of permutations satisfying $(\mathcal{A}_m)$ from other 3-tuples of permutations (in lower dimension in Proposition 4 and from linear permutations in Proposition 5). All these results lead to three new infinite families of bent functions (Theorems 1, 2 and 3). Next, we present a new infinite family of permutations constructed from permutations in higher dimension. We show that elements of this family are good candidates to construct 3-tuples of permutations satisfying $(\mathcal{A}_m)$. Finally, in Sect. 3.4, we show that the bent functions obtained here are generally not in the Maiorana-McFarland Class.

2   Several New Constructions for Bent Functions

Proposition 3. *Let $m, t \in \mathbb{Z}^+$ such that $t < m$ and t divides m. Let f be a mapping from $\mathbb{F}_{2^m}$ to $\mathbb{F}_{2^t}$. Let $L : \mathbb{F}_{2^m} \rightarrow \mathbb{F}_{2^m}$ be an $\mathbb{F}_{2^t}$-linear permutation of $\mathbb{F}_{2^m}$. Let $g : \mathbb{F}_{2^t} \rightarrow \mathbb{F}_{2^t}$ be a permutation. Assume $\alpha \in \mathbb{F}_{2^m}^*$ and $a \in \mathbb{F}_{2^t}^*$ such that α is an a-linear translator of f with respect to $\mathbb{F}_{2^t}$. Set $\phi : \mathbb{F}_{2^m} \rightarrow \mathbb{F}_{2^m}$ as*

$$\phi(x) := L(x) + L(\alpha)\left[g(f(x)) + \frac{f(x)}{a}\right]. \quad (3)$$

Then ϕ is a permutation polynomial of $\mathbb{F}_{2^m}$ and

$$\phi^{-1}(x) = L^{-1}(x) + \frac{\alpha}{a}\left[f(L^{-1}(x)) + g^{-1}\left(\frac{f(L^{-1}(x))}{a}\right)\right]. \tag{4}$$

Proof. Define $h : \mathbb{F}_{2^m} \rightarrow \mathbb{F}_{2^m}$ as

$$h(x) := x + \alpha\left(g(f(x)) + \frac{f(x)}{a}\right). \tag{5}$$

Note that the equality $y = x + \alpha\left(g(f(x)) + \frac{f(x)}{a}\right)$ leads to

$$f(y) = f\left(x + \alpha\left(g(f(x)) + \frac{f(x)}{a}\right)\right) = f(x) + a\left(g(f(x)) + \frac{f(x)}{a}\right) = ag(f(x))$$

from which one can obtain $f(x) = g^{-1}\left(\frac{f(y)}{a}\right)$. Therefore

$$x = y + \alpha\left(g(f(x)) + \frac{f(x)}{a}\right) = y + \frac{\alpha}{a}\left(f(y) + g^{-1}\left(\frac{f(y)}{a}\right)\right).$$

In other words, h is a permutation of $\mathbb{F}_{2^m}$ and its inverse map is

$$h^{-1}(x) = x + \frac{\alpha}{a}\left(g^{-1}\left(\frac{f(x)}{a}\right) + f(x)\right).$$

Finally, for any $x \in \mathbb{F}_{2^m}$, expression (5) and the definition of L implies that

$$L(h(x)) = L\left(x + \alpha\left(g(f(x)) + \frac{f(x)}{a}\right)\right) = L(x) + L(\alpha)\left(g(f(x)) + \frac{f(x)}{a}\right),$$

which means $\phi = L \circ h$. Being a composition of two permutations of $\mathbb{F}_{2^m}$, ϕ should also be a permutation of $\mathbb{F}_{2^m}$, and the inverse of ϕ should be $\phi^{-1} = h^{-1} \circ L^{-1}$ which gives the result

$$\phi^{-1}(x) = L^{-1}(x) + \frac{\alpha}{a}\left(g^{-1}\left(\frac{f(L^{-1}(x))}{a}\right) + f(L^{-1}(x))\right).$$

$\square$

Observe that expressions (3) and (4) depend linearly on α. Therefore, if we define three permutations ϕ_1, ϕ_2 and ϕ_3 in the form of the expression (3) by using the same L, f and g, but three-many pairwise distinct a-linear translators $\alpha_1, \alpha_2, \alpha_3$ of f, then $\psi := \phi_1 + \phi_2 + \phi_3$ would be

$$\psi(x) = L(x) + [L(\alpha_1) + L(\alpha_2) + L(\alpha_3)]\left[g(f(x)) + \frac{f(x)}{a}\right],$$

which is obviously a permutation by Propositions 1 and 3. Furthermore, for any $x \in \mathbb{F}_{2^m}$, $\phi_1^{-1}(x) + \phi_2^{-1}(x) + \phi_3^{-1}(x)$ becomes

$$L^{-1}(x) + \left(\frac{\alpha_1}{a} + \frac{\alpha_2}{a} + \frac{\alpha_3}{a}\right)\left[f(L^{-1}(x)) + g^{-1}\left(\frac{f(L^{-1}(x))}{a}\right)\right]$$

which is equal to $\psi^{-1}(x)$. Thus we conclude that ϕ_1, ϕ_2, ϕ_3 satisfy $(\mathcal{A}_m)$.

Theorem 1. *Let* $m, t \in \mathbb{Z}^+$ *such that* $t < m$ *and* t *divides* m. *Let* f *be a mapping from* $\mathbb{F}_{2^m}$ *to* $\mathbb{F}_{2^t}$. *Let* $L : \mathbb{F}_{2^m} \to \mathbb{F}_{2^m}$ *be an* $\mathbb{F}_{2^t}$-*linear permutation of* $\mathbb{F}_{2^m}$. *Let* $g : \mathbb{F}_{2^t} \to \mathbb{F}_{2^t}$ *be a permutation. Assume* α_1, α_2, $\alpha_3 \in \mathbb{F}_{2^m}^*$ *and* $a \in \mathbb{F}_{2^t}^*$ *such that* α_1, α_2, α_3 *are all pairwise distinct a-linear translators of* f *with respect to* $\mathbb{F}_{2^t}$. *Suppose* $\alpha_1 + \alpha_2 + \alpha_3 \neq 0$. *Set* $\rho : \mathbb{F}_{2^m} \to \mathbb{F}_{2^m}$ *as* $\rho(x) := g(f(x)) + \frac{f(x)}{a}$ *and* $\tilde{\rho} : \mathbb{F}_{2^m} \to \mathbb{F}_{2^m}$ *as* $\tilde{\rho}(x) := \frac{1}{a}\left(g^{-1}\left(\frac{f(x)}{a}\right) + f(x)\right)$. *Then,*

$$H(x,y) = Tr_1^m\big(xL(y)\big) + Tr_1^m\big(L(\alpha_1)x\rho(y)\big)Tr_1^m\big(L(\alpha_2)x\rho(y)\big) +$$
$$Tr_1^m\big(L(\alpha_1)x\rho(y)\big)Tr_1^m\big(L(\alpha_3)x\rho(y)\big) +$$
$$Tr_1^m\big(L(\alpha_2)x\rho(y)\big)Tr_1^m\big(L(\alpha_3)x\rho(y)\big) \quad (6)$$

is bent. Furthermore, its dual function $\tilde{H}$ *is given by*

$$\tilde{H}(x,y) = Tr_1^m\big(yL^{-1}(x)\big) + Tr_1^m\big(\alpha_1 y\tilde{\rho}(L^{-1}(x))\big)Tr_1^m\big(\alpha_2 y\tilde{\rho}(L^{-1}(x))\big) +$$
$$Tr_1^m\big(\alpha_1 y\tilde{\rho}(L^{-1}(x))\big)Tr_1^m\big(\alpha_3 y\tilde{\rho}(L^{-1}(x))\big) +$$
$$Tr_1^m\big(\alpha_2 y\tilde{\rho}(L^{-1}(x))\big)Tr_1^m\big(\alpha_3 y\tilde{\rho}(L^{-1}(x))\big). \quad (7)$$

Proof. Proposition 2 implies that the Boolean function $H : \mathbb{F}_{2^m} \times \mathbb{F}_{2^m} \to \mathbb{F}_2$ defined by (1) will be bent for any 3-tuples of ϕ_i satisfying $(\mathcal{A}_m)$. Define ϕ_1, ϕ_2, ϕ_3 in the same nature of (3) as follows:

$$\phi_i(x) := L(x) + L(\alpha_i)\left[g(f(x)) + \frac{f(x)}{a}\right], \text{ for all } i \in \{1, 2, 3\}.$$

Then these three permutations will satisfy $(\mathcal{A}_m)$ as proven above, so they can be used to define the bent function $H(x,y)$ given in (1).
For any $i \in \{1, 2, 3\}$, we have

$$Tr_1^m\big(x\phi_i(y)\big) = Tr_1^m\big(xL(y)\big) + Tr_1^m\big(L(\alpha_i)\rho(y)\big)$$

and so for any distinct values of $i, j \in \{1, 2, 3\}$, $Tr_1^m\big(x\phi_i(y)\big)Tr_1^m\big(x\phi_i(y)\big)$ will be equal to the product

$$[Tr_1^m\big(xL(y)\big) + Tr_1^m\big(L(\alpha_i)\rho(y)\big)]\,[Tr_1^m\big(xL(y)\big) + Tr_1^m\big(L(\alpha_j)\rho(y)\big)]$$

which is equal to the sum

$$Tr_1^m\big(xL(y)\big) + Tr_1^m\big(xL(y)\big)Tr_1^m\big(L(\alpha_j)\rho(y)\big)$$
$$+Tr_1^m\big(xL(y)\big)Tr_1^m\big(L(\alpha_i)\rho(y)\big) + Tr_1^m\big(L(\alpha_i)\rho(y)\big)Tr_1^m\big(L(\alpha_j)\rho(y)\big)$$

since $Tr_1^m : \mathbb{F}_{2^m} \to \mathbb{F}_2$. Hence we directly obtain (6) from the expression (1), by only defining ϕ_is as in (3). $\square$

Remark 1. It is easy to find non-trivial examples satisfying the conditions of Theorem 1 for all large m. For example, let $\mathbb{F}_{2^t}$ be a fixed (small) finite field and $a \in \mathbb{F}_{2^t}^*$. Let $g : \mathbb{F}_{2^t} \to \mathbb{F}_{2^t}$ be a fixed permutation of $\mathbb{F}_{2^t}$. An obvious choice would be $g(x) = x^j$ where $gcd(j, 2^t - 1) = 1$. Let $\alpha_1, \alpha_2, \alpha_3$ be three linearly independent elements of $\mathbb{F}_{2^m}$ over $\mathbb{F}_{2^t}$, which only requires that $m \geq 3t$. Let $\mathcal{W} \subseteq \mathbb{F}_{2^m}$ be the $\mathbb{F}_{2^t}$-linear span of α_1, α_2 and α_3. Let $\mathcal{V}$ be an $\mathbb{F}_{2^t}$-linear subspace of $\mathbb{F}_{2^m}$ such that $\mathbb{F}_{2^m} = \mathcal{V} \oplus \mathcal{W}$. Let $\mu : \mathcal{V} \to \mathbb{F}_{2^t}$ be an arbitrary map. For any $x \in \mathbb{F}_{2^m}$ there exist uniquely determined $v \in \mathcal{V}$, $c_1, c_2, c_3 \in \mathbb{F}_{2^t}$ such that $x = v + c_1\alpha_1 + c_2\alpha_2 + c_3\alpha_3$. Let $f : \mathbb{F}_{2^m} \to \mathbb{F}_{2^t}$ be the map defined as $f(x) = \mu(v) + a(c_1 + c_2 + c_3)$. It is not difficult to see that $\alpha_1, \alpha_2, \alpha_3$ are pairwise distinct a-linear translators of f.

Example 1. Consider the special case of Proposition 3: $m = 6$ and $t = 3$. Let ω be a primitive element of $\mathbb{F}_8$ satisfying the equation $\omega^3 + \omega + 1 = 0$, and ξ be a primitive element of $\mathbb{F}_{64}$ satisfying the equation $\xi^6 + \xi^4 + \xi^3 + \xi + 1 = 0$. Define $f : \mathbb{F}_{64} \to \mathbb{F}_8$, $L : \mathbb{F}_{64} \to \mathbb{F}_{64}$, $g : \mathbb{F}_8 \to \mathbb{F}_8$, a, and α_is as follows: $f(x) = x + x^8$, $L(x) = x$, $g(x) = x$, $a = \omega$, $\alpha_1 = \xi$, $\alpha_2 = \xi^8$ and $\alpha_3 = \xi^{56}$. Then α_1, α_2, α_3 are all a-linear translators of f with respect to $\mathbb{F}_8$ satisfying $\alpha_1 + \alpha_2 + \alpha_3 = \xi^7 \neq 0$. With this setup, $\phi_i : \mathbb{F}_{64} \to \mathbb{F}_{64}$ becomes

$$\phi_i(x) = x + \omega^2\alpha_i f(x),$$

where $i \in \{1, 2, 3\}$. We know by Proposition 3 that these ϕ_is are permutation polynomials of $\mathbb{F}_{64}$ satisfying $(\mathcal{A}_4)$. So, by Theorem 1, they can be used to define a bent function H from $\mathbb{F}_{64} \times \mathbb{F}_{64}$ to $\mathbb{F}_2$.

Let us observe that ϕ defined by (3) depends linearly on g as well as its inverse map depends linearly on g^{-1}. Thus, one can deduce from this observation the following secondary construction.

Proposition 4. *Let $m, t \in \mathbb{Z}^+$ such that $t < m$ and t divides m. Let f be a mapping from $\mathbb{F}_{2^m}$ to $\mathbb{F}_{2^t}$. Let $L : \mathbb{F}_{2^m} \to \mathbb{F}_{2^m}$ be an $\mathbb{F}_{2^t}$-linear permutation of $\mathbb{F}_{2^m}$. Let g_1, g_2 and g_3 be three permutations on $\mathbb{F}_{2^t}$ satisfying $(\mathcal{A}_t)$. Assume that $\alpha \in \mathbb{F}_{2^m}^*$ and $a \in \mathbb{F}_{2^t}^*$ such that α is an a-linear translator of f with respect to $\mathbb{F}_{2^t}$. Set $\phi_i : \mathbb{F}_{2^m} \to \mathbb{F}_{2^m}$, $i \in \{1, 2, 3\}$ as*

$$\phi_i(x) := L(x) + L(\alpha) \left[g_i(f(x)) + \frac{f(x)}{a} \right]. \tag{8}$$

Then ϕ_1, ϕ_2 and ϕ_3 satisfies $(\mathcal{A}_m)$.

Proof. Define $\psi : \mathbb{F}_{2^m} \to \mathbb{F}_{2^m}$ as $\psi := \phi_1 + \phi_2 + \phi_3$. Then

$$\psi(x) = L(x) + L(\alpha) \left[(g_1 + g_2 + g_3)(f(x)) + \frac{f(x)}{a} \right].$$

$g_1 + g_2 + g_3$ is a permutation for hypothesis since they satisfy $(\mathcal{A}_t)$. So, by Proposition 3, ψ must also be a permutation and

$$\psi^{-1}(x) = L^{-1}(x) + \frac{\alpha}{a} \left[f(L^{-1}(x)) + (g_1 + g_2 + g_3)^{-1} \left(\frac{f(L^{-1}(x))}{a} \right) \right].$$

On the other hand, from the explicit formulas of the permutations ϕ_1^{-1}, ϕ_2^{-1}, ϕ_3^{-1} given in Proposition 3, we obtain the equation

$$\phi_1^{-1}(x) + \phi_2^{-1}(x) + \phi_3^{-1}(x) = L^{-1}(x) +$$
$$\frac{\alpha}{a}\left[f(L^{-1}(x)) + (g_1^{-1} + g_2^{-1} + g_3^{-1})\left(\frac{f(L^{-1}(x))}{a}\right)\right].$$

Since $g_1^{-1} + g_2^{-1} + g_3^{-1} = (g_1 + g_2 + g_3)^{-1}$, we get the result $\psi^{-1} = \phi_1^{-1} + \phi_2^{-1} + \phi_3^{-1}$. $\qquad\square$

In the line of Theorem 1, one can then obtain from Proposition 2 the following construction of bent function.

Theorem 2. *Let $m, t \in \mathbb{Z}^+$ such that $t < m$ and t divides m. Let f be a mapping from $\mathbb{F}_{2^m}$ to $\mathbb{F}_{2^t}$. Let $L : \mathbb{F}_{2^m} \to \mathbb{F}_{2^m}$ be an $\mathbb{F}_{2^t}$-linear permutation of $\mathbb{F}_{2^m}$. Let g_1, g_2 and g_3 be three permutations of $\mathbb{F}_{2^t}$ satisfying $(\mathcal{A}_t)$. Let $a \in \mathbb{F}_{2^t}^*$. Let $\alpha \in \mathbb{F}_{2^m}^*$ be an a-linear translator of f with respect to $\mathbb{F}_{2^t}$. For $i \in \{1, 2, 3\}$, set $\rho_i : \mathbb{F}_{2^m} \to \mathbb{F}_{2^m}$ as $\rho_i(x) := g_i(f(x)) + \frac{f(x)}{a}$ and $\tilde{\rho}_i : \mathbb{F}_{2^m} \to \mathbb{F}_{2^m}$ as $\tilde{\rho}_i(x) := \frac{1}{a}\left(g_i^{-1}\left(\frac{f(x)}{a}\right) + f(x)\right)$. Then,*

$$H(x, y) = Tr_1^m\big(xL(y)\big) + Tr_1^m\big(L(\alpha)x\rho_1(y)\big)Tr_1^m\big(L(\alpha)x\rho_2(y)\big) +$$
$$Tr_1^m\big(L(\alpha)x\rho_1(y)\big)Tr_1^m\big(L(\alpha)x\rho_3(y)\big) + Tr_1^m\big(L(\alpha)x\rho_2(y)\big)Tr_1^m\big(L(\alpha)x\rho_3(y)\big) \tag{9}$$

is bent. Furthermore, its dual function $\tilde{H}$ is given by

$$\tilde{H}(x, y) = Tr_1^m\big(yL^{-1}(x)\big) + Tr_1^m\big(\alpha y\tilde{\rho}_1(L^{-1}(x))\big)Tr_1^m\big(\alpha y\tilde{\rho}_2(L^{-1}(x))\big) +$$
$$Tr_1^m\big(\alpha y\tilde{\rho}_1(L^{-1}(x))\big)Tr_1^m\big(\alpha y\tilde{\rho}_3(L^{-1}(x))\big) +$$
$$Tr_1^m\big(\alpha y\tilde{\rho}_2(L^{-1}(x))\big)Tr_1^m\big(\alpha y\tilde{\rho}_3(L^{-1}(x))\big). \tag{10}$$

Example 2. Consider the special case $m = 4$ and $t = 2$ of Theorem 2. Let ω be a root of the irreducible polynomial $x^2 + x + 1 \in \mathbb{F}_2[x]$ and ρ be a root of the irreducible polynomial $x^4 + x + 1 \in \mathbb{F}_2[x]$. Set $L : \mathbb{F}_{16} \to \mathbb{F}_{16}$, $f : \mathbb{F}_{16} \to \mathbb{F}_4$, $a \in \mathbb{F}_4^*$ and $\alpha \in \mathbb{F}_{16}^*$ as follows: $L(x) = x$, $f(x) = x + x^4$, $a = \omega$ and $\alpha = \rho^6$. Thus α is an a-linear translator of f with respect to $\mathbb{F}_4$. Also set $g_i : \mathbb{F}_4 \to \mathbb{F}_4$, where $i \in \{1, 2, 3\}$ as follows: $g_1(x) = x$, $g_2(x) = x^2$ and $g_3(x) = x^2 + 1$. (It is not hard to see that g_1, g_2, g_3 satisfy $(\mathcal{A}_2)$). With this setup, ϕ_1, ϕ_2 and ϕ_3 become permutations of $\mathbb{F}_{16}$ defined by $\phi_1(x) = (1 + \omega\rho^6)x + \omega\rho^6 x^4$, $\phi_2(x) = (1 + \omega^2\rho^6)x + \rho^6 x^2 + \omega^2\rho^6 x^4 + \rho^6 x^8$ and $\phi_3(x) = \rho^6 + (1 + \omega^2\rho^6)x + \rho^6 x^2 + \omega^2\rho^6 x^4 + \rho^6 x^8$. By Proposition 4, we know that these ϕ_is are satisfying $(\mathcal{A}_4)$, so they can be used to define a bent function H from $\mathbb{F}_{16} \times \mathbb{F}_{16}$ to $\mathbb{F}_2$ with the construction defined in Theorem 2.

Finally, observe that (3) depends linearly on L and (4) depends linearly on L^{-1} when g is an automorphism and f is a linear map. Therefore, one obtains the following result.

Proposition 5. *Let $m, t \in \mathbb{Z}^+$ such that $t < m$ and t divides m. Let $f : \mathbb{F}_{2^m} \to \mathbb{F}_{2^t}$ be a linear mapping. Let $g : \mathbb{F}_{2^t} \to \mathbb{F}_{2^t}$ be an automorphism on $\mathbb{F}_{2^t}$. Let $\alpha \in \mathbb{F}_{2^m}^*$ and $a \in \mathbb{F}_{2^t}^*$ such that α is an a-linear translator of f with respect to $\mathbb{F}_{2^t}$. For $i \in \{1, 2, 3\}$, let $L_i : \mathbb{F}_{2^m} \to \mathbb{F}_{2^m}$ be an $\mathbb{F}_{2^t}$-linear permutation of $\mathbb{F}_{2^m}$ and $\phi_i : \mathbb{F}_{2^m} \to \mathbb{F}_{2^m}$ be the mapping defined by*

$$\phi_i(x) := L_i(x) + L_i(\alpha) \left[g(f(x)) + \frac{f(x)}{a} \right]. \tag{11}$$

Suppose that L_1, L_2 and L_3 satisfies $(\mathcal{A}_m)$. Then ϕ_1, ϕ_2 and ϕ_3 satisfies $(\mathcal{A}_m)$.

Proof. Define $\psi : \mathbb{F}_{2^m} \to \mathbb{F}_{2^m}$ as $\psi := \phi_1 + \phi_2 + \phi_3$. Then

$$\psi(x) = (L_1 + L_2 + L_3)(x) + \left[(L_1 + L_2 + L_3)(\alpha) \right] \left[g(f(x)) + \frac{f(x)}{a} \right].$$

The fact that L_1, L_2 and L_3 are all $\mathbb{F}_{2^t}$-linear maps implies $L_1 + L_2 + L_3$ is also $\mathbb{F}_{2^t}$-linear. And, $L_1 + L_2 + L_3$ must be a permutation since L_1, L_2, L_3 satisfy $(\mathcal{A}_m)$. Furthermore, as being an automorphism between finite fields, g must be a permutation, too. Thus, we conclude that ψ is a permutation with inverse

$$\psi^{-1}(x) = (L_1 + L_2 + L_3)^{-1}(x) +$$
$$\frac{\alpha}{a} \left[f\big((L_1 + L_2 + L_3)^{-1}(x)\big) + g^{-1}\left(\frac{f\big((L_1 + L_2 + L_3)^{-1}(x)\big)}{a} \right) \right],$$

by Proposition 3. However, since $(L_1 + L_2 + L_3)^{-1} = L_1^{-1} + L_2^{-1} + L_3^{-1}$, we get

$$\psi^{-1}(x) = L_1^{-1}(x) + L_2^{-1}(x) + L_3^{-1}(x) +$$
$$\frac{\alpha}{a} \left[f\big(L_1^{-1}(x) + L_2^{-1}(x) + L_3^{-1}(x)\big) + g^{-1}\left(\frac{f\big(L_1^{-1}(x) + L_2^{-1}(x) + L_3^{-1}(x)\big)}{a} \right) \right].$$

On the other hand,

$$\phi_1^{-1}(x) + \phi_2^{-1}(x) + \phi_3^{-1}(x) = L_1^{-1}(x) + L_2^{-1}(x) + L_3^{-1}(x) +$$
$$\frac{\alpha}{a} \left[f(L_1^{-1}(x)) + f(L_2^{-1}(x)) + f(L_3^{-1}(x)) \right] +$$
$$\frac{\alpha}{a} \left[g^{-1}\left(\frac{f(L_1^{-1}(x))}{a} \right) + g^{-1}\left(\frac{f(L_2^{-1}(x))}{a} \right) + g^{-1}\left(\frac{f(L_3^{-1}(x))}{a} \right) \right].$$

Since g is a bijective automorphism and f is linear, we get $\psi^{-1} = \phi_1^{-1} + \phi_2^{-1} + \phi_3^{-1}$. $\qquad\square$

Similarly, by using Proposition 2, the above proposition leads then to the following construction of bent functions.

Theorem 3. *Let* $m, t \in \mathbb{Z}^+$ *such that* $t < m$ *and* t *divides* m. *Let* $f : \mathbb{F}_{2^m} \to \mathbb{F}_{2^t}$ *be a linear mapping. Let* $g : \mathbb{F}_{2^t} \to \mathbb{F}_{2^t}$ *be an automorphism of* $\mathbb{F}_{2^t}$. *Let* $\alpha \in \mathbb{F}_{2^m}^*$ *and* $a \in \mathbb{F}_{2^t}^*$ *such that* α *is an* a-*linear translator of* f *with respect to* $\mathbb{F}_{2^t}$. *For* $i \in \{1, 2, 3\}$. *Let* $L_i : \mathbb{F}_{2^m} \to \mathbb{F}_{2^m}$ *be an* $\mathbb{F}_{2^t}$-*linear permutation of* $\mathbb{F}_{2^m}$, $i \in \{1, 2, 3\}$. *Suppose that* L_1, L_2 *and* L_3 *satisfies* $(\mathcal{A}_m)$. *Set* $\rho : \mathbb{F}_{2^m} \to \mathbb{F}_{2^m}$ *as* $\rho(x) := g(f(x)) + \frac{f(x)}{a}$ *and* $\tilde{\rho} : \mathbb{F}_{2^m} \to \mathbb{F}_{2^m}$ *as* $\tilde{\rho}(x) := \frac{1}{a}\left(g^{-1}\left(\frac{f(x)}{a}\right) + f(x)\right)$. *Then,*

$$
\begin{aligned}
H(x, y) = &Tr_1^m(xL_1(y))Tr_1^m(xL_2(y)) + Tr_1^m(xL_1(y))Tr_1^m(xL_3(y)) + \\
&Tr_1^m(xL_2(y))Tr_1^m(xL_3(y)) + Tr_1^m(xL_1(y))Tr_1^m((L_2(\alpha) + L_3(\alpha))x\rho(y)) + \\
&Tr_1^m(xL_2(y))Tr_1^m((L_1(\alpha) + L_3(\alpha))x\rho(y)) + \\
&Tr_1^m(xL_3(y))Tr_1^m((L_1(\alpha) + L_2(\alpha))x\rho(y)) + \\
&Tr_1^m(L_1(\alpha)x\rho(y))Tr_1^m(L_2(\alpha)x\rho(y)) + \\
&Tr_1^m(L_1(\alpha)x\rho(y))Tr_1^m(L_3(\alpha)x\rho(y)) + \\
&Tr_1^m(L_2(\alpha)x\rho(y))Tr_1^m(L_3(\alpha)x\rho(y)) \quad (12)
\end{aligned}
$$

is bent. Furthermore, its dual function $\tilde{H}$ *is given by*

$$
\begin{aligned}
\tilde{H}(x, y) = &Tr_1^m(yL_1^{-1}(x))Tr_1^m(yL_2^{-1}(x)) + Tr_1^m(yL_1^{-1}(x))Tr_1^m(yL_3^{-1}(x)) + \\
&Tr_1^m(yL_2^{-1}(x))Tr_1^m(yL_3^{-1}(x)) + \\
&Tr_1^m(yL_1^{-1}(x))Tr_1^m(\alpha y(\tilde{\rho}(L_2^{-1}(x)) + L_3^{-1}(x))) + \\
&Tr_1^m(yL_2^{-1}(x))Tr_1^m(\alpha y(\tilde{\rho}(L_1^{-1}(x)) + L_3^{-1}(x))) + \\
&Tr_1^m(yL_3^{-1}(x))Tr_1^m(\alpha y(\tilde{\rho}(L_1^{-1}(x)) + L_2^{-1}(x))) + \\
&Tr_1^m(\alpha y\tilde{\rho}(L_1^{-1}(x)))Tr_2^m(\alpha y\tilde{\rho}(L_2^{-1}(x))) + \\
&Tr_1^m(\alpha y\tilde{\rho}(L_1^{-1}(x)))Tr_3^m(\alpha y\tilde{\rho}(L_3^{-1}(x))) + \\
&Tr_2^m(\alpha y\tilde{\rho}(L_2^{-1}(x)))Tr_3^m(\alpha y\tilde{\rho}(L_3^{-1}(x))). \quad (13)
\end{aligned}
$$

Example 3. Consider the special case of Theorem 3: $m = 4$ and $t = 2$. Let ρ be a primitive element of $\mathbb{F}_{16}$ satisfying the equation $\rho^4 + \rho + 1 = 0$ and ω be a primitive element of $\mathbb{F}_4$ satisfying the equation $\omega^2 + \omega + 1 = 0$. Set $f : \mathbb{F}_{16} \to \mathbb{F}_4$, $g : \mathbb{F}_4 \to \mathbb{F}_4$, $a \in \mathbb{F}_4^*$ and $\alpha \in \mathbb{F}_{16}^*$ as follows: $f(x) = x + x^4$, $g(x) = x$, $a = \omega$, and $\alpha = \rho$. Thus α is an a-linear translator of f with respect to $\mathbb{F}_4$. Also, for $i \in \{1, 2, 3\}$, set $L_i : \mathbb{F}_{16} \to \mathbb{F}_{16}$ as $L_1(x) = \rho^{14}x^8 + \rho^{13}x^4 + \rho^5 x^2 + \rho^{13}x$, $L_2(x) = \rho^{13}x^8 + \rho^4 x^4 + \rho^7 x^2 + \rho^{10}x$, and $L_3(x) = \rho^5 x^8 + \rho^2 x^4 + \rho^3 x^2 + \rho^8 x$. (This choice of L_1, L_2, L_3 makes them to satisfy $\mathcal{A}_4$.) With this setup, for $i \in \{1, 2, 3\}$, $\phi_i : \mathbb{F}_{16} \to \mathbb{F}_{16}$ defined as $\phi_1(x) = \rho^{14}x^8 + \rho^8 x^4 + \rho^5 x^2 + \rho^8 x$, $\phi_2(x) = \rho^{13}x^8 + \rho^7 x^4 + \rho^7 x^2 + \rho^{12}x$, and $\phi_3(x) = \rho^5 x^8 + \rho^6 x^4 + \rho^3 x^2 + \rho^{13}x$ should satisfy $\mathcal{A}_4$, by Proposition 5. Hence they can be used to define a bent function H from $\mathbb{F}_{16} \times \mathbb{F}_{16}$ to $\mathbb{F}_2$ with the construction defined in Theorem 3.

Proposition 6. *Let $m, t \in \mathbb{Z}^+$ be such that $t < m$ and t divides m. Let both f and h be mappings from $\mathbb{F}_{2^m}$ to $\mathbb{F}_{2^t}$. Let $\alpha_1, \alpha_2 \in \mathbb{F}_{2^m}^*$ and $a_1, a_2, b_1, b_2 \in \mathbb{F}_{2^t}^*$ be such that*

i. α_1 is an a_1-linear translator of f and b_1-linear translator of h with respect to $\mathbb{F}_{2^t}$,
ii. α_2 is an a_2-linear translator of f and b_2-linear translator of h with respect to $\mathbb{F}_{2^t}$.

Let g_1 and g_2 be both polynomials over $\mathbb{F}_{2^t}$. Assume $\psi : \mathbb{F}_{2^t} \times \mathbb{F}_{2^t} \to \mathbb{F}_{2^t} \times \mathbb{F}_{2^t}$ defined as

$$\psi(x, y) := \big(x + a_1 g_1(x) + a_2 g_2(y), \; y + b_1 g_1(x) + b_2 g_2(y)\big)$$

is a permutation. Set $\phi : \mathbb{F}_{2^m} \to \mathbb{F}_{2^m}$ as

$$\phi(x) := x + \alpha_1 g_1(f(x)) + \alpha_2 g_2(h(x)). \tag{14}$$

Then ϕ is a permutation polynomial of $\mathbb{F}_{2^m}$ and

$$\phi^{-1}(x) = x + \alpha_1(g_1 \circ \pi_1 \circ \psi^{-1})(f(x), h(x))$$
$$+ \alpha_2(g_2 \circ \pi_2 \circ \psi^{-1})(f(x), h(x)), \tag{15}$$

where $\pi_1 : \mathbb{F}_{2^t} \times \mathbb{F}_{2^t} \to \mathbb{F}_{2^t}$ and $\pi_2 : \mathbb{F}_{2^t} \times \mathbb{F}_{2^t} \to \mathbb{F}_{2^t}$ are projections defined as

$$\pi_1(x, y) := x \text{ and } \pi_2(x, y) := y.$$

Proof. Let $\phi(x) = y$ for some $y \in \mathbb{F}_{2^m}$. By definition,

$$x + \alpha_1 g_1(f(x)) + \alpha_2 g_2(h(x)) = y. \tag{16}$$

Taking f of both sides and using the fact that α_1 is an a_1-linear translator of f and α_2 is an a_2-linear translator of f, we obtain

$$f(x) + a_1 g_1(f(x)) + a_2 g_2(h(x)) = f(y). \tag{17}$$

Similarly, taking h of both sides in the Eq. (16) and using the fact that α_1 is an b_1-linear translator of h and α_2 is an b_2-linear translator of h, we also get

$$h(x) + b_1 g_1(f(x)) + b_2 g_2(h(x)) = h(y). \tag{18}$$

Combining Eqs. (17) and (18) gives us

$$\psi\big(f(x), h(x)\big) = \big(f(y), h(y)\big).$$

Since ψ is a permutation, ψ^{-1} exists and we have

$$\big(f(x), h(x)\big) = \psi^{-1}\big(f(y), h(y)\big).$$

By definition of π_1 and π_2, the Eq. (16) gives us

$$x = y + \alpha_1 g_1(\pi_1 \circ \psi^{-1}(f(y), h(y))) + \alpha_2 g_2(\pi_2 \circ \psi^{-1}(f(y), h(y))),$$

proving that ϕ is a permutation whose compositional inverse ϕ^{-1} is given by the above expression. $\qquad\square$

Now, by a similar approach, one can obtain a new family of bent functions as follows.

Theorem 4. *Let m, $t \in \mathbb{Z}^+$ such that $t < m$ and t divides m. Let both f and h be mappings from $\mathbb{F}_{2^m}$ to $\mathbb{F}_{2^t}$. For $i \in \{1, 2, 3\}$, let $\alpha_{1i}, \alpha_{2i} \in \mathbb{F}^*_{2^m}$ and $a_1, a_2, b_1, b_2 \in \mathbb{F}^*_{2^t}$ be such that*

i. α_{1i} is an a_1-linear translator of f and b_1-linear translator of h with respect to $\mathbb{F}_{2^t}$,

ii. α_{2i} is an a_2-linear translator of f and b_2-linear translator of h with respect to $\mathbb{F}_{2^t}$.

Let g_1 and g_2 be both polynomials over $\mathbb{F}_{2^t}$. Assume $\psi : \mathbb{F}_{2^t} \times \mathbb{F}_{2^t} \to \mathbb{F}_{2^t} \times \mathbb{F}_{2^t}$ defined as

$$\psi(x, y) := (x + a_1 g_1(x) + a_2 g_2(y), y + b_1 g_1(x) + b_2 g_2(y))$$

is a permutation. Then

$$H(x, y) = Tr_1^m(xy) + Tr_1^m(x\alpha_{11}g_1(f(y)))Tr_1^m(x\alpha_{12}g_1(f(y)))$$
$$+Tr_1^m(x\alpha_{11}g_1(f(y)))Tr_1^m(x\alpha_{13}g_1(f(y))) + Tr_1^m(x\alpha_{11}g_1(f(y)))Tr_1^m(x\alpha_{22}g_2(h(y)))$$
$$+Tr_1^m(x\alpha_{11}g_1(f(y)))Tr_1^m(x\alpha_{23}g_2(h(y))) + Tr_1^m(x\alpha_{12}g_1(f(y)))Tr_1^m(x\alpha_{13}g_1(f(y)))$$
$$+Tr_1^m(x\alpha_{12}g_1(f(y)))Tr_1^m(x\alpha_{21}g_2(h(y))) + Tr_1^m(x\alpha_{12}g_1(f(y)))Tr_1^m(x\alpha_{23}g_2(h(y)))$$
$$+Tr_1^m(x\alpha_{13}g_1(f(y)))Tr_1^m(x\alpha_{21}g_2(h(y))) + Tr_1^m(x\alpha_{13}g_1(f(y)))Tr_1^m(x\alpha_{22}g_2(h(y)))$$
$$+Tr_1^m(x\alpha_{21}g_2(h(y)))Tr_1^m(x\alpha_{22}g_2(h(y))) + Tr_1^m(x\alpha_{21}g_2(h(y)))Tr_1^m(x\alpha_{23}g_2(h(y)))$$
$$+Tr_1^m(x\alpha_{22}g_2(h(y)))Tr_1^m(x\alpha_{23}g_2(h(y))) \qquad (19)$$

is bent.

Proof. For $i \in \{1, 2, 3\}$, define $\phi_i : \mathbb{F}_{2^m} \to \mathbb{F}_{2^m}$ as follows:

$$\phi_i(y) = y + \alpha_{1i}g_1(f(y)) + \alpha_{2i}g_2(h(y)).$$

Then by Proposition 6, ϕ_1, ϕ_2 and ϕ_3 are all permutations. Define $\phi := \phi_1 + \phi_2 + \phi_3$. Then easy to verify that ϕ is also a permutation. Moreover; by Proposition 1 and Proposition 6, one can prove that $\phi^{-1} = \phi_1^{-1} + \phi_2^{-1} + \phi_3^{-1}$. So, these three permutations satisfy (A_m). By Proposition 2, we conclude that

$$H(x, y) = Tr_1^m(x\phi_1(y))Tr_1^m(x\phi_2(y)) + Tr_1^m(x\phi_1(y))Tr_1^m(x\phi_3(y)) +$$
$$Tr_1^m(x\phi_2(y))Tr_1^m(x\phi_3(y))$$

is bent. The result follows after several computations.   □

Moreover, we shall see in the following examples that it is not hard to find functions g_1 and g_2 leading to the construction of the permutation ψ of $\mathbb{F}_{2^t} \times \mathbb{F}_{2^t}$ defined as in Theorem 4.

Example 4. For $a_1 = a_2 = b_1 = b_2$, let $g_1(x) = g_2(x) = x^2$. So $\psi(x, y) = (x + cx^2 + cy^2, y + cx^2 + cy^2)$, for some $c \in \mathbb{F}_{2^t}$. Assume that

$$(x_1 + cx_1{}^2 + cy_1{}^2, y_1 + cx_1{}^2 + cy_1{}^2) = (x_2 + cx_2{}^2 + cy_2{}^2, y_2 + cx_2{}^2 + cy_2{}^2),$$

for some $(x_1, y_1), (x_2, y_2) \in \mathbb{F}_{2^t} \times \mathbb{F}_{2^t}$. Then we obtain the equation $x_1 + y_1 = x_2 + y_2$, so $x_1{}^2 + y_1{}^2 = x_2{}^2 + y_2{}^2$ which gives the result $(x_1, y_1) = (x_2, y_2)$.

Example 5. For $a_1 = a_2 = b_1 = b_2 = 1$, let $g_1(x) = x^2 + cx$ and $g_2(x) = x^2 + dx$, where c and d are two elements of $\mathbb{F}_{2^t}$ satisfying $c + d \neq 1$. Then, the ψ function defined in Theorem 4 becomes

$$\psi(x, y) = ((c + 1)x + x^2 + dy + y^2, cx + x^2 + (d + 1)y + y^2),$$

and by a similar way used in the previous example, it easy to conclude that ψ is a permutation.

3   Our New Families Are Not in the Maiorana-McFarland Class

In the world of bent functions, within all primary constructions, the Maiorana-McFarland gives the widest class by far. Therefore, to prove that several bent functions constructed by using the idea presented in the previous section will not be in the Maiorana-McFarland Class is quite important. Section 3.4 is reserved mainly for this purpose.

First, recall that the first derivative of a Boolean function f in the direction of $a \in \mathbb{F}_{2^n}$ is defined as $D_a f(x) = f(x) + f(x + a)$. A publicly know fact is stating that a bent function $H : \mathbb{F}_{2^m} \times \mathbb{F}_{2^m} \to \mathbb{F}_2$ is in the class of Maiorana-McFarland if and only if $D_{(b,0)} D_{(c,0)} H(x, y) = 0$, for all $b, c \in \mathbb{F}_{2^m}^*$. Let's start with the rough sketch of the proof of this fact for the curious reader: let $F : \mathbb{F}_{2^m} \times \mathbb{F}_{2^m} \to \mathbb{F}_2$ be a Boolean function. Then F is in the Maiorana-McFarland class if and only if

$$F(x, y) = Tr_1^m(x, \pi(y)) + h(y)$$

for some $h : \mathbb{F}_{2^m} \to \mathbb{F}_2$ and $\pi : \mathbb{F}_{2^m} \to \mathbb{F}_{2^m}$. Fix $y \in \mathbb{F}_{2^m}$, and let $x_1, x_2 \in \mathbb{F}_{2^m}$. Then

$$F(x_1 + x_2, y) = Tr_1^m(x_1, \pi(y)) + Tr_1^m(x_2, \pi(y)) + h(y),$$

which is equal to

$$F(x_1, y) + F(x_2, y) + h(y).$$

Therefore $F \in \mathcal{M}$ if and only if it is affine in x, for all y. Clearly, any affine function $G : \mathbb{F}_{2^m} \to \mathbb{F}_2$ can be defined by the property $D_b D_c G(x)|_{x=0} = 0$, $\forall b, c \in \mathbb{F}_{2^m}$. Functions $F : \mathbb{F}_{2^m} \times \mathbb{F}_{2^m} \to \mathbb{F}_2$ belonging to the class $\mathcal{M}$ can be therefore defined by the property

$$D_{(b,0)} D_{(c,0)}|_{x=0} F(x, y) = 0, \ \forall y \in \mathbb{F}_{2^m}.$$

Consider one of the bent functions $H : \mathbb{F}_{2^m} \times \mathbb{F}_{2^m} \to \mathbb{F}_2$ constructed in the previous section. All of these functions are of the following form:

$$H(x,y) = Tr_1^m\big(x\phi_1(y)\big)Tr_1^m\big(x\phi_2(y)\big) + Tr_1^m\big(x\phi_1(y)\big)Tr_1^m\big(x\phi_3(y)\big) + \\ Tr_1^m\big(x\phi_2(y)\big)Tr_1^m\big(x\phi_3(y)\big).$$

Taking the first derivative of H will result in

$$\begin{aligned}
D_{(c,0)}H(x,y) = {} & Tr_1^m\big(x\phi_1(y)\big)Tr_1^m\big(c\phi_2(y)\big) + Tr_1^m\big(c\phi_1(y)\big)Tr_1^m\big(x\phi_2(y)\big) \\
& + Tr_1^m\big(c\phi_1(y)\big)Tr_1^m\big(c\phi_2(y)\big) + Tr_1^m\big(x\phi_1(y)\big)Tr_1^m\big(c\phi_3(y)\big) \\
& + Tr_1^m\big(c\phi_1(y)\big)Tr_1^m\big(x\phi_3(y)\big) + Tr_1^m\big(c\phi_1(y)\big)Tr_1^m\big(c\phi_3(y)\big) \\
& + Tr_1^m\big(x\phi_2(y)\big)Tr_1^m\big(c\phi_3(y)\big) + Tr_1^m\big(c\phi_2(y)\big)Tr_1^m\big(x\phi_3(y)\big) \\
& \hphantom{{}+{}} + Tr_1^m\big(c\phi_2(y)\big)Tr_1^m\big(c\phi_3(y)\big),
\end{aligned}$$

and the second derivative of H at $x = 0$ will be equal to

$$\begin{aligned}
D_{(b,0)}D_{(c,0)}H(x,y) = {} & Tr_1^m\big(b\phi_1(y)\big)Tr_1^m\big(c\phi_2(y)\big) + Tr_1^m\big(c\phi_1(y)\big)Tr_1^m\big(b\phi_2(y)\big) \\
& + Tr_1^m\big(b\phi_1(y)\big)Tr_1^m\big(c\phi_3(y)\big) + Tr_1^m\big(c\phi_1(y)\big)Tr_1^m\big(b\phi_3(y)\big) \\
& + Tr_1^m\big(b\phi_2(y)\big)Tr_1^m\big(c\phi_3(y)\big) + Tr_1^m\big(c\phi_2(y)\big)Tr_1^m\big(b\phi_3(y)\big).
\end{aligned}$$

Therefore, in order to prove that $H \notin \mathcal{M}$, it is enough to find two elements $b, c \in \mathbb{F}_{2^m}^*$ satisfying

$$\begin{aligned}
& Tr_1^m\big(b\phi_1(y)\big)Tr_1^m\big(c\phi_2(y)\big) + Tr_1^m\big(c\phi_1(y)\big)Tr_1^m\big(b\phi_2(y)\big) \\
& + Tr_1^m\big(b\phi_1(y)\big)Tr_1^m\big(c\phi_3(y)\big) + Tr_1^m\big(c\phi_1(y)\big)Tr_1^m\big(b\phi_3(y)\big) \\
& + Tr_1^m\big(b\phi_2(y)\big)Tr_1^m\big(c\phi_3(y)\big) + Tr_1^m\big(c\phi_2(y)\big)Tr_1^m\big(b\phi_3(y)\big) \neq 0. \quad (20)
\end{aligned}$$

3.1   Families Obtained by the Construction Defined in Theorem 1

Recall the bent function H obtained in Example 1. A brief search on (b,c)-tuples will show that there exist various choices for them satisfying the Eq. (20), and $(b,c) = (\xi^4, \xi^2)$ is just one of them.

3.2   Families Obtained by the Construction Defined in Theorem 2

Consider the bent function H obtained in Example 2. Within a lot of (b,c)-tuples satisfying the Eq. (20), one can take $(b,c) = (\rho, \rho^3)$.

3.3   Families Obtained by the Construction Defined in Theorem 3

Recall the bent function H obtained in Example 3. One can see that the Eq. (20) is satisfied for many (b,c)-tuples, where $y = \rho^3$; and $(b, c) = (\rho^7, \rho^5)$ is just one of them.

3.4 Families Obtained by the Construction Defined in Theorem 4

Consider the special case of 4: $m = 4$ and $t = 2$. Let ρ be a primitive element of $\mathbb{F}_{16}$ satisfying the equation $\rho^4 + \rho + 1 = 0$ and ω be a primitive element of $\mathbb{F}_4$ satisfying the equation $\omega^2 + \omega + 1 = 0$. For $i \in \{1, 2, 3\}$ and $j \in \{1, 2\}$, set $f, h : \mathbb{F}_{16} \rightarrow \mathbb{F}_4$, $g_j : \mathbb{F}_4 \rightarrow \mathbb{F}_4$, a_j, $b_j \in \mathbb{F}_4^*$ and $\alpha_{ij} \in \mathbb{F}_{16}^*$ as follows: $f(x) = h(x) = x + x^4$, $g_1(x) = x^2 + \omega x$, $g_2(x) = x^2 + x$, $a_1 = a_2 = b_1 = b_2 = 1$, $\alpha_{11} = \rho$, $\alpha_{12} = \alpha_{31} = \rho^2$, $\alpha_{21} = \alpha_{32} = \rho^4$, and $\alpha_{22} = \rho^8$. Thus all of these α_{ij}'s will be 1-linear translators of f with respect to $\mathbb{F}_4$. With this setup, $\phi_i : \mathbb{F}_{16} \rightarrow \mathbb{F}_{16}$ defined as $\phi_1(x) = x + \alpha_{11}g_1(x + x^4) + \alpha_{12}g_2(x + x^4)$, $\phi_2(x) = x + \alpha_{21}g_1(x + x^4) + \alpha_{22}g_2(x + x^4)$ and $\phi_3(x) = x + \alpha_{31}g_1(x + x^4) + \alpha_{32}g_2(x + x^4)$ should satisfy $\mathcal{A}_4$, and so they can be used to define a bent function H from $\mathbb{F}_{16} \times \mathbb{F}_{16}$ to $\mathbb{F}_2$, by using the construction given in Theorem 4. For this bent function H, the choice $(b, c) = (\rho, \rho^3)$ satisfies the Eq. (20) within many (b, c)-choices satisfying it.

4 Conclusion

In this paper, we present new infinite families of permutations. We show that those families have the nice property that one can select three elements among them which allow to use the construction proposed in [7,8]. To this end, we have exploited the linear dependence of the expressions of those permutations on some coefficients to obtain several new infinite families of bent functions together with their dual functions. Furthermore, we show that our families are not in the Maiorana-McFarland class in general. It remains obviously to check the affine inequivalence of the proposed bent functions to the other known constructions.

References

1. Canteaut, A., Naya-Plasencia, M.: Structural weakness of mappings with a low differential uniformity. In: Conference on Finite Fields and Applications, Dublin, 13–17 July 2009
2. Carlet, C.: On bent and highly nonlinear balanced/resilient functions and their algebraic immunities. In: Fossorier, M.P.C., Imai, H., Lin, S., Poli, A. (eds.) AAECC 2006. LNCS, vol. 3857, pp. 1–28. Springer, Heidelberg (2006). doi:10.1007/11617983_1
3. Carlet, C., Mesnager, S.: Four decades of research on bent functions. J. Des. Codes Crypt. **78**(1), 5–50 (2016)
4. Dillon, J.: Elementary hadamard difference sets. PhD Thesis, University of Maryland (1974)
5. Koçak, N., Mesnager, S., Özbudak, F.: Bent and semi-bent functions via linear translators. In: Groth, J. (ed.) IMACC 2015. LNCS, vol. 9496, pp. 205–224. Springer, Cham (2015). doi:10.1007/978-3-319-27239-9_13
6. Kyureghyan, G.M.: Constructing permutations of finite fields via linear translators. J. Comb. Theory Ser. A **118**, 1052–1061 (2011)
7. Mesnager, S.: Several new infinite families of bent functions and their duals. IEEE Trans. Inf. Theory **60**(7), 4397–4407 (2014)

8. Mesnager, S.: Further constructions of infinite families of bent functions from new permutations and their duals. J. Crypt. Commun. (CCDS) **8**(2), 1–18 (2016). Springer
9. Mesnager, S.: A Note on Constructions of Bent Functions from Involutions. IACR, Cryptology ePrint Archive, 982 (extended version of ISIT 2016) (2015)
10. Mesnager, S.: On constructions of bent functions from involutions. In Proceedings of 2015 IEEE International Symposium on Information Theory, ISIT (2016)
11. Mesnager, S.: Bent Functions: Fundamentals and Results. Springer, New York (2016)
12. Mesnager, S., Cohen, G., Madore, D.: On existence (based on an arithmetical problem) and constructions of bent functions. In: Groth, J. (ed.) IMACC 2015. LNCS, vol. 9496, pp. 3–19. Springer, Cham (2015). doi:10.1007/978-3-319-27239-9_1
13. Rothaus, O.S.: On "bent" functions. J. Comb. Theory Ser. A **20**, 300–305 (1976)
14. Wang, Q., Johansson, T., Kan, H.: Some results on fast algebraic attacks and higher-order non-linearities. IET Inf. Secur. **6**, 41–46 (2012)

Bent Functions in $\mathcal{C}$ and $\mathcal{D}$ Outside the Completed Maiorana-McFarland Class

F. Zhang[1], E. Pasalic[2]($\boxtimes$), N. Cepak[3], and Y. Wei[4]

[1] School of Computer Science and Technology,
China University of Mining and Technology,
Xuzhou 221116, Jiangsu, People's Republic of China
zhfl203@cumt.edu.cn
[2] FAMNIT and IAM, University of Primorska, Koper, Slovenia
enes.pasalic6@gmail.com
[3] FAMNIT, University of Primorska, Koper, Slovenia
nastja.cepak@gmail.com
[4] Guilin University of Electronic Technology, Guilin, People's Republic of China
walker_wei@msn.com

Abstract. Two new classes of bent functions derived from the Maiorana-McFarland ($\mathcal{M}$) class, so-called $\mathcal{C}$ and $\mathcal{D}$, were introduced by Carlet [2] two decades ago. However, apart from the subclass $\mathcal{D}_0$, some explicit construction methods for these functions were not provided in [2]. Assuming the possibility of specifying a bent function f that belongs to one of these two classes (apart from $\mathcal{D}_0$), the most important issue is then to determine whether f is still contained in the known primary classes or lies outside their completed versions. In this article we partially solve this question by providing sufficient conditions on the permutation and related characteristic function (used to define f in these classes) so that f is provably outside the completed $\mathcal{M}$ class. To give some existence results, we employ recent results in [12] where some instances of bent functions in $\mathcal{C}$ were identified by providing specific permutations and related characteristic functions. More precisely, using our sufficient conditions that apply to both $\mathcal{C}$ and $\mathcal{D}$, it is shown that these identified classes of $\mathcal{C}$ functions described in [12] do not belong to the completed $\mathcal{M}$ class, whereas the question (which is more difficult) whether these functions are also outside the completed partial spread class remains open. We also propose some generic methods for specifying bent functions in $\mathcal{D}$ outside the completed Maiorana-McFarland class.

Keywords: Bent functions $\cdot$ $\mathcal{C}$ and $\mathcal{D}$ class $\cdot$ Maiorana-McFarland class

1 Introduction

Bent functions, as a particular subclass of Boolean functions, have many different and regular characterizations thus giving many interesting connections to other combinatorial objects such as difference sets, error correcting codes, sequences,

© Springer International Publishing AG 2017
S. El Hajji et al. (Eds.): C2SI 2017, LNCS 10194, pp. 298–313, 2017.
DOI: 10.1007/978-3-319-55589-8_20

and in general having a wide range of cryptographic applications. These discrete mathematical objects were introduced by Rothaus [13] and then later elaborated by Dillon [8] and McFarland [11]. The most significant impact of the latter two works is that essentially generic primary classes of bent functions could be deduced which is today known as partial spread ($\mathcal{PS}$) class due to Dillon and as Maiorana-McFarland ($\mathcal{M}$) class stemming from [11]. In 1993, two additional classes of bent functions derived by a suitable modification of bent functions in the $\mathcal{M}$ class, named as $\mathcal{C}$ and $\mathcal{D}$, were proposed by Carlet [2] and these classes fall under the term secondary constructions. Another generic and primary class $\mathcal{H}$ was proposed by Dobbertin [9] which includes both $\mathcal{M}$ and $\mathcal{PS}$. Both these primary classes and the secondary classes of Carlet greatly contribute to enumeration and classification of bent functions even though a complete solution to these problems seems to be elusive. In this context, it is worthy to emphasize the importance and the need of better understanding of a general construction method of bent functions due to Rothaus [13] because it is unclear to which primary class these functions belong to and after all it might be the case that the construction of Rothaus gives rise to provably new class of bent functions assuming that the initial bent functions are suitably chosen. For further details regarding the construction of Rothaus the interested reader is referred to [13]. For a survey of many other secondary constructions of bent functions the reader is referred to [3] whereas an exhaustive survey on bent functions can be found in [5].

The secondary classes of bent functions $\mathcal{C}$ and $\mathcal{D}$ are derived from the $\mathcal{M}$ class (see (1), (2) and property (C) below) by adding the indicator functions of suitably chosen vector subspaces to the functions in the $\mathcal{M}$ class. Nevertheless, apart from an explicit subclass denoted by $\mathcal{D}_0$, the bent conditions in terms of the selection of a vector subspace L and permutation π (used to define the initial function $f(x, y) = x \cdot \pi(y)$ in $\mathcal{M}$, where $x, y \in \mathbb{F}_2^n$) are rather hard to satisfy. This problem was recently addressed in [12] and the hardness of satisfying the property (C) (thus identifying a suitable permutation and related vector subspace) was confirmed true since for some classes of permutation polynomials there are no suitable linear subspaces of certain dimension for which the modification of $f \in \mathcal{M}$ would give a bent function $f^* \in \mathcal{C}$. On the other hand, for some other classes of permutations and associated linear subspaces of the same dimension it could be verified that indeed we get a bent function $f^* \in \mathcal{C}$. Thus, given the existence of bent functions $f^* \in \mathcal{C}$ the most fundamental issue is to determine whether these functions are essentially contained in the known primary classes (which gives nothing new in that case) or these functions potentially lie outside the known classes. It should be remarked that certain choices of the indicator functions used to define f^* from $f \in \mathcal{M}$ are provably non-efficient in this context, thus giving rise to bent functions f^* within the class $\mathcal{M}$.

In this article we provide sufficient conditions on the choice of the permutation π and the corresponding linear subspace so that a bent function f^* that belongs either to $\mathcal{C}$ or $\mathcal{D}$ is outside the completed $\mathcal{M}$ class. This is the first step towards a better understanding of classification of bent functions in these secondary classes which also opens up for further investigation concerning a more refined classification in terms of determining whether these functions are also outside the completed $\mathcal{PS}$ and $\mathcal{H}$ class (which is intrinsically more difficult due to the absence of efficient indicators for these classes). The derived sufficient conditions are relatively simple and they roughly speaking correspond to the existence of permutations without linear structures. Then, using the sufficient conditions that the bent functions in $\mathcal{C}$ or $\mathcal{D}$ do not belong to the completed $\mathcal{M}$ class we could show that some instances of bent functions in $\mathcal{C}$ identified in [12] are indeed outside the completed $\mathcal{M}$ class, thus answering positively the classification issue raised in [12]. Furthermore, some generic methods for specifying suitable monomial permutations are given for the purpose of generating bent functions in $\mathcal{D}$ outside the completed $\mathcal{M}$ class.

The rest of this article is organized as follows. In Sect. 2 we provide some basic definitions related to Boolean (and in particular bent) functions along with the exact definitions of $\mathcal{C}$ and $\mathcal{D}$ classes. Sufficient conditions that bent functions in $\mathcal{C}$ or $\mathcal{D}$ do not belong to the completed $\mathcal{M}$ class are given in Sect. 3. In Sect. 4, we demonstrate that some instances of bent functions in $\mathcal{C}$, identified in [12], do not belong to the completed $\mathcal{M}$ class. Sufficient conditions ensuring that bent functions in $\mathcal{D}$ are outside the completed $\mathcal{M}$ class are shown to be relatively easily satisfied in Sect. 5. Some concluding remarks are given in Sect. 6.

2   Preliminaries

Let $\mathbb{F}_2$ denote the binary field and let the n-dimensional vector space spanned over $\mathbb{F}_2$ be denoted by $\mathbb{F}_2^n = \{x = (x_1, \ldots, x_n) : x_i \in \mathbb{F}_2, \text{ for } i = 1, \ldots, n\}$. The extended Galois field of degree n over $\mathbb{F}_2$ is denoted by $\mathbb{F}_{2^n}$. Any function from $\mathbb{F}_2^n$ to $\mathbb{F}_2$ (or, equivalently from $\mathbb{F}_{2^n}$ to $\mathbb{F}_2$) is called a *Boolean function* on n variables and the set of all Boolean functions on n variables is denoted by $\mathfrak{B}_n$.

For a detailed study of Boolean functions we refer to Carlet [3,4], and Cusick and Stănică [7]. For the convenience of the reader, we recall some basic notions below. For any $x \in \mathbb{F}_2^n$, the (Hamming) *weight* of x is defined as the number of nonzero entries of x. The *algebraic normal form* (ANF) of a Boolean function $f \in \mathfrak{B}_n$ is

$$f(x_1, \ldots, x_n) = \sum_{a=(a_1,\ldots,a_n)\in\mathbb{F}_2^n} \mu_a x_1^{a_1} \cdots x_n^{a_n},$$

where $\mu_a \in \mathbb{F}_2$, for all $a \in \mathbb{F}_2^n$. The *algebraic degree* of f is $\deg(f) = \max_{a\in\mathbb{F}_2^n}\{\text{wt}(a) : \mu_a \neq 0\}$. The standard inner (dot) product of two vectors $u, x \in \mathbb{F}_2^n$ is defined as $u \cdot x := \sum_{i=1}^{n} u_i x_i$, for all. Once the basis of $\mathbb{F}_{2^n}$ over $\mathbb{F}_2$ is fixed we isomorphically identify $\mathbb{F}_2^n$ with $\mathbb{F}_{2^n}$.

We denote by $Tr(\cdot)$ the absolute trace on $\mathbb{F}_{2^n}$ and by $T_k^n(\cdot)$ the trace function from $\mathbb{F}_{2^n}$ to $\mathbb{F}_{2^k}$, where k divides n:

$$T_k^n(\beta) = \beta + \beta^{p^k} + \cdots + \beta^{p^{(n/k-1)k}}.$$

The *Walsh-Hadamard transform* of $f \in \mathcal{B}_n$ at $u \in \mathbb{F}_2^n$ is

$$W_f(u) = \sum_{x \in \mathbb{F}_2^n} (-1)^{f(x)}(-1)^{u \cdot x},$$

and the multiset $[W_f(u) : u \in \mathbb{F}_2^n]$ is said to be the *Walsh-Hadamard spectrum* of f. The *derivative* of $f \in \mathcal{B}_n$ at $a \in \mathbb{F}_2^n$, denoted by $D_a f$, is a Boolean function defined by

$$D_a f(x) = f(x+a) + f(x), \text{ for all } x \in \mathbb{F}_2^n.$$

Higher order derivatives of a Boolean function refer to k-dimensional vector subspaces, where $k > 1$. Suppose $\{a_1, a_2, \ldots, a_k\}$ is a basis of a k-dimensional subspace V of $\mathbb{F}_2^n$ (we write $\dim(V) = k$). The *k-th derivative* of f with respect to V, denoted by $D_V f$, is a Boolean function defined by

$$D_V f(x) = D_{a_k} D_{a_{k-1}} \ldots D_{a_1} f(x), \text{ for all } x \in \mathbb{F}_2^n.$$

It is to be noted that $D_V f$ is independent of the choice of the basis of V.

A Boolean function $f \in \mathcal{B}_n$, where n is an even positive integer, is said to be a *bent function* if $W_f(u) \in \{-2^{n/2}, 2^{n/2}\}$, for all $u \in \mathbb{F}_2^n$.

2.1 Bent Functions in $\mathcal{C}$ and $\mathcal{D}$

The Maiorana-McFarland class $\mathcal{M}$ is the set of m-variable ($m = 2n$) Boolean functions of the form

$$f(x, y) = x \cdot \pi(y) + g(y), \text{ for all } x, y \in \mathbb{F}_2^n,$$

where π is a permutation on $\mathbb{F}_2^n$, and g is an arbitrary Boolean function on $\mathbb{F}_2^n$. From this class Carlet derived the $\mathcal{C}$ class of bent functions that contains all functions of the form

$$f(x, y) = x \cdot \pi(y) + 1_{L^\perp}(x) \tag{1}$$

where L is any linear subspace of $\mathbb{F}_2^n$, $1_{L^\perp}$ is the indicator function of the space $L^\perp$, and π is any permutation on $\mathbb{F}_2^n$ such that:

(C) $\phi(a + L)$ is a flat (affine subspace), for all $a \in \mathbb{F}_2^n$, where $\phi := \pi^{-1}$.

The permutation ϕ and the subspace L are then said to satisfy property (C), for short (ϕ, L) *has property* (C).

Another class introduced by Carlet, called $\mathcal{D}$, is defined similarly as

$$f(x, y) = x \cdot \pi(y) + 1_{E_1}(x) 1_{E_2}(y) \tag{2}$$

with π a permutation on $\mathbb{F}_2^n$ and E_1, E_2 two linear subspaces of $\mathbb{F}_2^n$ such that $\pi(E_2) = E_1^\perp$.

Definition 1. *A class of bent functions* $\{f\} \in \mathcal{B}_n$ *is* complete *if it is globally invariant under the action of the general affine group (the group of all invertible matrices of size $n \times n$ over $\mathbb{F}_2$) and under the addition of affine functions. The* completed *class is the smallest possible class that properly includes the class under consideration.*

3 Sufficient Conditions for Functions in $\mathcal{C}$ and $\mathcal{D}$ to Be Outside $\mathcal{M}^{\#}$

A useful indicator for the purpose of establishing whether a given bent function belongs to the completed Maiorana-McFarland class ($\mathcal{M}^{\#}$) is given below.

Lemma 1 *[8, p. 102]. An m-variable bent function f, $m = 2n$, belongs to $\mathcal{M}^{\#}$ if and only if there exists an n-dimensional linear subspace V of $\mathbb{F}_2^m$ such that the second order derivatives*

$$D_\alpha D_\beta f(x) = f(x) \oplus f(x \oplus \alpha) \oplus f(x \oplus \beta) \oplus f(x \oplus \alpha \oplus \beta)$$

vanish for any $\alpha, \beta \in V$.

Using this criterion we firstly address the problem of deciding whether bent functions in $\mathcal{C}$ are outside the completed $\mathcal{M}$ class.

Theorem 1. *Let $m = 2n > 4$ be an even integer and let $f(x,y) = \pi(y) \cdot x \oplus 1_{L^{\perp}}(x)$, where L is any linear subspace of $\mathbb{F}_2^n$ and π is a permutation on $\mathbb{F}_2^n$ such that (π, L) has property (C). If π satisfies:*

1. $\dim(L) \geq 2$;
2. π has no nonzero linear structure;

then f does not belong to $\mathcal{M}^{\#}$.

Proof. Let $a^{(1)}, b^{(1)}, a^{(2)}, b^{(2)} \in \mathbb{F}_2^n$. We prove that f does not belong to $\mathcal{M}^{\#}$, by using Lemma 1. We need to show that there does not exist an n-dimensional subspace V such that

$$D_{(a^{(1)}, a^{(2)})} D_{(b^{(1)}, b^{(2)})} f = 0,$$

for any $(a^{(1)}, a^{(2)}), (b^{(1)}, b^{(2)}) \in V$.

The second derivative of f with respect to a and b can be written as,

$$D_{(a^{(1)}, a^{(2)})} D_{(b^{(1)}, b^{(2)})} f(x) = x \cdot (D_{a^{(2)}} D_{b^{(2)}} \pi(y)) \oplus a^{(1)} \cdot D_{b^{(2)}} \pi(y \oplus a^{(2)}) \quad (3)$$
$$\oplus b^{(1)} \cdot D_{a^{(2)}} \pi(y \oplus b^{(2)}) \oplus D_{a^{(1)}} D_{b^{(1)}} 1_{L^{\perp}}(x)$$

We denote the set $\{(x, 0_n) \mid x \in \mathbb{F}_2^n\}$ by Δ. We will distinguish two main cases depending on whether $V = \Delta$ or $V \neq \Delta$.

For $V = \Delta$, we can find two vectors $(a^{(1)}, 0_n), (b^{(1)}, 0_n) \in \Delta$ such that

$$D_{a^{(1)}} D_{b^{(1)}} 1_{L^{\perp}}(x) \neq 0$$

since $\dim(L) \geq 2$ (i.e., $\deg(1_{L^{\perp}}) \geq 2$). Further, we know

$$D_{(a^{(1)}, a^{(2)})} D_{(b^{(1)}, b^{(2)})} f(x) = D_{a^{(1)}} D_{b^{(1)}} 1_{L^{\perp}}(x) \neq 0.$$

Let now $V \neq \Delta$. We split the proof into three cases depending on the cardinality of $V \cap \Delta$. We set $V = \left\{ (v_1^{(1)}, v_2^{(1)}), (v_1^{(2)}, v_2^{(2)}), \dots, (v_1^{(2^n)}, v_2^{(2^n)}) \right\},$

1. For $|V \cap \Delta| = 1$, we have $v_2^{(i)} \neq v_2^{(j)}$ for any $i \neq j$. If there exist two vectors $v_2^{(i_1)}, v_2^{(j_1)}$ such that $v_2^{(i_1)} = v_2^{(j_1)}$, then $v_1^{(i_1)} = v_1^{(j_1)}$, (or $(v_1^{(i_1)} \oplus v_1^{(j_1)}, 0_n) \in V \cap \Delta$), that is, $(v_1^{(i_1)}, v_2^{(i_1)}) = (v_1^{(j_1)}, v_2^{(j_1)})$. Further, $|\{v_2^{(1)}, v_2^{(2)}, \ldots, v_2^{(2^n)}\}| = |V| = 2^n$, that is, $\{v_2^{(1)}, v_2^{(2)}, \ldots, v_2^{(2^n)}\} = \mathbb{F}_2^n$ (here, if $v_2^{(i_1)} = v_2^{(i_2)}$, they are called one element).

   Now, there are two cases to be considered.

   (a) If there exists one vector $\mathbf{v} = (v^{(1)}, v^{(2)}) \in V \backslash \{0_{2n}\}$ such that $v^{(1)} = 0_n$, we set $a = \mathbf{v}$. We know

   $$D_{a^{(1)}} 1_{L^\perp}(x) = 0.$$

   For the nonzero vector a, we have

   $$\deg(D_{a^{(2)}} \pi(y)) \geq 1$$

   since π has no nonzero linear structure (i.e., $\deg(\pi) \geq 2$). Further, since $\{v_2^{(1)}, v_2^{(2)}, \ldots, v_2^{(2^n)}\} = \mathbb{F}_2^n$, we are able to select $b \in V \backslash \{0_{2n}, a\}$ such that

   $$D_{a^{(2)}} D_{b^{(2)}} \pi(y) \neq 0_n.$$

   Thus, $D_{(a^{(1)}, a^{(2)})} D_{(b^{(1)}, b^{(2)})} f(x) = x \cdot (D_{a^{(2)}} D_{b^{(2)}} \pi(y)) \oplus b^{(1)} \cdot D_{a^{(2)}} \pi(y \oplus b^{(2)}) \neq 0$, since $D_{a^{(2)}} D_{b^{(2)}} \pi(y) \neq 0$ implies that $x \cdot (D_{a^{(2)}} D_{b^{(2)}} \pi(y))$ is not constant, i.e. depends on x.

   (b) If there does not exist a vector $\mathbf{v} = (v^{(1)}, v^{(2)}) \in V \backslash \{0_{2n}\}$ such that $v^{(1)} = 0_n$, then we have $|\{v_1^{(1)}, v_1^{(2)}, \ldots, v_1^{(2^n)}\}| = |V| = 2^n$ (that is, $\{v_1^{(1)}, v_1^{(2)}, \ldots, v_1^{(2^n)}\} = \mathbb{F}_2^n$) since V is a subspace and $|\{v_2^{(1)}, v_2^{(2)}, \ldots, v_2^{(2^n)}\}| = |V| = 2^n$. We set $a \in V \backslash \{0_{2n}\}$ such that $a^{(1)} \in L^\perp$. From the definition of indicator functions, we know

   $$D_{a^{(1)}} 1_{L^\perp}(x) = 0.$$

   Further, we have

   $$D_{a^{(1)}} D_{b^{(1)}} 1_{L^\perp}(x) = 0.$$

   Further, since $\{v_2^{(1)}, v_2^{(2)}, \ldots, v_2^{(2^n)}\} = \mathbb{F}_2^n$, we are able to select $b \in V \backslash \{0_{2n}, a\}$ such that

   $$D_{a^{(2)}} D_{b^{(2)}} \pi(y) \neq 0_n.$$

   Thus, $D_{(a^{(1)}, a^{(2)})} D_{(b^{(1)}, b^{(2)})} f(x) = x \cdot (D_{a^{(2)}} D_{b^{(2)}} \pi(y)) \oplus a^{(1)} \cdot D_{b^{(2)}} \pi(y \oplus a^{(2)}) \oplus b^{(1)} \cdot D_{a^{(2)}} \pi(y \oplus b^{(2)}) \neq 0$, since $D_{a^{(2)}} D_{b^{(2)}} \pi(y) \neq 0$ implies that $x \cdot (D_{a^{(2)}} D_{b^{(2)}} \pi(y))$ is not constant, i.e. depends on x.

   Hence, we have

   $$D_{(a^{(1)}, a^{(2)})} D_{(b^{(1)}, b^{(2)})} f(x) \neq 0$$

   for $|V \cap \Delta| = 1$.

2. For $|V \cap \Delta| \geq 2$, without loss of generality, let $(a^{(1)}, 0_n)(\neq 0_{2n}) \in V \cap \Delta$. Set $b \in V \backslash \{0_{2n}, a\}$, then $b^{(2)} \neq 0_n$. Thus,

$$D_a D_b f(x) = a^{(1)} \cdot D_{b^{(2)}} \pi(y) \oplus D_{a^{(1)}} D_{b^{(1)}} 1_{L^{\perp}}(x) \neq 0$$

since π has no nonzero linear structure.

Combining both cases $V = \Delta$ and $V \neq \Delta$ we deduce that f does not belong to $\mathcal{M}^{\#}$. □

A similar set of conditions on permutation π used in the definition of $\mathcal{D}$ class of bent functions can be deduced.

Theorem 2. *Let* $m = 2n > 6$ *be an even integer and let* $f(x, y) = \pi(y) \cdot x \oplus 1_{E_1}(x) 1_{E_2}(x)$, *where* π *is a permutation on* $\mathbb{F}_2^n$, *and* E_1, E_2 *are two linear subspaces of* $\mathbb{F}_2^n$ *such that* $\pi(E_2) = E_1^{\perp}$. *If* π *satisfies:*

1. $\dim(E_1) \geq 2$ *and* $\dim(E_2) \geq 2$;
2. π *has no nonzero linear structure;*
3. $\deg(\pi) \leq n - \dim(E_2)$,

then f *does not belong to* $\mathcal{M}^{\#}$.

The lengthy proof of Theorem 2 is given in the Appendix.

4   Some Examples of Bent Functions in $\mathcal{C}$ Outside $\mathcal{M}^{\#}$

In this section we apply the criterion derived in the previous section to those bent functions given in [12] that satisfy the property (C). Notice that the condition in Theorem 1 regards the condition imposed on $\pi(x)$ and not on $\phi(x) = \pi^{-1}(x)$ but this is of no relevance due the result of Charpin and Sarkar [6]. More precisely, it was shown that if F is a permutation then linear structures of F and F^{-1} are closely related and in particular the non-existence of linear structures for F implies the no-existence of linear structures for F^{-1}, see Lemma 2 in [6]. For convenience of the reader, we recall a few examples of bent functions satisfying the property (C), cf. [12].

Theorem 3 *[12]. Suppose* $\phi(x) = x^{2^r + 1}$, *for all* $x \in \mathbb{F}_{2^n}$, *where* $\gcd(r, n) = e$, n/e *is odd and* $\gcd(2^n - 1, 2^r + 1) = 1$.

(i) *Then* (ϕ, L) *(where* L *is a subspace of* $\dim(L) = 2$) *satisfies the (C) property if and only if* $L = \langle u, cu \rangle$ *where* $u \in \mathbb{F}_{2^n}^*$ *and* $1 \neq c \in \mathbb{F}_{2^e}^*$.
(ii) *We assume that* $e = \gcd(n, r) > 1$ *and* $L = \langle u_1, c_1 u_1, \ldots, c_{s-1} u_1 \rangle$, $\dim(L) = s$, $c_i \in \mathbb{F}_{2^e}^*$, $1 \leq i \leq s - 1$, $s \geq 2$, *and* $u_1 \in \mathbb{F}_{2^n}^*$. *Then* (ϕ, L) *satisfies the (C) property.*

The following example was also provided in [12], thus providing an infinite class of bent functions in $\mathcal{C}$ other than $\mathcal{D}_0$.

Example 1. *Let $n = 2p$ where p is any odd prime, $r = 2$ and $e = \gcd(n, r) = 2$. Since n/e is odd, it is known that $\gcd(2^r + 1, 2^n - 1) = 1$. Therefore $\phi(x) = x^{2^r+1}$ is a permutation on $\mathbb{F}_{2^n}$. Let ζ be a primitive element of $\mathbb{F}_{2^n}$. Therefore, $\lambda = \zeta^{\frac{2^n-1}{2^e-1}} = \zeta^{\frac{2^n-1}{3}}$ is a generator of $\mathbb{F}_{2^e}$. Suppose that the permutation $\pi(x) = \phi^{-1}(x) = x^\gamma$ where $\gamma(2^r + 1) \equiv 1 \pmod{2^n - 1}$. Given r and n, γ can be computed easily by the Euclidean algorithm. Consider the Maiorana-McFarland bent $f(x, y) = x \cdot \pi(y)$. According to Theorem 3 if we choose $L = \langle 1, \lambda \rangle$, then the function $f^*(x, y) = x \cdot \pi(y) + 1_{L^\perp}(x)$ is in $\mathcal{C}$. The bent function f^* can be explicitly written as*

$$
\begin{aligned}
f^*(x, y) &= Tr_1^n(xy^\gamma) + (Tr_1^n(x) + 1)(Tr_1^n(\lambda x) + 1) \\
&= Tr_1^n(xy^\gamma) + Tr_1^n(x)Tr_1^n(\lambda x) + Tr_1^n((1 + \lambda)x) + 1.
\end{aligned}
\tag{4}
$$

Using new tools presented in this article we can answer the question of whether the function f^* defined above is outside the completed $\mathcal{M}$ class.

Lemma 2. *For $r \neq 0$ the function f^* from Example 1 does not belong to the completed $\mathcal{M}$ class.*

Proof. Using Theorem 1 we need to prove that $\dim(L) \geq 2$ and that the permutation $\pi(x) = x^\gamma$ has no linear structures. Since $L = \langle 1, \lambda \rangle$, where λ is the generator of $\mathbb{F}_{2^e} = \mathbb{F}_{2^2}$, we have $\dim(L) = 2$. By Lemma 2 in [6], instead of considering $\pi(x)$ we show the non-existence of linear structures of $\phi(x) = x^{2^r+1}$.

Suppose the mapping $\phi(x)$ has a c-linear structure a, where $a, c \in \mathbb{F}_{2^n}^*$. Then

$$
(x + a)^{2^r+1} + x^{2^r+1} = c,
$$

which implies $x^{2^r} + a^{2^r-1}x + a^{2^r} + a^{-1}c = 0$, for every $x \in \mathbb{F}_{2^n}$. Taking $x = 0$ forces $a^{2^r} + a^{-1}c = 0$ and taking $x = 1$ forces $a^{2^r-1} = 1$. This leaves us with the equation $x^{2^r} + x = 0$ for every $x \in \mathbb{F}_{2^n}$, which implies $2^r \equiv 1 \mod (2^n - 1)$ and $r = 0$. It follows that for $r = 2$ the permutation π does not have linear structures and thus the function f^* from Example 1 does not belong to the completed $\mathcal{M}$ class. $\square$

Remark 1. *Note that when $r = 0$, the function $\phi(x) = x^{2^r+1} = x^2$ obviously has linear structures since it is a linear permutation, and is not covered by Theorem 3.*

Another class of so-called bilinear split permutations (considered originally in [1,10]) of the form

$$
\phi(x) = x(Tr_l^n(x) + ax),
\tag{5}
$$

where $n = kl$, $l > 1$, $a \in \mathbb{F}_{2^l} \backslash \mathbb{F}_2$ and $Tr_l^n(x) = \sum_{i=0}^{k-1} x^{2^{li}}$, was also analyzed in [12]. It was shown that when k is odd these permutations also give rise to bent functions satisfying (C).

Lemma 3. *The above defined function $\phi(x)$ has a linear structure if and only if $l = n$.*

Proof. Let b be a c-linear structure of $\phi(x) = x(Tr_l^n(x) + ax)$. Then

$$(x + b)(Tr_l^n(x + b) + a(x + b)) + x(Tr_l^n(x) + ax) = c$$
$$x(Tr_l^n(b) + ab) + b(Tr_l^n(x) + Tr_l^n(b) + ax + ab)) = c$$
$$xTr_l^n(b) + bTr_l^n(x) + bTr_l^n(b) + ab^2 = c$$
$$xTr_l^n(b) + bTr_l^n(x) + (bTr_l^n(b) + ab^2 + c) = 0,$$

for every $x \in \mathbb{F}_2^n$. Taking $x = 0$ forces $(bTr_l^n(b) + ab^2 + c) = 0$ and taking $x = 1$, since k is odd, implies that $Tr_l^n(b) = b$. We are left with the equation $Tr_l^n(x) = x$. This equation is valid for any $x \in \mathbb{F}_{2^n}$ if and only if $l = n$. □

Thus the bilinear permutations defined by (5) can be used in constructions of functions satisfying (C) and being outside the completed $\mathcal{M}$ class whenever we have a nontrivial factorization $n = kl$.

5   Bent Functions in $\mathcal{D}$ Outside $\mathcal{M}^*$

The set of sufficient conditions related to class $\mathcal{D}$ given in Theorem 2 is harder to satisfy than those related to class $\mathcal{C}$ so we have limited ourselves to the study of monomial permutations.

Proposition 1. *Let n be even. Then any non-linear monomial permutation $\pi(y) = y^d$, where $\deg(\pi) \leq n - 2$, satisfies the required conditions in Theorem 2 for the 2-dimensional vector subspace $E_2 = \langle \zeta^{\frac{2^n-1}{3}}, \zeta^{\frac{2(2^n-1)}{3}} \rangle$, where ζ is a primitive element of $\mathbb{F}_{2^n}$.*

Proof. Since n is even, $3 \mid 2^n - 1$ and furthermore E_2 is not only a vector subspace but also corresponds to a subfield $\{0, 1, \zeta^{\frac{2^n-1}{3}}, \zeta^{\frac{2(2^n-1)}{3}}\}$. This is because π is a monomial permutation and it must map every subfield to itself (multiplication being closed). Therefore, $\pi(E_2) = E_2 = E_1^\perp$. The permutation π is a non-linear monomial, therefore it does not have a linear structure. The condition $\deg(\pi) \leq n - \dim(E_2)$ is satisfied as well since $\deg(\pi) \leq n - 2$ and $\dim(E_2) = 2$. □

We illustrate this approach by providing an example for $n = 6$.

Example 2. *Let $n = 6$ and $d = 11$ (smaller d will be covered by Proposition 2 below). Since $(2^6 - 1, 11) = 1$ and the binary weight of 11 is 3, $\pi(x) = x^d$ is a cubic permutation. Using the programming package Magma, the vector space representation on $\mathbb{F}_2^6$ of the subspace $E_2 = \langle \zeta^{21}, \zeta^{42} \rangle$, where ζ is the generating element of the field $\mathbb{F}_{2^6}$, is :*

$$E_2 = \begin{cases} (0,0,0,0,0,0) \\ (1,0,0,0,0,0) \\ (1,1,1,1,0,0) \\ (0,1,1,1,0,0) \end{cases}.$$

Since $1^{11} = 1, (\zeta^{21})^{11} = \zeta^{42}$, and $(\zeta^{42})^{11} = \zeta^{21}$, the subspace E_2 is indeed mapped to itself. This gives us $E_2 = E_1^{\perp}$ and

$$E_1 = \left\langle \begin{matrix} (0,1,0,1,0,0) \\ (0,0,1,1,0,0) \\ (0,0,0,0,1,0) \\ (0,0,0,0,0,1) \end{matrix} \right\rangle.$$

Thus, all the requirements of Theorem 2 are satisfied and the permutation π gives rise to a bent function $f(x,y) = \pi(y) \cdot x \oplus 1_{E_1}(x)1_{E_2}(x)$ contained in $\mathcal{D}$ but outside the $\mathcal{M}^*$ class.

The next result partially overlaps with Proposition 1 but, as shown in Example 3, it also includes cases when n is odd.

Proposition 2. *Let $\pi(y) = y^d$ be a quadratic permutation over $\mathbb{F}_{2^n}$ ($n \geq 4$), where $d = 2^i + 2^j, i > j$, and $(2^n - 1, 2^i + 2^j) = 1$. Let also $E_2 = \langle \zeta^a, \zeta^b \rangle$ be a 2-dimensional linear subspace of $\mathbb{F}_2^n$, where ζ is a primitive element of $\mathbb{F}_{2^n}$. If*

$$(a - b)(2^i - 2^j) \equiv 0 \mod (2^n - 1)$$

then π satisfies all the conditions in Theorem 2.

Proof. Since π is a quadratic permutation monomial it has no linear structures. Because $n \geq 4$ and $\deg(\pi) = 2$, it also satisfies $\deg(\pi) \leq n - \dim(E_2)$. It remains to determine when the subspace E_2 is mapped to a subspace. Noting that $\zeta^a \mapsto \zeta^{ad}$ and $\zeta^b \mapsto \zeta^{bd}$, it is required that $\zeta^a + \zeta^b$ is mapped to $(\zeta^a + \zeta^b)^d = \zeta^{ad} + \zeta^{bd}$. Therefore

$$(\zeta^a + \zeta^b)^{2^i + 2^j} = \zeta^{a(2^i + 2^j)} + \zeta^{b(2^i + 2^j)}$$
$$\zeta^{a(2^i + 2^j)} + \zeta^{a2^i + b2^j} + \zeta^{b2^i + a2^j} + \zeta^{b(2^i + 2^j)} = \zeta^{a(2^i + 2^j)} + \zeta^{b(2^i + 2^j)}$$
$$\zeta^{a2^i + b2^j} = \zeta^{b2^i + a2^j}.$$

It follows that

$$a2^i + b2^j \equiv b2^i + a2^j \mod (2^n - 1),$$

which implies $(a - b)(2^i - 2^j) \equiv 0 \mod (2^n - 1)$, as stated. Thus, all three conditions imposed by Theorem 2 are satisfied. □

Remark 2. *It should be noted that given the set of parameters a, b, i and j satisfying the main condition in Proposition 2 we are still left with some freedom in choosing the subspace E_2 since the only constraint is on the fixed difference $a - b$ satisfying $(a - b)(2^i - 2^j) \equiv 0 \mod (2^n - 1)$. This gives multiple choices of a and b for specifying the elements ζ^a, ζ^b.*

It turns out that the conditions in Proposition 2 cannot be satisfied for relatively small n. It was confirmed (using the programming package Magma) that the smallest n for which a 2-dimensional subspace E_2 in Proposition 2 can be found is $n = 6$. Nevertheless, in order to also present a construction for odd n, we give below an example for $n = 9$.

Example 3. *Let $n = 9$ and $\pi(y) = y^9$, thus $i = 3, j = 0$. Then π is a quadratic permutation since $(2^9 - 1, 9) = 1$. Furthermore, $(a - b) = (2^9 - 1)/(2^3 - 2^0) = 73$. We choose $a = 74, b = 1$ and use Magma to get the vector space representation of the subspace $E_2 = \langle \zeta, \zeta^{74} \rangle$, where ζ is the generating element of the field $\mathbb{F}_{2^9}$:*

$$
E_2 = \left\{ \begin{array}{l} (0,0,0,0,0,0,0,0,0) \\ (1,1,0,0,1,1,0,1,0) \\ (0,1,0,0,0,0,0,0,0) \\ (1,0,0,0,1,1,0,1,0) \end{array} \right\}
$$

$$
E_1^\perp = \pi(E_2) = \left\{ \begin{array}{l} (0,0,0,0,0,0,0,0,0) \\ (1,1,1,0,1,1,0,0,1) \\ (1,0,0,0,1,0,0,0,0) \\ (0,1,1,0,0,1,0,0,1) \end{array} \right\}.
$$

One can readily check that all the requirements of Theorem 2 are satisfied.

Remark 3. *Finding non-monomial permutations that satisfy the conditions of Theorem 2 appears to be much harder and is still an open problem.*

6 Conclusions

Two secondary classes of bent functions, that possibly provide instances of bent functions outside the standard primary classes, was introduced by Carlet more than two decades ago and a single class named $\mathcal{D}_0$ was shown to be outside $\mathcal{PS}^*$ and $\mathcal{M}^*$. For the first time, by specifying sufficient conditions for these two classes $\mathcal{C}$ and $\mathcal{D}$ to be outside $\mathcal{M}^*$, we have been able to identify several infinite subclasses of bent functions that do not belong to the completed $\mathcal{M}$ class. The question whether these functions belong to other two primary classes remains open.

Acknowledgements. Fengrong Zhang is supported in part by National Science Foundation of China (Grant No. 61303263), and in part by the Fundamental Research Funds for the Central Universities (Grant No. 2015XKMS086), and in part by the China Post-doctoral Science Foundation funded project (Grant No. 2015T80600). Enes Pasalic is partly supported by the Slovenian Research Agency (research program P3- 0384 and research project J1-6720). Yongzhuang Wei is supported in part by the Natural Science Foundation of China (61572148), in part by the Guangxi Natural Science Found (2015GXNSFGA139007), in part by the project of Outstanding Young Teachers Training in Higher Education Institutions of Guangxi. Nastja Cepak is supported in part by the Slovenian Research Agency (research 25 program P3-0384 and Young Researchers Grant).

Appendix

Proof of Theorem 2:

Proof. Let $a^{(1)}, b^{(1)}, a^{(2)}, b^{(2)} \in \mathbb{F}_2^n$. We prove that f does not belong to $\mathcal{M}^{\#}$, by using Lemma 1. We need to show that there does not exist an $(\frac{n}{2})$-dimensional subspace V such that

$$D_{(a^{(1)}, a^{(2)})} D_{(b^{(1)}, b^{(2)})} f = 0,$$

for any $(a^{(1)}, a^{(2)}), (b^{(1)}, b^{(2)}) \in V$.

The second derivative of f with respect to a and b can be written as,

$$
\begin{aligned}
&D_{(a^{(1)}, a^{(2)})} D_{(b^{(1)}, b^{(2)})} f(x) \\
&= x \cdot (D_{a^{(2)}} D_{b^{(2)}} \pi(y)) \oplus a^{(1)} \cdot D_{b^{(2)}} \pi(y \oplus a^{(2)}) \\
&\quad \oplus b^{(1)} \cdot D_{a^{(2)}} \pi(y \oplus b^{(2)}) \oplus D_a D_b 1_{E_1}(x) 1_{E_2}(y) \\
&= x \cdot (D_{a^{(2)}} D_{b^{(2)}} \pi(y)) \oplus a^{(1)} \cdot D_{b^{(2)}} \pi(y \oplus a^{(2)}) \oplus b^{(1)} \cdot D_{a^{(2)}} \pi(y \oplus b^{(2)}) \quad (6) \\
&\quad \oplus 1_{E_1}(x) D_{a^{(2)}} D_{b^{(2)}} 1_{E_2}(y) \oplus 1_{E_2}(y \oplus a^{(2)}) D_{a^{(1)}} 1_{E_1}(x) \\
&\quad \oplus 1_{E_2}(y \oplus b^{(2)}) D_{b^{(1)}} 1_{E_1}(x) \oplus 1_{E_2}(y \oplus a^{(2)} \oplus b^{(2)}) D_{a^{(1)} \oplus b^{(1)}} 1_{E_1}(x).
\end{aligned}
$$

We denote the set $\{(x, 0_n) \mid x \in \mathbb{F}_2^n\}$ by Δ, and consider two cases $V = \Delta$ and $V \neq \Delta$.

1. For $V = \Delta$, we can find two vectors $(a^{(1)}, 0_n), (b^{(1)}, 0_n) \in \Delta$ such that

$$D_{a^{(1)}} D_{b^{(1)}} 1_{E_1}(x) \neq 0$$

since $\dim(E_1) \geq 2$. Further, we have

$$
\begin{aligned}
D_{(a^{(1)}, a^{(2)})} D_{(b^{(1)}, b^{(2)})} f(x) &= 1_{E_2}(y)(D_{a^{(1)}} 1_{E_1}(x) \oplus D_{b^{(1)}} 1_{E_1}(x) \\
&\quad \oplus D_{a^{(1)} \oplus b^{(1)}} 1_{E_1}(x)) \\
&= 1_{E_2}(y) D_{a^{(1)}} D_{b^{(1)}} 1_{E_1}(x) \neq 0.
\end{aligned}
$$

2. For $V \neq \Delta$, we split the proof into three cases depending on the cardinality of $V \cap \Delta$. We set $V = \left\{ (v_1^{(1)}, v_2^{(1)}), (v_1^{(2)}, v_2^{(2)}), \ldots, (v_1^{(2^n)}, v_2^{(2^n)}) \right\}$,

    (a) For $|V \cap \Delta| = 1$, we have $v_2^{(i)} \neq v_2^{(j)}$ for any $i \neq j$. If there exist two vectors $v_2^{(i_1)}, v_2^{(j_1)}$ such that $v_2^{(i_1)} = v_2^{(j_1)}$, then $v_1^{(i_1)} = v_1^{(j_1)}$, (or $(v_1^{(i_1)} \oplus v_1^{(j_1)}, 0_n) \in V \cap \Delta$), that is, $(v_1^{(i_1)}, v_2^{(i_1)}) = (v_1^{(j_1)}, v_2^{(j_1)})$. Further, $|\{v_2^{(1)}, v_2^{(2)}, \ldots, v_2^{(2^n)}\}| = |V| = 2^n$, that is, $\{v_2^{(1)}, v_2^{(2)}, \ldots, v_2^{(2^n)}\} = \mathbb{F}_2^n$ (here, if $v_2^{(i_1)} = v_2^{(i_2)}$, they are called one element). Thus, we can find two vectors $a, b \in V$ such that

$$D_{a^{(2)}} D_{b^{(2)}} 1_{E_2}(y) \neq 0$$

since $\dim(E_2) \geq 2$.

Now, there are four cases to be considered.

i. If $a^{(1)} = b^{(1)} = 0_n$, from (6), we have

$$D_{(a^{(1)},a^{(2)})}D_{(b^{(1)},b^{(2)})}f(x)$$
$$= x \cdot (D_{a^{(2)}}D_{b^{(2)}}\pi(y)) \oplus 1_{E_1}(x)D_{a^{(2)}}D_{b^{(2)}}1_{E_2}(y) \neq 0 \qquad (7)$$

since $\dim(E_1) + \dim(E_2) = n$ and $\dim(E_2) \geq 2$, that is, $\deg(1_{E_1}(x)) \geq 2$.

ii. If $a^{(1)} = 0_n, b^{(1)} \neq 0_n$, from (6), we have

$$D_{(a^{(1)},a^{(2)})}D_{(b^{(1)},b^{(2)})}f(x)$$
$$= x \cdot (D_{a^{(2)}}D_{b^{(2)}}\pi(y)) \oplus b^{(1)} \cdot D_{a^{(2)}}\pi(y \oplus b^{(2)})$$
$$\oplus 1_{E_1}(x)D_{a^{(2)}}D_{b^{(2)}}1_{E_2}(y)$$
$$\oplus 1_{E_2}(y \oplus b^{(2)})D_{b^{(1)}}1_{E_1}(x) \oplus 1_{E_2}(y \oplus a^{(2)} \oplus b^{(2)})D_{b^{(1)}}1_{E_1}(x)$$
$$= x \cdot (D_{a^{(2)}}D_{b^{(2)}}\pi(y)) \oplus b^{(1)} \cdot D_{a^{(2)}}\pi(y \oplus b^{(2)})$$
$$\oplus 1_{E_1}(x)D_{a^{(2)}}D_{b^{(2)}}1_{E_2}(y) \oplus D_{b^{(1)}}1_{E_1}(x)D_{a^{(2)}}1_{E_2}(y \oplus b^{(2)}).$$

We know $\dim(E_1) + \dim(E_2) = n$ and $\dim(E_2) \geq 2$, thus $\deg(1_{E_1}(x)) \geq 2$. Further, $\deg(1_{E_1}(x)) > \deg(D_{b^{(1)}}1_{E_1}(x))$. Thus, we have

$$D_{(a^{(1)},a^{(2)})}D_{(b^{(1)},b^{(2)})}f(x) \neq 0.$$

iii. If $a^{(1)} \neq 0_n, b^{(1)} = 0_n$, from (6), we have

$$D_{(a^{(1)},a^{(2)})}D_{(b^{(1)},b^{(2)})}f(x)$$
$$= x \cdot (D_{a^{(2)}}D_{b^{(2)}}\pi(y)) \oplus a^{(1)} \cdot D_{b^{(2)}}\pi(y \oplus a^{(2)}) \qquad (8)$$
$$\oplus 1_{E_1}(x)D_{a^{(2)}}D_{b^{(2)}}1_{E_2}(y) \oplus D_{a^{(1)}}1_{E_1}(x)D_{b^{(2)}}1_{E_2}(y \oplus a^{(2)}).$$

We know $\dim(E_1) + \dim(E_2) = n$ and $\dim(E_2) \geq 2$, thus $\deg(1_{E_1}(x)) \geq 2$. Further, $\deg(1_{E_1}(x)) > \deg(D_{a^{(1)}}1_{E_1}(x))$. Thus, we have

$$D_{(a^{(1)},a^{(2)})}D_{(b^{(1)},b^{(2)})}f(x) \neq 0.$$

iv. If $a^{(1)} \neq 0_n, b^{(1)} \neq 0_n$, from (6), we have

$$D_{(a^{(1)},a^{(2)})}D_{(b^{(1)},b^{(2)})}f(x) \neq 0.$$

Since $\dim(E_1) + \dim(E_2) = n$ and $\dim(E_2) \geq 2$, then $\deg(1_{E_1}(x)) \geq 2$. Furthermore, $\deg(1_{E_1}(x)) > \deg(D_{b^{(1)}}1_{E_1}(x))$, $\deg(1_{E_1}(x)) > \deg(D_{a^{(1)}}1_{E_1}(x))$ and $\deg(1_{E_1}(x)) > \deg(D_{a^{(1)} \oplus b^{(1)}}1_{E_1}(x))$.
Hence, we have

$$D_{(a^{(1)},a^{(2)})}D_{(b^{(1)},b^{(2)})}f(x) \neq 0$$

for $|V \cap \Delta| = 1$.

(b) For $|V \cap \Delta| = 2$, without loss of generality, let $(a^{(1)}, 0_n) \in V \cap \Delta$, $a^{(1)} \neq 0_n$.
We know $\{v_2^{(1)}, v_2^{(2)}, \ldots, v_2^{(2^n)}\}$ is a subspace of $\mathbb{F}_2^n$ which is denoted by $\mathcal{V}'$.

We first prove $\dim(\mathcal{V}') = n-1$ by showing that $|\{v_2^{(1)}, v_2^{(2)}, \ldots, v_2^{(2^n)}\}| = 2^{n-1}$, where we only count distinct vectors (e.g. if $v_2^{(i_1)} = v_2^{(i_2)}$ only one vector is counted). If $|\{v_2^{(1)}, v_2^{(2)}, \ldots, v_2^{(2^n)}\}| = 2^n$, then it is clear that V is not a subspace. If $|\{v_2^{(1)}, v_2^{(2)}, \ldots, v_2^{(2^n)}\}| < 2^{n-1}$, there must exist three vectors $v_2^{(i_1)} = v_2^{(i_2)} = v_2^{(i_3)}$, where $i_1 \neq i_2 \neq i_3$. Thus, we will have $(v_1^{(i_1)}, v_2^{(i_1)}) \oplus (v_1^{(i_2)}, v_2^{(i_2)}) \in V \cap \Delta$, $(v_1^{(i_1)}, v_2^{(i_1)}) \oplus (v_1^{(i_3)}, v_2^{(i_3)}) \in V \cap \Delta$ and $(v_1^{(i_3)}, v_2^{(i_3)}) \oplus (v_1^{(i_2)}, v_2^{(i_2)}) \in V \cap \Delta$, which contradicts the fact that $|V \cap \Delta| = 2$.

We now show that $|E_2 \cap \mathcal{V}'| \geq 1$ by using a well-known fact that

$$\dim(E_2 \cap \mathcal{V}') = \dim(E_2) + \dim(\mathcal{V}') - \dim(E_2 \boxplus \mathcal{V}'),$$

where $E_2 \boxplus \mathcal{V}' = \{\alpha \oplus \beta \mid \alpha \in E_2, \beta \in \mathcal{V}'\}$. Since by assumption $\dim(E_2) \geq 2$ and we have shown that $\dim(\mathcal{V}') = n-1$, then $\dim(E_2 \cap \mathcal{V}') \geq 1$.

We now choose one vector $b^{(2)}$ from $(\mathcal{V}' \cap E_2) \backslash \{0_n\}$, then $b^{(2)} \neq 0_n$ and $1_{E_2}(y) = 1_{E_2}(y \oplus b^{(2)})$ (since $b^{(2)} \in E_2$). Set $b = (b^{(1)}, b^{(2)}) \in V$. From (6), we have

$$
\begin{aligned}
& D_{(a^{(1)}, a^{(2)})} D_{(b^{(1)}, b^{(2)})} f(x) \\
&= a^{(1)} \cdot D_{b^{(2)}} \pi(y) \oplus 1_{E_2}(y) D_{a^{(1)}} 1_{E_1}(x) \qquad\qquad (9)\\
&\quad \oplus 1_{E_2}(y \oplus b^{(2)}) D_{b^{(1)}} 1_{E_1}(x) \oplus 1_{E_2}(y \oplus b^{(2)}) D_{a^{(1)} \oplus b^{(1)}} 1_{E_1}(x) \\
&= a^{(1)} \cdot D_{b^{(2)}} \pi(y) \oplus 1_{E_2}(y) D_{a^{(1)}} D_{b^{(1)}} 1_{E_1}(x).
\end{aligned}
$$

Now, there are three cases to be considered. If $D_{a^{(1)}} D_{b^{(1)}} 1_{E_1}(x) \neq const.$ or $D_{a^{(1)}} D_{b^{(1)}} 1_{E_1}(x) = 0$, then it is clear that

$$D_{(a^{(1)}, a^{(2)})} D_{(b^{(1)}, b^{(2)})} f(x) \neq 0$$

since π has no nonzero linear structure and $b^{(2)} \neq 0_n$.

If $D_{a^{(1)}} D_{b^{(1)}} 1_{E_1}(x) = 1$, then it is clear that

$$D_{(a^{(1)}, a^{(2)})} D_{(b^{(1)}, b^{(2)})} f(x) = a^{(1)} \cdot D_{b^{(2)}} \pi(y) \oplus 1_{E_2}(y) \neq 0$$

since $\deg(\pi) \leq n - \dim(E_2)$, that is, $\deg(a^{(1)} \cdot D_{b^{(2)}} \pi(y)) < n - \dim(E_2) = \deg(1_{E_2}(y))$.

(c) For $|V \cap \Delta| > 2$ (i.e., $|V \cap \Delta| \geq 4$), without loss of generality, let $a = (a^{(1)}, 0_n)(\neq 0_{2n}) \in V \cap \Delta$. Here, there are two cases to be considered.

i. If there exists one vector $v = (0_n, v_2) \in V \backslash \{0_{2n}\}$, then we set $b = v$. Further, using that $b^{(1)} = 0_n$, we have

$$D_{(a^{(1)}, a^{(2)})} D_{(b^{(1)}, b^{(2)})} f(x) = a^{(1)} \cdot D_{b^{(2)}} \pi(y) \oplus D_{b^{(2)}} 1_{E_2}(y) D_{a^{(1)}} 1_{E_1}(x).$$

If $D_{a^{(1)}} 1_{E_1}(x) \neq constant$ or $D_{a^{(1)}} 1_{E_1}(x) = 0$, then again

$$D_{(a^{(1)}, a^{(2)})} D_{(b^{(1)}, b^{(2)})} f(x) \neq 0,$$

since π has no nonzero linear structure.

We now show that $D_{a^{(1)}}1_{E_1}(x) = 1$ is impossible. We have that $D_{a^{(1)}}1_{E_1}(x) = 0$ if $a^{(1)} \in E_1$, or alternatively if $a^{(1)} \notin E_1$

$$\deg(D_{a^{(1)}}1_{E_1}(x)) = n - \dim(E_1) - 1,$$

since $E_1 \cup (a^{(1)} \oplus E_1)$ is a subspace of dimension $\dim(E_1) + 1$. Since $n - \dim(E_1) - 1 > 0$ and by assumption $\dim(E_1) < n - 1$, we have $D_{a^{(1)}}1_{E_1}(x) \neq 1$.

ii. Let $v = (v_1, v_2) \in V \backslash \{0_{2n}\}$. If we always have $v = (v_1, v_2)$ such that $v_1 \neq 0_n$ for every $v_2 \neq 0_n$, then we set $b = v \in V \backslash \{0_{2n}\}$ such that $v_2 \neq 0_n$. Further, we have

$$
\begin{aligned}
D_{(a^{(1)},a^{(2)})}D_{(b^{(1)},b^{(2)})}f(x) &= a^{(1)} \cdot D_{b^{(2)}}\pi(y) \oplus 1_{E_2}(y)D_{a^{(1)}}1_{E_1}(x) \\
&\oplus 1_{E_2}(y \oplus b^{(2)})D_{b^{(1)}}1_{E_1}(x) \oplus 1_{E_2}(y \oplus b^{(2)})D_{a^{(1)} \oplus b^{(1)}}1_{E_1}(x) \\
&= a^{(1)} \cdot D_{b^{(2)}}\pi(y) \oplus 1_{E_2}(y)D_{a^{(1)}}1_{E_1}(x) \\
&\oplus 1_{E_2}(y \oplus b^{(2)})D_{a^{(1)}}1_{E_1}(x \oplus b^{(1)}).
\end{aligned}
\tag{10}
$$

There are two cases to be considered.

If $b^{(2)} \in E_2$, then we have

$$
\begin{aligned}
D_{(a^{(1)},a^{(2)})}D_{(b^{(1)},b^{(2)})}f(x) &= a^{(1)} \cdot D_{b^{(2)}}\pi(y) \oplus 1_{E_2}(y)(D_{a^{(1)}}1_{E_1}(x) \\
&\oplus D_{a^{(1)}}1_{E_1}(x \oplus b^{(1)})) \neq 0,
\end{aligned}
$$

since $\deg(1_{E_2}(y)) > \deg(a^{(1)} \cdot D_{b^{(2)}}\pi(y))$.

If $b^{(2)} \notin E_2$, then we have three cases to be considered.

A. For $a^{(1)} \in E_1$ we have

$$D_{(a^{(1)},a^{(2)})}D_{(b^{(1)},b^{(2)})}f(x) = a^{(1)} \cdot D_{b^{(2)}}\pi(y) \neq 0.$$

B. For $a^{(1)} \notin E_1, b^{(1)} \in E_1$ we have

$$
\begin{aligned}
D_{(a^{(1)},a^{(2)})}D_{(b^{(1)},b^{(2)})}f(x) &= a^{(1)} \cdot D_{b^{(2)}}\pi(y) \\
&\oplus D_{b^{(2)}}1_{E_2}(y)D_{a^{(1)}}1_{E_1}(x) \neq 0,
\end{aligned}
$$

since $D_{a^{(1)}}1_{E_1}(x) \neq \text{constant}$.

C. For $a^{(1)} \notin E_1, b^{(1)} \notin E_1$ we have

$$
\begin{aligned}
D_{(a^{(1)},a^{(2)})}D_{(b^{(1)},b^{(2)})}f(x) &= a^{(1)} \cdot D_{b^{(2)}}\pi(y) \\
&\oplus D_{b^{(2)}}1_{E_2}(y)D_{a^{(1)}}1_{E_1}(x) \\
&\oplus 1_{E_2}(y \oplus b^{(2)})D_{a^{(1)}}D_{b^{(1)}}1_{E_1}(x) \neq 0,
\end{aligned}
$$

since $D_{a^{(1)}}1_{E_1}(x) \neq \text{constant}$ and furthermore $\deg(D_{a^{(1)}}1_{E_1}(x)) > \deg(D_{a^{(1)}}D_{b^{(1)}}1_{E_1}(x))$.

Combining items 1 and 2, we deduce that f does not belong to $M^{\#}$. $\square$

References

1. Blokhuis, A., Coulter, R.S., Henderson, M., O'Keefe, C.M.: Permutations amongst the Dembowski-Ostrom polynomials. In: Jungnickel, D., Niederreiter, H. (eds.) Proceedings of the Fifth International Conference on Finite Fields and Applications F_{q5}, pp. 37–42. Springer, Berlin (2001)
2. Carlet, C.: Two new classes of bent functions. In: Helleseth, T. (ed.) EURO-CRYPT 1993. LNCS, vol. 765, pp. 77–101. Springer, Heidelberg (1994). doi:10. 1007/3-540-48285-7_8
3. Carlet, C.: Boolean functions for cryptography and error correcting codes. In: Crama, Y., Hammer, P.L. (eds.) Boolean Models and Methods in Mathematics, Computer Science, and Engineering, pp. 257–397. Cambridge University Press, Cambridge (2010). Chapter of the monograph. http://www-roc.inria.fr/secret/Claude.Carlet/pubs.html
4. Carlet, C.: Vectorial Boolean functions for cryptography. In: Crama, Y., Hammer, P. (eds.) Boolean Methods and Models, pp. 398–469. Cambridge University Press, Cambridge (2010). http://www-roc.inria.fr/secret/Claude.Carlet/pubs.html
5. Carlet, C., Mesnager, S.: Four decades of research on bent functions. Des. Codes Crypt. **78**(1), 5–50 (2016)
6. Charpin, P., Sarkar, S.: Polynomials with linear structure and Maiorana-McFarland construction. IEEE Trans. Inform. Theory, IT **57**(6), 3796–3804 (2011)
7. Cusick, T.W., Stănică, P.: Cryptographic Boolean Functions and Applications. Elsevier-Academic Press, Cambridge (2009)
8. Dillon, J.F.: Elementary hadamard difference sets. In: Proceedings of 6th S.E. Conference of Combinatorics, Graph Theory, and Computing, Utility Mathematics, Winnipeg, pp. 237–249 (1975)
9. Dobbertin, H.: Construction of bent functions and balanced Boolean functions with high nonlinearity. In: Preneel, B. (ed.) FSE 1994. LNCS, vol. 1008, pp. 61–74. Springer, Heidelberg (1995). doi:10.1007/3-540-60590-8_5
10. Laigle-Chapuy, Y.: A note on a class of quadratic permutations over $\mathbb{F}_{2^n}$. In: Boztaş, S., Lu, H.-F.F. (eds.) AAECC 2007. LNCS, vol. 4851, pp. 130–137. Springer, Heidelberg (2007). doi:10.1007/978-3-540-77224-8_17
11. McFarland, R.L.: A family of noncyclic difference sets. J. Comb. Theory Ser. A **15**, 1–10 (1973)
12. Mandal, B., Stanica, P., Gangopadhyay, S., Pasalic, E.: An analysis of $\mathcal{C}$ class of bent functions. Fundamenta Informaticae **147**(3), 271–292 (2016)
13. Rothaus, O.S.: On bent functions. J. Comb. Theory Ser. A **20**, 300–305 (1976)

Quantum Algorithms Related to *HN*-Transforms of Boolean Functions

Sugata Gangopadhyay[1(✉)], Subhamoy Maitra[2], Nishant Sinha[1],
and Pantelimon Stănică[3]

[1] Department of Computer Science and Engineering,
Indian Institute of Technology Roorkee, Roorkee 247667, India
gsugata@gmail.com, nishantsinha.iitr@gmail.com
[2] Applied Statistics Unit, Indian Statistical Institute,
203, B.T. Road, Kolkata 700108, India
subho@isical.ac.in
[3] Department of Applied Mathematics, Naval Postgraduate School,
Monterey, CA 93943–5216, USA
pstanica@nps.edu

Abstract. *HN*-transforms, which have been proposed as generalizations of Hadamard transforms, are constructed by tensoring Hadamard and nega-Hadamard kernels in any order. We show that all the 2^n possible *HN*-spectra of a Boolean function in n variables, each containing 2^n elements (i.e., in total 2^{2n} values in transformed domain) can be computed in $O(2^{2n})$ time (more specific with little less than 2^{2n+1} arithmetic operations). We propose a generalization of Deutsch-Jozsa algorithm, by employing *HN*-transforms, which can be used to distinguish different classes of Boolean functions over and above what is possible by the traditional Deutsch-Jozsa algorithm.

Keywords: Boolean function · *HN*-transform , Deutsch-Jozsa algorithm

1 Introduction

Hadamard spectrum (or, Walsh-Hadamard spectrum) is possibly the most important tool in analyzing a Boolean function. This explains how a given Boolean function is correlated with each linear function and thus provides non-linearity as a summary data. High nonlinearity is an important property for the Boolean functions used in cryptographic primitives for resisting linear cryptanalysis [14] as well as correlation and fast correlation attacks [15,22]. Consider Boolean functions on n-variables. For n even, the functions with provably maximum nonlinearity $2^{n-1}-2^{\frac{n}{2}-1}$ exist [6] and such functions are called bent, though

---

S. Maitra is supported by the project "Cryptography & Cryptanalysis: How far can we bridge the gap between Classical and Quantum Paradigm", awarded by the Scientific Research Council of the Department of Atomic Energy (DAE-SRC), the Board of Research in Nuclear Sciences (BRNS).

S. El Hajji et al. (Eds.): C2SI 2017, LNCS 10194, pp. 314–327, 2017.
DOI: 10.1007/978-3-319-55589-8_21

the complete characterization of such functions is not yet known for $n > 8$. For n odd, consider the truth table of an n-variable function f constructed by the concatenation of the truth tables of two $(n-1)$-variable bent functions g and h, i.e., $f(x_0, x_1, \ldots, x_{n-1}) = x_0 g(x_1, \ldots, x_{n-1}) \oplus (x_0 \oplus 1) h(x_1, \ldots, x_{n-1})$ for all $(x_0, x_1, \ldots, x_{n-1}) \in \mathbb{F}_2^n$. One can then easily check that the nonlinearity of f is $2^{n-1} - 2^{\frac{n-1}{2}}$. This is famously known as the bent concatenation bound, which had been conjectured [10] to be the maximum attainable nonlinearity until disproved [18] in 1983. The maximum nonlinearity problem is directly related to coding theory also, since it corresponds to the covering radius of the first order Reed-Muller codes of block length 2^n.

There are several efficient methods in constructing Boolean functions with reasonably good cryptographic properties. However, commercial symmetric (stream or block) ciphers generally do not exploit Boolean functions on large number of variables. Instead, the trend is to use Boolean functions or S-Boxes on small number of variables (say 4 to 8) and then to introduce several rounds to obtain high confusion and diffusion. One can certainly regard the complete algorithm as a Boolean function on the key and IV bits, however, since we generally use between 80 to 256-bit key or IV, these Boolean functions are in reality very complicated to analyze. It is generally impossible to write the complete Truth Table (TT) or Algebraic Normal Form (ANF) of such functions. At the same time, it is well known that for randomly chosen Boolean functions the Hadamard spectrum values are concentrated around a low value [12] (i.e., their nonlinearities are high). However, it is not only the properties of the Boolean function as a whole that need to be studied. One may consider some sub-functions of the said Boolean function or the coefficient of certain monomials that may provide substantially high values of the Hadamard spectrum (i.e., low values of nonlinearity). Such a situation is needed for differential [11] or cube attacks [1,7] on heuristically designed stream ciphers. Thus identifying such high Hadamard spectrum values for a Boolean function (or its sub-functions) on large number of variables is an important question from cryptanalytic perspective. Apart from classical algorithms, quantum algorithms are also considered for approximating large spectrum values (and their positions). It has been observed [13] that in the quantum domain Deutsch-Josza algorithm [5] can create a superposition of states whose amplitudes are precisely the corresponding spectrum values.

The theory of linear approximations, which is based on Hadamard transform of the functions, has been generalized by Danielsen and Parker [3,4] as well as Riera and Parker [19,20], by introducing nega-Hadamard transforms leading to a class of generalized transforms, referred to as *HN*-transforms, combining Hadamard and nega-Hadamard kernels. It has been observed [19,20] that the quantum error correcting codes with optimal distance appear to have most flat spectra with respect to such transforms. In the context of *HN*-spectra, several results and constructions of Boolean functions and cryptographically strong S-Boxes had been studied in [4,8,16,17,19,21,24]. Surprisingly, while the *HN*-transform has been used for several purposes, its algorithmic issues have never been studied in detail. While it is natural that similar kind of ideas as for the traditional Hadamard transform might be applicable, there are specific details that need to be worked out. The algorithmic issues also provide several generalized techniques and characterizations related to Boolean functions.

In Sect. 3, we show that all the HN-spectra of an n-variable function can be simultaneously computed in time $O(2^{2n})$ as opposed to the naive estimate $O(n2^{2n})$ and we therefore design the corresponding algorithm. Note that the computation of the Hadamard (or Walsh-Hadamard transform) of a Boolean function on n-variables require $O(n2^n)$ time, by using the Fast Discrete Fourier Transform algorithm. As we will explain later, there are 2^n different HN-spectra, each containing 2^n elements. One of them is the well known Walsh spectrum. Similar to the algorithm of Hadamard spectrum, we may re-use the algorithm for each of the HN spectrum and that would require $O(2^n \cdot n2^n)$ time. However, while analyzing the algorithm for obtaining all the 2^n spectra, we note that the structure of the transforms are of such a nice pattern that this can be executed in $O(2^{2n})$ time, to be more specific, in exactly $2^{2n+1} - 2^{n+1}$ addition or subtraction operations. This is indeed a tight bound as 2^{2n} transformed values can be computed using $2^{2n+1} - 2^{n+1}$ arithmetic operations. Note that each transformed value, which depends on all the 2^n values of the Boolean function, can be obtained at an average cost of only 2 operations.

Next we consider quantum algorithms with respect to the HN-spectra. Suppose that we have an oracle access to a Boolean function f in n variables which is either constant or balanced. A classical algorithm will require $2^{n-1} + 1$ queries to determine whether f is constant or balanced. It is well known that Deutsch-Jozsa algorithm [5] solves this problem in a single query. In Sect. 4, we generalize the Deutsch-Jozsa algorithm by using HN-transforms and characterize larger classes of Boolean functions that can be distinguished by exploiting these transforms. We identify certain classes of quadratic symmetric functions that are related to these separations.

2   Preliminaries

Let $\mathbb{F}_2$ be the finite field with two elements and $\mathbb{Z}$ be the ring of integers. For any $n \in \mathbb{Z}^+$ (the set of positive integers), let $[n] = \{1, \ldots, n\}$. The Cartesian product of n copies of $\mathbb{F}_2$ is $\mathbb{F}_2^n = \{\mathbf{x} = (x_n, \ldots, x_1) : x_i \in \mathbb{F}_2, i \in [n]\}$ which is an n-dimensional vector space over $\mathbb{F}_2$ with respect to element-wise addition denoted by $\oplus$, scalar multiplication defined by $a\mathbf{x} = (ax_n, \ldots, ax_1)$, for all $a \in \mathbb{F}_2$ and $\mathbf{x} \in \mathbb{F}_2^n$. We define the inner product by $\mathbf{u} \cdot \mathbf{x} = \bigoplus_{i \in [n]} u_i x_i$ and intersection by $\mathbf{u} * \mathbf{x} = (u_n x_n, \ldots, u_1 x_1)$, for all $\mathbf{u} = (u_n, \ldots, u_1), \mathbf{x} = (x_n, \ldots, x_1) \in \mathbb{F}_2^n$. For any $\mathbf{v} = (v_n, \ldots, v_1) \in \mathbb{F}_2^n$, we can associate a unique integer $j = \sum_{i \in [n]} v_i 2^{i-1}$. When order is needed, we shall write $\mathbf{v} = \mathbf{u}_j$. The (Hamming) weight of a vector $\mathbf{v} \in \mathbb{F}_2^n$ is the integer sum $wt(\mathbf{v}) = \sum_{i \in [n]} v_i$. The (Hamming) distance between two vectors $\mathbf{u}, \mathbf{v} \in \mathbb{F}_2^n$ is $d(\mathbf{u}, \mathbf{v}) = wt(\mathbf{u} \oplus \mathbf{v})$.

Any function from $\mathbb{F}_2^n$ to $\mathbb{F}_2$ is said to be a Boolean function in n variables, whose set will be denoted by $\mathcal{B}_n$. The character form of $f \in \mathcal{B}_n$, $\chi_f(\mathbf{x}) =$

$(-1)^{f(\mathbf{x})}$, for all $\mathbf{x} \in \mathbb{F}_2^n$. Let M^T denote the transpose of a matrix M. We associate the column vectors (i.e., $2^n \times 1$ matrices) $\mathbf{f} = (f(\mathbf{u}_0), \dots, f(\mathbf{u}_{2^n-1}))^T$ and $\chi_{\mathbf{f}} = (\chi_f(\mathbf{u}_0), \dots, \chi_f(\mathbf{u}_{2^n-1}))^T$ to $f \in \mathcal{B}_n$. The vector $\mathbf{f}^T \in \mathbb{F}_2^{2^n}$ is said to be the truth table of f. The weight of a Boolean function f is $wt(f) = wt(\mathbf{f}^T)$. The Hamming distance between two Boolean functions $f, g \in \mathcal{B}_n$ is $d(f,g) = wt(\mathbf{f}^T \oplus \mathbf{g}^T)$. The algebraic normal form of $f \in \mathcal{B}_n$ is $f(\mathbf{x}) = \bigoplus_{\mathbf{a} \in \mathbb{F}_2^n} \mu_{\mathbf{a}} \prod_{i \in [n]} x_i^{a_i}$, where $\mu_{\mathbf{a}} \in \mathbb{F}_2$, for all $\mathbf{a} = (a_n, \dots, a_1) \in \mathbb{F}_2^n$. The algebraic degree of f, $\deg(f) = \max_{\mathbf{a} \in \mathbb{F}_2^n}\{wt(\mathbf{a}) : \mu_{\mathbf{a}} \neq 0\}$. The Boolean functions of the form $f(\mathbf{x}) = \bigoplus_{i \in [n]} a_i x_i \oplus a_0 = \mathbf{a} \cdot \mathbf{x} \oplus a_0$, where $a_i \in \mathbb{F}_2$ for all $i \in [n] \cup \{0\}$, are said to be affine functions. Affine functions are said to be linear if $\mu_0 = 0$. The set of affine functions and linear functions are denoted by $\mathcal{A}_n$ and $\mathcal{L}_n$, respectively.

2.1   *HN*-Transforms as a Generalization of Hadamard Transform

Recall that the tensor (sometimes, called Kronecker) product $A \otimes B$, where $A = (a_{ij})_{ij}, B = (b_{k\ell})_{k\ell}$ are $m \times n$, respectively, $p \times q$ matrices, is defined by

$$A \otimes B = \begin{bmatrix} a_{11}B & a_{12}B & \cdots & a_{1n}B \\ \vdots & \vdots & \ddots & \vdots \\ a_{m1}B & a_{m2}B & \cdots & a_{mn}B \end{bmatrix}.$$

The Hadamard and nega-Hadamard kernels $H = \frac{1}{\sqrt{2}}\begin{bmatrix} 1 & 1 \\ 1 & -1 \end{bmatrix}$, $N = \frac{1}{\sqrt{2}}\begin{bmatrix} 1 & i \\ 1 & -i \end{bmatrix}$, respectively, are unitary transformations over $\mathbb{C}^{\otimes 2} = \mathbb{C} \otimes \mathbb{C}$, where $\mathbb{C}$ is the field of complex numbers. The set of all tensor products

$$\{H, N\}^n = \left\{ \bigotimes_{i=n}^{1} K_i = K_n \otimes \cdots \otimes K_1 : K_i \in \{H, N\}, i \in [n] \right\}$$

is a subset (its cardinality is 2^n) of the set of all unitary transformations overs $(\mathbb{C}^2)^{\otimes n}$.

Definition 1. *Let $f \in \mathcal{B}_n$. Suppose $\mathbf{c} = (c_n, \dots, c_1) \in \mathbb{F}_2^n$ and $\mathcal{K}^{\mathbf{c}} \in \{H, N\}^n$ is such that $\mathcal{K}^{\mathbf{c}} = K_n \otimes \cdots \otimes K_1 = \bigotimes_{i=n}^{1} K_i$ where $K_i = \begin{cases} H \text{ if } c_i = 0, \\ N \text{ if } c_i = 1 \end{cases}$. For $0 \leq j \leq 2^n - 1$, we define,*

$$\mathcal{K}_f^{\mathbf{c}}(\mathbf{u}_j) = 2^{-\frac{n}{2}} \sum_{\mathbf{x} \in \mathbb{F}_2^n} (-1)^{f(\mathbf{x}) \oplus \mathbf{u}_j \cdot \mathbf{x}}\, i^{wt(\mathbf{c} * \mathbf{x})}, \tag{1}$$

which is referred to as the HN-transform of f at $\mathbf{u}_j$ with respect to $\mathcal{K}^{\mathbf{c}}$ (cf. [8]). The whole spectrum is denoted by $\mathcal{K}^{\mathbf{c}}\chi_{\mathbf{f}}$ and is referred as the HN-spectrum of f with respect to $\mathcal{K}^{\mathbf{c}}$.

For easy writing, let us denote $\mathbf{u}_0$ by $\mathbf{0}$ and $\mathbf{u}_{2^n-1}$ by $\mathbf{1}$. Then, $\mathcal{K}_f^{\mathbf{0}}(\mathbf{u}) = 2^{-\frac{n}{2}} \sum_{\mathbf{x} \in \mathbb{F}_2^n} (-1)^{f(\mathbf{x}) \oplus \mathbf{u} \cdot \mathbf{x}}$ and $\mathcal{K}_f^{\mathbf{1}}(\mathbf{u}) = 2^{-\frac{n}{2}} \sum_{\mathbf{x} \in \mathbb{F}_2^n} (-1)^{f(\mathbf{x}) \oplus \mathbf{u} \cdot \mathbf{x}} i^{wt(\mathbf{x})}$ are said to be the Hadamard and nega-Hadamard transforms of f at $\mathbf{u}$ and denoted by $\mathcal{H}_f(\mathbf{u})$ and $\mathcal{N}_f(\mathbf{u})$, respectively. For a detailed theory of Hadamard transform in the context of cryptographic Boolean functions we refer to [2,9,19,20,24].

3   The Complexity of Computing HN-Spectra

Given any function $f \in \mathcal{B}_n$ we can apply transformations from $\{H, N\}^n$ to obtain 2^n HN-spectra. Time complexity of computing each HN-spectrum is the same as the time complexity of computing the Hadamard spectrum of f, which is $O(n2^n)$, using the fast Hadamard transform algorithm. Thus, naively computing all HN-spectra will require $O(n2^{2n})$ time if we calculate each of them separately. In the following theorem we prove that this complexity can be improved.

Theorem 1. *The time complexity of computing HN-spectra of $f \in \mathcal{B}_n$ is $O(2^{2n})$.*

Proof. For $f \in \mathcal{B}_n$ there exist two functions $f_1, f_2 \in \mathcal{B}_{n-1}$ such that $f(\mathbf{x}, y) = (y \oplus 1) f_1(\mathbf{x}) \oplus y f_2(\mathbf{x})$, for all $\mathbf{x} \in \mathbb{F}_2^{n-1}$ and $y \in \mathbb{F}_2$. Then $2^{-\frac{n}{2}} \mathcal{K}_f^{(c_n, \mathbf{c})}(v, \mathbf{u}) =$

$$\sum_{\mathbf{x} \in \mathbb{F}_2^{n-1}} (-1)^{\mathbf{u} \cdot \mathbf{x} \oplus f_1(\mathbf{x})}{}_\imath wt(\mathbf{c} * \mathbf{x}) + (-1)^v {}_\imath wt(c_n) \sum_{\mathbf{x} \in \mathbb{F}_2^{n-1}} (-1)^{\mathbf{u} \cdot \mathbf{x} \oplus f_2(\mathbf{x})}{}_\imath wt(\mathbf{c} * \mathbf{x}), \qquad (2)$$

for all $(v, \mathbf{u}), (c_n, \mathbf{c}) \in \mathbb{F}_2 \times \mathbb{F}_2^{n-1}$. We denote by $T(n)$ the time complexity to compute all HN-spectra of any Boolean function in n variables. We show our result by finding a recurrence satisfied by $T(n)$. The computation of $2^{-\frac{n}{2}} \mathcal{K}_{f_1}^{\mathbf{c}}(\mathbf{u}) = \sum_{\mathbf{x} \in \mathbb{F}_2^{n-1}} (-1)^{\mathbf{u} \cdot \mathbf{x} \oplus f_1(\mathbf{x})}{}_\imath wt(\mathbf{c} * \mathbf{x})$, $2^{-\frac{n}{2}} \mathcal{K}_{f_2}^{\mathbf{c}}(\mathbf{u}) = \sum_{\mathbf{x} \in \mathbb{F}_2^{n-1}} (-1)^{\mathbf{u} \cdot \mathbf{x} \oplus f_2(\mathbf{x})}{}_\imath wt(\mathbf{c} * \mathbf{x})$, for all $\mathbf{u}, \mathbf{c} \in \mathbb{F}_2^{n-1}$, will therefore take $2T(n-1)$ time. For each $\mathbf{c} \in \mathbb{F}_2^{n-1}$, the computation of $2^{-\frac{n}{2}} \mathcal{K}_f^{(c_n, \mathbf{c})}(v, \mathbf{u})$ where $c_n \in \mathbb{F}_2$ and $(v, \mathbf{u}) \in \mathbb{F}_2 \times \mathbb{F}_2^{n-1}$ requires $4 \cdot 2^{n-1} = 2^{n+1}$ additions. If we vary $\mathbf{c}$ over $\mathbb{F}_2^{n-1}$ the total number of additions to compute all HN-spectra is $2^{n-1} \cdot 2^{n+1} = 2^{2n}$. Thus, we have the following first order recurrence relation:

$$T(n) = 2T(n-1) + 2^{2n},$$

which by iteration renders

$$T(n) = 2^{n-1} T(1) + 2^{2n} \sum_{i=0}^{n-2} \frac{1}{2^i} = 2^{n-1} T(1) + 2^{2n+1} - 2^{n+2} = O(2^{2n}),$$

and the theorem is shown. □

3.1   Fast HN-Transform Algorithm

Based on the above observations we design Algorithm 1 to efficiently compute HN-spectra of a Boolean function $f \in \mathcal{B}_n$. In Fig. 1 we demonstrate the steps of Algorithm 1 when $f \in \mathcal{B}_3$. It is clear from Fig. 1 that the total number of additions and subtractions required is $T(3) = 8 \times 2 + 8 \times 4 + 8 \times 8 = 2^3(2 + 2^2 + 2^3) = 2^4(2^3 - 1) = 2^{2(3)+1} - 2^{3+1} = 112$. In general $T(n) = 2^n(2 + 2^2 + \cdots + 2^{n-1} + 2^n) = 2^{2n+1} - 2^{n+1} = O(2^{2n})$, as discussed before.

Input: A Boolean function $f \in \mathcal{B}_n$, available in the form of the 2^n length array
$$\chi_{\mathbf{f}} = (\chi_f(\mathbf{u_0}), \dots, \chi_f(\mathbf{u}_{2^n-1}))$$
Output: All 2^n *HN*-spectra of f, each containing 2^n elements

1 Initialize a $2^n \times 2^n$ matrix h whose columns and rows are numbered from 0 to
   $2^n - 1$. The entry in the ith column and jth row is denoted by $h_{i,j}$.
2 $(h_{0,0}, h_{0,1}, \dots, h_{0,2^n-1}) \leftarrow (\chi_f(\mathbf{u_0}), \chi_f(\mathbf{u_1}), \dots, \chi_f(\mathbf{u}_{2^n-1}))$
3 **for** $j = 0$ **to** $n - 1$ **do**
4   **for** $\ell = 2^{j+1} - 1$ **downto** 0 **do**
5   **if** $\ell \equiv 0 \pmod 2$ **then**
6      $k = 0$
7   **while** $k < 2^n$ **do**
8   **for** $i = k$ **to** $k + 2^j - 1$ **do**
9        $tmp \leftarrow f_{\lfloor \frac{\ell}{2} \rfloor, i}$
10       $f_{\ell, i} \leftarrow tmp + f_{\lfloor \frac{\ell}{2} \rfloor, i+2^j}$
11       $f_{\ell, i+2^j} \leftarrow tmp - f_{\lfloor \frac{\ell}{2} \rfloor, i+2^j}$
12   **od**
13      $k \leftarrow k + 2^{j+1}$
14   **od**
15   **fi**
16   **if** $\ell \equiv 1 \pmod 2$ **then**
17     $k = 0$
18   **while** $k < 2^n$ **do**
19   **for** $i = k$ **to** $k + 2^j - 1$ **do**
20       $tmp \leftarrow f_{\lfloor \frac{\ell}{2} \rfloor, i}$
21       $f_{\ell, i} \leftarrow tmp + i f_{\lfloor \frac{\ell}{2} \rfloor, i+2^j}$
22       $f_{\ell, i+2^j} \leftarrow tmp - i f_{\lfloor \frac{\ell}{2} \rfloor, i+2^j}$
23   **od**
24      $k \leftarrow k + 2^{j+1}$
25   **od**
26   **fi**
27   **od**
28 **od**

Algorithm 1. Fast *HN*-transform algorithm.

3.2 *HN*-Transform and Quadratic Symmetric Functions on a Subspace Depending on c

In this section we describe the connection between *HN*-spectra and quadratic approximations of a Boolean function as discussed in Gangopadhyay, Pasalic and Stănică [8].

Consider any vector $\mathbf{c} \in \mathbb{F}_2^n$. The $(n-1)$-dimensional subspace orthogonal to $\mathbf{c}$ is $\mathbf{c}^\perp = \{\mathbf{x} \in \mathbb{F}_2^n : \mathbf{c} \cdot \mathbf{x} = 0\}$. Let $\ell_{\mathbf{c}} \in \mathcal{L}_n$ be defined by $\ell_{\mathbf{c}}(\mathbf{x}) = \mathbf{c} \cdot \mathbf{x}$, for all $\mathbf{x} \in \mathbb{F}_2^n$. Let $s \in \mathcal{B}_n$ be the symmetric quadratic bent function defined by

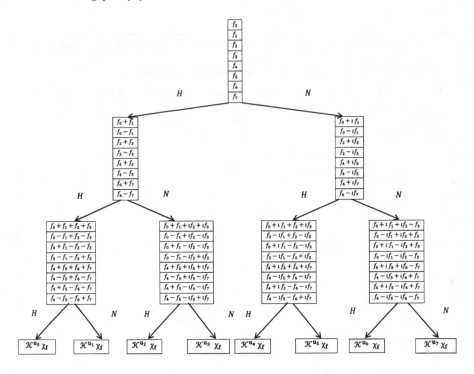

$\chi_f = (f_0, f_1, f_2, f_3, f_4, f_5, f_6, f_7)^T$. The vector $\mathbf{u}_j = (u_{j2}, u_{j1}, u_{j0})$ corresponds to the binary representation of $0 \le j \le 7$. $\mathcal{K}^{\mathbf{u}_j} \chi_f = K_2 \otimes K_1 \otimes K_0 \chi_f$ where $K_l = H$, if $u_{jl} = 0$ and $K_l = N$, if $u_{jl} = 1$, for all $0 \le i \le 2$. The normalizing factors of $\frac{1}{\sqrt{2}}$ and $\frac{1}{2}$ for the first and second levels, respectively, are not shown.

Fig. 1. Fast HN-transform algorithm for a function in $\mathcal{B}_n$.

$s(\mathbf{x}) = \bigoplus_{i<j} x_i x_j$, for all $\mathbf{x} \in \mathbb{F}_2^n$. For each $\mathbf{c} \in \mathbb{F}_2^n$ we define $s_\mathbf{c} \in \mathcal{B}_n$ by $s_\mathbf{c}(\mathbf{x}) = s(\mathbf{c}*\mathbf{x})$, for all $\mathbf{x}, \mathbf{c} \in \mathbb{F}_2^n$. We can think of $s_\mathbf{c}$'s as quadratic symmetric functions on the variables x_i's for which $c_i = 1$. Since (cf. [20,24]) $wt(\mathbf{c}*\mathbf{x}) \equiv 2s_\mathbf{c}(\mathbf{x}) + \mathbf{c}\cdot\mathbf{x}$ (mod 4), we obtain from (1)

$$2^{\frac{n}{2}}\mathcal{K}_f^\mathbf{c}(\mathbf{u}) = \sum_{\mathbf{x}\in\mathbf{c}^\perp} (-1)^{f(\mathbf{x})\oplus s_\mathbf{c}(\mathbf{x})\oplus\mathbf{u}\cdot\mathbf{x}} + \imath \sum_{\mathbf{x}\in\mathbb{F}_2^n\backslash\mathbf{c}^\perp} (-1)^{f(\mathbf{x})\oplus s_\mathbf{c}(\mathbf{x})\oplus\mathbf{u}\cdot\mathbf{x}}. \qquad (3)$$

Suppose that $f \in \mathcal{B}_n$ such that

$$\left| \sum_{\mathbf{x}\in\mathbf{c}^\perp} (-1)^{f(\mathbf{x})\oplus s_\mathbf{c}(\mathbf{x})\oplus\mathbf{u}\cdot\mathbf{x}} \right| = (-1)^{\epsilon_1(\mathbf{u},\mathbf{c})} \sum_{\mathbf{x}\in\mathbf{c}^\perp} (-1)^{f(\mathbf{x})\oplus s_\mathbf{c}(\mathbf{x})\oplus\mathbf{u}\cdot\mathbf{x}} \text{ and}$$

$$\left| \sum_{\mathbf{x}\in\mathbb{F}_2^n\backslash\mathbf{c}^\perp} (-1)^{f(\mathbf{x})\oplus s_\mathbf{c}(\mathbf{x})\oplus\mathbf{u}\cdot\mathbf{x}} \right| = (-1)^{\epsilon_2(\mathbf{u},\mathbf{c})} \sum_{\mathbf{x}\in\mathbb{F}_2^n\backslash\mathbf{c}^\perp} (-1)^{f(\mathbf{x})\oplus s_\mathbf{c}(\mathbf{x})\oplus\mathbf{u}\cdot\mathbf{x}},$$

where $\epsilon_1(\mathbf{u},\mathbf{c}), \epsilon_2(\mathbf{u},\mathbf{c}) \in \mathbb{F}_2$ and $\mathbf{c}, \mathbf{u} \in \mathbb{F}_2^n$. Then

$$\sum_{\mathbf{x}\in\mathbb{F}_2^n} (-1)^{f(\mathbf{x})\oplus s_\mathbf{c}(\mathbf{x})\oplus\epsilon_1(\mathbf{u},\mathbf{c})\oplus(\epsilon_1(\mathbf{u},\mathbf{c})\oplus\epsilon_2(\mathbf{u},\mathbf{c}))\mathbf{c}\cdot\mathbf{x}\oplus\mathbf{u}\cdot\mathbf{x}}$$

$$= \left| \sum_{\mathbf{x}\in\mathbf{c}^\perp} (-1)^{f(\mathbf{x})\oplus s_\mathbf{c}(\mathbf{x})\oplus\mathbf{u}\cdot\mathbf{x}} \right| + \left| \sum_{\mathbf{x}\in\mathbb{F}_2^n\setminus\mathbf{c}^\perp} (-1)^{f(\mathbf{x})\oplus s_\mathbf{c}(\mathbf{x})\oplus\mathbf{u}\cdot\mathbf{x}} \right| \qquad (4)$$

$$= |\Re(2^{\frac{n}{2}}\mathcal{K}_f^\mathbf{c}(\mathbf{u}))| + |\Im(2^{\frac{n}{2}}\mathcal{K}_f^\mathbf{c}(\mathbf{u}))|.$$

The Hamming distance between f and $s_\mathbf{c} \oplus \epsilon_1(\mathbf{u},\mathbf{c}) \oplus (\epsilon_1(\mathbf{u},\mathbf{c}) \oplus \epsilon_2(\mathbf{u},\mathbf{c}))\ell_\mathbf{c} \oplus \ell_\mathbf{u}$ is

$$2^{n-1} - \frac{1}{2}\left(|\Re(2^{\frac{n}{2}}\mathcal{K}_f^\mathbf{c}(\mathbf{u}))| + |\Im(2^{\frac{n}{2}}\mathcal{K}_f^\mathbf{c}(\mathbf{u}))|\right).$$

Given any Boolean function $f \in \mathcal{B}_n$, for all $\mathbf{c} \in \mathbb{F}_2^n$ we can obtain the spectra

$$\left[|\Re(2^{\frac{n}{2}}\mathcal{K}_f^\mathbf{c}(\mathbf{u}))| + |\Im(2^{\frac{n}{2}}\mathcal{K}_f^\mathbf{c}(\mathbf{u}))| : \mathbf{u} \in \mathbb{F}_2^n\right]. \qquad (5)$$

by computing the *HN*-spectra. We then find

$$\max_{\mathbf{c}\in\mathbb{F}_2^n} \max_{\mathbf{u}\in\mathbb{F}_2^n} \left[|\Re(2^{\frac{n}{2}}\mathcal{K}_f^\mathbf{c}(\mathbf{u}))| + |\Im(2^{\frac{n}{2}}\mathcal{K}_f^\mathbf{c}(\mathbf{u}))| : \mathbf{u} \in \mathbb{F}_2^n\right]. \qquad (6)$$

Suppose that the maximum value (6) is attained at $\mathbf{u}', \mathbf{c}' \in \mathbb{F}_2^n$. Then by using the *HN*-spectra the best possible quadratic approximation of f that we obtain is $s_{\mathbf{c}'} \oplus \epsilon_1(\mathbf{u}',\mathbf{c}') \oplus (\epsilon_1(\mathbf{u}',\mathbf{c}') \oplus \epsilon_2(\mathbf{u}',\mathbf{c}'))\ell_{\mathbf{c}'} \oplus \ell_{\mathbf{u}'}$.

Example 1. The 7-variable, 2-resilient functions with nonlinearity 56 are considered to be cryptographically strong functions and in [23, Table 4], all such rotation symmetric functions are listed. We have computed the spectra defined in (5), namely, $\left[|\Re(2^{\frac{n}{2}}\mathcal{K}_f^\mathbf{c}(\mathbf{u}))| + |\Im(2^{\frac{n}{2}}\mathcal{K}_f^\mathbf{c}(\mathbf{u}))| : \mathbf{u} \in \mathbb{F}_2^n\right]$ for all $\mathbf{c} \in \mathbb{F}_2^n$. Since these functions have nonlinearity the $\max_{\mathbf{u}\in\mathbb{F}_2^n} |2^{\frac{n}{2}}\mathcal{H}_f(\mathbf{u})| = 16$ for each function f in the list. Considering the *HN*-spectra for these functions we observe that for the first 12 functions $\max_{\mathbf{c}\in\mathbb{F}_2^n} \max_{\mathbf{u}\in\mathbb{F}_2^n} \left(|\Re(2^{\frac{n}{2}}\mathcal{K}_f^\mathbf{c}(\mathbf{u}))| + |\Im(2^{\frac{n}{2}}\mathcal{K}_f^\mathbf{c}(\mathbf{u}))| : \mathbf{u} \in \mathbb{F}_2^n\right) = 72$, and for the remaining functions

$$\max_{\mathbf{c}\in\mathbb{F}_2^n} \max_{\mathbf{u}\in\mathbb{F}_2^n} \left(|\Re(2^{\frac{n}{2}}\mathcal{K}_f^\mathbf{c}(\mathbf{u}))| + |\Im(2^{\frac{n}{2}}\mathcal{K}_f^\mathbf{c}(\mathbf{u}))| : \mathbf{u} \in \mathbb{F}_2^n\right) = 40.$$

This provides an example of how the *HN*-transforms enable us to obtain quadratic approximations efficiently and it is very clear that the second set of functions will have less correlation to the quadratic functions than the first ones.

Example 2. Parker [16] has computed the maximum of the square of the moduli of the 2^n times the *HN*-transformation values for several S-boxes including the AES S-box. This is related to peak-to-average ration (PAR) of the corresponding functions. In this example we consider the PRESENT S-box which is a permutation on $\mathbb{F}_2^4$. Let $\{f_i : i = 1,\ldots,15\}$ be its 15 non-zero component functions. For each $\mathbf{c} \in \mathbb{F}_2^4$ we compute

$$\max\{2^n|\mathcal{K}_f^\mathbf{c}(\mathbf{u})|^2 : \mathbf{u} \in \mathbb{F}_2^4\}.$$

Table 1. *HN*-spectra analysis of PRESENT *S*-box.

	c_0	c_1	c_2	c_3	c_4	c_5	c_6	c_7	c_8	c_9	c_{10}	c_{11}	c_{12}	c_{13}	c_{14}	c_{15}
f_1	64	32	64	32	64	32	128	64	32	16	32	16	32	16	64	32
f_2	64	32	40	40	40	40	32	32	40	40	32	32	32	32	40	40
f_3	64	32	40	40	40	40	32	32	32	64	40	72	40	72	32	64
f_4	64	32	40	40	40	40	32	32	40	40	32	32	32	32	40	40
f_5	64	64	40	72	40	72	32	64	40	72	32	64	32	64	40	72
f_6	64	32	40	40	40	40	32	32	32	32	40	40	40	40	32	32
f_7	64	64	40	72	40	72	32	64	40	72	32	64	32	64	40	72
f_8	64	32	32	32	32	64	32	32	40	40	40	40	40	72	40	40
f_9	64	32	32	16	32	16	64	32	32	16	64	32	64	32	128	64
f_{10}	64	32	32	64	32	32	32	32	40	40	40	72	40	40	40	40
f_{11}	64	32	32	32	32	64	32	32	40	40	40	40	40	72	40	40
f_{12}	64	32	32	16	32	16	64	32	64	32	32	16	32	16	32	16
f_{13}	64	32	32	64	32	32	32	32	40	40	40	72	40	40	40	40
f_{14}	64	32	40	40	40	40	32	32	32	32	40	40	40	40	32	32
f_{15}	64	32	40	40	40	40	32	32	32	64	40	72	40	72	32	64

Whether this provides us the best possible distribution of the *HN*-transformation values among all the permutations on $\mathbb{F}_2^4$ is an open question. In Table 1 we tabulate the values of $\max\{2^n|\mathcal{K}_f^c(\mathbf{u})|^2 : \mathbf{u} \in \mathbb{F}_2^4\}$ for each $\mathbf{c} \in \mathbb{F}_2^4$. For convenience we write $\mathbf{c}_j = (c_3, c_2, c_1, c_0)$ whenever $j = 2^3 c_3 + 2^2 c_2 + 2 c_1 + 1 c_0$. If F is the vector Boolean function corresponding to the PRESENT *S*-box then define $f_i = \mathbf{c}_i \cdot F$ for all $i = 0, 1, \ldots, 15$.

4 Extended Deutsch-Jozsa Algorithm

The extended Deutsch-Jozsa algorithm is pictorially represented in Fig. 2 and described in Algorithm 2. If we consider the specific case $H^{\otimes n}$ in place of $\mathcal{K}^c$, then we obtain the traditional Deutsch-Jozsa algorithm [5]. Given $f \in \mathcal{B}_n$ either constant or balanced, if the corresponding quantum bit oracle implementation $\mathcal{U}_f$ is available, Deutsch-Jozsa [5] provided a quantum algorithm that decides in a constant number of queries which one it is. One can simply describe Deutsch-Jozsa algorithm in terms of Hadamard spectrum values and it can be observed that

$$\sum_{\mathbf{z} \in \mathbb{F}_2^n} \sum_{\mathbf{x} \in \mathbb{F}_2^n} \frac{(-1)^{\mathbf{x} \cdot \mathbf{z} \oplus f(\mathbf{x})}}{2^n} |\mathbf{z}\rangle = \sum_{\mathbf{z} \in \mathbb{F}_2^n} 2^{-\frac{n}{2}} \mathcal{H}_f(\mathbf{z}) |\mathbf{z}\rangle,$$

i.e., the associated probability for the state $|\mathbf{z}\rangle$ is $2^{-\frac{n}{2}} \mathcal{H}_f(\mathbf{z})$. In this regard, we have the following technical result (see [13] for details).

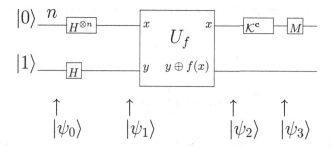

Fig. 2. Quantum circuit to implement extended Deutsch-Jozsa algorithm

---

Input: A Boolean function $f \in \mathcal{B}_n$, available in the form of the unitary
transformation $\mathcal{U}_f$
Output: n-bit pattern

1 Take an $(n+1)$ qubit state $|\psi_0\rangle = |0\rangle^{\otimes n}|1\rangle$;
2 Apply Hadamard Transform $H^{\otimes(n+1)}$ on $|\psi_0\rangle$ to get

$|\psi_1\rangle = \sum_{\mathbf{x}\in\mathbb{F}_2^n} \frac{|\mathbf{x}\rangle}{\sqrt{2^n}} \left[\frac{|0\rangle - |1\rangle}{\sqrt{2}}\right]$3; Apply $\mathcal{U}_f$ on $|\psi_1\rangle$ to get

$|\psi_2\rangle = \sum_{\mathbf{x}\in\mathbb{F}_2^n} \frac{(-1)^{f(\mathbf{x})}|\mathbf{x}\rangle}{\sqrt{2^n}} \left[\frac{|0\rangle - |1\rangle}{\sqrt{2}}\right]$4; Apply *HN*-Transform on the first
n qubits of $|\psi_2\rangle$ to obtain

$$|\psi_3\rangle = \sum_{\mathbf{z}\in\mathbb{F}_2^n} \sum_{\mathbf{x}\in\mathbb{F}_2^n} \frac{(-1)^{\mathbf{x}\cdot\mathbf{z}\oplus f(\mathbf{x})} \imath^{wt(\mathbf{c}*\mathbf{x})}|\mathbf{z}\rangle}{2^n} \left[\frac{|0\rangle - |1\rangle}{\sqrt{2}}\right];$$

5 Measurement at M: measure the first n qubits of $|\psi_3\rangle$ in
   computational basis;
6 After measurement, the state $\mathbf{v}$ such that $wt(\mathbf{v}) = 0$ or $\mathbf{v} = \mathbf{c}$ implies
   that the function is in $S_{\mathbf{c}}$, else it is in $T_{\mathbf{c}}$.

Algorithm 2. Extended Deutsch-Jozsa algorithm.

Proposition 1. *Given $f \in \mathcal{B}_n$, $\mathcal{D}_f|0\rangle^{\otimes n}$ produces a superposition of all states
$\mathbf{z} \in \mathbb{F}_2^n$ with the amplitude $2^{-\frac{n}{2}}\mathcal{H}_f(\mathbf{z})$ corresponding to each state $|\mathbf{z}\rangle$.*

In what follows, we trace the states through this circuit in the general case.
The input state is $|\psi_0\rangle = |0\rangle^{\otimes n}|1\rangle$. After applying $H^{\otimes(n+1)}$ and $\mathcal{U}_f$ successively
we obtain as before

$$|\psi_1\rangle = \sum_{\mathbf{x}\in\mathbb{F}_2^n} \frac{|\mathbf{x}\rangle}{\sqrt{2^n}} \left[\frac{|0\rangle - |1\rangle}{\sqrt{2}}\right] \text{ and } |\psi_2\rangle = \sum_{\mathbf{x}\in\mathbb{F}_2^n} \frac{(-1)^{f(\mathbf{x})}|\mathbf{x}\rangle}{\sqrt{2^n}} \left[\frac{|0\rangle - |1\rangle}{\sqrt{2}}\right],$$

respectively. Finally we apply the *HN*-transform $\mathcal{K}^c$ on the first n qubits of $|\psi_2\rangle$ to obtain

$$|\psi_3\rangle = \sum_{\mathbf{z}\in\mathbb{F}_2^n}\sum_{\mathbf{x}\in\mathbb{F}_2^n}\frac{(-1)^{\mathbf{x}\cdot\mathbf{z}\oplus f(\mathbf{x})}\imath^{wt(\mathbf{c}*\mathbf{x})}|\mathbf{z}\rangle}{2^n}\left[\frac{|0\rangle-|1\rangle}{\sqrt{2}}\right]$$

$$= \sum_{\mathbf{z}\in\mathbb{F}_2^n}2^{-\frac{n}{2}}\mathcal{K}_f^c(\mathbf{z})|\mathbf{z}\rangle\left[\frac{|0\rangle-|1\rangle}{\sqrt{2}}\right]. \tag{7}$$

Consider the sets $S_{\mathbf{c}} = \{s_{\mathbf{c}}(\mathbf{x}), 1\oplus s_{\mathbf{c}}(\mathbf{x}), s_{\mathbf{c}}(\mathbf{x})\oplus\ell_{\mathbf{c}}(x), 1\oplus s_{\mathbf{c}}(\mathbf{x})\oplus\ell_{\mathbf{c}}(\mathbf{x})\}$ and

$$T_{\mathbf{c}} = \left\{g\in\mathcal{B}_n : \sum_{\mathbf{x}\in\mathbf{c}^\perp=0}(-1)^{g(\mathbf{x})\oplus s_{\mathbf{c}}(\mathbf{x})} = \sum_{\mathbf{x}\in\mathbb{F}_2^n\backslash\mathbf{c}^\perp}(-1)^{g(\mathbf{x})\oplus s_{\mathbf{c}}(\mathbf{x})} = 0\right\}.$$

Theorem 2. *Suppose that $f\in\mathcal{B}_n$ is chosen from the set $S_{\mathbf{c}}\cup T_{\mathbf{c}}$ where*

$$S_{\mathbf{c}} = \{s_{\mathbf{c}}(\mathbf{x}), 1\oplus s_{\mathbf{c}}(\mathbf{x}), s_{\mathbf{c}}(\mathbf{x})\oplus\ell_{\mathbf{c}}(x), 1\oplus s_{\mathbf{c}}(\mathbf{x})\oplus\ell_{\mathbf{c}}(\mathbf{x})\}\ and$$

$$T_{\mathbf{c}} = \left\{g\in\mathcal{B}_n : \sum_{\mathbf{x}\in\mathbf{c}^\perp}(-1)^{g(\mathbf{x})\oplus s_{\mathbf{c}}(\mathbf{x})} = \sum_{\mathbf{x}\in\mathbb{F}_2^n\backslash\mathbf{c}^\perp}(-1)^{g(\mathbf{x})\oplus s_{\mathbf{c}}(\mathbf{x})} = 0,\right\},$$

for any $\mathbf{c}\in\mathbb{F}_2^n$. Applying the extended Deutsch–Jozsa algorithm on f, as above, and measure the first n qubits of $|\psi_3\rangle$ as obtained in (7), if we observe n-bit string $\mathbf{v}$ such that $wt(\mathbf{v})=0$ or $\mathbf{v}=\mathbf{c}$, then the function is in $S_{\mathbf{c}}$, otherwise the function is in $T_{\mathbf{c}}$.

Proof. Using Eq. (3) $|\psi_3\rangle$ is equal to

$$\sum_{\mathbf{z}\in\mathbb{F}_2^n}\frac{\sum_{\mathbf{x}\in\mathbf{c}^\perp}(-1)^{\mathbf{x}\cdot\mathbf{z}\oplus s_{\mathbf{c}}(\mathbf{x})\oplus f(\mathbf{x})} + \imath\sum_{\mathbf{x}\in\mathbb{F}_2^n\backslash\mathbf{c}^\perp}(-1)^{\mathbf{x}\cdot\mathbf{z}\oplus s_{\mathbf{c}}(\mathbf{x})\oplus f(\mathbf{x})}|\mathbf{z}\rangle}{2^n}\left[\frac{|0\rangle-|1\rangle}{\sqrt{2}}\right].$$

If $f\in S_{\mathbf{c}}$, then $f(\mathbf{x}) = s_{\mathbf{c}}(\mathbf{x})\oplus a_1\ell_{\mathbf{c}}(\mathbf{x})\oplus a_2$, where $(a_1,a_2)\in\mathbb{F}_2\times\mathbb{F}_2$. Putting $\mathbf{z}=0$

$$\sum_{\mathbf{x}\in\mathbf{c}^\perp}(-1)^{s_{\mathbf{c}}(\mathbf{x})\oplus f(\mathbf{x})} + \imath\sum_{\mathbf{x}\in\mathbb{F}_2^n\backslash\mathbf{c}^\perp}(-1)^{s_{\mathbf{c}}(\mathbf{x})\oplus f(\mathbf{x})}$$

$$= \sum_{\mathbf{x}\in\mathbf{c}^\perp}(-1)^{a_1\ell_{\mathbf{c}}(\mathbf{x})\oplus a_2} + \imath\sum_{\mathbf{x}\in\mathbb{F}_2^n\backslash\mathbf{c}^\perp}(-1)^{a_1\ell_{\mathbf{c}}(\mathbf{x})\oplus a_2}$$

$$= \sum_{\mathbf{x}\in\mathbf{c}^\perp}(-1)^{a_2} + \imath(-1)^{a_1}\sum_{\mathbf{x}\in\mathbb{F}_2^n\backslash\mathbf{c}^\perp}(-1)^{a_2} = (-1)^{a_2}2^{n-1}(1+(-1)^{a_1}).$$

Putting $\mathbf{z}=\mathbf{c}$, $\sum_{\mathbf{x}\in\mathbf{c}^\perp}(-1)^{\mathbf{x}\cdot\mathbf{c}\oplus s_{\mathbf{c}}(\mathbf{x})\oplus f(\mathbf{x})} + \imath\sum_{\mathbf{x}\in\mathbb{F}_2^n\backslash\mathbf{c}^\perp}(-1)^{\mathbf{x}\cdot\mathbf{c}\oplus s_{\mathbf{c}}(\mathbf{x})\oplus f(\mathbf{x})}$

$$= \sum_{\mathbf{x}\in\mathbf{c}^\perp}(-1)^{a_1\ell_{\mathbf{c}}(\mathbf{x})\oplus a_2} + \imath\sum_{\mathbf{x}\in\mathbb{F}_2^n\backslash\mathbf{c}^\perp}(-1)^{1\oplus a_1\ell_{\mathbf{c}}(\mathbf{x})\oplus a_2}$$

$$= \sum_{\mathbf{x}\in\mathbf{c}^\perp}(-1)^{a_2} + \imath(-1)^{a_1\oplus 1}\sum_{\mathbf{x}\in\mathbb{F}_2^n\backslash\mathbf{c}^\perp}(-1)^{a_2} = (-1)^{a_2}2^{n-1}(1+(-1)^{a_1\oplus 1}).$$

Thus, if $f \in S_{\mathbf{c}}$, then after measuring the first n qubits of $|\psi_3\rangle$ we will observe the $|\mathbf{0}\rangle$ or the $|\mathbf{c}\rangle$ state each with probability $\frac{1}{2}$. The probability is zero that any other state is observed.

On the other hand, if $f \in T_{\mathbf{c}}$ then

$$\sum_{\mathbf{x} \in \mathbf{c}^\perp} (-1)^{f(\mathbf{x}) \oplus s_{\mathbf{c}}(\mathbf{x})} = \sum_{\mathbf{x} \in \mathbb{F}_2^n \setminus \mathbf{c}^\perp} (-1)^{f(\mathbf{x}) \oplus s_{\mathbf{c}}(\mathbf{x})} = 0.$$

The probability amplitudes of the first n qubits of $|\psi_3\rangle$ for the states $|\mathbf{0}\rangle$ and $|\mathbf{c}\rangle$ are

$$\sum_{\mathbf{x} \in \mathbf{c}^\perp} (-1)^{s_{\mathbf{c}}(\mathbf{x}) \oplus f(\mathbf{x})} + \imath \sum_{\mathbf{x} \in \mathbb{F}_2^n \setminus \mathbf{c}^\perp} (-1)^{s_{\mathbf{c}}(\mathbf{x}) \oplus f(\mathbf{x})} = 0 + \imath 0 \text{ and}$$

$$\sum_{\mathbf{x} \in \mathbf{c}^\perp} (-1)^{\mathbf{x} \cdot \mathbf{c} \oplus s_{\mathbf{c}}(\mathbf{x}) \oplus f(\mathbf{x})} + \imath \sum_{\mathbf{x} \in \mathbb{F}_2^n \setminus \mathbf{c}^\perp} (-1)^{\mathbf{x} \cdot \mathbf{c} \oplus s_{\mathbf{c}}(\mathbf{x}) \oplus f(\mathbf{x})}$$

$$= \sum_{\mathbf{x} \in \mathbf{c}^\perp} (-1)^{s_{\mathbf{c}}(\mathbf{x}) \oplus f(\mathbf{x})} - \imath \sum_{\mathbf{x} \in \mathbb{F}_2^n \setminus \mathbf{c}^\perp} (-1)^{s_{\mathbf{c}}(\mathbf{x}) \oplus f(\mathbf{x})} = 0 + \imath 0,$$

respectively. Therefore, the observation of either the state $|\psi_3\rangle$ or the state $|\psi_3\rangle$ implies $f \in S_{\mathbf{c}}$, otherwise $f \in T_{\mathbf{c}}$.   □

Form this theorem we obtain Algorithm 2 which can distinguish Boolean functions from a larger set than the set of constant and balanced functions. It is to be noted that, if $wt(\mathbf{c}) = 0$, then we obtain the traditional Deutsch-Jozsa algorithm with all H gates.

5   Conclusion

In this paper, we have studied algorithms related to the *HN*-transform which is a generalization of the well known (Walsh-)Hadamard transform. First we presented an $O(2^{2n})$ algorithm to obtain all the values in the *HN*-spectra. Then we show that the Deutsch-Jozsa algorithm can be generalized considering the *HN*-transform. These results have application in cryptology, coding theory and related areas. While results related to *HN*-spectra have been investigated for more than a decade, a disciplined study of the related computing algorithms had not been attempted earlier, and that is the main goal of this paper.

References

1. Aumasson, J.-P., Dinur, I., Meier, W., Shamir, A.: Cube testers and key recovery attacks on reduced-round MD6 and trivium. In: Dunkelman, O. (ed.) FSE 2009. LNCS, vol. 5665, pp. 1–22. Springer, Heidelberg (2009). doi:10.1007/978-3-642-03317-9_1
2. Cusick, T.W., Stănică, P.: Cryptographic Boolean Functions and Applications, 2nd edn. Academic Press, San Diego (2017). 1st edn. (2009)

3. Danielsen, L.E., Parker, M.G.: Spectral orbits and peak-to-average power ratio of boolean functions with respect to the I, H, N^n transform. In: Helleseth, T., Sarwate, D., Song, H.-Y., Yang, K. (eds.) SETA 2004. LNCS, vol. 3486, pp. 373–388. Springer, Heidelberg (2005). doi:10.1007/11423461_28

4. Danielsen, L.E.: On connections between graphs, codes, quantum states, and Boolean functions. Ph.D. thesis, Department of Informatics, The Selmer Center, University of Bergen, Norway (2008)

5. Deutsch, D., Jozsa, R.: Rapid solution of problems by quantum computation. Proc. Roy. Soc. Lond. **A439**, 553–558 (1992)

6. Dillon, J.F.: Elementary Hadamard difference sets. Ph.D. thesis, University of Maryland (1974)

7. Dinur, I., Shamir, A.: Cube attacks on tweakable black box polynomials. In: Joux, A. (ed.) EUROCRYPT 2009. LNCS, vol. 5479, pp. 278–299. Springer, Heidelberg (2009). doi:10.1007/978-3-642-01001-9_16. See also: Cube Attacks on Tweakable Black Box Polynomials. http://eprint.iacr.org/2008/385.pdf

8. Gangopadhyay, S., Pasalic, E., Stănică, P.: A note on generalized bent criteria for boolean functions. IEEE Trans. Inf. Theor. **59**(5), 3233–3236 (2013)

9. Gangopadhyay, S., Gangopadhyay, A.K., Pollatos, S., Stănică, P.: Cryptographic Boolean functions with biased inputs. Crypt. Commun. Discrete Struct. Seq. **9**, 301–314 (2017). doi:10.1007/s12095-015-0174-1

10. Helleseth, T., Kløve, T., Mvkkeltveit, J.: On the covering radius of binary codes. IEEE Trans. Inf. Theor. **24**(5), 627–628 (1978)

11. Knudsen, L.R.: Truncated and higher order differentials. In: Preneel, B. (ed.) FSE 1994. LNCS, vol. 1008, pp. 196–211. Springer, Heidelberg (1995). doi:10.1007/3-540-60590-8_16

12. Litsyn, S., Shpunt, A.: On the distribution of Boolean function nonlinearity. SIAM J. Discrete Math. **23**(1), 79–95 (2008)

13. Maitra, S., Mukhopadhyay, P.: Deutsch-Jozsa algorithm revisited in the domain of cryptographically significant boolean functions. Int. J. Quantum Inf. **3**(2), 359–370 (2005)

14. Matsui, M.: Linear cryptanalysis method for DES cipher. In: Helleseth, T. (ed.) EUROCRYPT 1993. LNCS, vol. 765, pp. 386–397. Springer, Heidelberg (1994). doi:10.1007/3-540-48285-7_33

15. Meier, W., Staffelbach, O.: Fast correlation attacks on stream ciphers. In: Barstow, D., Brauer, W., Brinch Hansen, P., Gries, D., Luckham, D., Moler, C., Pnueli, A., Seegmüller, G., Stoer, J., Wirth, N., Günther, C.G. (eds.) EUROCRYPT 1988. LNCS, vol. 330, pp. 301–314. Springer, Heidelberg (1988). doi:10.1007/3-540-45961-8_28

16. Parker, M.G.: Generalised S-box nonlinearity. NESSIE Public Document, 11.02.03: NES/DOC/UIB/WP5/020/A

17. Parker, M.G., Pott, A.: On boolean functions which are bent and negabent. In: Golomb, S.W., Gong, G., Helleseth, T., Song, H.-Y. (eds.) SSC 2007. LNCS, vol. 4893, pp. 9–23. Springer, Heidelberg (2007). doi:10.1007/978-3-540-77404-4_2

18. Patterson, N.J., Wiedemann, D.H.: The covering radius of the $(2^{15}, 16)$ Reed-Muller code is at least 16276. IEEE Trans. Inf. Theor. **29**(3), 354–356 (1983). See also correction: IEEE Trans. Inf. Theor. **36**(2), 443 (1990)

19. Riera, C.: Spectral properties of Boolean functions, graphs and graph states. Ph.D. thesis, University of Bergen (2005)

20. Riera, C., Parker, M.G.: Generalized bent criteria for Boolean functions. IEEE Trans. Inf. Theor. **52**(9), 4142–4159 (2006)

21. Schmidt, K.-U., Parker, M.G., Pott, A.: Negabent functions in the Maiorana–McFarland class. In: Golomb, S.W., Parker, M.G., Pott, A., Winterhof, A. (eds.) SETA 2008. LNCS, vol. 5203, pp. 390–402. Springer, Heidelberg (2008). doi:10.1007/978-3-540-85912-3_34
22. Siegenthaler, T.: Decrypting a class of stream ciphers using ciphertext only. IEEE Trans. Comput. **34**(1), 81–85 (1985)
23. Stănică, P., Maitra, S.: Rotation symmetric Boolean functions - count and cryptographic properties. Disc. Appl. Math. **156**, 1567–1580 (2008)
24. Stănică, P., Gangopadhyay, S., Chaturvedi, A., Kar-Gangopadhyay, A., Maitra, S.: Investigations on bent and negabent functions via the nega-Hadamard transform. IEEE Trans. Inf. Theor. **58**(6), 4065–4072 (2012)

Explicit Characterizations for Plateaued-ness of p-ary (Vectorial) Functions

Claude Carlet[1,3], Sihem Mesnager[1,3,4($\boxtimes$)], Ferruh Özbudak[5,6],
and Ahmet Sınak[2,6,7]

[1] Department of Mathematics, University of Paris VIII, Saint-Denis, France
{claude.carlet,smesnager}@univ-paris8.fr
[2] LAGA, UMR 7539, CNRS, University of Paris VIII, Saint-Denis, France
[3] LAGA, UMR 7539, CNRS, University of Paris XIII, Villetaneuse, France
[4] Telecom ParisTech, Paris, France
[5] Department of Mathematics, Middle East Technical University, Ankara, Turkey
{ozbudak,sahmet}@metu.edu.tr
[6] Institute of Applied Mathematics,
Middle East Technical University, Ankara, Turkey
[7] Department of Mathematics and Computer Sciences,
Necmettin Erbakan University, Konya, Turkey

Abstract. Plateaued (vectorial) functions have an important role in the sequence and cryptography frameworks. Given their importance, they have not been studied in detail in general framework. Several researchers found recently results on their characterizations and introduced new tools to understand their structure and to design such functions. In this work, we mainly extend some of the observations made in characteristic 2 and given in (Carlet, IEEE Trans. Inf. Theor. **61**(11), 6272–6289, 2015) to arbitrary characteristic. We first extend to arbitrary characteristic the characterizations of plateaued (vectorial) Boolean functions by the auto-correlation functions, next their characterizations in terms of the second-order derivatives, and finally their characterizations via the moments of the Walsh transform.

Keywords: Vectorial functions · p-ary functions · Bent functions · Plateaued functions

1 Introduction

Boolean bent functions were introduced by Rothaus [16] in the 1970s and then generalized to arbitrary characteristic by Kumar et al. [9]. Since bent functions can never be balanced, Carlet (1993) introduced in [4] the superclass of the class of bent functions whose elements are the so-called partially-bent functions. As a further extension, Zheng and Zhang introduced in [17] the notion of Boolean plateaued functions whose absolute Walsh transform has at most one nonzero value. Moreover, Boolean plateaued functions were generalized to arbitrary characteristic: the so-called p-ary plateaued functions from $\mathbb{F}_{p^n}$ to $\mathbb{F}_p$ (see for instance

© Springer International Publishing AG 2017
S. El Hajji et al. (Eds.): C2SI 2017, LNCS 10194, pp. 328–345, 2017.
DOI: 10.1007/978-3-319-55589-8_22

[6,13]). The vectorial p-ary plateaued functions from $\mathbb{F}_{p^n}$ to $\mathbb{F}_{p^m}$ are the functions whose component functions are p-ary plateaued. This notion covers the vectorial bent functions. Since plateaued (vectorial) functions have a significant role in the sequence and cryptography frameworks, several researchers have widely studied those functions (see for instance [1–3,5,6,8,10–13,18]). More precisely, new characterizations of p-ary plateaued functions by using the moment of the Walsh transform and the second-order derivative were provided (2014) in [10], and then those characterizations were completed and extended in [12,13]. Independently, Carlet [2,3] introduced several characterizations of plateaued (vectorial) Boolean functions by means of the first-order and second-order derivatives, autocorrelation functions and power moments of the Walsh transform values. The aim of this paper is mainly to extend some characterizations of plateaued (vectorial) Boolean functions given in [2,3] to arbitrary characteristic.

The paper is structured as follows. Section 2 sets the necessary background. In Sect. 3, we extend to arbitrary characteristic the characterizations of plateaued (vectorial) Boolean functions given in [2,3] and give more characterizations, by means of the autocorrelation function. Section 4 generally extends the characterizations of plateaued (vectorial) Boolean functions given in [2,3] to arbitrary characteristic as well as providing the characterizations of plateaued (vectorial) p-ary functions, in terms of the first-order and second-order derivatives. Section 5 extends to arbitrary characteristic the characterizations of plateaued (vectorial) Boolean functions given in [2,3], and introduces further results by means of the power moments of the Walsh transform.

2 Preliminaries

For any set E, $\#E$ denotes the size of E and $E^{\star} = E \backslash \{0\}$. Let $\mathbb{C}$ be the field of complex numbers. Given a complex number $z \in \mathbb{C}$, $|z|$ and $\bar{z}$ denote the absolute value and the conjugate of z, respectively. Let p be a prime number and n be a positive integer. The finite field with p^n elements is denoted by $\mathbb{F}_{p^n}$, which can be viewed as an n-dimensional vector space over $\mathbb{F}_p$, and it is denoted by $\mathbb{F}_p^n$. The trace function $\mathrm{Tr} : \mathbb{F}_{p^n} \to \mathbb{F}_p$ is defined as

$$\mathrm{Tr}_{p^n/p}(x) = \sum_{i=0}^{n-1} x^{p^i} = x + x^p + x^{p^2} + \cdots + x^{p^{n-1}},$$

which is called *the absolute trace* of $x \in \mathbb{F}_{p^n}$. Let f be a function from $\mathbb{F}_p^n$ to $\mathbb{F}_p$. We can give a corresponding complex-valued function χ_f from $\mathbb{F}_p^n$ to $\mathbb{C}$ defined as $\chi_f(x) = \xi_p^{f(x)}$ for all $x \in \mathbb{F}_p^n$ where $\xi_p = e^{(2\pi\sqrt{-1})/p}$ is a complex primitive p-th root of unity. The Walsh transform of f is the Fourier transform $\widehat{\chi_f}$ from $\mathbb{F}_p^n$ to $\mathbb{C}$ of χ_f defined as

$$\widehat{\chi_f}(\omega) = \sum_{x \in \mathbb{F}_p^n} \xi_p^{f(x) - \omega \cdot x}$$

for all $\omega \in \mathbb{F}_p^n$ where "." denotes an inner product (for instance, the usual inner product) in $\mathbb{F}_p^n$. We can take $\omega \cdot x = \mathrm{Tr}_{p^n/p}(\omega x)$ if $\mathbb{F}_p^n$ is identified with $\mathbb{F}_{p^n}$. The Walsh support of f is defined as the set $\{\omega \in \mathbb{F}_p^n : \widehat{\chi}_f(\omega) \neq 0\}$. A p-ary function f can be recovered from $\widehat{\chi}_f$ by the inverse transform:

$$\xi_p^{f(x)} = \frac{1}{p^n} \sum_{b \in \mathbb{F}_{p^n}} \widehat{\chi}_f(b) \xi_p^{b \cdot x}.$$

A p-ary function f is called *plateaued* if its absolute Walsh transform takes only one nonzero value (and also possibly the value 0). More precisely, f is p-ary *bent* if $|\widehat{\chi}_f(\omega)|^2 = p^n$ for all $\omega \in \mathbb{F}_{p^n}$, and f is said to be p-ary s-*plateaued* (i.e., *plateaued of amplitude μ*) if $|\widehat{\chi}_f(\omega)|^2 \in \{0, \mu^2\}$ for all $\omega \in \mathbb{F}_{p^n}$ (where $\mu^2 = p^{n+s}$ for an integer s with $0 \leq s \leq n$, and μ is called the amplitude of plateaued p-ary function). By definition, p-ary bent functions are p-ary 0-plateaued functions and hence, all results about p-ary plateaued functions in this paper are valid for p-ary bent functions.

A function F from $\mathbb{F}_p^n$ to $\mathbb{F}_p^m$ is called an (n, m)-p-ary function or vectorial p-ary function. The component functions of F are in the form $f_\lambda = \lambda \cdot F$ for $\lambda \in \mathbb{F}_p^m \backslash \{0\}$ defined as $f_\lambda(x) = \lambda \cdot F(x)$ for all $x \in \mathbb{F}_p^n$, where "." denotes an inner product in $\mathbb{F}_p^m$. The vector spaces $\mathbb{F}_p^n$ and $\mathbb{F}_p^m$ can be identified with the Galois fields $\mathbb{F}_{p^n}$ and $\mathbb{F}_{p^m}$ of orders p^n and p^m, respectively. Then the component functions $f_\lambda, \lambda \in \mathbb{F}_{p^m}^\star$, of F are defined as $f_\lambda(x) = \mathrm{Tr}_{p^m/p}(\lambda F(x))$ for all $x \in \mathbb{F}_{p^n}$. A p-ary function F is called *vectorial plateaued* if its component functions $f_\lambda, \lambda \in \mathbb{F}_{p^m}^\star$, are p-ary plateaued with possibly different amplitudes. In particular, F is called *vectorial s-plateaued* (i.e., *plateaued with single amplitude*) p-ary function if its component functions $f_\lambda, \lambda \in \mathbb{F}_{p^m}^\star$, are p-ary s-plateaued (see for instance [13]). Notice that vectorial p-ary bent functions are vectorial p-ary 0-plateaued functions.

Remark 1. A vectorial p-ary function is plateaued if and only if all its component functions are p-ary plateaued with possibly different amplitudes. Moreover, a vectorial p-ary function is plateaued with single amplitude if and only if all its component functions are p-ary plateaued of the same amplitude. Those facts are frequently used in the sequel.

For a nonnegative integer i, the even power moment of the Walsh transform of f is given as

$$S_i(f) = \sum_{\omega \in \mathbb{F}_{p^n}} |\widehat{\chi}_f(\omega)|^{2i}$$

with the convention $S_0(f) = p^n$. It is a well known fact that $S_1(f) = p^{2n}$ is known as the *Parseval identity*. By the Parseval identity, we have (see for instance [10]):

Lemma 1. *Let $f : \mathbb{F}_{p^n} \to \mathbb{F}_p$ be p-ary s-plateaued where s is an integer with $0 \leq s \leq n$. Then for $\omega \in \mathbb{F}_{p^n}$, $|\widehat{\chi}_f(\omega)|^2$ takes p^{n-s} times the value p^{n+s} and $p^n - p^{n-s}$ times the value 0.*

For every nonnegative integers A and i, we have

$$\sum_{\omega \in \mathbb{F}_{p^n}} \left(|\widehat{\chi_f}(\omega)|^2 - A \right)^2 |\widehat{\chi_f}(\omega)|^{2i} = S_{i+2}(f) - 2AS_{i+1}(f) + A^2 S_i(f) \geq 0. \quad (1)$$

For a positive integer i and $A = p^{n+s}$ with an integer $1 \leq s \leq n$, the inequality (1) becomes an equality if and only if f is p-ary s-plateaued. For instance for $i = 1$, f is p-ary s-plateaued if and only if $S_3(f) + p^{4n+2s} = 2p^{n+s}S_2(f)$.

We finally introduce the following notations. A p-ary function f is said to be *balanced* over $\mathbb{F}_p$ if $\#\{x \in \mathbb{F}_{p^n} : f(x) = k\} = p^{n-1}$ for each $k \in \mathbb{F}_p$, i.e., f takes every value of $\mathbb{F}_p$ the same number, p^{n-1} times. The first-order derivative of f at $a \in \mathbb{F}_{p^n}$ is the map $\mathcal{D}_a f : \mathbb{F}_{p^n} \to \mathbb{F}_p$ defined as $\mathcal{D}_a f(x) = f(x + a) - f(x)$ for $x \in \mathbb{F}_{p^n}$. The second-order derivative of f at $(a, b) \in \mathbb{F}_{p^n}^2$ is given as $\mathcal{D}_b \mathcal{D}_a f(x) = f(x + a + b) - f(x + a) - f(x + b) + f(x)$ for $x \in \mathbb{F}_{p^n}$. For $(a, b) \in \mathbb{F}_{p^n}^2$, one can readily see $\mathcal{D}_b \mathcal{D}_a f(x) = \mathcal{D}_a \mathcal{D}_b f(x)$ for $x \in \mathbb{F}_{p^n}$. Similarly, the first-order derivative of vectorial F at $a \in \mathbb{F}_{p^n}$ is defined as

$$\mathcal{D}_a F(x) = F(x + a) - F(x),$$

and the second-order derivative of F at $(a, b) \in \mathbb{F}_{p^n}^2$ is given as $\mathcal{D}_b \mathcal{D}_a F(x) = F(x + a + b) - F(x + a) - F(x + b) + F(x)$ for $x \in \mathbb{F}_{p^n}$. The autocorrelation function of f is a map from $\mathbb{F}_{p^n}$ to $\mathbb{C}$ defined as

$$a \mapsto \mathcal{F}(\mathcal{D}_a f) = \sum_{x \in \mathbb{F}_{p^n}} \xi_p^{f(x+a)-f(x)},$$

which can be denoted by $\varDelta_f(a)$ for all $a \in \mathbb{F}_{p^n}$. Let $G_1, G_2 : \mathbb{F}_{p^n} \to \mathbb{C}$ be two functions. The convolution of G_1 and G_2 is a map from $\mathbb{F}_{p^n}$ to $\mathbb{C}$ defined as

$$(G_1 \otimes G_2)(u) = \sum_{x \in \mathbb{F}_{p^n}} G_1(x) G_2(u - x)$$

for all $u \in \mathbb{F}_{p^n}$ (see for instance [14, Definition 10.1.18]).

3   Characterizations by the Autocorrelation Functions

In this section, we begin extending the characterizations of plateaued (vectorial) Boolean functions to arbitrary characteristic by presenting further characterizations of those functions in arbitrary characteristic by the autocorrelation functions.

We start with the known elementary materials related to the Fourier transform, convolution and autocorrelation function, which are useful to prove the main results in this section.

Let $G : \mathbb{F}_{p^n} \to \mathbb{C}$ be a function, and let $\widehat{G}(v) = \sum_{u \in \mathbb{F}_{p^n}} G(u) \xi_p^{-u \cdot v}$ be its Fourier transform for $v \in \mathbb{F}_{p^n}$. Then for all $u \in \mathbb{F}_{p^n}$

$$\widehat{\widehat{G}}(u) = p^n G(-u). \quad (2)$$

By (2), one can find the known relation $G(-u) = \frac{1}{p^n} \sum_{v \in \mathbb{F}_{p^n}} \widehat{G}(v) \xi_p^{-v \cdot u}$ for all $u \in \mathbb{F}_{p^n}$. Therefore, $G(u) = 0$ for all $u \in \mathbb{F}_{p^n}$ if and only if $\widehat{G}(v) = 0$ for all $v \in \mathbb{F}_{p^n}$. Let $G_1, G_2 : \mathbb{F}_{p^n} \to \mathbb{C}$, then

$$G_1(u) = G_2(u),\ \forall u \in \mathbb{F}_{p^n} \iff \widehat{G}_1(v) = \widehat{G}_2(v),\ \forall v \in \mathbb{F}_{p^n}. \tag{3}$$

We recall the convolution theorem of Fourier analysis (see for instance [14, Theorem 10.1.19]).

Proposition 1 *[14]. Let $G_1, G_2 : \mathbb{F}_{p^n} \to \mathbb{C}$. Then we have $\widehat{G_1 \otimes G_2} = \widehat{G}_1 \widehat{G}_2$. Moreover, $\widehat{G}_1 \otimes \widehat{G}_2 = p^n \widehat{G_1 G_2}$.*

Let $f : \mathbb{F}_{p^n} \to \mathbb{F}_p$ be a p-ary function and let $\Delta_f(a) = \sum_{x \in \mathbb{F}_{p^n}} \xi_p^{f(x+a)-f(x)}$ be its autocorrelation function for $a \in \mathbb{F}_{p^n}$. By (2), we have

$$\widehat{\widehat{\Delta_f}}(a) = p^n \Delta_f(-a) \tag{4}$$

for all $a \in \mathbb{F}_{p^n}$. As readily seen, for all $a \in \mathbb{F}_{p^n}$

$$\overline{\Delta_f}(a) = \sum_{x \in \mathbb{F}_{p^n}} \xi_p^{f(x)-f(x+a)} = \Delta_f(-a) \tag{5}$$

where we used the bijective change of variable $x \mapsto x - a$. The Walsh transform $\widehat{\chi}_f$ of f satisfies

$$|\widehat{\chi_f}(b)|^2 = \sum_{x,y \in \mathbb{F}_{p^n}} \xi_p^{f(x)-f(y)-b \cdot (x-y)} = \sum_{x,y \in \mathbb{F}_{p^n}} \xi_p^{f(x+y)-f(y)-b \cdot x} = \widehat{\Delta_f}(b) \tag{6}$$

for all $b \in \mathbb{F}_{p^n}$ where in the second equality we used the bijective change of variable $x \mapsto x + y$. Hence, combining (4), (5) and (6), we have $|\widehat{\widehat{\chi_f}}(a)|^2 = p^n \overline{\Delta_f}(a)$ for all $a \in \mathbb{F}_{p^n}$. Moreover, by Proposition 1, the Fourier transform of $|\widehat{\chi}_f|^4$ is obtained as

$$\widehat{|\widehat{\chi_f}|^2 |\widehat{\chi_f}|^2} = p^{-n} \left(\widehat{|\widehat{\chi_f}|^2} \otimes \widehat{|\widehat{\chi_f}|^2} \right) = p^{-n} \left(p^n \overline{\Delta_f} \otimes p^n \overline{\Delta_f} \right) = p^n \left(\overline{\Delta_f} \otimes \overline{\Delta_f} \right). \tag{7}$$

We first characterize p-ary plateaued functions by considering the Fourier transforms of their absolute Walsh transforms. By definition of p-ary plateaued functions, we can say that f is p-ary plateaued of amplitude μ if and only if the two functions $|\widehat{\chi}_f|^4$ and $\mu^2 |\widehat{\chi}_f|^2$ are equal; equivalently, by (3) their Fourier transforms are equal. We now present the following characterization of plateaued p-ary functions.

Theorem 1. *Let $f : \mathbb{F}_{p^n} \to \mathbb{F}_p$. Then, f is p-ary plateaued of amplitude μ if and only if for all $x \in \mathbb{F}_{p^n}$*

$$\sum_{a \in \mathbb{F}_{p^n}} \Delta_f(a) \Delta_f(x - a) = \mu^2 \Delta_f(x). \tag{8}$$

Let $F : \mathbb{F}_{p^n} \to \mathbb{F}_{p^m}$ and let f_λ, $\lambda \in \mathbb{F}_{p^m}^\star$, be the component functions of F. Then, F is p-ary plateaued with single amplitude if and only if for all $x \in \mathbb{F}_{p^n}$ and $\lambda \in \mathbb{F}_{p^m}^\star$, we have $\sum_{a \in \mathbb{F}_{p^n}} \Delta_{f_\lambda}(a) \Delta_{f_\lambda}(x - a) = \mu^2 \Delta_{f_\lambda}(x)$.

Proof. As stated above, f is p-ary plateaued of amplitude μ if and only if the two functions $\overline{\Delta_f} \otimes \overline{\Delta_f}$ and $\mu^2 \overline{\Delta_f}$ are equal; equivalently, $(\Delta_f \otimes \Delta_f)(x) = \mu^2 \Delta_f(x)$ for all $x \in \mathbb{F}_{p^n}$ by (5). This proves the first assertion. By Remark 1, the second statement is a direct consequence of the first statement.   □

Now let us compute the Fourier transform of $|\widehat{\chi_f}|^6$, by Proposition 1,

$$\widehat{|\widehat{\chi_f}|^2 |\widehat{\chi_f}|^4} = p^{-n} \left(\widehat{|\widehat{\chi_f}|^2} \otimes \widehat{|\widehat{\chi_f}|^4} \right) = p^n \left(\overline{\Delta_f} \otimes \overline{\Delta_f} \otimes \overline{\Delta_f} \right)$$

where we used (7) in the last equality. We can present the following result, which may be less practical than Theorem 1.

Theorem 2. *Let* $f : \mathbb{F}_{p^n} \to \mathbb{F}_p$. *Then,* f *is* p-ary plateaued of amplitude μ if and only if for all $x \in \mathbb{F}_{p^n}$

$$\sum_{a,b \in \mathbb{F}_{p^n}} \Delta_f(a) \Delta_f(b) \Delta_f(x - a - b) = \mu^2 \sum_{c \in \mathbb{F}_{p^n}} \Delta_f(c) \Delta_f(x - c).$$

Let $F : \mathbb{F}_{p^n} \to \mathbb{F}_{p^m}$ *and let* f_λ, $\lambda \in \mathbb{F}_{p^m}^\star$, *be the component functions of* F. *Then,* F *is* p-ary plateaued with single amplitude if and only if for all $x \in \mathbb{F}_{p^n}$ *and* $\lambda \in \mathbb{F}_{p^m}^\star$,

$$\sum_{a,b \in \mathbb{F}_{p^n}} \Delta_{f_\lambda}(a) \Delta_{f_\lambda}(b) \Delta_{f_\lambda}(x - a - b) = \mu^2 \sum_{c \in \mathbb{F}_{p^n}} \Delta_{f_\lambda}(c) \Delta_{f_\lambda}(x - c).$$

Proof. As in the proof of Theorem 1, f is p-ary plateaued of amplitude μ if and only if the two functions $|\widehat{\chi_f}|^6$ and $\mu^2 |\widehat{\chi_f}|^4$ are equal; equivalently, from (3) and (5), $(\Delta_f \otimes \Delta_f \otimes \Delta_f)(x) = \mu^2 (\Delta_f \otimes \Delta_f)(x)$ for all $x \in \mathbb{F}_{p^n}$. By Remark 1, the proof of the second statement is a direct consequence of the first statement.   □

In order to characterize vectorial plateaued p-ary functions whose component functions may have different amplitudes, we need to get rid of the μ^2 in (8). Then putting $x = 0$ in (8), we have $\sum_{a \in \mathbb{F}_{p^n}} |\Delta_f(a)|^2 = \mu^2 \Delta_f(0) = \mu^2 p^n$ by (5) since $\Delta_f(0) = p^n$. Hence we can directly derive from Theorem 1 the following result.

Theorem 3. *Let* $f : \mathbb{F}_{p^n} \to \mathbb{F}_p$. *Then,* f *is* p-ary plateaued if and only if for all $x \in \mathbb{F}_{p^n}$,

$$p^n \sum_{a \in \mathbb{F}_{p^n}} \Delta_f(a) \Delta_f(x - a) = \sum_{a \in \mathbb{F}_{p^n}} |\Delta_f(a)|^2 \Delta_f(x).$$

Let $F : \mathbb{F}_{p^n} \to \mathbb{F}_{p^m}$, *and* f_λ, $\lambda \in \mathbb{F}_{p^m}^\star$, *be the component functions of* F. *Then,* F *is* p-ary plateaued if and only if for all $x \in \mathbb{F}_{p^n}$ *and* $\lambda \in \mathbb{F}_{p^m}^\star$,

$$p^n \sum_{a \in \mathbb{F}_{p^n}} \Delta_{f_\lambda}(a) \Delta_{f_\lambda}(x - a) = \sum_{a \in \mathbb{F}_{p^n}} |\Delta_{f_\lambda}(a)|^2 \Delta_{f_\lambda}(x).$$

We can rewrite Theorem 3 as follows. A p-ary function f is plateaued if and only if for all $x \in \mathbb{F}_{p^n}$ (by the bijective change of variable $a \mapsto a - b$)

$$p^n \sum_{a,b,c \in \mathbb{F}_{p^n}} \xi_p^{-f(a)+f(b)+f(c)-f(-a+b+c+x)}$$
$$= \sum_{a,b,c,d \in \mathbb{F}_{p^n}} \xi_p^{-f(a)+f(b)+f(c)-f(-a+b+c)+f(d)-f(d+x)}.$$

For vectorial $F : \mathbb{F}_{p^n} \to \mathbb{F}_{p^m}$, we can write this by considering $f = \lambda \cdot F$ for $\lambda \in \mathbb{F}_{p^m}^*$. Applying the Fourier transform, by (3) their Fourier transforms are equal, and hence we deduce (see the proof of Theorem 5) that:

Corollary 1. *A vectorial function F is plateaued if and only if for all $x \in \mathbb{F}_{p^n}$ and $v \in \mathbb{F}_{p^m}$ $p^n \#\{(a, b, c) \in \mathbb{F}_{p^n}^3 : -F(a)+F(b)+F(c)-F(-a+b+c+x) = v\} = \#\{(a, b, c, d) \in \mathbb{F}_{p^n}^4 : -F(a)+F(b)+F(c)-F(-a+b+c)+F(d)-F(d+x) = v\}.$*

One can obtain the following characterization of p-ary bent (vectorial) functions by means of the autocorrelation function (see for instance [9, Property 4]).

Corollary 2. *Let $f : \mathbb{F}_{p^n} \to \mathbb{F}_p$. Then we have*

$$\sum_{a \in \mathbb{F}_{p^n}} |\Delta_f(a)|^2 \geq p^{2n}, \tag{9}$$

with an equality if and only if f is p-ary bent. Let $F : \mathbb{F}_{p^n} \to \mathbb{F}_{p^m}$, and let f_λ, $\lambda \in \mathbb{F}_{p^m}^*$, be the component functions of F. Then F is vectorial p-ary bent if and only if $\sum_{a \in \mathbb{F}_{p^n}} |\Delta_{f_\lambda}(a)|^2 = p^{2n}$ for all $\lambda \in \mathbb{F}_{p^m}^*$.

Proof. Recall that $\Delta_f(a) = \sum_{x \in \mathbb{F}_{p^n}} \xi_p^{f(x+a)-f(x)}$ for $a \in \mathbb{F}_{p^n}$. For any function f, $\Delta_f(0) = p^n$ and $|\Delta_f(a)| \geq 0$ for all $a \in \mathbb{F}_{p^n}^*$. Hence, the bound in (9) holds for every function, and it is achieved by bent p-ary functions because of the fact that f is p-ary bent if and only if $\Delta_f(a) = 0$ for all $a \in \mathbb{F}_{p^n}^*$. Thanks to the well known fact that a vectorial p-ary function is bent if and only if all its components are p-ary bent, the proof of the second statement is a direct consequence of the first statement.   □

4   Characterizations in Terms of the Derivatives

This section generally extends to arbitrary characteristic the characterizations of plateaued (vectorial) Boolean functions in terms of the first-order and second-order derivatives given in [2,3]. More precisely, we characterize power plateaued p-ary functions and vectorial plateaued p-ary functions whose components are unbalanced. We finally extend the notion of strongly-plateaued functions.

The first characterization of Boolean plateaued functions in terms of the second-order derivatives was provided in [5], and has been extended to arbitrary characteristic in [2,13].

Theorem 4 *[13]. Let $f : \mathbb{F}_{p^n} \to \mathbb{F}_p$. Set $\theta_f(x) = \sum_{a,b \in \mathbb{F}_{p^n}} \xi_p^{\mathcal{D}_b \mathcal{D}_a f(x)}$ for $x \in \mathbb{F}_{p^n}$. Then, f is p-ary plateaued of amplitude μ if and only if $\theta_f(x) = \mu^2$ for all $x \in \mathbb{F}_{p^n}$. In particular, f is p-ary bent if and only if $\theta_f(x) = p^n$ for all $x \in \mathbb{F}_{p^n}$.*

The extension of [2, 3, Theorem 1] to arbitrary characteristic was given in [12].

Theorem 5 *[12]. Let $F, G : \mathbb{F}_{p^n} \to \mathbb{F}_{p^m}$. For $v \in \mathbb{F}_{p^m}$ and $x \in \mathbb{F}_{p^n}$, let us denote by $N_F(v; x)$ the size of the set $\{(a, b) \in \mathbb{F}_{p^n}^2 : \mathcal{D}_b \mathcal{D}_a F(x) = v\}$. Then $N_F(v; x) = \#\{(a, b) \in \mathbb{F}_{p^n}^2 : \mathcal{D}_a F(b) - \mathcal{D}_a F(x) = v\}$ for all $v \in \mathbb{F}_{p^m}$ and $x \in \mathbb{F}_{p^n}$. Moreover,*

- *F is p-ary plateaued if and only if for all $v \in \mathbb{F}_{p^m}$, $N_F(v; x)$ is independent of $x \in \mathbb{F}_{p^n}$.*
- *F is p-ary plateaued with single amplitude if and only if there exist two integers u_1 and u_2 such that $N_F(0; x) = u_1$ and $N_F(v; x) = u_2$ for all $v \in \mathbb{F}_{p^m}^\star$ and $x \in \mathbb{F}_{p^n}$.*
- *For two plateaued p-ary functions F and G, if $N_F(v; x) = N_G(v; x)$ for all $v \in \mathbb{F}_{p^m}$, which means that F and G have the same distribution for $\mathcal{D}_b \mathcal{D}_a F(x)$ and $\mathcal{D}_b \mathcal{D}_a G(x)$, then the component functions f_λ and g_λ, $\lambda \in \mathbb{F}_{p^m}^\star$, are s_λ-plateaued p-ary functions.*

The following link between the second-order derivative and the fourth power moment of the Walsh transform was given in [10] (for characteristic 2, see [1]).

Proposition 2. *Let $f : \mathbb{F}_{p^n} \to \mathbb{F}_p$. Then, $S_2(f) = p^n \sum_{a,b,x \in \mathbb{F}_{p^n}} \xi_p^{\mathcal{D}_b \mathcal{D}_a f(x)}$.*

We can derive from Theorems 4 and 5 and Proposition 2 the following result.

Corollary 3. *Let $f : \mathbb{F}_{p^n} \to \mathbb{F}_p$. Set $\theta_f(x) = \sum_{a,b \in \mathbb{F}_{p^n}} \xi_p^{\mathcal{D}_b \mathcal{D}_a f(x)}$ for $x \in \mathbb{F}_{p^n}$. Then, f is p-ary plateaued if and only if $S_2(f) = p^{2n} \theta_f(x)$ for all $x \in \mathbb{F}_{p^n}$. Let $F : \mathbb{F}_{p^n} \to \mathbb{F}_{p^m}$, and let f_λ, $\lambda \in \mathbb{F}_{p^m}^\star$, be the component functions of F. Then F is p-ary plateaued if and only if for each $\lambda \in \mathbb{F}_{p^m}^\star$, $S_2(f_\lambda) = p^{2n} \theta_{f_\lambda}(x)$ for all $x \in \mathbb{F}_{p^n}$. In particular, F is p-ary plateaued with single amplitude if and only if, additionally, $S_2(f_\lambda)$ does not depend on λ for $\lambda \neq 0$.*

Proof. By Proposition 2, $S_2(f) = p^n \sum_{x \in \mathbb{F}_{p^n}} \theta_f(x)$. Assume that f is p-ary plateaued. By Theorem 4, $S_2(f) = p^n p^n \theta_f(x)$ for all $x \in \mathbb{F}_{p^n}$. Conversely, for all $x \in \mathbb{F}_{p^n}$ we have $\theta_f(x) = \theta$ where $\theta = p^{-2n} S_2(f)$. By Theorem 4, f is p-ary plateaued.

By Remark 1, the second statement is a direct consequence of the first statement.

The last assertion, with the above arguments, can be completed from Theorems 4 and 5 and Proposition 2. $\qquad \square$

Notice that for all $a, b, c \in \mathbb{F}_{p^n}$ we have $\mathcal{D}_a \mathcal{D}_b f_\lambda(c) = \lambda \cdot \mathcal{D}_a \mathcal{D}_b F(c)$ where $f_\lambda = \lambda \cdot F$ for $\lambda \in \mathbb{F}_{p^m}^\star$. By Proposition 2 and Corollary 3, F is p-ary plateaued if and only if for all $x \in \mathbb{F}_{p^n}$ and $\lambda \in \mathbb{F}_{p^m}^\star$

$$\sum_{a,b,c \in \mathbb{F}_{p^n}} \xi_p^{\lambda \cdot \mathcal{D}_a \mathcal{D}_b F(c)} = p^n \sum_{a,b \in \mathbb{F}_{p^n}} \xi_p^{\lambda \cdot \mathcal{D}_a \mathcal{D}_b F(x)},$$

equivalently, applying the Fourier transform, by (3) for all $x \in \mathbb{F}_{p^n}$ and $v \in \mathbb{F}_{p^m}$

$$\#\{(a,b,c) \in \mathbb{F}_{p^n}^3 : D_a D_b F(c) = v\} = p^n \#\{(a,b) \in \mathbb{F}_{p^n}^2 : D_a D_b F(x) = v\}, \quad (10)$$

that is, for all $v \in \mathbb{F}_{p^m}$, $\#\{(a,b) \in \mathbb{F}_{p^n}^2 : D_a D_b F(x) = v\}$ is independent of $x \in \mathbb{F}_{p^n}$. Thus, Corollary 3 can also be derived only from Theorem 5.

Remark 2. Notice that if we add an affine function to F, then plateaued-ness of F is preserved because it does not change the value of the second-order derivative of F. On the other hand, adding a quadratic function to F changes this value since the distribution of the second-order derivative of F is dependent on x in general. We indicate this in the following results which generalize those of [2].

Corollary 4. *Let $F : \mathbb{F}_{p^n} \to \mathbb{F}_{p^m}$. Then, F is p-ary plateaued if and only if for all $x \in \mathbb{F}_{p^n}$, there exists a permutation ϕ_x of $\mathbb{F}_{p^n}^2$ defined as $\phi_x(a,b) = (a_x, b_x)$ such that $D_b D_a F(x) = D_{b_x} D_{a_x} F(0)$; or equivalently, there exists a permutation ψ_x of $\mathbb{F}_{p^n}^2$ defined as $\psi_x(a,b) = (a'_x, b'_x)$ such that $D_a F(b) - D_a F(x) = D_{a'_x} F(b'_x) - D_{a'_x} F(0)$.*

Proof. For $v \in \mathbb{F}_{p^m}$ and $x \in \mathbb{F}_{p^n}$, let us define the following sets

$$\{(a,b) \in \mathbb{F}_{p^n}^2 : D_b D_a F(x) = v\} \text{ and } \{(a_x, b_x) \in \mathbb{F}_{p^n}^2 : D_{b_x} D_{a_x} F(0) = v\}. \quad (11)$$

Assume that F is p-ary plateaued. By Theorem 5, for each $x \in \mathbb{F}_{p^n}$ the sizes of the sets in (11) are equal for all $v \in \mathbb{F}_{p^m}$, then for all $x \in \mathbb{F}_{p^n}$ there exists a permutation ϕ_x of $\mathbb{F}_{p^n}^2$ from the first set (defined for some value of v and $x \neq 0$) to the second set (defined for the same value of v and for $x = 0$) in (11) defined as $\phi_x(a,b) = (a_x, b_x)$. Conversely, because of the permutation ϕ_x, for all $v \in \mathbb{F}_{p^m}$ and $x \in \mathbb{F}_{p^n}$, the sizes of the sets in (11) are equal. By Theorem 5, F is p-ary plateaued. The proof is complete.

For the second statement, we consider the following sets

$$\{(a,b) \in \mathbb{F}_{p^n}^2 : D_a F(b) - D_a F(x) = v\} \text{ and} \quad (12)$$
$$\{(a'_x, b'_x) \in \mathbb{F}_{p^n}^2 : D_{a'_x} F(b'_x) - D_{a'_x} F(0) = v\} \quad (13)$$

for $v \in \mathbb{F}_{p^m}$ and $x \in \mathbb{F}_{p^n}$. By Theorem 5, using the above arguments, F is p-ary plateaued if and only if for all $x \in \mathbb{F}_{p^n}$, there exists a permutation ψ_x of $\mathbb{F}_{p^n}^2$ mapping from the set in (12) to the set in (13) defined as $\psi_x(a,b) = (a'_x, b'_x)$. $\square$

Recall that $D_a F(b) - D_a F(x) = D_a D_{b-x} F(x)$ for $a, b, x \in \mathbb{F}_{p^n}$. Hence we have $\psi_x(a,b) = \phi_x(a, b-x)$ since $D_{a'_x} F(b'_x) - D_{a'_x} F(0) = D_a F(b) - D_a F(x) = D_a D_{b-x} F(x) = D_{a''_x} D_{b''_x} F(0) = D_{a''_x} F(b''_x) - D_{a''_x} F(0)$ where $\psi_x(a,b) = (a'_x, b'_x)$ and $\phi_x(a, b-x)$ is denoted by (a''_x, b''_x).

Remark 3. Notice that the simple permutation $\phi_x(a,b) = (a,b)$ for all $a, b \in \mathbb{F}_{p^n}$ correlates with quadratic functions. Actually, F admits such associated ϕ_x if and only if $D_b D_a F(c) = D_b D_a F(0)$ at $(a,b) \in \mathbb{F}_{p^n}^2$ for all $c \in \mathbb{F}_{p^n}$, that is, $D_c D_b D_a F(0) = 0$ at $(a,b,c) \in \mathbb{F}_{p^n}^3$, which means that it is a quadratic function.

Corollary 5. *Let $F : \mathbb{F}_{p^n} \to \mathbb{F}_{p^m}$ be p-ary plateaued, and for all $x \in \mathbb{F}_{p^n}$, let ϕ_x be a permutation defined by $\phi_x(a, b) = (a_x, b_x)$ as in Corollary 4. Let $G : \mathbb{F}_{p^n} \to \mathbb{F}_{p^m}$ be a function such that $\mathcal{D}_b \mathcal{D}_a G(x) = \mathcal{D}_{b_x} \mathcal{D}_{a_x} G(0)$ at $(a, b) \in \mathbb{F}_{p^n}^2$ for all $x \in \mathbb{F}_{p^n}$. Then, $F + G$ is p-ary plateaued.*

Proof. By Corollary 4, for all $x \in \mathbb{F}_{p^n}$, we have $\mathcal{D}_b \mathcal{D}_a F(x) = \mathcal{D}_{b_x} \mathcal{D}_{a_x} F(0)$ where $(a_x, b_x) = \phi_x(a, b)$. Then, for all $x \in \mathbb{F}_{p^n}$, $\mathcal{D}_b \mathcal{D}_a (F + G)(x) = \mathcal{D}_b \mathcal{D}_a F(x) + \mathcal{D}_b \mathcal{D}_a G(x) = \mathcal{D}_{b_x} \mathcal{D}_{a_x} F(0) + \mathcal{D}_{b_x} \mathcal{D}_{a_x} G(0) = \mathcal{D}_{b_x} \mathcal{D}_{a_x} (F + G)(0)$ where $(a_x, b_x) = \phi_x(a, b)$. Thus, $F + G$ is p-ary plateaued. $\qquad\square$

Remark 4. We derive from the above results that in general $F + G$ may not be p-ary plateaued when F is p-ary plateaued and G is quadratic. For a quadratic function G, although we have $\mathcal{D}_b \mathcal{D}_a G(x) = \mathcal{D}_b \mathcal{D}_a G(0)$ (see Remark 3), $\mathcal{D}_b \mathcal{D}_a G(x)$ may not equal $\mathcal{D}_{b_x} \mathcal{D}_{a_x} G(0)$ for some $x \in \mathbb{F}_{p^n}$ where $(a_x, b_x) = \phi_x(a, b)$ for the associated permutation ϕ_x of F.

We now investigate power functions on $\mathbb{F}_{p^n}$ in terms of the first-order derivatives. Power functions are widely studied because of their interesting algebraic and combinatorial properties, and their applications in sequence design, coding theory and cryptography.

Corollary 6. *Let F be a power function on $\mathbb{F}_{p^n}$ defined as $F(x) = x^d$. For $v, x \in \mathbb{F}_{p^n}$, let $N_F(v; x)$ be the size of the set $\{(a, b) \in \mathbb{F}_{p^n}^2 : \mathcal{D}_a F(b) - \mathcal{D}_a F(x) = v\}$. Then for all $v, x, \lambda \in \mathbb{F}_{p^n}$ with $\lambda \neq 0$,*

$$N_F(v; x) = \#\{(a, b) \in \mathbb{F}_{p^n}^2 : \mathcal{D}_a F(b) - \mathcal{D}_a F(x/\lambda) = v/\lambda^d\}. \tag{14}$$

In particular, for all $v \in \mathbb{F}_{p^n}$, $N_F(v; 0) = N_F(v/\lambda^d; 0)$ for any $\lambda \in \mathbb{F}_{p^n}^\star$. Moreover

i. F is p-ary plateaued if and only if $N_F(v; 1) = N_F(v; 0)$ for all $v \in \mathbb{F}_{p^n}$.
ii. F is p-ary plateaued with single amplitude if and only if $N_F(0; 1) = N_F(0; 0)$ and there exists an integer u such that $N_F(v; 1) = N_F(v; 0) = u$ for all $v \in \mathbb{F}_{p^n}^\star$.

If F is p-ary plateaued and $\gcd(d, p^n - 1) = 1$, then it has single amplitude.

Proof. For all $\lambda \in \mathbb{F}_{p^n}$ with $\lambda \neq 0$, by the bijective change of variable $a \mapsto \lambda a$ and $b \mapsto \lambda b$, we have $\#\{(a, b) \in \mathbb{F}_{p^n}^2 : \mathcal{D}_a F(b) - \mathcal{D}_a F(x) = v\} = \#\{(a, b) \in \mathbb{F}_{p^n}^2 : \mathcal{D}_{\lambda a} F(\lambda b) - \mathcal{D}_{\lambda a} F(x) = v\}$. For all $a, b, x, \lambda \in \mathbb{F}_{p^n}$ with $\lambda \neq 0$, we can easily see $\mathcal{D}_{\lambda a} F(\lambda b) = (\lambda b + \lambda a)^d - (\lambda b)^d = \lambda^d \mathcal{D}_a F(b)$ and $\mathcal{D}_{\lambda a} F(x) = \lambda^d \mathcal{D}_a F(x/\lambda)$. Hence, (14) holds for all $v, x, \lambda \in \mathbb{F}_{p^n}$ with $\lambda \neq 0$.
In particular, for $x = 0$ in (14), we have $\#\{(a, b) \in \mathbb{F}_{p^n}^2 : \mathcal{D}_a F(b) - \mathcal{D}_a F(0) = v\} = \#\{(a, b) \in \mathbb{F}_{p^n}^2 : \mathcal{D}_a F(b) - \mathcal{D}_a F(0) = v/\lambda^d\}$, that is, $N_F(v; 0) = N_F(v/\lambda^d; 0)$ for all $v, \lambda \in \mathbb{F}_{p^n}$ with $\lambda \neq 0$.
We now prove (i). By (14), for all $v \in \mathbb{F}_{p^n}$ we have (by taking $\lambda = x$ for $x \neq 0$)

$$N_F(v; x) = \#\{(a, b) \in \mathbb{F}_{p^n}^2 : \mathcal{D}_a F(b) - \mathcal{D}_a F(1) = v/x^d\}. \tag{15}$$

Assume that $N_F(v; 1) = N_F(v; 0)$ for all $v \in \mathbb{F}_{p^n}$. Then we have $\#\{(a, b) \in \mathbb{F}_{p^n}^2 : \mathcal{D}_a F(b) - \mathcal{D}_a F(1) = v/x^d\} = \#\{(a, b) \in \mathbb{F}_{p^n}^2 : \mathcal{D}_a F(b) - \mathcal{D}_a F(0) = v/x^d\}$, which equals $\#\{(a, b) \in \mathbb{F}_{p^n}^2 : \mathcal{D}_a F(b) - \mathcal{D}_a F(0) = v\}$ from the second statement. Then, for all $v \in \mathbb{F}_{p^n}$, $N_F(v; x) = N_F(v; 0)$ for all $x \in \mathbb{F}_{p^n}^\star$ by (15). Hence, F is p-ary plateaued by Theorem 5. Conversely, the other direction is clear from Theorem 5.

Next we prove (ii). The second item of Theorem 5 says that F is p-ary plateaued with single amplitude if and only if there exist two integers u_1 and u_2 such that $N_F(0; x) = u_1$ and $N_F(v; x) = u_2$ for all $x \in \mathbb{F}_{p^n}$ and $v \in \mathbb{F}_{p^n}^\star$. Assume that $N_F(0; 1) = N_F(0; 0)$ and there exists an integer u such that $N_F(v; 1) = N_F(v; 0) = u$ for all $v \in \mathbb{F}_{p^n}^\star$. From the proof of (i), we have $N_F(v; x) = N_F(v; 0)$ for all $v, x \in \mathbb{F}_{p^n}$ with $x \neq 0$. Combining them, $N_F(v; x) = u$ for all $v, x \in \mathbb{F}_{p^n}$ with $v \neq 0$ and $N_F(0; x)$ is independent of $x \in \mathbb{F}_{p^n}$. Hence, by Theorem 5, F is p-ary plateaued with single amplitude. Conversely, the other direction is clear from Theorem 5.

Finally we prove the last assertion. Assume that F is p-ary plateaued. By (i), $N_F(v; 1) = N_F(v; 0)$ for all $v \in \mathbb{F}_{p^n}$. From the second assertion, $N_F(v; 0) = N_F(v/\lambda^d; 0)$ for all $v, \lambda \in \mathbb{F}_{p^n}$ with $\lambda \neq 0$. Then, $N_F(v; 1) = N_F(v/\lambda^d; 0)$ for all $v, \lambda \in \mathbb{F}_{p^n}$ with $\lambda \neq 0$. For $v = 0$, it is obvious that $N_F(0; 1) = N_F(0; 0)$. For $v \in \mathbb{F}_{p^n}^\star$, if we set $v = 1$ and using the fact $\gcd(d, p^n - 1) = 1$, then $\lambda \mapsto 1/\lambda^d$ is a permutation of $\mathbb{F}_{p^n}^\star$. Then we get $N_F(1, 1) = N_F(v, 0)$ for all $v \in \mathbb{F}_{p^n}^\star$, that is, $N_F(v, 0) = N_F(v; 1)$ does not depend on $v \in \mathbb{F}_{p^n}^\star$. Hence, plateaued F has single amplitude from (ii). $\qquad\square$

Remark 5. With the above notations, for a power function $F(x) = x^d$, in general we have $N_F(v; 1) \neq N_F(v/\lambda^d; 1)$ for $v, \lambda \in \mathbb{F}_{p^n}$ with $\lambda \neq 0$. However, the equality case is necessary for plateaued-ness.

We now consider plateaued-ness property of vectorial p-ary functions whose component functions are all unbalanced. Recall that a function is balanced if and only if its Walsh transform vanishes at zero input.

Theorem 6. *Let $F : \mathbb{F}_{p^n} \to \mathbb{F}_{p^m}$, and let f_λ, $\lambda \in \mathbb{F}_{p^m}^\star$, be unbalanced functions. For $v \in \mathbb{F}_{p^m}$ and $x \in \mathbb{F}_{p^n}$, let $N_F(v; x)$ be the size of the set $\{(a, b) \in \mathbb{F}_{p^n}^2 : \mathcal{D}_a \mathcal{D}_b F(x) = v\}$. Then, F is p-ary plateaued if and only if for all $v \in \mathbb{F}_{p^m}$ and $x \in \mathbb{F}_{p^n}$ we have*

$$N_F(v; x) = \#\{(a, b) \in \mathbb{F}_{p^n}^2 : F(a) - F(b) = v\}. \tag{16}$$

In particular, F is p-ary plateaued with single amplitude if and only if for all $v \in \mathbb{F}_{p^m}$ and $x \in \mathbb{F}_{p^n}$ (16) holds and it is independent of $v \in \mathbb{F}_{p^m}^\star$.

Proof. Assume that F is p-ary plateaued. Since $f_\lambda = \lambda \cdot F$, $\lambda \in \mathbb{F}_{p^m}^\star$, are all unbalanced p-ary plateaued of amplitude μ_λ, we have $\widehat{\chi}_{f_\lambda}(0) \neq 0$ for all $\lambda \in \mathbb{F}_{p^m}^\star$ (and also for $\lambda = 0$), and hence $\mu_\lambda^2 = |\widehat{\chi}_{f_\lambda}(0)|^2$. For $\lambda \in \mathbb{F}_{p^m}$, since $|z|^2 = z\bar{z}$ for $z \in \mathbb{C}$, we can easily see

$$|\widehat{\chi}_{f_\lambda}(0)|^2 = \sum_{a, b \in \mathbb{F}_{p^n}} \xi_p^{\lambda \cdot (F(a) - F(b))}. \tag{17}$$

Recall that $\mathcal{D}_a \mathcal{D}_b f_\lambda(x) = \lambda \cdot (\mathcal{D}_a \mathcal{D}_b F(x))$ for all $a, b, x \in \mathbb{F}_{p^n}$ and $\lambda \in \mathbb{F}_{p^m}$. Then, by Theorem 4, for all $x \in \mathbb{F}_{p^n}$ and $\lambda \in \mathbb{F}_{p^m}$ we have

$$G(\lambda; x) = \sum_{a,b \in \mathbb{F}_{p^n}} \xi_p^{\lambda \cdot \mathcal{D}_a \mathcal{D}_b F(x)} = \sum_{a,b \in \mathbb{F}_{p^n}} \xi_p^{\lambda \cdot (F(a) - F(b))} \tag{18}$$

where the second equality follows from (17). By (3), for all $x \in \mathbb{F}_{p^n}$ and $v \in \mathbb{F}_{p^m}$, the Fourier transforms of the equal functions in (18) are equal:

$$\widehat{G}(v; x) = \sum_{\lambda \in \mathbb{F}_{p^m}} \sum_{a,b \in \mathbb{F}_{p^n}} \xi_p^{\lambda \cdot (\mathcal{D}_a \mathcal{D}_b F(x) - v)} = \sum_{\lambda \in \mathbb{F}_{p^m}} \sum_{a,b \in \mathbb{F}_{p^n}} \xi_p^{\lambda \cdot (F(a) - F(b) - v)}, \tag{19}$$

equivalently, $\widehat{G}(v; x) = p^m \#\{(a, b) \in \mathbb{F}_{p^n}^2 : \mathcal{D}_a \mathcal{D}_b F(x) = v\} = p^m \#\{(a, b) \in \mathbb{F}_{p^n}^2 : F(a) - F(b) = v\}$. Hence, the assertion holds.

Conversely, assume that for all $x \in \mathbb{F}_{p^n}$ and $v \in \mathbb{F}_{p^m}$ (16) holds, that is, (19) holds. By (3), for all $x \in \mathbb{F}_{p^n}$ and $\lambda \in \mathbb{F}_{p^m}$, (18) holds, that is, by (17), $G(\lambda; x) = |\widehat{\chi}_{f_\lambda}(0)|^2$, which is nonzero since f_λ, $\lambda \in \mathbb{F}_{p^m}^\star$, are all unbalanced. Then, for all $\lambda \in \mathbb{F}_{p^m}^\star$, $G(\lambda; x)$ does not depend on $x \in \mathbb{F}_{p^n}$. By Theorem 4, f_λ, $\lambda \in \mathbb{F}_{p^m}^\star$, is p-ary plateaued, and hence, F is p-ary plateaued.

Finally we prove the last statement. Theorem 4 says that F is p-ary plateaued with single amplitude if and only if $G(\lambda; x)$ in (18) does not depend on $x \in \mathbb{F}_{p^n}$ nor λ for $\lambda \neq 0$; equivalently by (3), $\widehat{G}(v; x)$ in (19) does not depend on $x \in \mathbb{F}_{p^n}$ nor of v for $v \neq 0$. Hence, with the above arguments, this completes the proof. $\qquad\square$

We derive from Theorem 6 and [12, Theorem 8 and Corollary 4] the following result.

Corollary 7. *Let $F : \mathbb{F}_{p^n} \to \mathbb{F}_{p^m}$ be a function whose component functions are all unbalanced. Let s be an integer with $0 \leq s \leq n$. Then F is vectorial p-ary s-plateaued if and only if for all $v \in \mathbb{F}_{p^m}^\star$ we have $\#\{(a, b) \in \mathbb{F}_{p^n}^2 : F(a) - F(b) = v\} = p^{2n-m} - p^{n+s-m}$. In particular, F is vectorial p-ary bent if and only if $\#\{(a, b) \in \mathbb{F}_{p^n}^2 : F(a) = F(b)\} = p^{2n-m} + p^n - p^{n-m}$. Moreover, if F is vectorial p-ary s-plateaued, $\#\{(a, b) \in \mathbb{F}_{p^n}^2 : F(a) = F(b)\} = p^{2n-m} + p^{n+s} - p^{n+s-m}$.*

In the following, we study a particular case of p-ary plateaued (vectorial) functions: when the value distribution of $b \mapsto \mathcal{D}_a \mathcal{D}_b F(x)$ is independent of $x \in \mathbb{F}_{p^n}$ for each fixed value of a although the value distribution of $\mathcal{D}_b \mathcal{D}_a F(x)$ when $(a, b) \in \mathbb{F}_{p^n}^2$ is independent of $x \in \mathbb{F}_{p^n}$ in Theorem 5.

Definition 1. *Let $F : \mathbb{F}_{p^n} \to \mathbb{F}_{p^m}$. Then, F is called vectorial p-ary strongly-plateaued if for all $a \in \mathbb{F}_{p^n}$ and $v \in \mathbb{F}_{p^m}$, the size of the set $\{b \in \mathbb{F}_{p^n} : \mathcal{D}_a \mathcal{D}_b F(x) = v\}$ is independent of $x \in \mathbb{F}_{p^n}$. In particular, $f : \mathbb{F}_{p^n} \to \mathbb{F}_p$ is called p-ary strongly-plateaued if for all $a \in \mathbb{F}_{p^n}$ and $v \in \mathbb{F}_p$, the size of the set $\{b \in \mathbb{F}_{p^n} : \mathcal{D}_a \mathcal{D}_b f(x) = v\}$ is independent of $x \in \mathbb{F}_{p^n}$.*

Remark 6. By Theorem 5, any p-ary strongly-plateaued function is p-ary plateaued. Moreover, a vectorial p-ary function is strongly-plateaued if and only if its component functions are p-ary strongly-plateaued.

Theorem 7. *Let* $F : \mathbb{F}_{p^n} \to \mathbb{F}_{p^m}$. *For all* $a, x \in \mathbb{F}_{p^n}$ *and* $v \in \mathbb{F}_{p^m}$ *we have*
$$\#\{b \in \mathbb{F}_{p^n} : \mathcal{D}_a \mathcal{D}_b F(x) = v\} = \#\{b \in \mathbb{F}_{p^n} : \mathcal{D}_a F(b) - \mathcal{D}_a F(x) = v\}.$$

Proof. For all $a, b, x \in \mathbb{F}_{p^n}$, (by the bijective change of variable $b \mapsto b - x$), we have $\mathcal{D}_a \mathcal{D}_b F(x) = \mathcal{D}_a \mathcal{D}_{b-x} F(x) = F(a + b) - F(x + a) - F(b) + F(x) = \mathcal{D}_a F(b) - \mathcal{D}_a F(x)$. This completes the proof.   □

The notion of p-ary strongly-plateaued is closely connected to p-ary partially-bent.

Proposition 3. *Let* $f : \mathbb{F}_{p^n} \to \mathbb{F}_p$. *Then* f *is* p-*ary strongly-plateaued if and only if* f *is* p-*ary partially-bent.*

Proof. By the fact (see for instance [7]) that f is p-ary partially-bent if and only if the derivative $\mathcal{D}_a f$ is either balanced or constant for all $a \in \mathbb{F}_{p^n}$, then f is p-ary partially-bent if and only if for all $v \in \mathbb{F}_{p^m}$ and $a \in \mathbb{F}_{p^n}$, $\#\{b \in \mathbb{F}_{p^n} : \mathcal{D}_a f(b) = \mathcal{D}_a f(x) + v\}$ is independent of $x \in \mathbb{F}_{p^n}$; equivalently, f is p-ary strongly-plateaued by Theorem 7.   □

Proposition 4. *A vectorial* p-*ary function is strongly-plateaued if and only if its all component functions are* p-*ary partially-bent. In particular,* p-*ary bent and quadratic (vectorial) functions are* p-*ary strongly-plateaued (vectorial) functions.*

Proof. The first assertion follows from Remark 6 and Proposition 3. Because of the well known fact that p-ary bent and quadratic (vectorial) functions are p-ary partially-bent (vectorial) functions (see for instance [7]), the last assertion follows from the first statement.   □

5 Characterizations by Using Power Moments of the Walsh Transform

In 2014, new characterizations of p-ary plateaued functions, and in 2015 different characterizations of plateaued (vectorial) Boolean functions were provided in [2,3,10] in terms of the Walsh transform. Those characterizations are completed and extended in this section. We first derive from Theorem 4 the following characterization of p-ary plateaued (vectorial) functions.

Theorem 8. *Let* $f : \mathbb{F}_{p^n} \to \mathbb{F}_p$. *Then* f *is* p-*ary plateaued if and only if for all* $\alpha \in \mathbb{F}_{p^n}^\star$, *we have*

$$\sum_{\omega \in \mathbb{F}_{p^n}} \widehat{\chi}_f(\alpha + \omega) \overline{\widehat{\chi}_f(\omega)} \, |\widehat{\chi}_f(\omega)|^2 = 0. \tag{20}$$

Let $F : \mathbb{F}_{p^n} \to \mathbb{F}_{p^m}$ *and let* f_λ, $\lambda \in \mathbb{F}_{p^m}^\star$, *be the component functions of* F. *Then,* F *is* p-*ary plateaued if and only if* $\sum_{\omega \in \mathbb{F}_{p^n}} \widehat{\chi}_{f_\lambda}(\alpha + \omega) \overline{\widehat{\chi}_{f_\lambda}(\omega)} \, |\widehat{\chi}_{f_\lambda}(\omega)|^2 = 0$ *for all* $\alpha \in \mathbb{F}_{p^n}^\star$ *and* $\lambda \in \mathbb{F}_{p^m}^\star$. *In particular,* F *is* p-*ary plateaued with single amplitude if and only if, additionally,* $\sum_{\omega \in \mathbb{F}_{p^n}} |\widehat{\chi}_{f_\lambda}(\omega)|^4$ *does not depend on* λ *for* $\lambda \neq 0$.

Proof. By definition of $\widehat{\chi}_f$, for all $\alpha \in \mathbb{F}_{p^n}^\star$ (20) is equivalent to:

$$\sum_{\omega,x,y,z,t\in\mathbb{F}_{p^n}} \xi_p^{f(x)-(\alpha+\omega)\cdot x-f(y)+\omega\cdot y+f(z)-\omega\cdot z-f(t)+\omega\cdot t} = 0,$$

that is, to: $\displaystyle\sum_{\omega,x,y,z,t\in\mathbb{F}_{p^n}} \xi_p^{f(x)-f(y)+f(z)-f(t)-\omega\cdot(x-y+z-t)-\alpha\cdot x} = 0$, equivalently, to:

$$\sum_{x,y,z\in\mathbb{F}_{p^n}} \xi_p^{f(x)-f(y)+f(z)-f(x-y+z)-\alpha\cdot x} = 0$$

since $\sum_{\omega\in\mathbb{F}_{p^n}} \xi_p^{\omega\cdot(x-y+z-t)}$ is null if $x - y + z - t \neq 0$, that is, (by the bijective change of variables: $y = x+a$ and $z = x+a+b$) to: $\sum_{x,a,b\in\mathbb{F}_{p^n}} \xi_p^{\mathcal{D}_b\mathcal{D}_a f(x)-\alpha\cdot x} = 0$. This last expression equals the Fourier transform of $x \mapsto \sum_{a,b\in\mathbb{F}_{p^n}} \xi_p^{\mathcal{D}_b\mathcal{D}_a f(x)}$ at $\alpha \in \mathbb{F}_{p^n}^\star$. We know (see for instance [1]) that the Fourier transform of a function from $\mathbb{F}_{p^n}$ to $\mathbb{C}$ vanishes at any nonzero input if and only if the function is constant. Theorem 4 says that there exists $\theta \in \mathbb{N}$ such that $\sum_{a,b\in\mathbb{F}_{p^n}} \xi_p^{\mathcal{D}_b\mathcal{D}_a f(x)} = \theta$ for all $x \in \mathbb{F}_{p^n}$ if and only if f is p-ary plateaued. Thus the assertion holds.

By Remark 1, the second statement is a direct consequence of the first statement. The last assertion follows from Theorems 4 and 5 and Proposition 2.   □

Corollary 8. *Let* $f : \mathbb{F}_{p^n} \to \mathbb{F}_p$. *Then* f *is* p-ary plateaued if and only if for all $x \in \mathbb{F}_{p^n}$

$$\sum_{\omega\in\mathbb{F}_{p^n}} |\widehat{\chi}_f(\omega)|^4 = p^n \sum_{\omega\in\mathbb{F}_{p^n}} \xi_p^{f(x)-\omega\cdot x}\overline{\widehat{\chi}_f(\omega)}\,|\widehat{\chi}_f(\omega)|^2. \qquad (21)$$

Let $F : \mathbb{F}_{p^n} \to \mathbb{F}_{p^m}$, *and let* f_λ, $\lambda \in \mathbb{F}_{p^m}^\star$, *be the component functions of* F. *Then* F *is* p-ary plateaued if and only if for all $x \in \mathbb{F}_{p^n}$ and $\lambda \in \mathbb{F}_{p^m}^\star$

$$\sum_{\omega\in\mathbb{F}_{p^n}} |\widehat{\chi}_{f_\lambda}(\omega)|^4 = p^n \sum_{\omega\in\mathbb{F}_{p^n}} \xi_p^{f_\lambda(x)-\omega\cdot x}\overline{\widehat{\chi}_{f_\lambda}(\omega)}|\widehat{\chi}_{f_\lambda}(\omega)|^2. \qquad (22)$$

In particular, F *is* p-ary plateaued with single amplitude if and only if for all $x \in \mathbb{F}_{p^n}$ and $\lambda \in \mathbb{F}_{p^m}^\star$ (22) holds and it is independent of $\lambda \neq 0$.

Proof. Assume that f is p-ary plateaued of amplitude μ. By Lemma 1, the left-hand side of (21) is equal to $p^{2n}\mu^2$. For all $x \in \mathbb{F}_{p^n}$, the right-hand side of (21) equals

$$p^n\mu^2 \sum_{\omega\in\mathbb{F}_{p^n}} \xi_p^{f(x)-\omega\cdot x}\overline{\widehat{\chi}_f(\omega)} = p^n\mu^2 \sum_{y\in\mathbb{F}_{p^n}} \xi_p^{f(x)-f(y)} \sum_{\omega\in\mathbb{F}_{p^n}} \xi_p^{\omega\cdot(y-x)} = p^n p^n\mu^2.$$

Thus for all $x \in \mathbb{F}_{p^n}$, (21) holds. Conversely, assume that (21) holds for all $x \in \mathbb{F}_{p^n}$. That is, for all $x \in \mathbb{F}_{p^n}$, a function $G : \mathbb{F}_{p^n} \to \mathbb{C}$ defined as $x \mapsto G(x) = \sum_{\omega\in\mathbb{F}_{p^n}} \xi_p^{f(x)-\omega\cdot x}\overline{\widehat{\chi}_f(\omega)}|\widehat{\chi}_f(\omega)|^2$ is constant. Then its Fourier transform

$$\widehat{G}(\alpha) = \sum_{x \in \mathbb{F}_{p^n}} G(x) \xi_p^{-\alpha \cdot x} = \sum_{\omega \in \mathbb{F}_{p^n}} \sum_{x \in \mathbb{F}_{p^n}} \xi_p^{f(x) - x \cdot (\alpha + \omega)} \overline{\widehat{\chi}_f(\omega)} \, |\widehat{\chi}_f(\omega)|^2$$
$$= \sum_{\omega \in \mathbb{F}_{p^n}} \widehat{\chi}_f(\alpha + \omega) \overline{\widehat{\chi}_f(\omega)} \, |\widehat{\chi}_f(\omega)|^2$$

is null at any $\alpha \in \mathbb{F}_{p^n}^*$. Hence f is p-ary plateaued by Theorem 8.

By Remark 1, the second statement is a direct consequence of the first statement.

The last assertion, with the above arguments, follows from Theorems 4 and 5 and Proposition 2. □

Considering $f_\lambda = \lambda \cdot F$ for $\lambda \in \mathbb{F}_{p^m}^*$, for all $x \in \mathbb{F}_{p^n}$ and $\lambda \in \mathbb{F}_{p^m}^*$ the equality (22) is equivalent to:

$$\sum_{\omega, a, b, c, d \in \mathbb{F}_{p^n}} \xi_p^{\lambda \cdot (F(a) - F(b) + F(c) - F(d)) - \omega \cdot (a - b + c - d)}$$
$$= p^n \sum_{\omega, a, b, c \in \mathbb{F}_{p^n}} \xi_p^{\lambda \cdot (F(x) - F(a) + F(b) - F(c)) - \omega \cdot (x - a + b - c)},$$

equivalently,

$$\sum_{a, b, c \in \mathbb{F}_{p^n}} \xi_p^{\lambda \cdot (F(a) - F(b) + F(c) - F(a - b + c))} = p^n \sum_{a, b \in \mathbb{F}_{p^n}} \xi_p^{\lambda \cdot (F(x) - F(a) + F(b) - F(x - a + b))},$$

that is, (by the bijective change of variables: $a \mapsto a + b + c$ and $b \mapsto b + c$ in the left-hand side, and $a \mapsto a + x$ and $b \mapsto a + b + x$ in the right-hand side) we have

$$\sum_{a, b, c \in \mathbb{F}_{p^n}} \xi_p^{\lambda \cdot (\mathcal{D}_b \mathcal{D}_a F(c))} = p^n \sum_{a, b \in \mathbb{F}_{p^n}} \xi_p^{\lambda \cdot (\mathcal{D}_b \mathcal{D}_a F(x))},$$

which is equivalent to (10). Namely, the characterizations given by Corollaries 3 and 8 are equivalent. This link between the Walsh transform and second-order derivative of p-ary functions can be formalized as follows.

Proposition 5. *Let $f : \mathbb{F}_{p^n} \to \mathbb{F}_p$. Then, for all $x \in \mathbb{F}_{p^n}$*

$$\sum_{\omega \in \mathbb{F}_{p^n}} \xi_p^{f(x) - \omega \cdot x} \overline{\widehat{\chi}_f(\omega)} \, |\widehat{\chi}_f(\omega)|^2 = p^n \sum_{a, b \in \mathbb{F}_{p^n}} \xi_p^{\mathcal{D}_a \mathcal{D}_b f(x)}. \tag{23}$$

Proof. By definition $\widehat{\chi}_f$, for all $x \in \mathbb{F}_{p^n}$, the left-hand side of (23) equals

$$\sum_{\omega, a, b, c \in \mathbb{F}_{p^n}} \xi_p^{f(x) - f(a) - f(b) + f(c) + \omega \cdot (a + b - c - x)} = p^n \sum_{a, b \in \mathbb{F}_{p^n}} \xi_p^{f(x) - f(a) - f(b) + f(a + b - x)}$$

since $\sum_{\omega \in \mathbb{F}_{p^n}} \xi_p^{-\omega \cdot (x - a - b + c)}$ is null if $c \neq a + b - x$. For all $x \in \mathbb{F}_{p^n}$, by the bijective change of variables: $a \mapsto a + x$ and $b \mapsto b + x$, it is equal to the right-hand side of (23). Hence, the proof is complete. □

We now present further characterizations of plateaued p-ary functions in terms of the even power moments of the Walsh transform. We shall recover some results given by Mesnager [10]. According to (1), for $i \geq 1$ and a nonnegative integer A, f is p-ary plateaued of amplitude μ if and only if

$$S_i(f)A^2 - 2S_{i+1}(f)A + S_{i+2}(f) = 0, \tag{24}$$

where $A = \mu^2 > 0$. Then, the reduced discriminant $S_{i+1}(f)^2 - S_{i+2}(f)S_i(f) \leq 0$ of the above equation can not be positive, and it is zero if and only if f is p-ary plateaued. Moreover, this can be derived from the *Cauchy-Schwarz Inequality* (see for instance [15]), which states that for $(x_1, x_2, \ldots, x_m), (y_1, y_2, \ldots, y_m) \in \mathbb{R}^m$ or $\mathbb{C}^m$,

$$\left(\sum_{k=1}^{m} |x_k y_k| \right)^2 \leq \sum_{k=1}^{m} |x_k|^2 \sum_{k=1}^{m} |y_k|^2,$$

with an equality if and only if for all $k \in \{1, \ldots, m\}$, $|x_k|^2 = d|y_k|^2$ for some $d \in \mathbb{R}^+$. Applying the Cauchy-Schwarz Inequality for $x_k = |\widehat{\chi}_f(\omega)|^i$ and $y_k = |\widehat{\chi}_f(\omega)|^{i+2}$, we have

$$\left(\sum_{\omega \in \mathbb{F}_{p^n}} |\widehat{\chi}_f(\omega)|^{2i+2} \right)^2 \leq \sum_{\omega \in \mathbb{F}_{p^n}} |\widehat{\chi}_f(\omega)|^{2i} \sum_{\omega \in \mathbb{F}_{p^n}} |\widehat{\chi}_f(\omega)|^{2i+4},$$

that is, $S_{i+1}(f)^2 \leq S_i(f)S_{i+2}(f)$ for $i \geq 1$. The inequality is an equality if and only if for all $\omega \in \mathbb{F}_{p^n}$ $|\widehat{\chi}_f(\omega)|^{2i} = d|\widehat{\chi}_f(\omega)|^{2i+4}$ for some $d \in \mathbb{R}^+$ where $i \geq 1$; equivalently, $|\widehat{\chi}_f(\omega)|^2$ is constant (with also possibly the value 0), that is, f is p-ary plateaued. Notice that the equality case is equivalent to [10, Theorem 1].

Theorem 9. *Let $f : \mathbb{F}_{p^n} \to \mathbb{F}_p$. Then for all $i \geq 1$, we have*

$$S_{i+1}(f)^2 \leq S_{i+2}(f)S_i(f), \tag{25}$$

with an equality if and only if f is p-ary plateaued. Let $F : \mathbb{F}_{p^n} \to \mathbb{F}_{p^m}$ and let f_λ, $\lambda \in \mathbb{F}_{p^m}^\star$, be the component functions of F. Then for all $i \geq 1$, we have

$$\sum_{\lambda \in \mathbb{F}_{p^m}^\star} S_{i+1}(f_\lambda)^2 \leq \sum_{\lambda \in \mathbb{F}_{p^m}^\star} S_{i+2}(f_\lambda)S_i(f_\lambda), \tag{26}$$

equivalently,

$$\sum_{\lambda \in \mathbb{F}_{p^m}^\star} S_{i+1}(f_\lambda) \leq \sum_{\lambda \in \mathbb{F}_{p^m}^\star} \sqrt{S_{i+2}(f_\lambda)S_i(f_\lambda)}, \tag{27}$$

with an equality if and only if F is p-ary plateaued.

Proof. As stated above, the first assertion is a direct consequence of the Cauchy-Schwarz Inequality and its equality case. In the second statement, the inequalities

(26) and (27) can be easily obtained by (25). To prove equality cases, notice that by (25) we have

$$S_{i+2}(f_\lambda)S_i(f_\lambda) - S_{i+1}(f_\lambda)^2 \geq 0$$

for all $\lambda \in \mathbb{F}_{p^m}^\star$. Thanks to the well known fact that a sum of nonnegative terms is zero if and only if each term is zero, then the inequality (26) or (27) becomes an equality if and only if f_λ, $\lambda \in \mathbb{F}_{p^m}^\star$, are all p-ary plateaued by (25); equivalently, F is p-ary plateaued. $\square$

In particular, from (24), for $i = 1$ and $A > 0$, f is p-ary plateaued if and only if $S_1(f)A^2 - 2S_2(f)A + S_3(f) = 0$. The reduced discriminant $S_2(f)^2 - S_3(f)S_1(f) \leq 0$ of the above equation can not be positive and it is zero if and only if f is p-ary plateaued.

Corollary 9. *Let* $f : \mathbb{F}_{p^n} \to \mathbb{F}_p$. *Then* $S_2(f)^2 \leq p^{2n}S_3(f)$, *with an equality if and only if f is p-ary plateaued. Let* $F : \mathbb{F}_{p^n} \to \mathbb{F}_{p^m}$ *and let* f_λ, $\lambda \in \mathbb{F}_{p^m}^\star$, *be the component functions of* F. *Then we have*

$$\sum_{\lambda \in \mathbb{F}_{p^m}^\star} S_2(f_\lambda)^2 \leq p^{2n} \sum_{\lambda \in \mathbb{F}_{p^m}^\star} S_3(f_\lambda),$$

equivalently, $\sum_{\lambda \in \mathbb{F}_{p^m}^\star} S_2(f_\lambda) \leq p^n \sum_{\lambda \in \mathbb{F}_{p^m}^\star} \sqrt{S_3(f_\lambda)}$, *with an equality if and only if F is p-ary plateaued.*

6   Conclusion

Plateaued functions have attracted attention since their introduction in the literature because of their role in diverse domains of Boolean and vectorial functions for sequences and cryptography like correlation immune functions and orthogonal arrays (since plateaued functions offer the best possible compromise between resiliency order and nonlinearity), APN functions and S-boxes (since plateaued APN functions in odd dimension are almost bent), and because, like partially-bent functions, they represent a natural class for generalizing in the same time bent functions and quadratic functions, but they constitute a larger class than partially-bent functions, including also all semi-bent and near-bent functions. But their structure is still more difficult to characterize and, little is known about those functions already in characteristic 2 and still more in arbitrary characteristic. The objective of this paper was to provide several various tools to handle the plateaued-ness property of p-ary (vectorial) functions in order to clarify their structure. To this end, several explicit characterizations were given, extending the recent results of Carlet [2,3] (valid in characteristic 2).

Acknowledgment. The fourth author is supported by the Scientific and Technological Research Council of Turkey (TÜBITAK)-BIDEB 2214-A program.

References

1. Carlet, C.: Boolean functions for cryptography and error correcting codes. In: Crama, Y., Hammer, P.L. (eds.) Boolean Models and Methods in Mathematics, Computer Science, and Engineering, pp. 257–397. Cambridge University Press, Cambridge (2010)
2. Carlet, C.: Boolean and vectorial plateaued functions, and APN functions. IEEE Trans. Inf. Theor. **61**(11), 6272–6289 (2015)
3. Carlet, C.: On the properties of vectorial functions with plateaued components and their consequences on APN functions. In: El Hajji, S., Nitaj, A., Carlet, C., Souidi, E.M. (eds.) C2SI 2015. LNCS, vol. 9084, pp. 63–73. Springer, Cham (2015). doi:10.1007/978-3-319-18681-8_5
4. Carlet, C.: Partially-bent functions. Des. Code Crypt. **3**(2), 135–145 (1993)
5. Carlet, C., Prouff, E.: On plateaued functions and their constructions. In: Johansson, T. (ed.) FSE 2003. LNCS, vol. 2887, pp. 54–73. Springer, Heidelberg (2003). doi:10.1007/978-3-540-39887-5_6
6. Çesmelioglu, A., Meidl, W.: A construction of bent functions from plateaued functions. Des. Code Crypt. **66**(1–3), 231–242 (2013)
7. Çesmelioglu, A., Meidl, W., Topuzoglu, A.: Partially bent functions and their properties. In: Larcher, G., Pillichshammer, F., Winterhof, A., Xing, C. (eds.) Applications of Algebra and Number Theory, pp. 22–40. Cambridge University Press, Cambridge (2014)
8. Hyun, J.Y., Lee, J., Lee, Y.: Explicit criteria for construction of plateaued functions. IEEE Trans. Inf. Theor. **62**(12), 7555–7565 (2016)
9. Kumar, P.V., Scholtz, R.A., Welch, L.R.: Generalized bent functions and their properties. J. Comb. Theor. **A–40**, 90–107 (1985)
10. Mesnager, S.: Characterizations of plateaued and bent functions in characteristic p. In: Schmidt, K.-U., Winterhof, A. (eds.) SETA 2014. LNCS, vol. 8865, pp. 72–82. Springer, Cham (2014). doi:10.1007/978-3-319-12325-7_6
11. Mesnager, S.: Bent Functions: Fundamentals and Results. Springer, Cham (2016)
12. Mesnager, S., Özbudak, F., Sınak, A.: On the p-ary (cubic) bent and plateaued (vectorial) functions. Des. Code Crypt. (2017, submitted)
13. Mesnager, S., Özbudak, F., Sınak, A.: Results on characterizations of plateaued functions in arbitrary characteristic. In: Pasalic, E., Knudsen, L.R. (eds.) Balkan-CryptSec 2015. LNCS, vol. 9540, pp. 17–30. Springer, Cham (2016). doi:10.1007/978-3-319-29172-7_2
14. Mullen, G.L., Panario, D.: Handbook of finite fields. CRC Press, New York (2013)
15. Rudin, W.: Principles of Mathematical Analysis. McGraw-Hill, New York (1964)
16. Rothaus, O.S.: On bent functions. J. Comb. Theor. A. **20**, 300–305 (1976)
17. Zheng, Y., Zhang, X.-M.: Plateaued functions. In: Varadharajan, V., Mu, Y. (eds.) ICICS 1999. LNCS, vol. 1726, pp. 284–300. Springer, Heidelberg (1999). doi:10.1007/978-3-540-47942-0_24
18. Zheng, Y., Zhang, X.-M.: Relationships between bent functions and complementary plateaued functions. In: Song, J.S. (ed.) ICISC 1999. LNCS, vol. 1787, pp. 60–75. Springer, Heidelberg (2000). doi:10.1007/10719994_6

A New Dynamic Code-Based Group Signature Scheme

Berenger Edoukou Ayebie$^{(\boxtimes)}$, Hafsa Assidi, and El Mamoun Souidi

Laboratory of Mathematics, Computer Science and Applications, Faculty of Sciences, Mohammed V University in Rabat, BP 1014 RP, Rabat, Morocco
berenger.ayebie@gmail.com, assidihafsa@gmail.com, emsouidi@gmail.com

Abstract. Group signature is a cryptographic primitive where a user can anonymously sign a message on behalf of group users. The dynamic case in group signature is more interesting than the static one. The general idea of this scheme consists in finding a collision between two ciphertexts using two different Quasi-cyclic Moderate Density Parity-Check (QC-MDPC) matrices in McEliece cryptosystem. We use a variation of AGS Zero-Knowledge protocol to prove the possession of the secret key and then we use the Fiat Shamir transformation to turn it into a signature. The public key and signature sizes are constants and independent of group users size and are shorter than those presented in the literature for 80 bits security level. Furthermore the proposed group signature scheme presents several advantages: it is a dynamic group signature based on error correcting code assumptions which are supposed resistant to quantum computing.

Keywords: Dynamic group signature · Code-based cryptography · QC-MDPC codes · McEliece cryptosystem · General decoding problem

1 Introduction

Group signature schemes have been introduced by Chaum and van Heyst [3], in order to provide revocable anonymity to a signer, who is allowed to sign on behalf of a group. In such a scheme an authority is able, in exceptional cases, to "open" any group signature and thus recover the actual signer. The properties of group signature schemes make them a very important cryptographic primitive, with several applications in real-life scenarios such as e-voting schemes, e-bidding, digital right management systems, controlled anonymous printing services and other domains.

For many years, several group signatures have been introduced namely the ACJT [4], which was the first provably secure coalition-resistant scheme under the Strong RSA and DDH assumptions. Boneh, Boyen and Shacham in [5] and later Camenisch and Lysyanskaya in [6] proposed very efficient group signature schemes using bilinear maps. Independently, Nguyen and Safavi-Naini presented in [7] another group signature scheme using bilinear maps. We note that all these schemes were analyzed in the random oracle model [8].

© Springer International Publishing AG 2017
S. El Hajji et al. (Eds.): C2SI 2017, LNCS 10194, pp. 346–364, 2017.
DOI: 10.1007/978-3-319-55589-8_23

Bellare *et al.* in [9] gave formal definitions of the security properties of group signatures, they proposed the first scheme provably secure in the standard model (while totally unpractical). Independently, Kiayias and Yung [10] (and later [11]) defined a security model. Bellare *et al.* [12] extended the [9] model to the case of dynamic groups.

All the aforementioned group signature schemes are based on number theory assumptions. However, number-theoretic based cryptography will not resist to the quantum computer. Recently, the research for post-quantum group signatures is quite active as shows these publications [2,13–16,18]. The majority of these works are based on lattice assumptions, while in code-based cryptography we denote three group signature schemes. The two first schemes presented in Asiacrypt 2015 [17,18] in Secrypt 2016 are based on BMW (Bellare, Micciancio and Warinschi) model [9]. The last one is based on a slight relaxation of the BSZ (Bellare, Shi and Zang) model [12] and is proposed in [2].

In this paper, we present a new dynamic code-based group signature scheme which is an improvement of [2]. It is based on two main ideas: the first one consists in using as Zero-Knowledge argument system the AGS (Aguilar, Gaborit and Shrek) identification scheme [1], this choice implies lower signature and public key sizes. For example, to achieve a 80 bits security level, the public key and signature sizes are 2.5 Ko and 1.32 Mo respectively in our scheme instead of 2.5 Mo and 20 Mo in [2].

We explain in the following points how we reduce the signature and public key length:

- We use two double circulant matrices (concatenation of an identity matrix and a circulant matrix) as public key. Therefore, we can generate such matrices only from the first row of the circulant matrix since the others are obtained by applying a shift.
- AGS identification scheme [1] have low communication data than all existing code based identification schemes. In addition, the cheating probability is around $\frac{1}{2}$ witch implies less number of rounds. Thus the signature size is reduced.

The second difference between our construction and that presented in [2] consists to find cipher collision instead of syndrome collision in the joining protocol. In [2], the authors generate trapdoor matrix to make syndrome collision. Notice that in Hamming metric, the one existing trapdoor matrix is CFS signature instance [27]. To generate cipher collision, we use QC-MDPC McEliece variant where in [19] the authors assume that QC-MDPC decoding algorithm is fast and simple to implement.

This paper is organized as follows: we set in Sect. 2 some notations and preliminaries concerning code based cryptography, AGS Zero-Knowledge identification protocol and Fiat-Shamir paradigm. In Sect. 3, we give a security model of group signature in dynamic case. Then, in Sect. 4 we explain our proposed group signature scheme and we prove its security in Sect. 5 with specific parameters. We conclude in Sect. 6.

2 Preliminaries

In this section, we first provide the notations that will be used all along this work. Secondly, we give background in code-based cryptography. We present also the QC-MDPC McElice variant and the AGS Zero-Knowledge identification protocol.

2.1 Notation

Let μ denotes some randomness and the symbol $\|$ is for concatenation. $v[r]$ denotes the $r-th$ coordinate of the vector v and s_r denotes the $r-th$ symbol of the string s. $\mathbb{F}_q$ denotes the finite field of cardinality q. $\mathcal{M}_{m \times n}(\mathbb{F}_q)$ denotes matrices over $\mathbb{F}_q$ of m rows and n columns. S_w^n is the set of vectors of weight w lying in $\mathbb{F}_2^n$. $\mathcal{H}$ denotes a generic random oracle. Let $h : \mathbb{F}_2^\star \to \mathbb{F}_2^n$, $h' : \mathbb{F}_2^\star \to \mathcal{M}_{k \times n}(\mathbb{F}_2)$ and $h'' : \mathbb{F}_2^\star \to \mathbb{F}_2^\ell$ be random oracles model. For protocols, we denote by $\mathcal{P}$ and $\mathcal{V}$ respectively the prover and the verifier. We use coding theory notation, where G and H respectively denote generator and parity check matrices of a code. Let $H \in \mathcal{M}_{k \times n}(\mathbb{F}_q)$ and $x \in \mathbb{F}_q^n$. The product $H \cdot x^T$ is called the syndrome of x and $wt(x)$ refers to the Hamming weight of x. We define 1^λ as a string of ones λ times. We denote by $a \xleftarrow{\$} S$ if a is chosen uniformly at random from the finite set S.

We define $Pr[A = a]$ as being the probability to have the event $A = a$. Let $Exp_{B,A}^{C-b}(\lambda)$ and $Adv_{B,A}^C(\lambda)$ define respectively the experiment where adversary $\mathcal{A}$ attack the properties C of the scheme B, where the objective is to output the bit b and the advantage of the adversary $\mathcal{A}$ in the experiment $Exp_{B,A}^{C-b}(\lambda)$.

We define ρ_r as a left shift rotation of r position applied to vector u as follows:

$$\rho_r : \mathbb{F}_2^n \to \mathbb{F}_2^n$$
$$u \mapsto \rho_r(u)$$

2.2 Code-Based Cryptography

Now we give some necessary notions in code-based cryptography for the well understanding of our work.

The Syndrome decoding problem and the General Decoding problem, that we recall hereafter, are two problems based on coding theory proved to be NP-complete in [20].

Problem 1 (Syndrome Decoding Problem). The $SD(n, k, \omega)$ problem is formulated as follows: let n, k and ω be integers, given an uniformly random matrix $H \in \mathcal{M}_{k \times n}(\mathbb{F}_2)$ and a uniformly random syndrome $y \in \mathbb{F}_2^k$, find a vector $s \in \mathbb{F}_2^n$ such that $wt(s) \le \omega$ and $H \cdot s^\top = y^\top$.

Problem 2 (General Decoding Problem GD). The $GD(n, n-k, \omega)$ problem is defined as follows: let k, n and w be integers, and let (G, x, ω) be a triple consisting of a matrix $G \in \mathcal{M}_{n-k \times n}(\mathbb{F}_2)$, vector $x \in \mathbb{F}_2^n$ and an integer $\omega < n$. Find a vector $m \in \mathbb{F}_2^{n-k}$ and a vector $e \in \mathbb{F}_2^n$ such that $wt(e) \le \omega$ and $x = m \cdot G + e$.

Double circulant matrices. We say that H is a double circulant matrix if $H = [I_p|A]$ where I_p is the identity matrix of size p and A is a circulant matrix of length p, which means a $p \times p$ matrix generated from its first row $a = (a_0, \cdots, a_{p-1})$

$$A = \begin{pmatrix} a_0 & a_1 & \dots & a_{p-1} \\ a_{p-1} & a_0 & \dots & a_{p-2} \\ \cdot & \cdot & \dots & \cdot \\ \cdot & \cdot & \dots & \cdot \\ \cdot & \cdot & \dots & \cdot \\ a_1 & a_2 & \dots & a_0 \end{pmatrix} \tag{1}$$

(n, p, w)**-QC-MDPC code construction.** We use specially a (n, p, w)-QC-MDPC codes where $n = n_0 p$. This means that the parity-check matrix has the form: $H = [H_0|H_1|\cdots|H_{n_0-1}]$, where H_i is a $p \times p$ circulant block.

We define the first row of H by picking a random vector of length $n = n_0 p$ and weight w. The other $p-1$ rows are obtained from the $p-1$ quasi-cyclic shifts of this first row. Each block H_i will have a row weight w_i such that $w = \sum_{i=0}^{n_0-1} w_i$.

A generator matrix G in row reduced echelon form can be easily derived from the H_i's blocks. Assuming the rightmost block H_{n_0-1} is non-singular, we construct a generator-matrix as follows:

$$G = \begin{pmatrix} & & (H_{n_0-1}^{-1}.H_0)^T \\ & & (H_{n_0-1}^{-1}.H_1)^T \\ & I_{n-p} & \cdot \\ & & \cdot \\ & & \cdot \\ & & (H_{n_0-1}^{-1}.H_{n_0-2})^T \end{pmatrix} \tag{2}$$

QC-MDPC McEliece Encryption Scheme: Using QC-MDPC code, Misoczki et al. present in [19] the QC-MDPC version of McEliece cryptosystem. For instance, to achieve a 128 bits of security we just need 9857 bits for the public key which is very compact unlike the original version of McEliece using Goppa codes.

2.3   AGS Zero-Knowledge Identification Protocol

In 2011 Aguilar *et al.* published a new identification scheme AGS [1] improving the Veron's protocol [22] which is a variation of Stern protocol [23]. The improvement consists in reducing the cheating probability from $\frac{2}{3}$ to $\frac{1}{2}$ asymptotically exploiting the structure of random double circulant codes. In fact, decreasing the cheating probability for one round affect directly the number of rounds needed to reach a certain level of security. For example, to achieve a 80 bits security level for signature we have to execute the AGS protocol 88 rounds whereas in Stern [23] case we need 140 rounds.

Remark 1. It is proved in [1] that the AGS identification protocol verifies the Zero-Knowledge, completeness and the soundness properties and thus a malicious prover cannot be authenticated with probability much higher than $\frac{1}{2}$.

Fiat-Shamir paradigm. It's possible to transform an identification protocol into a signature scheme through the Fiat-Shamir paradigm [24]. The general difference between authentication and signature is the number of characters involved. In fact, an authentication protocol consists of an interaction between a prover and a verifier while a signature needs only one signer. Fiat and Shamir propose an idea consisting at the generation of the challenge with a random oracle. We explain in the following how it works: firstly, a signer generates all the commitments at once. By applying an oracle to these elements, he obtains challenges that are used to compute the answers to send to the verifier. We use a hash function as a random oracle and the challenges are deduced from the hash value of the message and the commitments.

The Fiat-Shamir paradigm has been proved secure for the three-pass protocols in [25] and more recently for five-pass protocols in [26].

3   Group Signature

In this paper, we are interested in the dynamic group signature case. For this reason, we follow the definitions presented in [21].

Definition 1. *A group signature scheme* $GS = (Setup, Join, Sign, Verif, Open)$ *is a sequence of protocols such as:*

- *$Setup(1^\lambda)$: this algorithm generates global public parameters of the system params, the group public key gpk and the group manager secret key gmsk encompassing the opening key, skO;*
- *$Join(\mathcal{U}_i)$: this is an interactive protocol between a user $\mathcal{U}_i$ and the group manager. At the end of the protocol, the user obtains a secret signing key $sk[i]$. The group manager adds the new user $\mathcal{U}_i$ and updates skO;*
- *$Sign(gpk, sk[i], m; \mu)$: to sign a message m, the user uses his secret key $sk[i]$ and some randomness μ to output a signature σ valid under the group public key gpk;*
- *$Verif(gpk, m, \sigma)$: anybody should be able to verify the validity of the signature σ on the message m with respect to gpk. It thus outputs 1 if the signature is valid, and 0 otherwise;*
- *$Open(skO, gpk, m, \sigma)$: for a valid signature σ with respect to gpk, the group manager can provide the signer identity. It thus outputs the user $\mathcal{U}_i$.*

Following [21], to be claimed secure, a group signature scheme has to prove its correctness and fulfill three properties: anonymity($anon$), traceability(tr) and non frameability(nf). For more details on security requirement of group signature, see [21].

Correctness: This notion guarantees that honest users should be able to generate valid signatures and the opener should then be able to revoke anonymity of the signers.

Unforgeability: Informally, the unforgeability notion guarantees that no one can produce a valid signature that cannot be opened in convincing way (traceability) and that no one can produce a signature on behalf of some group member (non-frameability).

In the following experiments, to join the group, an adversary($\mathcal{A}$) runs the joinP-oracle (passive join). $\mathcal{A}$ creates an honest user for whom he does not know the secret key: the index i is added to the HU (Honest Users) list.

For users whose secret keys are known to the adversary, we let the adversary play on their behalf. For honest users, the adversary can interact with them granted some oracles:

- $corrupt(i)$, if $i \in HU$, provides the secret key $sk[i]$ of this user. The adversary can now control it. The index i is then moved from HU to the list of corrupted users CU;
- $sign(i, m)$, if $i \in HU$, plays as the honest user $\mathcal{U}_i$ would do in the signature process. Then i is appended to the list $S[m]$.

We also define the open-oracle which on input (m, σ) returns $Open(skO, gpk, m, \sigma)$.

Algorithm 1. Unforgeability Notions

(a) Experiment $Exp_{GS,\mathcal{A}}^{tr}(\lambda)$

1. $(gpk, gmsk, skO) \leftarrow Setup(1^\lambda)$
2. $(m, \sigma) \leftarrow \mathcal{A}(gpk : joinP, corrupt, sign, open)$
3. if $Verif(gpk, m, \sigma) = 0$, return 0
4. if $\exists j \notin CU \cup S[m]$,
   $\quad\quad Open(gmsk, gpk, m, \sigma) = j$
   return 1
   else return 0
   $$Adv_{GS,\mathcal{A}}^{tr}(\lambda) = Pr[Exp_{GS,\mathcal{A}}^{tr}(\lambda) = 1]$$

(b) Experiment $Exp_{GS,\mathcal{A}}^{nf}(\lambda)$

1. $(gpk, gmsk, skO) \leftarrow Setup(1^\lambda)$
2. $(m, \sigma) \leftarrow \mathcal{A}(gpk, gmsk : joinP, corrupt, sign, open)$
3. if $Verif(gpk, m, \sigma) = 0$, return 0
4. if $\exists i \in HU \setminus S[m]$, $Open(gmsk, gpk, m, \sigma) = i$
   return 1
   else return 0
   $$Adv_{GS,\mathcal{A}}^{nf}(\lambda) = Pr[Exp_{GS,\mathcal{A}}^{nf}(\lambda) = 1]$$

Traceability and Non-frameability. Traceability (see Algorithm 1(a)) says that nobody should be able to produce a valid signature that cannot be opened in a convincing way. Furthermore, non-frameability (see Algorithm 1(b)) guarantees that no dishonest player (even the authorities, i.e. the Group Manager $\mathcal{GM}$, hence the keys when $gmsk$ is provided to the adversary) will be able to frame an honest user: an honest user that does not sign a message m should not be convincingly declared as a possible signer. Non-frameability also shows that the group manager cannot cheat. Thus, we say that:

- GS is traceable if, for any polynomial adversary $\mathcal{A}$, the advantage $Adv_{GS,A}^{tr}(\lambda)$ is negligible.
- GS is non-frameable if, for any polynomial adversary $\mathcal{A}$, the advantage $Adv_{GS,A}^{nf}(\lambda)$ is negligible.

In both games, the adversary generates a signature σ on a message m of its choice. In the latter game, the adversary itself can play the role of the opener trying to frame an honest user i.

Anonymity. Given two of honest users i_0 and i_1, the adversary should not have any significant advantage in guessing which one of them have issued a valid signature.

Algorithm 2. Anonymity notion

Experiment $Exp_{GS,\mathcal{A}}^{anon-b}(\lambda)$

1. $(gpk, gmsk, skO) \leftarrow (1^\lambda)$
2. $(m, i_0, i_1) \leftarrow \mathcal{A}(FIND, gpk : joinP, corrupt, sign)$
3. $\sigma \leftarrow Sign(gpk, i_b, m, sk[i])$
4. $b' \leftarrow \mathcal{A}(GUESS, \sigma : joinP, corrupt, sign)$
5. if $i_0 \notin HU$ or $i_1 \notin HU$ return 0
6. $return\ b'$

$$Adv_{GS,\mathcal{A}}^{anon}(\lambda) = Pr[Exp_{GS,\mathcal{A}}^{anon-1}(\lambda) = 1] - Pr[Exp_{GS,\mathcal{A}}^{anon-0}(\lambda) = 1]$$

The adversary can interact with honest users as before (with sign and corrupt) but the challenge signature is generated using the interactive signature protocol $Sign$. The adversary plays the role of corrupted users whereas honest users are activated to play their roles.

GS is anonymous if, for any polynomial adversary $\mathcal{A}$, the advantage $Adv_{GS,A}^{anon}(\lambda)$ is negligible.

4 The Proposed Dynamic Group Signature

In this section, we use notations defined in Sect. 3 concerning group signature. We first propose a variation of AGS identification protocol then, we present a high level overview of our scheme. Finally, we describe precisely the operations of the different algorithms required in the proposed group signature scheme.

4.1   Setup Algorithm

The following algorithm aims to output global parameters to generate our code (n, k, w), group signature public key $gpk = (\begin{pmatrix} R \\ Q \end{pmatrix}, w)$, group manager secret key $gmsk = (H, skO)$. We initialize the secret key of the opener skO to empty set.

Algorithm 3. Setup algorithm

Require: 1^{λ}
Ensure: $params, gpk$
   $(\lambda, k, n, w) \in \mathbb{N}^4 \leftarrow params(1^{\lambda})$
   $(Q, H) \in \mathcal{M}_{k \times n}(\mathbb{F}_2) \times \mathcal{M}_{(n-k) \times n}(\mathbb{F}_2) \leftarrow QC - MDPC(params)$
   $R \in \mathcal{M}_{k \times n}(\mathbb{F}_2)$ a generator matrix of a $QC - MDPC(params)$ codes
   $gpk \leftarrow (\begin{pmatrix} R \\ Q \end{pmatrix}, w)$
   $skO \leftarrow \emptyset$
   $gmsk \leftarrow (H, skO)$

4.2   AGS Variant Protocol with Collision

To build our variation of AGS protocol with ciphertext collision, the idea consists in using two related instances of AGS protocol as described below.

- The matrices R, Q and H are generated as explained in the setup algorithm (Algorithm 3) such that R and Q are two different generator matrices of QC-MDPC code of the same parameters. Let w_1, w_2 and w_3 be three integers that will constitute the weight of the error vectors and all of them are bellow to the capacity correction t of codes generated by R and Q.
- The user construct the first instance by choosing randomly $(e_1, m_1) \in \mathbb{F}_2^n \times \mathbb{F}_2^k$ such that $wt(e_1) = w_1$ then, he calculates $c_1 = m_1 \cdot R + e_1$. The user must also choose $(f_2, m_2) \in \mathbb{F}_2^n \times \mathbb{F}_2^k$ such that $wt(f_2) = w' < w_2$, calculate $\bar{c}_2 = m_2 \cdot Q + f_2$ and sends $c_1, \bar{c}_2$ to the group manager.
- The group manager receives $c_1, \bar{c}_2$ and applies the efficient decoding algorithm, of the QC-MDPC code using the secret matrix H, ψ_H to $\bar{c}_2$ ($\psi_H(\bar{c}_2)$ to get m_2). The next step consist in: the generation of $e_2 \in \mathbb{F}_2^n$ of weight w_2, computing $c_2 = m_2 \cdot Q + e_2$ and e_3 as follows $e_3 = c_1 + c_2$. He sets the opener secret key as follows $sko[i] = (e_2, e_3, m_2)$ and sends to the user the second part of his secret key (e_2, m_2).
- After receiving (e_2, m_2) from the manager, the user must verify if m_2 has been well calculated and sets his secret key as follow: $sk[i] = (e_1 + e_2, (m_1 \| m_2))$ with the following condition $c_1 + c_2 + e_3 = 0$.

$$c_1 + c_2 + e_3 = 0 \Longleftrightarrow e_1 + m_1 \cdot R + e_2 + m_2 \cdot Q + e_3 = 0 \qquad (3)$$
$$\Longleftrightarrow e_1 + e_2 + e_3 + m_1 \cdot R + m_2 \cdot Q = 0$$
$$\Longleftrightarrow e_1 + e_2 + e_3 + (m_1 \| m_2) \cdot \begin{pmatrix} R \\ Q \end{pmatrix} = 0$$

We note $e = e_1 + e_2 + e_3$, $m = (m_1 \| m_2)$ and $G = \begin{pmatrix} R \\ Q \end{pmatrix}$

Algorithm 4. The AGS variant

1. $\mathcal{P}$ randomly chooses $(u, v) \in \mathbb{F}_2^k \times \mathbb{F}_2^k$ and permutations (π, π') of $\{1, 2, ..., n\}$. Then $\mathcal{P}$ sends to $\mathcal{V}$ the commitments c_1, d_1, c_2 and d_2 such that: $c_1 = h(\pi)$, $d_1 = h(\pi')$, $c_2 = h(\pi(u \cdot R))$ and $d_2 = h(\pi'(v \cdot Q))$
   $rnd \leftarrow \{\pi', \pi, u, v\}$
2. $\mathcal{V}$ sends a value $0 \leq r \leq k - 1$ (number of shifted positions) to $\mathcal{P}$.
3. $\mathcal{P}$ build $\rho_r(e_1)$, $\rho_r(e_2)$, $\rho_r(e_3)$ and sends the last part of the commitment: $c_3 = h(\pi(u \cdot R + \rho_r(e_1) + v \cdot Q + \rho_r(e_2) + \rho_r(e_3)))$ and
   $d_3 = h(\pi'(u \cdot R + \rho_r(e_1) + v \cdot Q + \rho_r(e_2) + \rho_r(e_3)))$
   $\alpha \leftarrow \{c_1, d_1, c_2, d_2, c_3, d_3\}$
4. $\mathcal{V}$ sends $b \in \{0, 1\}$ to $\mathcal{P}$.
5. Two possibilities:
   - if $b = 0$: $\mathcal{P}$ reveals $\rho_r(u + \rho_r(m_1))$, $\rho_r(v + \rho_r(m_2))$, π, π'.
     $ans \leftarrow \{\rho_r(u + \rho_r(m_1)), \rho_r(v + \rho_r(m_2)), \pi, \pi'\}$
   - if $b = 1$: $\mathcal{P}$ reveals $\pi(u \cdot R)$, $\pi'(v \cdot Q)$, $\pi(\rho_r(e_1))$ and $\pi'(\rho_r(e_2))$, $\pi(\rho_r(e_2) + \rho_r(e_3) + v \cdot Q)$ and $\pi'(\rho_r(e_1) + \rho_r(e_3) + u \cdot R)$.
     $ans \leftarrow \{\pi(u \cdot R), \pi'(vQ), \pi(\rho_r(e_1)), \pi'(\rho_r(e_2)), \pi(\rho_r(e_2) + \rho_r(e_3) + v \cdot Q), \pi'(\rho_r(e_1) + \rho_r(e_3) + u \cdot R)\}$
6. $\mathcal{V}$ computes $c := check(\alpha, b, ans)$ where $check$ is defined by:
   - if $b = 0$: $\mathcal{V}$ verifies that c_1, d_1, c_3 and d_3 have been honestly computed.
   - if $b = 1$: $\mathcal{V}$ verifies that c_2, d_2, c_3 and d_3 have been honestly computed. and that the weight of $\pi(\rho_r(e_1))$ and $\pi'(\rho_r(e_2))$ are w_1 and w_2 respectively.
   if all checks passed
   $c \leftarrow 1$
   Else
   $c \leftarrow 0$
7. $\mathcal{V}$ outputs *Accept* if $c = 1$ and *Reject* otherwise
8. Additional check for $\mathcal{V}$ who knows m_2 $ac := addcheck(m_2, \alpha, \beta, ans)$:
   - If $b = 0$ $\mathcal{V}$ checks if $d_2 = h(\pi'(\rho_r(m_2)Q + (v + \rho_r(m_2))Q))$.
     - If d_2 is well calculated $ac \leftarrow 1$ else $ac \leftarrow 0$
   - else their is no additional checks to verify.

The Eq. (3) becomes

$$e + m \cdot G = 0$$

with $e \in \mathbb{F}_2^n$, $m \in \mathbb{F}_2^{2k}$ and $G \in \mathcal{M}_{2k \times n}(\mathbb{F}_2)$.
private key: (e, m) with e of weight $wt(e) \leq w_1 + w_2 + w_3$, of length n and $m \in \mathbb{F}_2^{2k}$.
public key: (G, w_1, w_2, w_3) with G a matrix of size $2k \times n$ and $e + m \cdot G = 0$.

Remark 2. In the case where $b = 0$, c_1 and d_1 can be easily calculated and

$$c_3 = h(\pi((u + \rho_r(m_1)) \cdot R + (v + \rho_r(m_2)) \cdot Q))$$
$$= h(\pi(u \cdot R + v \cdot Q + m_1 \cdot R + m_2 \cdot Q))$$
$$= h(\pi(u \cdot R + v \cdot Q + \rho_r(e_1) + \rho_r(e_2) + \rho_r(e_3)))$$

By analogy, we verify d_3. In addition, the opener can do additional checks by verifying $d_2 = h(\pi'(\rho_r(m_2)Q + (v + \rho_r(m_2))Q))$ (the opener have m_2 and Q is public).

In the case where $b = 1$, the verifier can check c_2 and d_2 by applying h on $\pi(u \cdot R)$ and $\pi'(v \cdot Q)$ respectively. We compute c_3 and d_3 as follows: $c_3 = h(\pi'(\pi(u \cdot R) + \pi(\rho_r(e_1)) + \pi(\rho_r(e_2) + \rho_r(e_3) + v \cdot Q)))$ and $d_3 = h(\pi'(v \cdot Q) + \pi'(\rho_r(e_2)) + \pi'(\rho_r(e_1) + \rho_r(e_3) + u \cdot R))$. We note that in our AGS variant protocol, we keep the same properties of Zero-Knowledge, correctness and soundness as the original AGS identification scheme and even the same cheating probability.

4.3   High Level Overview

In our signature scheme, we consider the following actors:

Group manager ($\mathcal{GM}$) represents the authority and he is able to run the setup algorithm (Algorithm 3) by giving as input a security parameter, to add new members to the group (by the join protocol) and to revoke the anonymity of the signers (by open algorithm).

Group members or users who sign on behalf of the group (by executing Sign algorithm).

Outsiders users who do not belong to the group but can verify a signature granted the group public key gpk (by running Algorithm 7).

The signature scheme occur in three steps: firstly, the user $\mathcal{U}$ asks the manager $\mathcal{GM}$ for joining the group and then he gets his secret signing key; secondly, the user sign on behalf of the group using his secret key and finally, the group manager $\mathcal{GM}$ is able to recover the signer's identity.

First step: the user $\mathcal{U}$ chooses randomly $e_1 \in \mathbb{F}_2^n$ of weight w_1, $m_1 \in \mathbb{F}_2^k$, $m_2 \in \mathbb{F}_2^k$ and $f_2 \in \mathbb{F}_2^n$ such that $wt(f_2) < w_2$. He compute $c_1 = m_1 \cdot R + e_1$ and $\bar{c}_2 = m_2 \cdot Q + f_2$. The user $\mathcal{U}$ sends c_1 and $\bar{c}_2$ to the $\mathcal{GM}$ who generate $e_2 \in \mathbb{F}_2^n$ of weight w_2, $m_2 \in \mathbb{F}_2^k$ and $e_3 \in \mathbb{F}_2^n$ such that $c_1 = m_1 \cdot R + e_1 = m_2 \cdot Q + e_2 + e_3$ (the collision as explained in Algorithm 5), the user's secret key is $(e_1 + e_2, m_1 \| m_2)$. The $\mathcal{GM}$ has access only to (e_2, e_3, m_2), while (e_1, m_1) is known only by the user himself.

Second step: every group user possesses a secret key verifying

$$(e_1 + e_2 + e_3) + (m_1 \| m_2) \cdot \begin{pmatrix} R \\ Q \end{pmatrix} = 0$$

The possession of such secret key will be proved in Zero-Knowledge way using the AGS variant identification protocol.

Third Step: the open algorithm is running by the group manager $\mathcal{GM}$ using the second part of the user's secret key.

4.4   Join Scheme

The Join protocol (Algorithm 5) is an interactive protocol between the group manager $\mathcal{GM}$ and the user $\mathcal{U}_i$. Firstly the user randomly chooses two pair $(e_1^{(i)}, m_1^{(i)}) \in \mathbb{F}_2^n \times \mathbb{F}_2^k$ and $(f_2^{(i)}, m_2^{(i)}) \in \mathbb{F}_2^n \times \mathbb{F}_2^k$ such that $wt(f_2^{(i)}) < w_2$ and $wt(e_1^{(i)}) = w_1$. He sends $c_1^{(i)} = m_1^{(i)} \cdot R + e_1^{(i)}$ and $\overline{c}_2^{(i)} = m_2^{(i)} \cdot Q + f_2^{(i)}$ to $\mathcal{GM}$ where $R, Q \in \mathcal{M}_{k \times n}(\mathbb{F}_2)$ are two generators matrix of two QC-MDPC codes of same parameters. Secondly the $\mathcal{GM}$ uses $H \in \mathcal{M}_{(n-k) \times n}(\mathbb{F}_2)$ from his $gmsk$ and Q from gpk to make cipher collision and compute $(e_2^{(i)}, e_3^{(i)}, m_2^{(i)}) \in S_{w_2}^n \times \mathbb{F}_2^n \times \mathbb{F}_2^k$ such that $c_1^{(i)} = m_2^{(i)} \cdot Q + e_2^{(i)} + e_3^{(i)}$ as explained in the Subsect. 4.1.

Algorithm 5. Join interactive scheme

---

1. $\mathcal{U}_i$ chooses $(e_1^{(i)}, m_1^{(i)}) \xleftarrow{\$} \mathbb{F}_2^n \times \mathbb{F}_2^k$, $(f_2^{(i)}, m_2^{(i)}) \in \mathbb{F}_2^n \times \mathbb{F}_2^k$ such that $wt(f_2^{(i)}) < w_2$, $wt(e_1) = w_1$, then send $c_1^{(i)} = m_1^{(i)} \cdot R + e_1^{(i)}$ and $\overline{c}_2^{(i)} = m_2^{(i)} \cdot Q + f_2^{(i)}$ to $\mathcal{GM}$
2. $\mathcal{GM}$ computes:
   - $m_2^{(i)} \cdot Q = \psi_H(\overline{c}_2^{(i)})$
   - retrieve $m_2^{(i)}$ from the first k positions of $m_2^{(i)} \cdot Q$
   - generate $e_2^{(i)} \xleftarrow{\$} S_{w_2}^n$
   - compute $c_2^{(i)} = m_2^{(i)} \cdot Q + e_2^{(i)}$ and $e_3^{(i)} = c_1^{(i)} + c_2^{(i)}$
   if $\exists i_0 : skO[i_0] = (e_2^{(i)}, e_3^{(i)}, m_2^{(i)})$ return false
   else $skO[i] = (e_2^{(i)}, e_3^{(i)}, m_2^{(i)})$ and send $e_2^{(i)}, m_2^{(i)}$ to $\mathcal{U}_i$
3. if the computed $m_2^{(i)}$ is different to the received $m_2^{(i)}$ return false
   else $sk[i] = (e_1^{(i)} + e_2^{(i)}, m_1^{(i)} \parallel m_2^{(i)})$

4.5   Signature Algorithm

This algorithm outputs a signature σ of a message m valid under gpk using secret key $sk[i]$ of user $\mathcal{U}_i$. The signature is obtained by applying the Fiat-Shamir transformation to our AGS Zero Knowledge identification variant. We repeat the variant l_λ times until achieving cheating probability close to 0 corresponding to λ security level. To sign a message m (Algorithm 6), a group member $\mathcal{U}_i$ produces a transcript $Tr = (\alpha, \beta, ans)$ of the AGS variant protocol (Algorithm 4) executed on public key gpk and secret key $sk[i]$ simulating the interaction through the use of a random oracle h''. In the round r, $rnd[r]$ are the random elements generated by $gen(\mu)$ ($gen(\mu)$ is a function that generate random elements in the AGS variant protocol). To run the AGS variant, $\alpha[r]$ are the commitment calculated as described in Algorithm 4 the value of $\beta \in \{0,1\}^l$ depends on the random oracle h'' (we choose h'' such that $l \geq l_\lambda$) and $ans[r]$ are responses obtained in the AGS variant protocol.

Algorithm 6. Sign$(gpk, sk[i], m, l_\lambda, \mu)$

$\alpha \leftarrow \emptyset, rnd \leftarrow \emptyset, ans \leftarrow \emptyset$ and $r \leftarrow 0$

1. **While**$(r < l_\lambda)$
   $rnd[r] \leftarrow gen(\mu)$
   $\alpha[r] \leftarrow com(gpk, sk[i], rnd[r])$
   $r \leftarrow r + 1$
2. $\beta = h''(m, \alpha)$
3. $r = 0$
4. **While** $r < l_\lambda$
   $ans[r] \leftarrow resp(sk[i], \beta_r, rnd[r])$
   $r \leftarrow r + 1$
5. Output (m, σ) where $\sigma = (\alpha, ans)$

4.6 Verification Algorithm

To verify a signature σ of a message m generate by Algorithm 7, the verifier should firstly split the signature as $\sigma = (\alpha, ans)$. Then he generates β from m and α using the random oracle h'' to verify the integrity of the message m and the membership of the user. Thereafter, we check the responses $ans[r]$ corresponding to $\alpha[r]$ and β_r for each iteration. If all checks passed, the signature is valid if not the verification fails.

Algorithm 7. Verif$(gpk, m, l_\lambda, \sigma)$

1. Split $\sigma = (\alpha, ans)$
   $\beta = h''(m, \alpha)$
2. $r \leftarrow 0$
   While $(r < l_\lambda)$
   $c = check(\alpha[r], \beta_r, ans[r])$
   if$(c == 0)$
   Output 0
   $r \leftarrow r + 1$
3. Output 1

4.7 Open Algorithm

We note that the manager secret key skO consists on a part of users secret key. We have for all users i, $skO[i] = (e_2^{(i)}, e_3^{(i)}, m_2^{(i)})$. We note also that a group manager can check more commitments than a classical verifier. Consequently $\mathcal{GM}$ may check additional requirements at Step 6 of our AGS variant protocol (Algorithm 4).

In the case $b = 0$, $\mathcal{GM}$ can check

$$d_2 = h(\sigma'(\rho_r(m_2)Q + (v + \rho_r(m_2))Q))$$

in addition to c_1, d_1, c_3, d_3 as described in Remark 2.

Algorithm 8. Open algorithm

Require: skO, gpk, m, σ
Ensure: i : the signer identity
  Split $\sigma = (\alpha, ans)$
  $\beta = h''(m, \alpha)$
  $i \leftarrow 1$
  while $i <$ *numberofUsers* **do**
    $r \leftarrow 0$
    while $r < \ell_\lambda$ **do**
      $c \leftarrow addCheck(skO[i], \alpha[r], \beta_r, ans[r])$
      if $c = 0$ **then**
        $i \leftarrow i + 1$
        $r \leftarrow 0$
      else if $c = 1$ and $r = \ell_\lambda - 1$ **then**
        return i
      else
        $i \leftarrow i + 1$
      end if
    end while
  end while
  return *False*

We define the function $addcheck(skO[i], \alpha[r], \beta_r, ans[r])$ which is similar to the function $check(\alpha[r], \beta_r, ans[r])$ but checks additional requirements as described above, we note that only $\mathcal{GM}$ can run the function $addcheck$ using skO.

Given a signature $\sigma = (\alpha, ans)$ on a message m, the group manager (who execute the open Algorithm 8) acts like in algorithm $Verify$ but now with additional checks. If there exists a user i such as all iterations on r are successful, the signer can only be i.

Moreover, $addcheck(skO[i], \alpha[r], \beta_r, ans[r])$ outputs 1 with a probability around to $\frac{1}{2}$ when i is not the real signer of the message m. Because only the case $\beta_r = 1$ who can pass (their is no additional check to do). We recall that the cheating probability in our AGS variant protocol is asymptotically equal to $\frac{1}{2}$, thus l_λ rounds required for algorithms $Sign$ and $Verify$ ensure an opening probability close to 0.

5   Security Analysis and Practical Results

5.1   Security Analysis

In this part we study the security requirements as it was defined in Algorithms 1 and 2 of our signature scheme. We start with anonymity properties and finish with Soundness properties. Our methodology is similar to the one used in the extended version of [2].

Anonymity: We now study the anonymity property.

Theorem 1. *If there exists an adversary A that can break the anonymity property of the scheme, then there exists an adversary B that can break the Zero-Knowledge property of our AGS variant.*

Proof. Let A be an adversary against the anonymity of our scheme with advantage ε. We will prove that ε is negligible using the Zero-Knowledge property of our AGS variant. We consider the following sequence of games: $\mathcal{G}_\star$, $\mathcal{G}_0$ and $\mathcal{G}_1$.

- $\mathcal{G}_\star$: the challenger run $Setup(1^\lambda)$ and gives gpk to A who has also access to oracles joinP, corrupt, sign and open. This game is the same as the real anonymity experiment (Algorithm 2), A outputs (i_0, i_1, m), but the challenger generates a simulated signature $\sigma^\star$ by programming the random oracle $\mathcal{H}_\lambda$ accordingly.
- $\mathcal{G}_0$: in this game the challenger and the adversary A has the same way and access to oracles as in game $\mathcal{G}_\star$. Now the challenger and the adversary execute the real anonymity experiment (Algorithm 2) for $b = 0$. A outputs (i_0, i_1, m) and the challenger outputs $\sigma_0 = Sign(gpk; sk[i_0]; m; \mu)$.
- $\mathcal{G}_1$: in this game the challenger and the adversary A has the same way and access to oracles as in game $\mathcal{G}_\star$. Now the challenger and the adversary execute the real anonymity experiment (Algorithm 2) for $b = 1$. A output (i_0, i_1, m) and the challenger outputs $\sigma_1 = Sign(gpk; sk[i_1]; m; \mu)$.

Given that our AGS version is Zero-Knowledge, in first time, we have that, $\sigma_\star$ is statistically close to σ_0 and in second time, we have that, $\sigma_\star$ is statistically close to σ_1. It then follows that $Exp_{GS,A}^{anon-1}(\lambda)$ and $Exp_{GS,A}^{anon-0}(\lambda)$ are indistinguishable so $Adv_{GS,A}^{anon}(\lambda) = Pr[Exp_{GS,A}^{anon-1}(\lambda) = 1] - Pr[Exp_{GS,A}^{anon-0}(\lambda) = 1]$ (where $Exp_{GS,A}^{anon-1}(\lambda)$ and $Exp_{GS,A}^{anon-0}(\lambda)$ are as defined in Algorithm 2) is negligible. ∎

Soundness: the soundness of our group signature scheme allows us to prove the traceability and the non-frameability.

We suppose that the adversary A can forge a signature on user $\mathcal{U}_{j^*}$ which is considered as an uncorrupted user. The adversary A produces a signature

$$\sigma = (cmt; rsp) = (cmt[1], \cdots, cmt[l_\lambda], rsp[1], \cdots, rsp[l_\lambda])$$

such as $Verif(gpk, m, \sigma) = 1$. A does not have any knowledge on the user secret key $sk[j^*]$ where l_λ is the number of iteration required in the AGS variation to achieve a certain security level.

Producing such as forgery by A means that he have been able to successfully run l_λ iterations of AGS variation protocol without knowing a valid secret whereas the cheating probability is close to $\frac{1}{2}$. Then A has either broken the soundness of the AGS variation or enabled the design of a knowledge extractor reaping benefits of the forgery to produce a valid solution to the general syndrome decoding problem.

The traceability and non-frameability are two notions closely related, consequently we treat both of them simultaneously. In both cases, the adversary A that can produce a valid forgery σ verifying:

$$Verif(gpk, m, \sigma) = 1$$

360   B.E. Ayebie et al.

However, breaking traceability implies for $\mathcal{A}$ to produce σ such as the group manager could not trace it back to any group member. More explicitly, to attack the traceability on user $\mathcal{U}_{j^*}$, $\mathcal{A}$ should produce a forgery σ on message m such that:

$$Verif(gpk, m, \sigma) = 1 \text{ and } Open(skO, gpk, m, \sigma) = 0 \tag{4}$$

while the non-frameability requires to produce a signature that does trace back to an actual group member. The adversary produce σ verifying:

$$Verif(gpk, m, \sigma) = 1 \text{ and } Open(skO, gpk, m, \sigma) = j^* \tag{5}$$

With the evidence that $\sigma \neq Sign(gpk, sk[j^*], m; \mu)$.

Theorem 2. *If there exists an adversary $\mathcal{A}$ against the traceability (resp the non frameability) of the scheme, then we can build an adversary $\mathcal{B}$ that can either break a general decoding problem GD, or the Simulation-Soundness of the AGS variation protocol.*

Proof. Assume that their exists an adversary $\mathcal{A}$ against the traceability of our group signature with success probability equal to ε. We use a sequence of games to show that if $\mathcal{A}$ is efficient which means he can break the traceability property (rep the non-frameability) then it's possible for an adversary $\mathcal{B}$ to solve a difficult problem with non negligible probability related to ε.

$\mathcal{G}_0$: this is the original game of the traceability (non frameability) as defined in Algorithm 1a (resp in Algorithm 1b) when the adversary $\mathcal{B}$ runs algorithm $Setup(1^\lambda)$, he chooses a user $\mathcal{U}_{j^*}$ on witch $\mathcal{A}$ will be challenged. Then $\mathcal{B}$ provides $gpk = (\begin{pmatrix} R \\ Q \end{pmatrix}, \omega)$ (resp $gpk = (\begin{pmatrix} R \\ Q \end{pmatrix}, \omega)$ and skO) to $\mathcal{A}$ which has access to the following oracles: $joinP$, $Sign$, $Open$ and $corrupt$ (resp $joinP$, $Sign$ and $corrupt$ but not $Open$ since he has skO). For any query of $\mathcal{A}$, $\mathcal{B}$ responds honestly but the game aborts if $\mathcal{A}$ tries to corrupt $\mathcal{U}_{j^*}$.

At some point $\mathcal{A}$ produces a forgery (m, σ) such that for all $j \in CU$, the signature on m were never queried and the probability for $\mathcal{A}$ to have $Verify(gpk, m, \sigma) = 1$ is equal to ε.

$\mathcal{G}_1$: this game is similar to the previous one $\mathcal{G}_0$ $\mathcal{A}$ still knows gpk (resp gpk and skO) with the same oracle access. It differ from $\mathcal{G}_0$ by the following: whenever $\mathcal{A}$ queries oracle sign on user $\mathcal{U}_{j^*}$, $\mathcal{B}$ generates a simulated valid signature. Like previously, the game aborts if $\mathcal{A}$ tries to corrupt $\mathcal{U}_{j^*}$.

At some point, $\mathcal{A}$ produces a forgery (m, σ) under the condition that for all $j \in CU$, signatures on m were never queried and signatures honestly produced on behalf of $\mathcal{U}_{j^*}$ in game $\mathcal{G}_0$ are indistinguishable from random ones produced in this game.

Consequently, for the adversary $\mathcal{A}$, games $\mathcal{G}_0$ and $\mathcal{G}_1$ are indistinguishable and the probability for $\mathcal{A}$ to have $Verif(gpk, m, \sigma) = 1$ is still ε.

Under the soundness of AGS variation protocol, we now treat separately traceability (game $\mathcal{G}_1^{trac}$) and non-frameability (game $\mathcal{G}_1^{nf}$).

$\mathcal{G}_1^{trac}$: $\mathcal{A}$ has been given a forgery σ verifying (4). Since the AGS variation protocol is sound, $\mathcal{B}$ could thus apply a knowledge extractor algorithm on σ to generate a (e^*, m^*) passing AGS variation protocol $(m^* \cdot G + e^* = 0)$. During this game $\mathcal{A}$ queried for oracle corrupt on all users except for $\mathcal{U}_{j^*}$ meaning that he may have obtained many secret keys solving the instance of general decoding $(G, \omega_1 + \omega_2 + \omega_3, 0)$. It mean that $\mathcal{A}$ has obtained the following GD instance $(G, \omega_1 + \omega_2 + \omega_3, 0, usk[i]_{i \in CU})$. If $(e^*, m^*) \notin usk[i] i \in CU$, $\mathcal{A}$ has then enabled $\mathcal{B}$ to find solution to the previous GD instance with probability directly related to ε, else, $\mathcal{A}$ replays the game until $\mathcal{B}$ gets a new solution to the aforesaid GD instance.

$\mathcal{G}_1^{nf}$: in the non frameability game, $\mathcal{A}$ has a forgery σ verifying (5). Since $\mathcal{A}$ has the skO it means that he has a knowledge of a part of the user's secret key because $usk[i] = (e_1^{(i)} + e_2^{(i)}, m_1^{(i)} \| m_2^{(i)})$ and $\mathcal{A}$ has the $(e_2^{(i)}, e_3^{(i)}, m_2^{(i)})$ for all i, we have that $\mathcal{A}$ can compute $c_1^{(i)} = m_2^{(i)} \cdot Q + e_2^{(i)} + e_3^{(i)} = (m_1^{(i)} \cdot R + e_1^{(i)})$.

Using the *corrupt* oracle, $\mathcal{A}$ can learn the entire $(e_i, m_i) = (e_1^{(i)} + e_2^{(i)} + e_3^{(i)}, m_1^{(i)} \| m_2^{(i)})$ for every user (different from $\mathcal{U}_{j^*}$) he might corrupt. In fact, $\mathcal{A}$ has obtained the following unsolved GD instances $(R, c_1^{(i)}, \omega_1)_{i \in HU}$ containing the particular one $(R, c_1^{(j^*)}, \omega_1)$. Now, under the soundness of AGS variation protocol, $\mathcal{B}$, by programming the ROM, exploits σ to get $(e^*, m^*) = (e_1^* + e_2^* + e_3^*, m_1^* \| m_2^*)$ from which he can issues signatures verifying (5) just like σ. In particular, it means that applying algorithm *Open* on signatures issued with (e^*, m^*) returns j^*. It leads to $(e_2^*, e_3^*, m_2^*) = (e_2^{(j^*)}, e_3^{(j^*)}, m_2^{(j^*)})$ and then that (e_1^*, m_1^*) is a solution to the GSD instance $(R, c_1^{(j^*)}, \omega_1)$.

At the end of game $\mathcal{G}_1$, $\mathcal{B}$ has either broken the soundness of AGS variation protocol or been able to solve a computational problem with non negligible probability related to ε. This concludes the proof. ∎

5.2 Practical Results

To instantiate our scheme, we propose the following parameters:

- In [28], Qian Guo et al. show that it's possible to successfully recovers the secret key of QC-MDPC instance for 80 bit security proposed by Misoczki et al. in [19]. Consequently, it's recommended to use this proposed instance of QC-MDPC for 128 bit security: $n = 19714$, $p = 9857$, $w = 142$, and $t = 134$.
- We generate R and Q two QC-MDPC matrices for the same parameters (n, p, w, t). Generating each one of these matrices require only 9857 bits and $w = 142$ can be stocked using 8 bits which implies a public key of size
$$size_{gpk} = (2 \times (n - p) + 8)\text{-bit} = 19750 \text{ bit} = 2.5 \text{ KB since } gpk = (\binom{R}{Q}, w).$$
- To achieve a $\lambda = 80$ bits security level in our AGS variation protocol, the number of rounds must be equal to $l_\lambda = 88$. We parse a signature σ as follows: $\sigma = (cmt[1], .., cmt[l_\lambda], rsp[1], .., rsp[l_\lambda])$. For each round $1 \leq i \leq l_\lambda$ $cmt[i] = (c_1, d_1, c_2, d_2, c_3, d_3)$ and we choose a hash function that returns a 160 bit then

the size of each $cmt[i]$ is $size_{cmt[i]} = 6 \times 160$ bits $= 960$ bits.

Now we calculate the size of $rsp[i]$ such $1 \leq i \leq l\lambda$, for this reason we calculate the size of permutations as follow:

$size_{perm} = 2 \times m \times 2^m$-bit where $m = \frac{\lceil log_2(2p) \rceil}{2}$ which implies a permutation of size $size_{perm} = 4096$ bits.

We distinguish 2 cases:
- The first case: where the challenge $b = 0$, the size of the response is $size_{rsp0} = 2 \times p + 2 \times size_{perm} = 27906$ bits
- The second case: where the challenge $b = 1$, the size of the response is $size_{rsp1} = 6 \times n = 118284$ bits

The size of a signature $\sigma = (cmt[1], .., cmt[l_\lambda], rsp[1], .., rsp[l_\lambda])$ is $size_\sigma = l_\lambda \times (size_{cmt[i]} + max(size_{rsp0}, size_{rsp1})) = 10493472$ bits, which means a signature of size 1.32 MB.

Comparing our results with the first dynamic group signature based on coding theory [2] for 80 bits security level, Our schema is more efficient in terms of public key and signature sizes which are respectively equal to 2.5 MB and 20 MB in [2] and only 2.5 KB and 1.32 MB in our scheme.

Our construction is not based on trapdoor matrix, to make syndrome collision. Notice that in hamming metric, the one existing trapdoor matrix is CFS signature instance [27]. We propose an other approach in the join protocol based on QC-MDPC McEliece to make cipher collision instead of syndrome collision.

6   Conclusion

In this work, we have proposed a new dynamic group signature scheme based on coding theory assumptions which improve the first code based group signature presented in [2]. In this proposed scheme, group public key and signatures sizes remain independent of group length but just depending on the security level. Our technique consist in using the QC-MDPC version of McEliece cryptosystem in the join protocol and AGS identification scheme as Zero-Knowledge argument system unlike using the trapdoor matrix and Stern identification scheme as described in [2]. Even if in [28] the authors proposed a key recovery attack for 80 bits security level, our scheme remain still secure because we use parameters for 128 bits security level. We have started to develop this idea before the publication of [28]. These choices make the join protocol very fast and decreases the public key and signature length typically from 2.5 MB in the original scheme to 2.5 KB in our one for the public key and from 20 MB in the original scheme to 1.32 MB in our scheme for the signature. In future work we will try to add the revocation properties and try to turn this scheme to a List signature scheme.

References

1. Aguilar Melchor, C., Gaborit, P., Schrek, J.: A new zero-knowledge code based identification scheme with reduced communication. In: 2011 IEEE Information Theory Workshop, pp. 648–652. IEEE Press, Paraty (2011)
2. Alamélou, Q., Blazy, O., Cauchie, S., Gaborit, P.: A code-based group signature scheme. In: WCC 2015. LNCS, vol. 942, pp. 260–285. Springer, Heidelberg (2015)
3. Chaum, D., Heyst, E.: Group signatures. In: Davies, D.W. (ed.) EUROCRYPT 1991. LNCS, vol. 547, pp. 257–265. Springer, Heidelberg (1991). doi:10.1007/3-540-46416-6_22
4. Ateniese, G., Camenisch, J., Joye, M., Tsudik, G.: A practical and provably secure coalition-resistant group signature scheme. In: Bellare, M. (ed.) CRYPTO 2000. LNCS, vol. 1880, pp. 255–270. Springer, Heidelberg (2000). doi:10.1007/3-540-44598-6_16
5. Boneh, D., Boyen, X., Shacham, H.: Short group signatures. In: Franklin, M. (ed.) CRYPTO 2004. LNCS, vol. 3152, pp. 41–55. Springer, Heidelberg (2004). doi:10.1007/978-3-540-28628-8_3
6. Camenisch, J., Lysyanskaya, A.: Signature schemes and anonymous credentials from bilinear maps. In: Franklin, M. (ed.) CRYPTO 2004. LNCS, vol. 3152, pp. 56–72. Springer, Heidelberg (2004). doi:10.1007/978-3-540-28628-8_4
7. Nguyen, L., Safavi-Naini, R.: Efficient and provably secure trapdoor-free group signature schemes from bilinear pairings. In: Lee, P.J. (ed.) ASIACRYPT 2004. LNCS, vol. 3329, pp. 372–386. Springer, Heidelberg (2004). doi:10.1007/978-3-540-30539-2_26
8. Bellare, M., Rogaway, P.: Random oracles are practical: a paradigm for designing efficient protocols. In: Proceedings of the 1st CCS, pp. 62–73 (1993)
9. Bellare, M., Micciancio, D., Warinschi, B.: Foundations of group signatures: formal definitions, simplified requirements, and a construction based on general assumptions. In: Biham, E. (ed.) EUROCRYPT 2003. LNCS, vol. 2656, pp. 614–629. Springer, Heidelberg (2003). doi:10.1007/3-540-39200-9_38
10. Kiayias, A., Yung, M.: Extracting group signatures from traitor tracing schemes. In: Biham, E. (ed.) EUROCRYPT 2003. LNCS, vol. 2656, pp. 630–648. Springer, Heidelberg (2003). doi:10.1007/3-540-39200-9_39
11. Kiayias, A., Yung, M.: Group Signatures with Efficient Concurrent Join. In: Cramer, R. (ed.) EUROCRYPT 2005. LNCS, vol. 3494, pp. 198–214. Springer, Heidelberg (2005). doi:10.1007/11426639_12
12. Bellare, M., Shi, H., Zhang, C.: Foundations of group signatures: the case of dynamic groups. In: Menezes, A. (ed.) CT-RSA 2005. LNCS, vol. 3376, pp. 136–153. Springer, Heidelberg (2005). doi:10.1007/978-3-540-30574-3_11
13. Laguillaumie, F., Langlois, A., Libert, B., Stehlé, D.: Lattice-based group signatures with logarithmic signature size. In: Sako, K., Sarkar, P. (eds.) ASIACRYPT 2013. LNCS, vol. 8270, pp. 41–61. Springer, Heidelberg (2013). doi:10.1007/978-3-642-42045-0_3
14. Langlois, A., Ling, S., Nguyen, K., Wang, H.: Lattice-based group signature scheme with verifier-local revocation. In: Krawczyk, H. (ed.) PKC 2014. LNCS, vol. 8383, pp. 345–361. Springer, Heidelberg (2014). doi:10.1007/978-3-642-54631-0_20
15. Ling, S., Nguyen, K., Wang, H.: Group signatures from lattices: simpler, tighter, shorter, ring-based. In: Katz, J. (ed.) PKC 2015. LNCS, vol. 9020, pp. 427–449. Springer, Heidelberg (2015). doi:10.1007/978-3-662-46447-2_19

16. Gordon, S.D., Katz, J., Vaikuntanathan, V.: A group signature scheme from lattice assumptions. In: Abe, M. (ed.) ASIACRYPT 2010. LNCS, vol. 6477, pp. 395–412. Springer, Heidelberg (2010). doi:10.1007/978-3-642-17373-8_23

17. Ezerman, M.F., Lee, H.T., Ling, S., Nguyen, K., Wang, H.: A provably secure group signature scheme from code-based assumptions. In: Iwata, T., Cheon, J.H. (eds.) ASIACRYPT 2015. LNCS, vol. 9452, pp. 260–285. Springer, Heidelberg (2015). doi:10.1007/978-3-662-48797-6_12

18. Assidi, H., Ayebie, E.B., Souidi, E.M.: A code-based group signature scheme with shorter public key length. In: ICETE 2016: SECRYPT, vol. 4, pp. 432–439. SciTePress, Lisbon, July 2016

19. Misoczki, R., Tillich, J.-P., Sendrier, N., Barreto, P.S.: MDPC-McEliece: new Mceliece variants from moderate density parity-check codes. In: IEEE International Symposium on Information Theory Proceedings (ISIT), pp. 2069–2073 (2013)

20. Berlekamp, E.R., McEliece, R.J., Van Tilborg, H.C.: On the inherent intractability of certain coding problems. IEEE Trans. Inf. Theor. **24**(3), 384–386 (1978)

21. Blazy, O.: Preuves de connaissance interactives et non-interactives. Part 1, Chap. 3. Ph.D. thesis, University Paris VII - Denis Diderot, September 2012

22. Véron, P.: Improved identification schemes based on error-correcting codes. Appl. Algebra Eng. Commun. Comput. **8**(1), 57–69 (1996)

23. Stern, J.: A method for finding codewords of small weight. In: Cohen, G., Wolfmann, J. (eds.) Coding Theory 1988. LNCS, vol. 388, pp. 106–113. Springer, Heidelberg (1989). doi:10.1007/BFb0019850

24. Fiat, A., Shamir, A.: How to prove yourself: practical solutions to identification and signature problems. In: Odlyzko, A.M. (ed.) CRYPTO 1986. LNCS, vol. 263, pp. 186–194. Springer, Heidelberg (1987). doi:10.1007/3-540-47721-7_12

25. Poincheval, D., Stern, J.: Security arguments for digital signatures and blind signatures. J. Cryptology **13**(3), 361–396 (2000)

26. Yousfi Alaoui, S.M., Dagdelen, Ö., Véron, P., Galindo, D., Cayrel, P.-L.: Extended security arguments for signature schemes. In: Mitrokotsa, A., Vaudenay, S. (eds.) AFRICACRYPT 2012. LNCS, vol. 7374, pp. 19–34. Springer, Heidelberg (2012). doi:10.1007/978-3-642-31410-0_2

27. Courtois, N.T., Finiasz, M., Sendrier, N.: How to achieve a McEliece-based digital signature scheme. In: Boyd, C. (ed.) ASIACRYPT 2001. LNCS, vol. 2248, pp. 157–174. Springer, Heidelberg (2001). doi:10.1007/3-540-45682-1_10

28. Guo, Q., Johansson, T., Stankovski, P.: A key recovery attack on MDPC with CCA security using decoding errors. In: Cheon, J.H., Takagi, T. (eds.) ASIACRYPT 2016. LNCS, vol. 10031, pp. 789–815. Springer, Heidelberg (2016). doi:10.1007/978-3-662-53887-6_29

A Secure Cloud-Based IDPS Using Cryptographic Traces and Revocation Protocol

Hind Idrissi[✉], Mohammed Ennahbaoui, Said El Hajji, and El Mamoun Souidi

Laboratory of Mathematics, Computing and Applications (LabMIA),
Faculty of Sciences, Mohammed-V University in Rabat, Rabat, Morocco
hind.idr@gmail.com, ennahbaoui.mohamed@gmail.com, elhajji.said@gmail.com,
emsouidi@gmail.com

Abstract. Cloud computing is a revolutionary information technology, that aims to provide reliable, customized and quality of service guaranteed environments, where virtualized and dynamic data are stored and shared among cloud users. Thanks to its significant benefits such as: on demand resources and low maintenance costs, cloud computing becomes a trend in the area of new technologies that facilitates communication and access to information. Despite the aforementioned facts, the distributed and open nature of this paradigm makes privacy and security of the stored resources a major challenge, that limits the use and agreement of cloud computing in practice. Among the strong security policies adopted to address this problem, there are Intrusion Detection and Prevention Systems (IDPS), that enable the cloud architecture to detect anomalies through monitoring the usage of stored resources, and then reacting prevent their expansion. In this paper, we propose a secure, reliable and flexible IDPS mainly based on autonomous mobile agents, that are associated with tracing and revocation protocol. While roaming among multiple cloud servers, our mobile agent is charged with executing requested tasks and collecting needed information. Thus, on each cloud server a "cryptographic trace" is produced in which all behaviors, results and data involved in the execution are recorded, which allow to identify any possible intrusions and hence predict a response to prevent them or end their processing, through using a server revocation technique based on trust threshold.

Keywords: Cloud computing · IDPS · Mobile agent · Cryptographic traces · Revocation protocol

1 Introduction

Cloud Computing is a rapidly expanding paradigm that brings a revolution in IT world through using the Internet services. These services that consist of applications and databases deployed in large centralized data centers, are delivered to end users on demand rather than being maintained in a large and expensive IT infrastructure. Cloud computing was defined by the NIST [1] as an emerging

© Springer International Publishing AG 2017
S. El Hajji et al. (Eds.): C2SI 2017, LNCS 10194, pp. 365–382, 2017.
DOI: 10.1007/978-3-319-55589-8_24

computing approach enabling ubiquitous, convenient and on-demand network access to shared resources (e.g., data, servers, applications, and services), that can be rapidly provisioned and released with minimal management effort or service provider interaction. Reliable and flexible computations on demand, available and easy accessible data/services, in addition to the wide storage capacities associated with high quality of services, are among the several benefits of the cloud that attract customers and make it a viable commercial option, particularly for small companies and startups which will potentially reduce their costs when paying only for what they really use.

Even though cloud computing shows a significant widespread according to its recent design of IT hardware, it is confronted to the substantial problem of security. With its open and distributed framework mainly based on resource virtualization, global replication and migration, cloud computing increasingly seduces potential attackers. In 2011 [2], the Amazons Elastic Computer Cloud service is used by a hacker to attack Sonys online entertainment systems, by registering and opening an Amazon account and using it anonymously. Such attack compromised more than 100 million customer accounts, the largest data breach in the U.S. One of the most common requirements for cloud security is an Intrusion Detection and Prevention System (IDPS), which can be efficient to early screen malicious entities, track their untrustworthy behaviors and prevent the damages they can cause to the systems. An IDPS was also defined by the NIST [3] as a software or hardware device, that has all the capabilities of an intrusion detection system (IDS) to potentially identify an attack and notify appropriate personnel immediately, and can also attempt to stop possible threats or at least prevent them from succeeding, so that they can be contained.

In this paper, a robust IDPS relying on mobile agent technology is proposed to ensure a safe interaction model where communication between the cloud provider and its related cloud storage servers are secure and reliable. The use of this technology will allow us to benefit from the autonomy, pro-activity, mobility and flexibility of its entities in order to provide the cloud with perceptive and talented solutions. Mobile agents [4] represent a particular category of software entities with the capacity to move across different platforms and execute the requested tasks independently from the environment where they are landed, as they transport their own resources including the code, the data to deploy and an execution state.

Our approach begins once the cloud provider receives requests from one or multiple cloud users. Then, it creates a mobile agent with specified constraints, and which will be charged with the migration across the cloud storage servers to gather or retrieve data, achieve actions and calculations, benefit from services and resources and finally return back with results. Along its round-trip, the mobile agent adopts two major mechanisms combined with cryptographic primitives (asymmetric encryption, digital signature and hash function) to ensure confidentiality and integrity of data associated. The first mechanism consists of generating cryptographic traces, where all behaviors, actions and computations performed either by the agent or the hosting server are recorded, so that any anomaly can be easily detected through verifying and analyzing these traces. The second one is a prevention mechanism based on revocation technique, where a trust threshold is assigned to each server to define its degree of maliciousness according to many factors and proofs. Once this threshold is reached, the named server is added to the black database of malicious servers hosted by a specified authority.

The reminder of the paper is organized as follows. Section 2 provides a statement of the security problems in cloud computing, especially those between the Cloud Provider and the storage servers. A brief review on the related works is provided in Sect. 3. In Sect. 4 we give an overview of the system architecture, then we describe the proposed IDPS with the mechanisms employed. An evaluation using CloudSim tool is exposed in Sect. 5, and the obtained results are compared to a basic mathematical model, in terms of security, reliability and efficiency. Finally, further discussion and perspectives are mooted in conclusion.

2   Security Issues

While cloud computing shows a large set of advantages more appealing than ever, it also brings new and challenging security threats to the outsourced data. This is relatively due to the concept of virtualization and the physical absence of resources, which makes the overall cloud architecture not fully controlled/managed and thus not safe.

We begin with an architecture description of cloud data storage services illustrated in Fig. 1. This latter consists of four different entities: data owner, cloud user (CU), cloud service provider (CSP) controlling and monitoring numerous storage servers (SS), and trusted third authority (TPA) that has the capabilities to assess cloud storage security on behalf of a data owner. When a CU

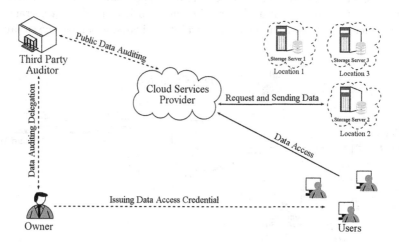

Fig. 1. The generic architecture of cloud computing

submits storage or computation service requests, the CSP charges its administration server to spread the request among its different SS located on different geographical areas. According to [1], among the essential features that cloud computing provides are defined:

- *Resource Pooling:* the provider's computing resources are pooled to serve multiple consumers, using a multi-tenant model, with different physical and virtual resources dynamically assigned and reassigned according to consumer demand. There is a sense of location independence in that the customer generally has no control or knowledge over the exact location of the provided resources but may be able to specify location at a higher level of abstraction (e.g., country, state, or data-center). Examples of resources include storage, processing, memory, and network bandwidth.
- *Broad Network Access:* capabilities are available over the network and accessed through standard mechanisms that promote use by heterogeneous thin or thick client platforms (e.g., mobile phones, tablets, laptops, and workstations).

These two features are extremely vulnerable to many attacks: Masquerading, Man-In-The-Middle, etc., since they evoke many security issues as illustrated in Fig. 2. In this context, multiple scenarios are considered:

- An intruder may intercept the communication between the CPU and its relevant cloud servers. He can proceed to the alteration of the exchanged data in order to compromise their confidentiality and integrity.
- An attacker can harm to the CUs privacy through leaking their confidential data to others.
- An attacker may corrupt one or multiple SSs with the aim to monitor them and initiate various cheating attacks.

Within the scope of this article, we focus on how to ensure secure cloud data storage services. We consider both malicious outsiders and a semi-trusted storage server SS as potential adversaries interrupting cloud data storage services. Malicious outsiders can be economically motivated and have the capability

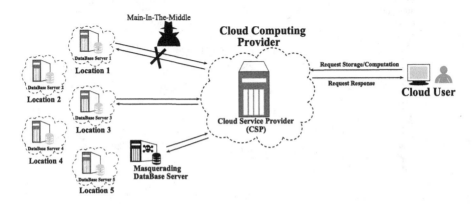

Fig. 2. Security issues in cloud computing architecture

to attack cloud storage servers, subsequently they are able to pollute or delete owner's data while remaining undetected. The CS is semi-trusted in the sense that most of the time it behaves properly and does not deviate from the prescribed protocol execution. However, for its own benefit the CS might neglect to keep or deliberately delete rarely accessed data files that belong to ordinary cloud owners. Moreover, the CS may decide to hide the data corruptions caused by server hacks or Byzantine failures to maintain its reputation.

In this context, many efforts have been devoted to investigate the security issue of Cloud Computing. Indeed, several solutions have been proposed in the literature, where IDPSs are supplied as very important and invaluable tools. Towards this, Gupta et al. [5] propose an intrusion detection and prevention approach that focuses on a non-conventional technique for securing cloud network from malicious insiders and outsiders, using network profiling to describe network behavior of cloud users and the attack patterns that needs to be looked over. In [6], Tapakula et al. have proposed an IDPS for cloud environment based on a virtual machine monitor (called hypervisor) to protect the system from different types of attacks in the infrastructure layer (IaaS). However, it has not provided a prevention solution face to high severe attacks over the system. Recently, Autonomic Computing drew researchers attention for CIDPS with minimal human intervention. The authors of [7] have proposed a customizable defense system, called VMFence, in a virtualization-based cloud computing environment. It is deployed with distributed intrusion prevention system and a file integrity monitoring tool (FIMT) endowed with a high efficiency of detection and response. This allows the cloud provider to configure detection rules for each domain, according to the type of service running in each virtual machine, while the file modification information is collected in real time. Smith et al. [8] presented an autonomic mechanism for anomaly detection in a cloud computing environment, with uniform format analysis and size reduction for data, as well as learnt how to detect the nodes which have abnormal behavior and act differently from others in an unsupervised mode. In the work of Alsafi et al. [9], they discuss a method where the use of multi-threading techniques provides a more efficient method for improving the performance of an IDPS in a cloud environment. Their method provides the system with the ability to handle a large number of data packet flows through an analysis module that filters out the bad packets and a reporting module that produces reports on security and accuracy of data. Dastjerdi et al. [10] have proposed an application of mobile agents in IDPS to provide flexible, scalable, and a cost effective system for the cloud environment. However, the inefficient sharing of knowledge among the mobile agents makes the robustness of the system not supported and the scalability not ensured.

3   Proposed IDPS for Cloud Security

In this section, a thorough description of our Cloud-based IDPS is provided, which is essentially made up of two main parts. In the first part, a robust detection mechanism is proposed, where mobile agents generate a behaviors record in the form of a cryptographic trace on each visited cloud server. The second part presents a prevention policy that punishes malicious entities through a server revocation technique based on trust threshold. The detection and prevention in our proposed system are complementary and dependent, as they are basically reliant to the same sensible parameters in the cryptographic traces obtained through the trip of the mobile agent, to perform the tasks that the CU requests.

3.1   Cryptographic Traces for Intrusion Detection

In this section, we describe the proposed detection method, which is mainly based on the mechanism of execution tracing associated to the mobile agents. Thanks to its numerous features, mobility of agents provides an agile, dynamic and operational aspects to the interactions with cloud servers, since the rate of the exchanges is lower which significantly lessens the network traffic. As illustrated in Fig. 3, in a normal scenario, the CSP receives multiple requests for computation or storage services from different CUs. According to this, a mobile agent is created and attributed a list of the IP addresses corresponding to the hosts in its itinerary, as well as the jobs to be carried out on each one. Then, the mobile agent analyzes the given positions to follow the shortest path, and moves among the specified cloud servers to execute its tasks, such that the obtained results are securely accumulated until it returns back to the CSP. Moreover, it is supposed that the TPA afforded the CSP and each cloud server a public and private key in order to deal with the computations and verification with the agent.

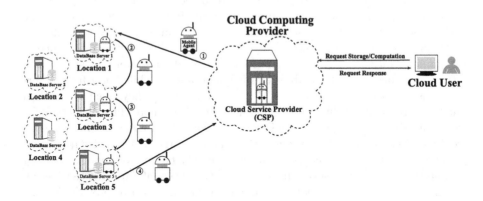

Fig. 3. The use of mobile agents in the interactions of the cloud computing

Before migration, the agent is also assigned some essential credentials given in Table 1, to authenticate the visited cloud servers according to their IP addresses. Thus, we make use of an enhanced version of Diffie-Hellman key exchange protocol inspired from [13], where the digital signature algorithm is integrated to fix security issues related to Man-in-the-Middle attacks. Figure 4 illustrates the authentication process, that generates a common session key at the end. This process employs the cryptographic generator ISAAC+[14] to produce randoms

Table 1. Authentication credentials assigned to the mobile agent by the CSP

Credential	Description
p, t	Random odd primes, where $t < p - 1$ and $q = t \dfrac{p-1}{2}$
g	Generator of the field F_p
X_{MA}	Random private key, with $1 < X_{MA} < q$
PK_{CSP}	Public key of the CSP
$H(.)$	256-bits hash function (SHA-3)

Mobile Agent (MA)	**Cloud Server (CS$_n$)**
a: random integer $\quad \xrightarrow{p,q,g,H}$	
	X_{CS_n}: random private key $< q$
	Y_{CS_n}: public key, with:
	$Y_{CS_n} = g^{X_{CS_n}} mod p$
	b: random integer
$V_{MA} = g^a mod p,$	
$W_{MA} = PK_{CSP}^a mod p$	
$R_{MA} = V_{MA} mod q \quad \xrightarrow{V_{MA}, W_{MA}}$	
	$V_{CS_n} = g^b mod p,$
	$W_{CS_n} = Y_{CS_n}^b mod p$
$\xleftarrow{V_{CS_n}, W_{CS_n}}$	$R_{CS_n} = V_{CS_n} mod q$
$K_{ab} = W_{CS_n}^a mod p,$	$K_{ba} = W_{MA}^b mod p,$
$K_{ba} = V_{CS_n}^{a \times X_{MA}} mod p$	$K_{ab} = V_{MA}^{b \times X_{CS_n}} mod p$
IP_h: address in the Host list	IP_{CS_n}: address of the platform
	Signature:
	$S_{CS_n} = b^{-1} \times H((V_{CS_n} \parallel K_{ba} \parallel K_{ab}$
$\xleftarrow{S_{CS_n}}$	$\parallel IP_{CS_n}) + X_{CS_n} \times R_{CS_n}) mod q$
Verify: $S_{CS_n} == S_{MA}$, where:	
$S_{MA} = a^{-1} \times H((V_{MA} \parallel K_{ab} \parallel K_{ba}$	
$\parallel IP_h) + X_{MA} \times R_{MA}) mod q$	

Common Shared Key
$\mathbf{SK_n = K_{ab} \times K_{ba} mod p = g^{vw \times (X_{MA} + X_{CS_n})} mod p}$

Fig. 4. Authentication process between the mobile agent and the cloud server

and the hash function SHA-3 [15] to elaborate signatures. The following are the main steps in this mechanism:

1. The mobile agent "MA" generate a random integer "a" and sends the authentication credentials (p, q, g, H) assigned by its owner to the cloud server

"CS_n". Then, MA computes the values: V_{MA} as the exponentiation of the generator g by the random a, W_{MA} as the exponentiation of the public key of the cloud service provider (as the agent owner) PK_{CSP} by a and R_{MA} as the modulo of V_{MA} by q. Finally, MA forwards V_{MA} and W_{MA} to "CS_n", while R_{MA} is kept secret.

2. Once receiving the authentication credentials, the "CS_n" chooses a random private key $X_{CS_n} < q$ and computes its public key PK_{CS_n} as the exponentiation of g by X_{CS_n}. Then, "CS_n" generates a random integer "b" and computes the values: V_{CS_n}, W_{CS_n} (using the "CS_n" public key) and R_{CS_n} in the same way the agent did. Afterwards, CS_n sends V_{CS_n} and W_{CS_n} to "MA", while R_{CS_n} is kept secret.

3. The "CS_n" calculates the exponentiation by b of the previously received W_{MA} and notates it as K_{ba}. Besides, the exponentiation by $b \times X_{CS_n}$ of the received value V_{MA} is computed and noted as K_{ab}. Then, using the IP address of the cloud server platform, the "CS_n" computes its own signature as: $S_{CS_n} = b^{-1} \times H((V_{CS_n}||K_{ba}||K_{ab}||IP_{CS_n}) + X_{CS_n} \times R_{CS_n})modq$. Finally, this signature is forwarded to the mobile agent.

4. Similarly, the "MA" computes the values K_{ab} and K_{ba} using the random a, the private key X_{CS_n} and (V_{CS_n}, W_{CS_n}) previously received from the "CS_n". Thus, the signature S_{MA} relevant to the "MA" is calculated using the IP address contained in the afforded Host IP List. If this signature is equal to that provided by the "CS_n", then this latter is said to be authenticated and a shared session key can be computed by both parties as: $SK_n = K_{ab} \times K_{ba}modp$.

The use of ephemeral secrets a and b chosen by the both sides provides two important properties. The first one is *forward secrecy* that prevents the disclosure of any of the previous session keys even if the long-term private key of any party is exposed. The second one is *key freshness* as neither of the authenticating parties can predetermine the value of the session key, since he would not know the ephemeral secret of the other party. In addition, the use of the CSP's public key in computations facilitates traceability and saves time and energy of producing a public key for the agent at each cloud sever visited. Thereby, the 256-bits length session key (SK_n) obtained at the cloud server (n) will be used to compute the cryptographic trace on that platform.

Being inspired from the work of [12], a trace is associated to the execution of the mobile agent on each visited server. It is noted $T = <ui, S>$ and contains the signature of the executed black statements (instructions of the agent code using information from external environments), in addition to a unique identifier of the trace. This process allows the owner of the agent to verify its execution and depict suspicious behaviors or operations carried out by malicious intruders. Indeed, the temporal correctness of the agent execution is a crucial point to guarantee that the connected cloud servers work in synchronized timing. Thus, we introduce Network Time Protocol (NTP) [11] to enforce the clocks of different cloud servers related to the CSP to perform in sync with it. This makes the control and the harmonization of log entries more straightforward when an event occurred across multiple servers.

A cryptographic trace is generated as indicated in the java function illustrated in Fig. 5. It contains a generic DSA signature [17] performed using the public key of the CSP, which makes this latter the only one able to decrypt it by means of its private key. In addition to the unique identifier (ui), the identities (ID_s: sender, $host$: current host, ID_next: intended host), the session key ($SK\_S(i)$), the timestamp (ts) and the required task ($Task$), the generic signature includes two nested fields:

```
public Trace<ui,S> Generate_CryptoTrace( Agent A, CloudServer S(i)){
  ID_s = A.getSender();
  ID_ma = A.getAgentIdentity();
  host = S(i).getIdentity();
  crd = A.getCredentials;
  PK_csp = crd.getCSPkey();
  ID_next = crd.getNextHost(S(i));
  SK_S(i) = this.getSessionKey();
  Task= A.getRequestTask();
  BS = class.execute(A).getBlackStatement();
  RS= class.execute(A).getResults();
  // current timestamp
  java.sql.Date ts = new java.sql.Timestamp(Calendar.getInstance().
                            getTime().getTime());
  // unique identifier
  ui = UUID.randomUUID();
  Signature dsa1 = Signature.getInstance("SHA256withDSA", "SUN");
  dsa1.initSign(PK_csp);
  Signature dsa2 = Signature.getInstance("SHA256withDSA", "SUN");
  dsa2.initSign(SK_S(i));
  if (i==1)
  T(S(i))= <ui, dsa1.sign(dsa1.update(ui, SK_S(i), ID_s, host, ID_next
             , ts, Task, AES_Encrypt(SK_S(i), (BS,RS)), dsa2.sign(dsa2.
             update(ID_s, ID_ma, Hash(ts, BS, RS) ) ) ) )>;
    else
  T(S(i))= <ui, dsa1.sign(dsa1.update( T(S(i-1)), ui, SK_S(i), ID_s,
             host, ID_next, ts, Task, AES_Encrypt( SK_S(i), (BS,RS)),
             dsa2.sign(dsa2.update(ID_s, ID_ma, Hash(ts, BS, RS),
             Hash (T(S(i-1))) ) ) ) )>;
  return T(S(i));
}
```

Fig. 5. Java pseudo-code of the cryptographic trace generation

- A symmetric encryption, using AES [16] with the session key $SK_S(i)$ of 256 bits length, of the black statements and the results obtained through the execution of requested tasks.
- The DSA signature, using $SK\_S(i)$, of the agent's identity and its sender, as well as a SHA-3 hash of the timestamp, black statements and the results of execution.

Our system is also provided with a chaining mechanism that links the traces produced during the trip of the mobile agent. Thereby, each trace being produced on a cloud server (current host) will be enclosed while generating the trace on the following server (intended host), either in the generic signature or in the nested signature which joins the trace hash. Since for the first destination in the itinerary list there is no prior trace, the chaining process begins at the second cloud server being visited ($i = 2$). Once returning back to the CSP, the mobile agent presents the final results along with the collected traces. Then, a verification of each trace $T(S(i))$ is performed following the steps below:

1. using its private key, the CSP decrypts the generic signature, verifies the identities involved and checks the freshness of the timestamp (ts). Then, decrypts the cipher using the session key $SK\_S(i)$, in order to get the black statements and the results in clear.
2. a SHA-3 hash of the previous trace provided by the mobile agent is calculated: $h1 = Hash(T(S(i-1)))$.
3. using the current session key $SK\_S(i)$, the CSP decrypts the nested signature and verifies the identity of the agent and its sender.
4. the CSP computes its own hash of the timestamp along with the black statements and the results, that are provided in the cipher $Hash(ts, BS, RS)$. Then, this hash is compared to that contained in the third field of the nested signature.
5. the CSP computes the hash of the given trace in the first field of the generic signature: $h2 = Hash(T(S(i-1)))$. Then h1 and h2 are compared to the hash in the last field of the nested signature.

Once the verification is successfully carried out, the cloud server is classified as trustworthy for this session, else it is considered suspicious and the CSP proceeds to its prevention policy.

3.2   Revocation-Based Trust Threshold for Intrusion Prevention

As a matter of fact, the prevention policy we adopt for our system consists in a double faced method: ending the activity of the malicious server in that session, and punishing it through decreasing its level of trust. When the CSP verifies the trace of the mobile agent's execution on a server $S(i)$, and finds that one or more comparisons do not match, then this server is qualified as suspicious. At this stage, the CSP sends a message to the named server asking it for forwarding the specified trace. The core of this message and its reply provided by $S(i)$ is shown in Fig. 6, such that the identity of each one is involved as identifier for its message.

```
From CSP to S(i):
M_(csp-to-si)=Sign_(sk_(si)) ( Hash(ID_(si)), Forward-Trace())

From S(i) to CSP:
M_(si-to-csp)=Sign_(pk_(csp))( Hash(ID_csp), Hash(ID_(si)),
                               M_(csp-to-si),T(S(i)))
```

Fig. 6. The core of the message and reply for Forward-Trace() transaction

Once the CSP receives the reply for its request and decrypts the involved signature using its private key, it proceeds to the verifications as indicated in Fig. 7:

1. Checks the hash containing its identity ID_csp, then verifies that the value $Hash(ID\_(si))$ is the same in both messages $M\_(csp-to-si)$ and $M\_(si-to-csp)$.
2. Decrypts the nested $M\_(csp-to-si)$ using the session key and verifies that it contains the forward request.
3. Computes the hash of the trace provided $Hash(T(S(i)))$ and verifies that it matches with the hash given by the server $S(i+1)$

```
Verification_Flow (suspicious Server (S_i)){

Request(Forward-Trace()) ===> (S_i)

While (timestamp){

    if (Not-Received())
       {Revocation ((S_i), Punishment(level=1));}

    else{
                /*     Verification  1 */

       if (Hash(ID_(S_i)).NotEqualTo (extract-ID_Hash(
                                M_(csp-to-si),M_(si-to-csp))))
          {Revocation ((S_i), Punishment(level=2));}

       else{
                  /*     Verification  2 */

          if (Hash(T(S_i)). NotEqualTo(extract-Trace_Hash(M_(si-to-csp))))
             {Revocation ((S_i), Punishment(level=3));}

          else{
                     /*     Verification  3 */

             Agent. re-Execute();
             if (Agent. getCurrentTrace(). NotEqualTo( T(S_i)))
                {Revocation ((S_i), Punishment(level=4));}

             else{
                     Print ("The Cloud Server is Honest")
}}
}}
}}
```

Fig. 7. Pseudo-code describing the verification flow of a cloud server maliciousness

It is sufficient to not receive a reply for the request or at least one of these verifications does not match, so that the CSP can start its revocation protocol in concordance with a Trust Authority (TA). Otherwise, in case a response is received such that all verifications are unverified, then the CSP charges the same mobile agent with moving to the server $S(i)$ in order to re-execute the same tasks. The returned results are verified to see if the execution agrees with the forwarded trace. If the traces match, then the server is qualified as honest, else the revocation protocol is triggered.

The revocation protocol has two instances according to the trust level of the cloud server. A trust level (TL) is represented by a decimal number between 0 and 1 and it is allocated by the Trust Authority (TA), which is the only one who possesses the right to modify it. According to this (TL), the revocation may be conditional or permanent, with regard to a threshold (d) decided by the TA for each cloud server registered in its database.

The conditional revocation protocol is processed when the cloud server has one TL that did not reach yet the threshold (d), and it is described as follows:

1. The CSP informs the TA that the server $S(i)$ is suspicious and sends to it the proofs it has.
2. The TA verifies that the TL of the named server is less than the (d), and then requests the trace directly from $S(i)$.
3. If the cloud server forwards the trace, the TA proceeds to a verification of all data involved in this trace, then it re-executes the agent again and compares the given trace with that provided by the CSP.
   - if they match, then the server is considered to be honest and the CSP is warned for dishonest attitude.
   - else, the server is provisionally revoked and punished by subtracting a percentage of its TL according to a punishment scale.
4. If the server did not present the requested trace, then it is revoked with decreasing its TL until sending the trace or proving its good intention.

Concerning the permanent revocation protocol, it becomes activated when the TL of the suspicious server exceeds the threshold (d). It is defined in the steps bellow:

1. The CSP informs the TA that the server $S(i)$ is malicious and sends all its proofs to demonstrate that it conducts incorrect behaviors.
2. The TA verifies that the TL of the named server is greater than (d).
3. The TA verifies narrowly all information provided by the CSP, beginning by the timestamp freshness, the identities of communicating parties and the session key produced. Whether one of these data is incorrect, the TA stops the revocation and warns the CSP for invalid information. Else, verification is pursued.
4. The TA computes the hash of the given trace in the CSP proofs $Hash(T(S(i)))$ with the hash included in the trace of the server $S(i + 1)$. If they match, it sends the mobile agent to be re-executed and verifies if the obtained trace agrees with the others.
   - if they agree, this means that the CSP claimed a revocation basing falsified proofs. In this case, the TA imposes a sanction to the CSP for its dishonest attitude.
   - else, the server is permanently revoked with decreasing its TL.

Table 2. Technical characteristics of the cloud platforms

Characteristic	Value
OS	Windows, Ubuntu, MacOs
Processor	Core i7 at 2.7 GHz
RAM	4 Go
Bandwidth	1000 Mbit/s

Table 3. Technical characteristics of the mobile agent

Characteristic	Value
Storage memory	512 Mbits
Processor	GenuineIntel 800 MHz
Communication protocol	HTTP-based-MTP
Bandwidth	100 Mbit/s

In the case a server is permanently revoked and classified in a black list, we consider a time interval before revising the behavior of that server to see if it may become honest. This timing interval is chosen according to many constraints: application field, sensitivity of data, category of users, etc. For example, in the field of e-commerce, the data involved are highly sensitive since it relies on monetary transactions and the users are categorized in the top classes of confidentiality (Banks, Stock Exchanges, Businessman, etc.). Thus, every millisecond of activity ban costs a lot for the server and all its dependents. Hence, after being revoked for certain period, the behaviors and activities of the server can be resumed with a maximum of restrictions and caution. The server needs to pass this procedure t times chosen randomly, such as $5 \leq t \leq 15$. Once this step is successfully performed, the server begins to accumulate a new threshold while preserving the same restrictions until it comes to prove its real honesty through reaching its old trust level.

4 Performance Analysis

In this section, we evaluate the detection and prevention performances of our Cloud-based IDPS to prove its reliability and efficiency. For that purpose, a basic cloud environment was implanted using the simulation toolkit CloudSim [18]. Besides, many virtual machines (VMs), with the characteristics denoted in Table 2, are assigned on real and heterogeneous hosts in order to represent the CSP and the related cloud servers. Indeed, the use of mobile agents needs the integration of an agent framework in each machine. Thus, we make use of JADE 4.3.3 FIPA-compliant agent platform [19], which is configured on Eclipse and charged with receiving, executing and dispatching our mobile agent. This latter has the characteristics denoted in Table 3.

The experimentations being conducted initiate five datacenters hosting increasingly 2, 4, 6, 10 and 15 cloud servers, as virtual machines (VMs). It is worth to mention that CloudSim involves two fundamental entities: *the brokers* responsible for administering the operations on the VMs, and *the cloudlets* that define the tasks to be executed by the cloud servers. In the context of our solution, a list of cloudlets assembling the solicited tasks is submitted by the broker, and attributed to the agent before moving. Figure 8 shows an example

```
public void processEvent(SimEvent ev) {
    int num_user = 1;
    Calendar calendar = Calendar.getInstance();
    CloudSim.init(num_user, calendar);
    Datacenter DC4 = createDatacenter("Datacenter_04", 10);
    DatacenterBroker broker = (DatacenterBroker)createBroker(
                                                 "DCBroker");
    broker.submitVmList(createVM(broker.getId(), 10, 1));
    broker.submitCloudletList(createCloudlet(broker.getId(), 7, 1);
    CSPAgent.Tasks= broker.getCloudletList;

    CloudSim.startSimulation();
    CSPAgent.move();
// ...
}
```

Fig. 8. Java pseudo-code of the proposed configuration in CloudSim

Table 4. Running time (in S) of the authentication and trace generation regarding the increase of cloud servers

Nb of servers	Authentication	Trace generation
2	0.024	0.088
7	0.079	0.312
15	0.166	0.654
30	0.325	1.42
60	0.643	2.65

of Java pseudo-code used in CloudSim to establish the proposed configuration. The evaluation of our IDPS for cloud emphasizes on three important criteria: response time, network load and detection rates. This is achieved while making a comparison with the mathematical model presented by Braun et al. [20], where a basic IDS architecture is adopted to ensure communication within the cloud. We choose to evaluate our approach compared to the work of [20], because this latter provides a prototype of toolkit totally based on mobile agents, called Tracy Mobile Agent, with layered architecture. Moreover, having a very small imperative core where all basic functions for starting and managing agent life-cycle are included and added as services, gives great advantages for clouds and imposes less load on processing systems. Tracy Mobile Agent still in use until now as an emergent toolkit that variety of approaches, mainly detection ones, adopt for conception and comparison.

4.1 Response Time

When the agent requires to migrate across various cloud servers with the aim to perform manifold tasks, it is strongly needed, before computing the overall response time expressed by the system, to calculate the overhead of security added through performing the authentication mechanism and the trace generation. Table 4 gives this overhead regarding the expansion of cloud servers in

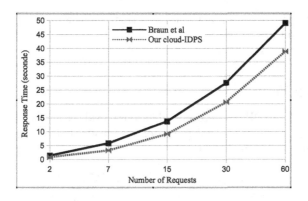

Fig. 9. Response time of our cloud-IDPS compared to [20]

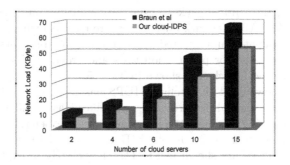

Fig. 10. Network load of our cloud-IDPS compared to [20]

the itinerary of the agent. Knowing that only one migration from one server to another takes about 156 ms, it is clearly noticed that the authentication and trace generation represent a very low percentage of the overall time spent during the agent round-trip.

A comparison of response time between our cloud-based IDPS and the work of [20], regarding the expansion of requested tasks, is shown in Fig. 9. By virtue of this latter, we prove that adopting mobile agent technology for cloud environments is beneficial as it reduces response time of about 34% and provides a flexible and secure layout.

4.2   Network Load

Our cloud-based IDPS shows very challenging performances in terms of network load, compared to Braun et al. [20]. This is illustrated in Fig. 10, where we remark that the use of mobile agents provided with security features demonstrates a lower network load of about 30%.

While analyzing the given results, we remark the appearance of an optimum load highly expressed when the mobile agent visits a number of cloud servers not beyond six. Hence, it is greatly advised that the itinerary of each dispatched mobile agent contains no more than six destinations, for better reliability and effectiveness in the execution of tasks.

4.3 Security Performance

The efficiency of an IDPS is determined through evaluating its capacity to make correct attack detection. For that purpose, we consider two commonly used rates to quantify the detection performance:

$$\text{Detection rate (DR)}: \frac{\text{Nb of detected attacks}}{\text{Nb of attacks}}$$

$$\text{False Alarm rate (FAR)}: \frac{\text{Nb of false alarms}}{\text{Nb of alarms}}$$

For capturing and sending packets (especially malicious ones), we make use of the java library JPCAP (Network Packet Capture Facility) with JADE. Besides, real attacks are simulated and injected in the running environments using MetaSploit tool [21] dedicated for penetration tests and creation of secure solutions. Examples of the used attacks are shown in Table 5.

It is well acknowledged that reliable IDPS has to express a high detection rate, while keeping the false alarm rate as reduced as possible. Table 6 lists the results of injecting various attacks to evaluate the behavior of our IDPS, compared to the system of Braun et al. Besides, the detection rates specific for each kind of attack are shown in Fig. 11. They clearly prove the robustness

Table 5. Simulated attacks to evaluate detection performance of our IDPS

Attack	Description
DoS/DDoS	Using a spoofed address, the attacker applies a TCP SYN flooding to burden the cloud server and make it unavailable.
Reused IP	The attacker burdens the server VM to force its disconnection and convince the centralized management component to allocate it for him.
Nmap TCP scan	Performs a scan of a remote machine to determine the available ports that can be exploited to gain shell access of the server hosting the mobile agent.

Table 6. Performance of our IDPS, in terms of attack detection, compared to Braun et al. [20]

Number of cloud servers	2	4	6	10	15	30
Number of injected attacks	1	2	3	6	10	17
Attacks detected: Braun et al.	0	0	1	2	4	9
Attacks detected: our IDPS	1	2	3	6	10	17

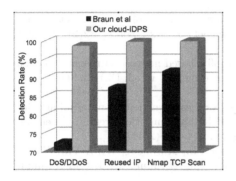

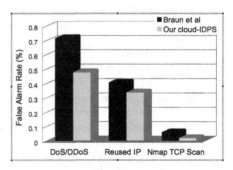

Fig. 11. Detection rate of our solution compared to [20]

Fig. 12. False alarm rate of our solution compared to [20]

and effectiveness of our IDPS in detecting intrusions, through highly significant detection rates that are enhanced of about 15% more than Braun et al.

The given results are further supported by the assessment of the false alarm rate as indicated in Fig. 12, where we notice a notably low false alarm rate of our IDPS versus a substantial rate for the system of Braun et al. Hence, our cloud-based IDPS demonstrates very promising features as security tool for cloud environment.

5   Conclusion

In this paper, a new approach for intrusion detection and prevention in cloud computing is proposed. We were concerned by persistently depicting security vulnerabilities through introducing autonomous mobile agents able to trace the execution of tasks on multiple cloud servers. These traces undergo a thorough verification associated to an agile prevention policy. Depending on a trust threshold, malicious servers are provisionally or permanently revoked, while decreasing their trust level for each suspicious behavior. In practice, our IDPS demonstrates low rates in terms of response time and network load against a high detection performance, compared to the system of Braun et al. In future, we envisage to extend our IDPS to cover the different layers and infrastructures of the cloud, as well as introducing an access control model able to collaborate with mobile agents.

References

1. Mell, P., Grance, T.: The NIST definition of cloud computing (2011)
2. Galante, J., Kharif, O., Alpeyev, P.: Sony network breach shows Amazon clouds appeal for hackers (2011)
3. Scarfone, K., Mell, P.: Guide to Intrusion Detection and Prevention Systems (IDPS). NIST Special Publication, 800, p. 94 (2007)

4. Gavalas, D., Tsekouras, G.E., Anagnostopoulos, C.: A mobile agent platform for distributed network and systems management. J. Syst. Softw. **82**(2), 355–371 (2009)
5. Gupta, S., Kumar, P., Abraham, A.: A profile based network intrusion detection and prevention system for securing cloud environment. Int. J. Distrib. Sens. Netw. **2013**, 1–12 (2013)
6. Tupakula, U., Varadharajan, V., Akku, N.: Intrusion detection techniques for infrastructure as a service cloud. In: IEEE International Conference on Dependable, Autonomic and Secure Computing, pp. 744–751 (2011)
7. Jin, H., Xiang, G., Zou, D., Wu, S., Zhao, F., Li, M., Zheng, W.: A VMM-based intrusion prevention system in cloud computing environment. J. Supercomput. **66**(3), 1133–1151 (2013)
8. Smith, D., Guan, Q., Fu, S.: An anomaly detection framework for autonomic management of compute cloud systems. In: 34th Annual Computer Software and Applications Conference Workshops (COMPSACW), Seoul, pp. 376–381 (2010)
9. Alsafi, H.M., Abduallah, W.M., Pathan, A.S.K.: IDPS: an integrated intrusion handling model for cloud computing environment. Int. J. Comput. Inf. Technol. (IJCIT) **4**(1), 1–16 (2012)
10. Dastjerdi, A.V., Bakar, K.A., Tabatabaei, S.G.H.: Distributed intrusion detection in clouds using mobile agents. In: Third International Conference on Advanced Engineering Computing and Applications in Sciences, Sliema, pp. 175–180 (2010)
11. Mills, D., Martin, J., Burbank, J., Kasch, W.: Network time protocol version 4: protocol and algorithms specification no. RFC5905 (2010)
12. Vigna, G.: Cryptographic traces for mobile agents. In: Vigna, G. (ed.) Mobile Agents and Security. LNCS, vol. 1419, pp. 137–153. Springer, Heidelberg (1998). doi:10.1007/3-540-68671-1_8
13. Phan, R.W.: Fixing the integrated Diffie-Hellman-Dsa key exchange protocol. Commun. Lett. IEEE **9**(6), 570–572 (2005)
14. Aumasson, J.: On the pseudo-random generator ISAAC. IACR Cryptology ePrint Archive 2006, p. 438 (2006)
15. Jaffar, A., Martinez, J.C.: Detail power analysis of the SHA-3Hashing algorithm candidates on Xilinx Spartan-3E. Int. J. Comput. Electr. Eng. **5**(4), 410–413 (2013)
16. Announcing the Advanced Encryption Standard (AES). FIPS Publication 197, NIST (2001)
17. Gallagher, P.: Digital signature standard (DSS). Federal Information Processing Standards Publication, FIPS PUB, 186–3 (2009)
18. Calheiros, R.N., Ranjan, R., Beloglazov, A., DeRose, C.A.F., Buyya, R.: CloudSim: a toolkit for modeling and simulation of cloud computing environments and evaluation of resource provisioning algorithms. Softw. Pract. Experience **41**(1), 23–50 (2010). Wiley publishers
19. Bellifemine, F., Poggi, A., Rimassa, G.: JADE: a FIPA2000-compliant agent development environment. In: The 5th International Conference on Autonomous Agents, pp. 216–217. ACM, Montreal (2001)
20. Braun, P., Rossak, R.: Mobile Agents: Basic Concepts, Mobility Models and the Tracy Toolkit. Elsevier, San Francisco (2005)
21. Maynor, D.: Metasploit Toolkit for Penetration Testing, Exploit Development, and Vulnerability Research. Elsevier, San Francisco (2011)

Author Index